Thermophilic Microorganisms

Edited by

Fu-Li Li

Qingdao Institute of Bioenergy and Bioprocess Technology
Chinese Academy of Sciences
Qingdao
China

Caister Academic Press

Caister Academic Press
Norfolk, UK

www.caister.com

British Library Cataloguing-in-Publication Data
A catalogue record for this book is available from the British Library

ISBN: 978-1-910190-13-5 (hardback)
ISBN: 978-1-910190-14-2 (ebook)

Cover design adapted from Figure 2.3.

Contents

Contributors

Garabed Antranikian
Institute of Technical Microbiology
Hamburg University of Technology (TUHH)
Hamburg
Germany

antranikian@tuhh.de

Sara E. Blumer-Schuette
Department of Chemical and Biomolecular Engineering
North Carolina State University
Raleigh, NC
USA

blumerschuette@oakland.edu

Jonathan M. Conway
Department of Chemical and Biomolecular Engineering
North Carolina State University
Raleigh, NC
USA

jmconway@ncsu.edu

Qiu Cui
Shandong Provincial Key Laboratory of Energy Genetics
Qingdao Institute of Bioenergy and Bioprocess Technology
Chinese Academy of Sciences
Qingdao
China

cuiqiu@qibebt.ac.cn

Skander Elleuche
Institute of Technical Microbiology
Hamburg University of Technology (TUHH)
Hamburg
Germany

skander.elleuche@tuhh.de

Wakao Fukuda
Department of Biotechnology
College of Life Sciences
Ritsumeikan University
Kusatsu
Japan

wfukuda@sk.ritsumei.ac.jp

Tadayuki Imanaka
Department of Biotechnology
College of Life Sciences
Ritsumeikan University
Kusatsu
Japan

imanaka@sk.ritsumei.ac.jp

Sonoko Ishino
Department of Bioscience and Biotechnology
Graduate School of Bioresource and Bioenvironmental Sciences
Kyushu University
Fukuoka
Japan

sonoko@agr.kyushu-u.ac.jp

Yoshizumi Ishino
Department of Bioscience and Biotechnology
Graduate School of Bioresource and Bioenvironmental Sciences
Kyushu University
Fukuoka
Japan

ishino@agr.kyushu-u.ac.jp

Shi-Qi Ji
Shandong Provincial Key Laboratory of Energy Genetics
Key Laboratory of Biofuel
Qingdao Institute of Bioenergy and Bioprocess Technology
Chinese Academy of Sciences
Qingdao
China

jisq@qibebt.ac.cn

Robert M. Kelly
Department of Chemical and Biomolecular Engineering
North Carolina State University
Raleigh, NC
USA

rmkelly@ncsu.edu

Laura L. Lee
Department of Chemical and Biomolecular Engineering
North Carolina State University
Raleigh, NC
USA

lllee@ncsu.edu

Duochuan Li
Department of Mycology Biology
Shandong Agricultural University
Tai'an
China

lidc20@sdau.edu.cn

Fu-Li Li
Qingdao Institute of Bioenergy and Bioprocess Technology
Chinese Academy of Sciences
Qingdao
China

lifl@qibebt.ac.cn

Xuegong Li
State Key Laboratory of Microbial Metabolism
Shanghai Jiao Tong University
Shanghai
China

xuegongli@sidsse.ac.cn

Ya-Jun Liu
Shandong Provincial Key Laboratory of Energy Genetics
Qingdao Institute of Bioenergy and Bioprocess Technology
Chinese Academy of Sciences
Qingdao
China

liuyj@qibebt.ac.cn

Kesen Ma
Department of Biology
University of Waterloo
Waterloo, ON
Canada

kma@uwaterloo.ca

Dong-Dong Meng
Shandong Provincial Key Laboratory of Energy Genetics
Key Laboratory of Biofuel
Qingdao Institute of Bioenergy and Bioprocess Technology
Chinese Academy of Sciences
Qingdao
China

mengdd@qibebt.ac.cn

Tairo Oshima
Institute of Environmental Microbiology
Kyowa-kako Co.
Machida
Tokyo
Japan

tairo.oshima@kyowa-kako.co.jp

Christian Schäfers
Institute of Technical Microbiology
Hamburg University of Technology (TUHH)
Hamburg
Germany

christian.schaefers@tuhh.de

Ching Tse
Department of Biology
University of Waterloo
Waterloo, ON
Canada

C6TSE@uwaterloo.ca

Xiang Xiao
State Key Laboratory of Microbial Metabolism
Shanghai Jiao Tong University
Shanghai
China

xoxiang@sjtu.edu.cn

Kun-Di Zhang
Shandong Provincial Key Laboratory of Energy Genetics
Key Laboratory of Biofuel
Qingdao Institute of Bioenergy and Bioprocess Technology
Chinese Academy of Sciences
Qingdao
China

zhangkd@qibebt.ac.cn

Yu Zhang
State Key Laboratory of Microbial Metabolism
Shanghai Jiao Tong University
Shanghai
China

zhang.yusjtu@sjtu.edu.cn

Jeffrey V. Zurawski
Department of Chemical and Biomolecular Engineering
North Carolina State University
Raleigh, NC
USA

jvzuraws@ncsu.edu

Current Books of Interest

Microalgae: Current Research and Applications	2016
Arboviruses: Molecular Biology, Evolution and Control	2016
Shigella: Molecular and Cellular Biology	2016
Aquatic Biofilms: Ecology, Water Quality and Wastewater Treatment	2016
Alphaviruses: Current Biology	2016
Flow Cytometry in Microbiology: Technology and Applications	2015
Probiotics and Prebiotics: Current Research and Future Trends	2015
Epigenetics: Current Research and Emerging Trends	2015
Corynebacterium glutamicum: From Systems Biology to Biotechnological Applications	2015
Advanced Vaccine Research Methods for the Decade of Vaccines	2015
Antifungals: From Genomics to Resistance and the Development of Novel Agents	2015
Bacteria-Plant Interactions: Advanced Research and Future Trends	2015
Aeromonas	2015
Antibiotics: Current Innovations and Future Trends	2015
Leishmania: Current Biology and Control	2015
Acanthamoeba: Biology and Pathogenesis (2nd edition)	2015
Microarrays: Current Technology, Innovations and Applications	2014
Metagenomics of the Microbial Nitrogen Cycle: Theory, Methods and Applications	2014
Pathogenic *Neisseria*: Genomics, Molecular Biology and Disease Intervention	2014
Proteomics: Targeted Technology, Innovations and Applications	2014
Biofuels: From Microbes to Molecules	2014
Human Pathogenic Fungi: Molecular Biology and Pathogenic Mechanisms	2014
Applied RNAi: From Fundamental Research to Therapeutic Applications	2014
Halophiles: Genetics and Genomes	2014
Molecular Diagnostics: Current Research and Applications	2014
Phage Therapy: Current Research and Applications	2014
Bioinformatics and Data Analysis in Microbiology	2014
The Cell Biology of Cyanobacteria	2014
Pathogenic *Escherichia coli*: Molecular and Cellular Microbiology	2014
Campylobacter Ecology and Evolution	2014
Burkholderia: From Genomes to Function	2014
Myxobacteria: Genomics, Cellular and Molecular Biology	2014
Next-generation Sequencing: Current Technologies and Applications	2014
Omics in Soil Science	2014

Full details at www.caister.com

Preface

Thermophiles thrive in various environments in both marine and terrestrial habitats. The ability of microorganisms to proliferate under extreme conditions is of widespread importance in microbial physiology, biological evolution, the ecological cycle and industry biotechnology. The discovery of thermophilic microorganisms and their enzyme systems has created new opportunities for various industrial applications over the past decades. Temperature is one of the most important factors controlling the adaptation and evolution of organisms, and high-temperature environments are of special interest for scientists, as they reveal the extremes to which evolution has been pushed.

In this book, leading scientists in this field highlight the current achievements of the most updated topic areas. The diversity and ecological roles of thermophiles, biochemical properties of thermostable biocatalysts and their application, the role of polyamines and viruses in thermophiles, DNA replication and metabolic engineering of thermophiles are all covered. Extensive focus is given to industrial applications of thermostable catalysts including alcohol dehydrogenase, glycoside hydrolase, protease and lipases. In addition, authors discussed technical challenges and future development trends.

International experts in this field from Canada, China, Germany, Japan and the USA collaborated on this book. Thank you for all your valuable contribution. In addition, I give my thanks to Caister Academic Press. I hope and I do believe that the book will be useful to students, scientists and engineers who are interested in extreme microbial research.

Dr Fu-Li Li

Ecology and Genetics of Deep-sea Thermophiles

1

Xuegong Li, Yu Zhang and Xiang Xiao

Abstract

Hydrothermal vent is a typical deep-sea ecosystem hosting the largest reservoir of archaea and many bacteria. Through a mixing of hot venting fluid and cold sea water under high hydrostatic pressure, a special environment was created with a steep gradient in terms of temperature, redox potential, pH and substrate concentration. In the past 30 years, the metagenome based analysis has showed at least 20 archaeal orders/lineages and 13 bacterial orders habituating in hydrothermal venting area, among which only a few have cultivatable represents due to the technique difficulties on incubation. Thus, to construct the genetic systems on the model strains become especially necessary. The hydrothermal vent systems and the (hyper) thermophiles are irreplaceable biological materials for a better understanding of life strategy and exploiting for industry use.

Introduction: hydrothermal vent and the microbial ecology therein

Since the discovery in 1977, deep-sea hydrothermal vents have always been a special ecosystem for searching thermophile. Some people believe hydrothermal vent is the cradle of early life, one important support for this is that hydrothermal vents are the largest reservoirs for archaeal diversity (Auguet *et al.*, 2009). To date, three types of hydrothermal vents had been identified at mid-ocean ridges: Type I, e.g. East Pacific Rise (9–10°N) and TAG (Mid-Atlantic Ridge), is the most commonly reported type with relatively high fluid temperature up to 407°C and relatively low dissolved CH_4 concentration that result in low CH_4:Mn ratios in their overlying plumes. Type II, e.g. Rainbow (36°N MAR), is with high fluid temperature and high dissolved H_2, CH_4, and Fe concentrations in the vent fluids that result in high CH_4:Mn ratios in their overlying plumes and serpentinization of ultramatic rocks. Type III, like Lost City (30°N MAR) is low fluid temperatures ~40–90°C with also serpentinization of ultramatic rocks (German *et al.*, 2010). According to the structure of chimney and the colour of the outflow, hydrothermal vents are also described as black smoker and white smoker. In Type I and Type II hydrothermal vents, the black smokers are formed when sulfate and metal sulfides precipitate. In Type III hydrothermal vent, the white smokers are formed with high porosity that favours outwards percolation of hot fluids from all surfaces. The large portion of fluids is disperse more wildly in hydrothermal plumes that rise over 1000 m off seafloor and travel thousands of kilometres from their source (Flores *et al.*, 2012).

In hydrothermal vent area, because of the mixing of reducing fluids and oxic sea water, a steep environmental gradient of temperature, pH, redox potential and metal is created, which offers a variety of habitat for diverse microorganisms. During the past decades, six fields of hydrothermal vents (Juan de Fuca, Guaymas Basin, Mid-Atlantic ridge, Eastern Lau spreading center, Lost City, East Pacific Rise EPR9°) have been extensively studied using omics-based techniques, geo-chips or cloning libraries (Table 1.1). Culture dependent and culture independent approaches have exposed a vast diversity of bacteria and archaea in hydrothermal vent

Table 1.1 Microbial community structure of different vent fields

Vent field	Sites	Depth	Sample	Microbial community	
				Bacteria	Archaea
Mid-Cayman Rise (German *et al.*, 2010)	Piccard	4150–4650 m	Fluid	γ-Proteobacteria ε-Proteobacteria	ND
	Europa	2100 m	Water	ε-Proteobacteria γ-Proteobacteria	ND
Eastern Lau Spreading Center (Flores *et al.*, 2012)	Kilo Moana	2606–2640 m	Sediment	*Helicobacteraceae* *Nautiliaceae* Desulfurobacteriaceae *Desulfobulbaceae* *Deferribacterales*	*Desulfurococcaceae* *Thermofilaceae* *Thermoproteaceae* *Thermococcaceae* *Pyrodictiaceae*
	Tow Cam	2690–2728 m		*Helicobacteraceae* *Nautiliaceae*	*Thermoproteaceae* *Thermofilaceae* *Pyrodictiaceae* *Thermococcaceae* *Desulfurococcaceae* *Archaeoglobaceae*
	Tahi Moana	2214–2260 m		*Helicobacteraceae* *Nautiliaceae* *Thermodesulfobacteriaceae* *Desulfobulbaceae*	*Thermofilaceae* *Pyrodictiaceae* *Desulfurococcaceae* *Thermococcaceae*
	ABE	2104–2163 m		*Helicobacteraceae* *Nautiliaceae* *Desulfobulbaceae* *Thermodesulfobacteriaceae*	*Thermofilaceae* *Desulfurococcaceae* *Thermoproteaceae* *Desulfurococcaceae* *Thermococcaceae*
	Tui Malila	1839–1928 m		*Nautiliaceae* *Helicobacteraceae* *Desulfobulbaceae*	*Thermofilaceae* *Thermococcaceae* *Pyrodictiaceae* *Methanomicrobiales* *Thermoproteaceae*
	Mariner	1877–1951 m		*Nautiliaceae* *Actinomycetales* *Helicobacteraceae* *Desulfobulbaceae*	*Caldisphaeraceae* *Thermoplasmatales* *Thermococcaceae* *Thermofilaceae* *Thermoproteaceae*
Juan de Fuca Ridge (Wang *et al.*, 2009)	4143-1		Chimney (outer section)	*Rhodobacteraceae* *Geobacteraceae* *Helicobacteraceae* *Campylobacteraceae* *Nitrospiraceae* *Flexibacteraceae* *Flavobacteriaceae* *Planctomycetaceae*	*Methanosarcinaceae* *Methanococcaceae* *Thermococcaceae*
Guaymas Basin (He *et al.*, 2013)	4558-6	2013 m	Chimney (outer section)	*Helicobacteraceae* *Thermodesulfobacteriaceae* *Colwelliaceae* *Thiotrichaceae* *Desulfovibrionaceae* *Thermotogaceae*	*Archaeoglobaceae* *Thermococcaceae* *Desulfobacteraceae* *Desulfurococcaceae*
Guaymas Basin (Dick and Tebo, 2010)		1700–2000 m	Fluid	*Methylococcaceae* *Methylophilaceae* *Alteromonadales* *Pseudoalteromonadaceae* *Vibrionaceae* *Rhodobacteraceae*	Marine *Crenarchaeota* group I *Nitrosopumilaceae*

Table 1.1 continued

				Microbial community	
Vent field	Sites	Depth	Sample	Bacteria	Archaea
Mid-Atlantic Ridge (Flores *et al.*, 2011)	Lucky Strike	1600–1075 m	Sediment	ε-Proteobacteria γ-Proteobacteria δ-Proteobacteria α-Proteobacteria *Aquificae* *Caldisericaceae* *Deferribacteres* *Thermodesulfobacteria* Thermotogae Firmicutes Bacteroidetes	*Desulfurococcaceae* *Thermococcaceae* *Archaeoglobaceae* *Nanoarchaea* *Pyrodictiaceae* *Thermoproteaceae* *Korarchaeum*
	Rainbow	2270–2330 m	Sediment	ε-Proteobacteria γ-Proteobacteria δ-Proteobacteria α-Proteobacteria *Aquificae* *Caldisericaceae* *Deferribacteres* *Thermodesulfobacteria* Thermotogae Firmicutes Bacteroidetes	*Desulfurococcaceae* *Thermococcaceae* *Thermofilaceae* *Archaeoglobaceae* *Pyrodictiaceae* *Thermoproteaceae* *Korarchaeum* *Nanoarchaea* *Methanopyraceae* *Methanococcaceae* *Methanocaldococcaceae*
Lost City Hydrothermal Field (Brazelton *et al.*, 2010)	Sample1		Chimney	*Flavobacteriales* *Sphingobacteriales* *Caldilineales* *Deferribacterales* *Rhizobiales* *Rhodobacterales* *Burkholderiales* *Desulfobacterales* *Desulfovibrionales* *Campylobacterales* *Thiotrichales*	*Crenarchaeota* *Euryarchaeota* *Korarchaeota*
	Sample2			*Flavobacteriales* *Sphingobacteriales* *Caldilineales* *Nitrospirales* *Rhizobiales* *Rhodobacterales* *Burkholderiales* *Desulfobacterales* *Desulfovibrionales* *Campylobacterales* *Methylococcales* *Thiotrichales* *Acholeplasmatales*	*Euryarchaeota*
	Sample3, 4			*Acidobacteriales* *Bacteroidales* *Flavobacteriales* *Sphingobacteriales* *Caldilineales* *Deferribacterales* *Nitrospirales* *Rhizobiales* *Rhodobacterales* *Burkholderiales* *Desulfobacterales* *Desulfovibrionales* *Campylobacterales* *Nautiliales* *Methylococcales* *Thiotrichales* *Acholeplasmatales*	*Crenarchaeota* *Euryarchaeota*

ecosystems. These studies have identified that Epsilonproteobacteria is the dominant bacterial lineage associated with vent deposits, while other bacterial groups, such as Aquificales, Thermales, Thermotogales, Deltaproteobacteria and Thermodesulfurobacteriacaeae, are also often detected. Archaea associated with active vent deposits with cultivable relatives are thermophiles belonging to Archaeoglobales, Thermococcales, Aciduliprofundales, Methanococcales and Desulfurococcales (Fig. 1.1). Besides, numerous archaeal lineages with no relatives in culture have also been reported (Wang *et al.*, 2009). By applying the ecological concept of 'indicator species', there are five archaea lineages are exclusive indicators to hydrothermal vent, and most of them located close to the root of the phylogenetic tree, suggesting it would be the earliest habitat colonized by Archaea (Auguet *et al.*, 2009). In the following pages, we are about to show several studies in different deep sea hydrothermal venting fields to bring up a general understanding on this special type of ecosystems.

Dynamics of deep-sea hydrothermal vent community in Juan de Fuca

For the obstacle of sampling and monitoring *in situ*, few researches have revealed the process of the establishment of nature ecosystem. During the expedition to Juan de Fuca Ridge in 2006, a cone-shaped cap was placed on the top of 310°C small chimney for 15 days. The inner part and outer part were separated and compared with a mature chimney rock 4143-1 nearby. Only 4143-1 had enough DNA for 16S rDNA PCR amplification. The bacterial community in 4143-1 was composed of γ-, ε-, and δ-Proteobacteria, Nitrospirae, Bacteroidetes and Planctomycetes, among which the γ-Proteobacteria was the most dominant while the ε-Proteobacteria followed. The archaeal community in 4143-1 was relatively

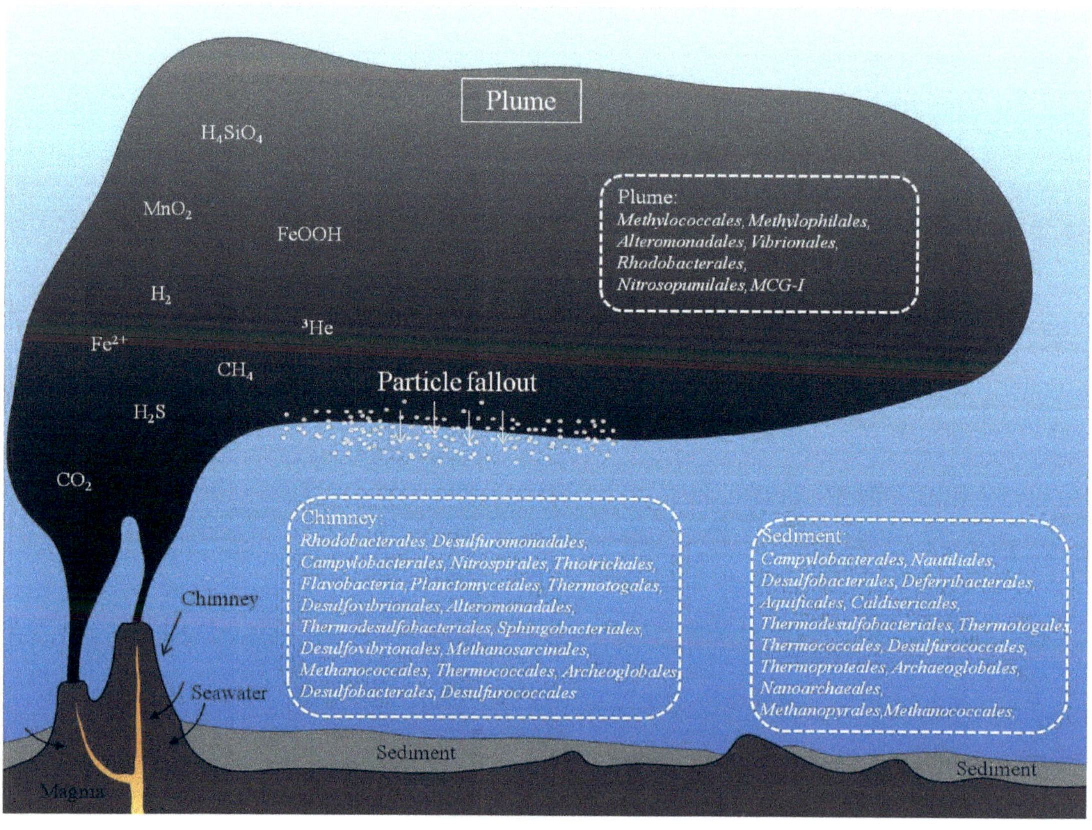

Figure 1.1 Microbial community structure in hydrothermal vent ecosystems.

simple and contained exclusively Euryarchaea. Unidentified Euryarchaea cluster and Thermococcus were predominant in the library (51% and 42%, respectively). The physiology and metabolic properties for most of the archaea in 4143-1 could not be determined, because they are distinct from the known species. Whole-community genome amplification was carefully performed to obtain enough DNA from new artificial chimney for microarray hybridization. On the GeoChip, there were a total of 8371 genes from bacteria and 594 genes from archaea. After hybridization, there were 113 genes detected in inner part and 504 genes in outer part, whereas 5414 functional genes were detected in 4143-1. Among the genes detected, those involved in organic contaminant degradation were most abundant in all 3 samples, followed by genes involved in carbon degradation and metal resistance. The inner part contained significantly higher numbers of genes involved in carbon fixation and nitrification than the other 2 samples, whereas fewer genes for methane oxidation and sulfate reduction were present in inner part than in the other samples. Significant variability of functional genes related to carbon, sulfur, and nitrogen metabolism was observed during the development of vent chimneys (Wang *et al.*, 2009).

Guaymas Basin field with high organic sediments

The Guaymas Basin (Gulf of California) is a unique hydrothermal vent field with high-temperature thick sediments (100–500 m), 2–4% organic matter. These sediments were formed by precipitation from the highly productive surface waters and terrigenous input (Von Damm *et al.*, 1985). The metagenome of an oil-immersed chimney in Guaymas Basin was analysed to demonstrate the metabolic potential and ecological functions of the inhabited microbial community (He *et al.*, 2013). At the phylum level, Proteobacteria (25.6%), Euryarchaeota (24.3%), Crenarchaeota (3.6%), Firmicutes (2.1%), Thermotogae (1.9%), and Bacteroidetes (1.6%) had the top matches. The majority of sequences in this chimney metagenome originated from the families Archaeoglobaceae, Thermococcaceae, and Desulfobacteraceae. Thermophilic sulfate reducing microorganisms, including bacteria (Thermodesulfobacteriaceae) and archaea (Archaeoglobaceae) were found predominant Mn^{2+} oxidation rate was 200 times faster and two groups uncultured Gammaproteobacteria and Betaproteobacteria phylogenetically clustered as methanotrophs were significantly more abundant. However, their over-represented in plumes (2–5×) is not nearly sufficient to accounts for the difference in Mn^{2+} oxidation rate (up to 200×), and unknown Mn^{2+} oxidizer (Dick and Tebo, 2010) and/or higher metabolic efficiency need to be identified.

Geochemical constrain on microbial community structure of different fields

It is still unclear that how the hydrothermal fluids chemistry controls the diversity and physiology of adjacent microorganisms. An efforts was reported along Mid Atlantic ridges: Flores and his colleagues examined the archaeal and bacterial communities of 12 samples collected from Lucky Strike and Rainbow vent fields along the Mid-Atlantic Ridge (MAR) using multiplexed barcoded pyrosequencing of V4 region of the 16S rRNA genes (Flores *et al.*, 2011). At the ultramafically hosted Rainbow site, the high concentrations of aqueous H_2 allow for methanogens to flourish. Conversely, the basalt-hosted Lucky Strike vent fluids have very low aqueous H_2 concentrations. Overall, Archaeal families shared between all vent samples include the thermophilic Desulfurococcaceae, Thermococcaceae, and Thermofilaceae. Other thermophilic lineages shared by most sampled vents include the mixotrophic Archaeoglobaceae, the acidophilic fermentative 'Aciduliprofundales' (DHVE2) and the Nanoarchaea. For bacteria, Epsilonproteobacteria are the most dominant class in half of the Rainbow samples and all of the Lucky Strike samples, within which the moderately thermophilic *Caminibacter* and *Nautilia*, and thermophilic *Nitratifrator* and *Sulfurovum* were detected in all samples although abundances differed. Of the samples not dominated by Epsilonproteobacteria, two were dominated by Gammaproteobacteria and one by unclassified sequences related to the Thermotogae. The hydrogen-oxidizing sulfate-reducing *Thermodesulfatator* and *Thermodesulfobacterium* were detected in all Rainbow samples but only detected in one of the Lucky Strike deposits (Flores *et al.*, 2011).

Eastern Lau Spreading Center has more geological diverse venting systems, owing to the presence of both convergent and divergent plate boundaries. In general, vent fluids of south field (Kilo Moana, Tow Cam, ABE and Tui Malila) are characterized by composition similar to those observed at typical Middle Oceanic Ridge. From north to south, these vent fields exhibit a systematic decrease of temperature accompanied by an increase in pH and deceases in the concentration of aqueous H_2, H_2S and Fe. Meanwhile, Mariner are substantially different from the others in composition with higher concentration of CO_2, H_2S, Fe and other transition metals and are hotter and more acidic relative to the northern vents. For 30 hydrothermal vent samples in Eastern Lau Spreading Center, the most abundant bacterial OTU was *Sulfurovum* (20.24%). *Nitratifrator* (4.20%), *Desulfolobas* (3.02%), and *Hippea* (0.47%) were the only other OTUs found in all 30 samples. With the exception of Kilo Moana and Mariner, the majority of sequences detected in all samples were classified as Epsilonproteobacteria. *Caminibacter, Lebetimonas, Nautilia, Nitratifractor, Nitratiruptor, Sulfurimonas* and *Sulfurovum* were associated with most samples although the dominant genera varied. In the case of Mariner, *Nautilia* were prevalent, while other taxa commonly observed as Alphaproteobacteria, Aquificales, Caldisericia, Chloroflexi, Deferribacteres, *Deinococcus/Thermus*, Deltaproteobacteria, Gammaproteobacteria, Thermodesulfobacteria, and Thermotogae. For archaea, the single most abundant OTU was *Thermofilum*. Only Thermococcales was observed in all samples. The comparisons from different vent fields showed that the communities associated with Mariner sample are distinct. It was still difficult to identify a main environment factor shaping the difference of community structure (Flores *et al.*, 2012).

Cradle of early life? The alkaline Lost City hydrothermal field (LCHF)

The early ocean was believed to be acid with pH 3.5–6; thus, the alkaline vent system (pH 9–10), like LCHF, could create an interface with a proton gradient sufficient to drive the first metabolism (Bernhardt and Tate, 2012). Lost City also differs from most magma-driven systems in that it is very long-lived. Modelling results indicate that the system could remain active for up to 1 million years, with individual chimneys active for at least 300 years. The serpentinization reaction produces alkaline (pH 9–11) fluids that are rich in calcium (up to 30 mmol/kg), low in metals, and have near zero concentrations of CO_2. From the four carbonate chimney samples, Brazelton *et al.* identified 817 OTU at the 3% distance threshold (Brazelton *et al.*, 2010). Notably, the most aged sample (^{230}Th age = 1245 ± 257 years) contained significantly fewer OTUs than the younger samples (^{230}Th age = 34–128 years). A single OTU Methanosarcinales comprised > 90% of all tag sequence from the young samples. The oldest sample was also dominated by one single OTU, which is similar to the ANME-1 group of anaerobic methanotrophic archaea. Besides, other archaeal clades, such as methanogens and ammonia oxidizers were present in the youngest, hottest chimney sample and in hot fluids sample, indicated that they probably represent thermophilic organisms. Most of the diversity present in the bacteria, e.g. sequence assigned to Gammaproteobacteria, Alphaproteobacteria, Firmicutes, and Chloroflexi, was from the youngest sample. In short, samples of similar ages contain similar communities with few exceptions, and this trend is true for dominant as well as rare organisms (Brazelton *et al.*, 2010).

Microbial community structure at East Pacific Rise (EPR)

Environmental samples from four vent sites (Tica, Bio9/P-vent, Marker 28 and v-vent) along EPR were collected for 16S rRNA gene sequence analysis. Parts of the samples were collected near macrofaunal communities and the others were collected near the diffuse flow vents. Marine Group I (MGI) within Thaumarchaeota are largely present (up to 60% of OTU) in both surrounding seawater and diffuse flow water samples, while *Sulfurovum* sp. within Epsilonproteobacteria are the dominant in the water samples near macrofaunal communities. The Oceanospirillales within *Gammaproteobacteria* comprise about 7%

OTU in EPR off-axis bacterioplankton. Other detected groups were Marine Group II, III, Thermococcaceae, SAR406, Methylococcales, Desulfobacteriales, Arcobacter, Campylobacter, Campylobacter, Sulfurimonas (Campbell *et al.*, 2013).

Another study on a white smoker, which was collected from a fluid seep area with a maximum measured temperature of 350°C at the top of the spire in EPR, showed that, within the bacteria domain, members of Epsilonproteobacteria, the Bacteroidetes phylum, the Thermales and Aquificales, were retrieved most frequently. Also Deltaproteobacteria and Gammaproteobacteria could be detected. Among the Archaea, clones representing archaeal orders with (hyper) thermophilic culture species (Thermococcales, Archaeoglobales, Desulfurococcales, Thermoproteales, Methanococcales) predominated. Also DSVG and DHV2 were detected (Kormas *et al.*, 2006).

The ecological global patterns of hydrothermal vents

A comparative genomic analysis demonstrated that the Juan de Fuca metagenome clustered most closely with that from a carbonate chimney from Lost City. Both were highly enriched in genes for mismatch repair and homologous recombination, to cope with the extreme conditions that have potential deleterious effects on the genome (Xie *et al.*, 2010). Later, the complete clustering of functional gene categories with a comparative metagenomic approach also showed that this Guaymas chimney metagenome was clustered most closely with a chimney metagenome from Juan de Fuca. All chimney samples were enriched with genes involved in recombination and repair, chemotaxis and flagellar assembly, highlighting their roles in coping with the fluctuating extreme deep-sea environments. A high proportion of transposase was observed in all the metagenomes from deep-sea chimneys, supporting the previous hypothesis that horizontal gene transfer may be common in the deep-sea vent chimney biosphere (He *et al.*, 2013).

Genetic tools on hyperthermophiles

Hyperthermophiles with optimal growth temperature above 80°C are wildly distributed in geothermal environments including hot-spring, volcano, and deep-sea hydrothermal vent (Stetter, 2006). These microorganisms are located in the deepest branches in the phylogenetic trees and deemed as a good subject for addressing evolutionary problems (Huber and Stetter, 1998). Moreover, hyperthermophiles live on unique metabolic pathway and biochemical reaction, thus are believed to be potential resources for mining highly thermostable enzymes (Atomi, 2005; de Miguel Bouzas *et al.*, 2006; Demirjian *et al.*, 2001). In the past 30 years, the metagenome based analysis has showed at least 20 archaeal orders/lineages and 13 bacterial orders habituating in hydrothermal venting area, among which only a few have cultivatable represents due to the technique difficulties on incubation. Thus, to construct the genetic systems on the model strains become especially necessary. Thermococcales is a group of archaea that are indicative as well as the most isolated in hydrothermal vent ecosystems (Nakagawa *et al.*, 2005). We will summarize the key techniques applied in constructing genetic systems on Thermococcales (Table 1.2).

Transformation

For most of the reported Thermococcales (including *Thermococcus kodakarensis* KOD1, *Thermococcus onnurineus* NA1, *Thermococcus barophilus* MP, and *Pyrococcus furiosus*), $CaCl_2$ method is suitable to deliver exogenous DNA. However, such method fails to work on the transformation of *Pyrococcus abyssi* (Lucas *et al.*, 2002). To break this limitation, a polyethylene glycol (PEG)-mediated spheroplast protocol was developed. In addition, it is notable that some Thermococcales possess natural competence for DNA uptake, such as *T. kodakarensis* KOD1 (Sato *et al.*, 2005) and *P. furiosus* COM1 (Lipscomb *et al.*, 2011). This character simplifies the procedure of genetic manipulation in such strains.

Resistance selectable marker

Most of the antibiotics do not work for archaea especially for the hyperthermophilic ones

Table 1.2 Genetic tools for Thermococcales (modified from Leigh *et al.*, 2011)

	T. kodakarensis	*P. abyssi*	*P. furiosus*	*T. barophilus*
Defined media	Yes	Yes	Yes	Yes
DNA delivery	Transformation	PEG-mediated transformation (Lucas *et al.*, 2002)	Transformation	Transformation
Replicative shuttle vectors	pLC70 (Santangelo and Reeve, 2008)	pYS2 (Lucas *et al.*, 2002)	pYS3 (Waege *et al.*, 2010)	–
Antibiotic selectable marker	Simvastatin (Matsumi *et al.*, 2007)	–	Simvastatin	Simvastatin (Thiel *et al.*, 2014)
Auxotrophic selectable markers	Uracil auxotroph (ΔpyrF) (Sato *et al.*, 2003) Tryptophan auxotroph (ΔtrpE) (Sato *et al.*, 2003) Agmatine auxotroph (ΔpdaD) (Santangelo and Reeve, 2010) Histidine auxotroph (ΔhisD) (Sato *et al.*, 2005)	Uracil auxotroph (ΔpyrE)	Uracil auxotroph (ΔpyrF)	Uracil auxotroph (ΔpyrF) Histidine auxotroph (ΔhisB)
Counter selection	5-FOA (*pyrF*) 6-Methyl purine (*TK0664*) (Santangelo and Reeve, 2010)	–	5-FOA	5-FOA
Markerless gene knockout	*TrpE-TK0664*	–	hmg_{pf}-*pyrF* (Lipscomb *et al.*, 2011)	hmg_{pf}-*pyrF* (Thiel *et al.*, 2014)
Reporter genes	*TK1761* (β-galactosidase) (Santangelo *et al.*, 2008)	–	–	–
Regulated gene expression	fbp (*TK2164*) promoter (Kanai *et al.*, 2007)		fbp (*PF0613*) promoter (Waege *et al.*, 2010)	
Protein overexpression	gdh (*TK1431*), TK2279 and csg (*TK0895*) promoters		gdh (*PF1602*) promoter	gdh (*PF1602*) promoter

because of the stability problems during high-temperature incubation. An alternative selectable marker for archaeal cells is statins. Archaeal lipids are synthesized by using mevalonate pathway, where HMG-CoA is converted into mevalonate catalysed by HMG-CoA reductase (Grochowski *et al.*, 2006). Statins (e.g. simvastatin and mevinolin) are competitive inhibitors to the biosynthesis of archaeal lipids. As proven in *Haloferax volcanii*, the overexpressed HMG-CoA reductase could conferred host cell resistance against mevinolin (Leigh *et al.*, 2011). To use simvastatin as a selectable marker, the native promoter region of HMG-CoA reductase in *T. kodakarensis* KOD1 was replaced by the strong promoter region of glutamate dehydrogenase and HMG-CoA reductase was overexpressed resulting in a high transcription of cassette P_{gdh}-*hmg*. This makes transformants selectable in presence of simvastatin, even in nutrient-rich medium. Using this cassette, the sugar transporter system of *T. kodakarensis* KOD1 was disrupted successfully (Matsumi *et al.*, 2007). The same overexpressed system has been successfully applied in *P. furiosus* (Waege *et al.*, 2010), *T. onnurineus* NA1 (Kim *et al.*, 2010), and *T. barophilus* MP (Thiel *et al.*, 2014).

Auxotrophic selectable marker

5-FOA as counter-selection agent was employed to isolated uracil auxotrophs. Orotate phosphoribosyltransferase and orotidine 5′-monophosphate decarboxylase catalysed the last two steps of uracil biosynthesis, which encoded by *pyrE* and *pyrF*, respectively. Only mutants with *pyrE/F* deficient can normally grow in the medium supplemented with 5-FOA. Based on the selection of 5-FOA, uracil auxotrophic mutants of *P. abyssi* GE9 (PA9-13 and PA9-20) were isolated and characterized with *pyrE*-deficient. To use uracil auxotroph as genetic host, a shuttle vector pYS2 was constructed by fused endogenous plasmid pGT5 from *P. abyssi* GE5 with *E. coli* pLitmus38 plasmid, and a *pyrE* from *Sulfolobus acidocaldarius* was introduced for complementation experiments. After transforming PA9-13 and PA9-20 with redesigned pYS2 by a PEG-spheroplast method, the transformants exhibit uracil prototrophy that can grow normally on the medium without uracil supplement (Lucas *et al.*, 2002).

To extent the use of this genetic system in other Thermococcales, a disruption plasmid was designed by double-crossover homologous recombination. A complete deletion of *pyrF* gene has resulted in uracil auxotrophic mutant from *T. kodakarensis* KOD1 (Sato *et al.*, 2005), which avoided the reverse mutation and other unknown mutations caused by UV irradiation. Using the *pyrF* as a positive selection marker and *pyrF*$^-$ strain as host, *trpE* from *T. kodakarensis* KOD1 has been disrupted by homologous recombination, then generated tryptophan auxotrophs (Sato *et al.*, 2003, 2005). Moreover, *trpE* was also used as a new selectable marker for genetic manipulation of *T. kodakarensis* KOD1. Take △*trpE* strain as host, *TK0475* encoding 3-hexulose-6-phosphate synthase and 6-phospho-3-hexuloisomerase was disrupted through the complementation with *trpE*. Then, the ribulose monophosphate pathway of KOD1 was investigated (Orita *et al.*, 2006).

6-MP is analogues of adenine that can inhibit growth by disrupting RNA synthesis, thus a good counter selection marker in genetic system of Thermococcales. Disruption of *TK0664* from KOD1 is resistant to 6-MP. *TK0664*–6-MP has been successfully used for genetic manipulation of *T. kodakarensis* KOD1 (Santangelo and Reeve, 2010). In addition, *hisD* gene (encoding histidinol dehydrogenase) involved in histidine biosynthesis was another candidate of selection marker in *T. kodakarensis* KOD1. *TK0149* gene encoding arginine decarboxylase (pdaD) from KOD1 has been disrupted and it resulted in agmatine auxotrophy (Sato *et al.*, 2005). Using *pdaD* as selection marker, transformants can be positive selection in nutrient rich medium (Santangelo and Reeve, 2010).

Markerless disruption system

To perform the gene disruption in Thermococcales under the limitation of selectable marker, a markerless disruption system was proposed. Using double-deletion (△*trp*E, △*TK0664*) strain from *T. kodakarensis* KOD1, a selection/counter selection markerless system based on *trpE*-6MP^S cassette was developed (Santangelo and Reeve, 2010). For single-crossover, the transformants were selected on the defined medium without tryptophan supplement. Then, transformants were further selected on the medium containing 6-MP and tryptophan. Plasmid was lost and a double-crossover event occurred. Use this system, any nonessential gene of *T. kodakarensis* KOD1 can be deleted without harbouring scar in the genome. Using similar strategy, a markerless cassette was designed for *T. barophilus*, which using simvastatin and 5-FOA as selection and counter selection marker, respectively (Thiel *et al.*, 2014). Also for *P. furiosus* COM1, a P_{gdh}-*pyrF* markerless cassette was constructed and the cytoplasmic hydrogenases were deleted (Lipscomb *et al.*, 2011).

Gene expression in Thermococcales

Hyperthermophiles produce some genes coding unique enzymes which can catalyse uncommon biochemical reactions and are stable at very high temperature. Although some pioneer works have demonstrated the expression of these gene in mesophilic bacteria, it is still of great advantage to express them in archaea host for a higher efficiency (Farkas *et al.*, 2013). Strong promoters are commonly used in Thermococcales gene expression, such as promoter of glutamate dehydrogenase (P_{gdh}) from *T. kodakarensis* KOD1 (Matsumi *et al.*, 2007) and *P. furiosus* (Waege *et al.*, 2010), promoter of the histone gene (P_{hmtB}) from *Methanothermobacter thermautotrophicus* (Santangelo and Reeve, 2010), promoter of the sugar nucleotidyltransferase gene *TK2279* (P_{2279}) from *T. kodakarensis* KOD1 (Santangelo and Reeve, 2008) and promoter of cell surface glycoprotein gene (P_{csg}) from *T. kodakarensis* KOD1 (Mueller *et al.*, 2009). Using P_{gdh} as a regulator, HMG-CoA reductases have been overexpressed in many hyperthermophiles and conferred host strains resistance to against simvastatin. In addition, the *pyrF* and β-glycosidase gene has been expressed in *T. kodakarensis* KOD1 under the control of strong promoter P_{hmtB}. Moreover, α-1,4-glucan phosphorylases from *Sulfolobus solfataricus* has been expressed in the *T. kodakarensis* KOD1 by the control of P_{csg}, which fails to express in *E. coli* host after many times trying out.

Thermococcales was also used as the engineered cells for gene expression and protein secretion. A chitinase gene from *T. kodakarensis* KOD1 was redesigned by fusing a signal peptide region and a strong promoter P_{csg}. Then, chitinase was overexpressed under the control of strong promoter P_{csg} and secreted to culture supernatant by the help of signal peptide (Takemasa *et al.*, 2011). Such expression and secretion systems are powerful tools for a better understanding of Thermococcales and the exploiting of the enzyme resource in hyperthermophiles.

References

Atomi, H. (2005). Recent progress towards the application of hyperthermophiles and their enzymes. Curr. Opin. Chem. Biol. *9*, 166–173.

Auguet, J.-C., Barberan, A., and Casamayor, E.O. (2009). Global ecological patterns in uncultured Archaea. ISME J. *4*, 182–190.

Bernhardt, H.S., and Tate, W.P. (2012). Primordial soup or vinaigrette: did the RNA world evolve at acidic pH. Biol. Direct *7*, 1–12.

Brazelton, W.J., Ludwig, K.A., Sogin, M.L., Andreishcheva, E.N., Kelley, D.S., Shen, C.-C., Edwards, R.L., and Baross, J.A. (2010). Archaea and bacteria with surprising microdiversity show shifts in dominance over 1,000-year time scales in hydrothermal chimneys. Proc. Natl. Acad. Sci. U.S.A. *107*, 1612–1617.

Campbell, B.J., Polson, S.W., Allen, L.Z., Williamson, S.J., Lee, C.K., Wommack, K.E., and Cary, S.C. (2013). Diffuse flow environments within basalt-and sediment-based hydrothermal vent ecosystems harbor specialized microbial communities. Front. Microbiol. *4*, 1–15.

Demirjian, D.C., Morís-Varas, F., and Cassidy, C.S. (2001). Enzymes from extremophiles. Curr. Opin. Chem. Biol. *5*, 144–151.

Dick, G.J., and Tebo, B.M. (2010). Microbial diversity and biogeochemistry of the Guaymas Basin deep-sea hydrothermal plume. Environ. Microbiol. *12*, 1334–1347.

Farkas, J.A., Picking, J.W., and Santangelo, T.J. (2013). Genetic techniques for the archaea. Annual review of genetics *47*, 539–561.

Flores, G.E., Campbell, J.H., Kirshtein, J.D., Meneghin, J., Podar, M., Steinberg, J.I., Seewald, J.S., Tivey, M.K., Voytek, M.A., and Yang, Z.K. (2011). Microbial community structure of hydrothermal deposits from geochemically different vent fields along the Mid-Atlantic Ridge. Environ. Microbiol. *13*, 2158–2171.

Flores, G., Shakya, M., Meneghin, J., Yang, Z., Seewald, J., Geoff Wheat, C., Podar, M., and Reysenbach, A.L. (2012). Inter-field variability in the microbial communities of hydrothermal vent deposits from a back-arc basin. Geobiology *10*, 333–346.

German, C.R., Bowen, A., Coleman, M.L., Honig, D.L., Huber, J.A., Jakuba, M.V., Kinsey, J.C., Kurz, M.D., Leroy, S., McDermott, J.M., *et al.* (2010). Diverse styles of submarine venting on the ultraslow spreading Mid-Cayman Rise. Proc. Natl. Acad. Sci. U.S.A. *107*, 14020–14025.

Grochowski, L.L., Xu, H., and White, R.H. (2006). Methanocaldococcus jannaschii uses a modified mevalonate pathway for biosynthesis of isopentenyl diphosphate. J. Bacteriol. *188*, 3192–3198.

He, Y., Xiao, X., and Wang, F. (2013). Metagenome reveals potential microbial degradation of hydrocarbon coupled with sulfate reduction in an oil-immersed chimney from Guaymas Basin. Front. Microbiol. *4*, 1–13.

Huber, H., and Stetter, K.O. (1998). Hyperthermophiles and their possible potential in biotechnology. J. Biotechnol. *64*, 39–52.

Kanai, T., Akerboom, J., Takedomi, S., van de Werken, H.J., Blombach, F., van der Oost, J., Murakami, T., Atomi, H., and Imanaka, T. (2007). A global transcriptional regulator in Thermococcus kodakaraensis controls the expression levels of both glycolytic and gluconeogenic enzyme-encoding genes. J. Biol. Chem. *282*, 33659–33670.

Kim, Y.J., Lee, H.S., Kim, E.S., Bae, S.S., Lim, J.K., Matsumi, R., Lebedinsky, A.V., Sokolova, T.G., Kozhevnikova, D.A., and Cha, S.-S. (2010). Formate-driven growth coupled with H2 production. Nature *467*, 352–355.

Kormas, K.A., Tivey, M.K., Von Damm, K., and Teske, A. (2006). Bacterial and archaeal phylotypes associated with distinct mineralogical layers of a white smoker spire from a deep-sea hydrothermal vent site (9 degrees N, East Pacific Rise). Environ. Microbiol. *8*, 909–920.

Leigh, J.A., Albers, S.V., Atomi, H., and Allers, T. (2011). Model organisms for genetics in the domain Archaea: methanogens, halophiles, Thermococcales and Sulfolobales. FEMS Microbiol. Rev. *35*, 577–608.

Lipscomb, G.L., Stirrett, K., Schut, G.J., Yang, F., Jenney, F.E., Scott, R.A., Adams, M.W., and Westpheling, J. (2011). Natural competence in the hyperthermophilic archaeon Pyrococcus furiosus facilitates genetic manipulation: construction of markerless deletions of genes encoding the two cytoplasmic hydrogenases. Appl. Environ. Microbiol. *77*, 2232–2238.

Lucas, S., Toffin, L., Zivanovic, Y., Charlier, D., Moussard, H., Forterre, P., Prieur, D., and Erauso, G. (2002). Construction of a shuttle vector for, and spheroplast transformation of, the hyperthermophilic archaeon Pyrococcus abyssi. Appl. Environ. Microbiol. *68*, 5528–5536.

Matsumi, R., Manabe, K., Fukui, T., Atomi, H., and Imanaka, T. (2007). Disruption of a sugar transporter gene cluster in a hyperthermophilic archaeon using a host-marker system based on antibiotic resistance. J. Bacteriol. *189*, 2683–2691.

de Miguel Bouzas, T., Barros-Velázquez, J., and Gonzalez Villa, T. (2006). Industrial applications of hyperthermophilic enzymes: a review. Protein Peptide Lett. *13*, 645–651.

Mueller, M., Takemasa, R., Schwarz, A., Atomi, H., and Nidetzky, B. (2009). "Short-chain" α-1, 4-glucan phosphorylase having a truncated N-terminal domain: Functional expression and characterization of the enzyme from *Sulfolobus solfataricus*. BBA-Proteins Proteom. *1794*, 1709–1714.

Nakagawa, S., Takai, K., Inagaki, F., Chiba, H., Ishibashi, J.i., Kataoka, S., Hirayama, H., Nunoura, T., Horikoshi, K., and Sako, Y. (2005). Variability in microbial community and venting chemistry in a sediment-hosted backarc hydrothermal system: Impacts of subseafloor phase-separation. FEMS Microbiol. Ecol. *54*, 141–155.

Orita, I., Sato, T., Yurimoto, H., Kato, N., Atomi, H., Imanaka, T., and Sakai, Y. (2006). The ribulose monophosphate pathway substitutes for the missing pentose phosphate pathway in the archaeon Thermococcus kodakaraensis. J. Bacteriol. *188*, 4698–4704.

Santangelo, T.J., and Reeve, J.N. (2008). Shuttle vector expression in Thermococcus kodakaraensis: contributions of cis elements to protein synthesis in a hyperthermophilic archaeon. Appl. Environ. Microbiol. *74*, 3099–3104.

Santangelo, T.J., and Reeve, J.N. (2010). Thermococcus kodakarensis genetics: TK1827-encoded β-glycosidase, new positive-selection protocol, and targeted and repetitive deletion technology. Appl. Environ. Microbiol. *76*, 1044–1052.

Santangelo, T.J., Cubonova, L., Matsumi, R., Atomi, H., Imanaka, T., and Reeve, J.N. (2008). Polarity in archaeal operon transcription in Thermococcus kodakaraensis. J. Bacteriol. *190*, 2244–2248.

Sato, T., Fukui, T., Atomi, H., and Imanaka, T. (2003). Targeted gene disruption by homologous recombination in the hyperthermophilic archaeon Thermococcus kodakaraensis KOD1. J. Bacteriol. *185*, 210–220.

Sato, T., Fukui, T., Atomi, H., and Imanaka, T. (2005). Improved and versatile transformation system allowing multiple genetic manipulations of the hyperthermophilic archaeon Thermococcus kodakaraensis. Appl. Environ. Microbiol. *71*, 3889–3899.

Stetter, K.O. (2006). History of discovery of the first hyperthermophiles. Extremophiles *10*, 357–362.

Takemasa, R., Yokooji, Y., Yamatsu, A., Atomi, H., and Imanaka, T. (2011). Thermococcus kodakarensis as a host for gene expression and protein secretion. Appl. Environ. Microbiol. *77*, 2392–2398.

Thiel, A., Michoud, G., Moalic, Y., Flament, D., and Jebbar, M. (2014). Genetic manipulations of the hyperthermophilic piezophilic archaeon Thermococcus barophilus. Appl. Environ. Microbiol. *80*, 2299–2306.

Von Damm, K.V., Edmond, J.t., Measures, C., and Grant, B. (1985). Chemistry of submarine hydrothermal solutions at Guaymas Basin, Gulf of California. Geochim. Cosmochim. Acta *49*, 2221–2237.

Waege, I., Schmid, G., Thumann, S., Thomm, M., and Hausner, W. (2010). Shuttle vector-based transformation system for Pyrococcus furiosus. Appl. Environ. Microbiol. *76*, 3308–3313.

Wang, F., Zhou, H., Meng, J., Peng, X., Jiang, L., Sun, P., Zhang, C., Van Nostrand, J.D., Deng, Y., and He, Z. (2009). GeoChip-based analysis of metabolic diversity of microbial communities at the Juan de Fuca Ridge hydrothermal vent. Proc. Natl. Acad. Sci. U.S.A. *106*, 4840–4845.

Xie, W., Wang, F., Guo, L., Chen, Z., Sievert, S.M., Meng, J., Huang, G., Li, Y., Yan, Q., and Wu, S. (2010). Comparative metagenomics of microbial communities inhabiting deep-sea hydrothermal vent chimneys with contrasting chemistries. ISME J. *5*, 414–426.

Diversity of Thermophilic Microorganisms and Their Roles in the Carbon Cycle

2

Shi-Qi Ji, Dong-Dong Meng, Kun-Di Zhang and Fu-Li Li

Abstract

Thermophiles are a large category of microorganisms that show optimum growth at temperatures of 50°C or higher. These microbes thrive in various environments in both marine and terrestrial habitats. The ability of microorganisms to proliferate under extreme conditions is of widespread importance in microbial physiology, ecological cycle, industry and evolution. Thermophiles play a complex role in ecosystem maintenance, particularly with respect to carbon cycle and biomass deconstruction. These ecological processes are carried out by various thermophilic microorganisms. Temperature is one of the most important factors controlling the activities and evolution of organisms, and limited species diversity found in extreme environments suggests that many organisms lack the capacity for successful adaptation to these environments. High-temperature environments are of special interest, in that they reveal the extremes to which evolution has been pushed. In this chapter, the diversity of thermophiles and their distinct strategies in carbon cycling and biomass degradation are discussed.

Introduction

The optimum growth temperatures of thermophiles range between 55°C and 80°C, and that of hyperthermophiles, at or above 80°C. Thermophilic microbes are also classified into moderate thermophiles (T_{opt} of 45–55°C), true thermophiles (T_{opt} of 55–75°C) and extremophiles (T_{opt} of>75°C) (Brock, 1986). The temperature boundary for defining thermophiles is not absolute, and in general, is proposed in accordance to certain considerations. The lower temperature boundary for thermophiles is that above which most mesophiles and Eukarya show weak or negative growth. On the other hand, for hyperthermophiles, the lower temperature boundary is that which distinguishes these microorganisms from thermophiles. To date, the upper temperature boundary of thermophiles has been observed to exceed 121°C, and their isolation has been carried out from samples at temperatures of over 350°C (Canganella and Wiegel, 2014; Prieur, 2001).

The microorganisms prevalent in high-temperature environments are typically anaerobic or microaerophilic, due to the low solubility of O_2 at high temperatures and the presence of reducing gases which render several of these biotopes anoxic (VanFossen *et al.*, 2008). Thermophilic microorganisms show a wide distribution, and are common in hot springs, shallow submarine hydrothermal systems, abyssal hot-vent environments, hot coal refuse piles and geothermal oil reservoirs (Blochl *et al.*, 1995). Although habitats with high temperatures support the growth and reproduction of a variety of thermophiles, the distribution of thermophiles is not limited to such geothermal areas.

Global carbon cycling involves the activities of both microorganisms and macroorganisms. Due to the acceleration of CO_2 inputs into the atmosphere by human industrial revolution, the carbon cycle has long been a subject of scientific interest. Carbon is cycled through all Earth's major carbon reservoirs including the atmosphere, the land, the oceans and other aquatic environment, sediments and rocks, and biomass. Through the largest carbon reservoir is in the sediment and rocks of

Earth's crust, but the turnover time is so long, which shows insignificant impacts on a human time scale. From the viewpoint of living organisms, a large amount of organic carbon is found in land plants and aquatic plants and algae. The most rapid means of global transfer of carbon is the CO_2 of the atmosphere. Carbon dioxide is removed by photosynthetic organisms from the atmosphere primarily and is returned to the atmosphere by respiration of animals and chemoorganotrophic microorganisms via fermentation, anaerobic respiration, or aerobic respiration (Madigan and Martinko, 2006).

Lignocellulose and macroalgae constitute two of the most abundant representative biomass types from terrestrial and marine environments, and are considered as promising feedstocks for bioconversion to biofuel and other bio-based products. Lignocellulosic biomass has broad prospects and has aroused intense interest. Wood and other lignocellulosic biomass are composed of cellulose (insoluble fibres of β-1,4-glucan), hemicellulose (non-cellulosic polysaccharides including xylan, mannan and glucan) and lignin (complex polyphenolic structure) (Kumar *et al.*, 2008). The precisely arranged crystalline cellulose core of cell-wall microfibrils is highly resistant to assault on its structural sugars by microbes and members of the animal kingdom (Himmel *et al.*, 2007). A major limitation to the active utilization of lignocellulose is its crystalline non-reactivity, and in particular, its resistance to hydrolysis. Recent reports highlight macroalgae as a promising feedstock for bioconversion (Wei *et al.*, 2013). The wide distribution of macroalgae in the ocean, particularly in the coastal areas, facilitates the culture and collection of these organisms. Brown algae, a scalable and sustainable seaweed, does not contain lignin, giving this organism a distinct advantage over lignocellulosic biomass. Direct conversion of macroalgae bypasses the need for physicochemical pre-treatment prior to fermentation, thereby reducing process cost. Numerous types of polysaccharides are present in seaweeds, and their structural information is shown in Fig. 2.1 (Wei *et al.*, 2013).

The process of carbon cycle involves the degradation of these complex carbohydrates into small molecules by hydrolysis. Thermophiles use a wide range of carbon sources, and possess the capacity for conversion of biomass to CO_2 or dissolved organic matter. Cellulolytic communities and species, such as the thermophilic bacterium *Clostridium thermocellum* and the extremely thermophilic *Caldicellulosiruptor* sp., have been identified from a wide variety of terrestrial sources, including bio-compost and soil. Hyperthermophilic archaea (T_{opt} of ≥80°C) utilize α- and β-linked glucans such as starch, barley glucan, laminarin and chitin, while hyperthermophilic bacteria (T_{opt} of ≥80°C) utilize not only glucans but also hemicelluloses such as xylan and mannan. The range of substrates utilized by a microorganism depends on its habitat and strain characteristics. A few terrestrial microorganisms have the capacity to utilize microcrystalline cellulose. On the other hand, cellulolytic microorganisms in marine environments have received less attention. The high salinity of the marine environment is likely to have facilitated the evolution of different cellulose-degrading mechanisms compared to the terrestrial organisms. The degradation of seaweed is usually performed by bacterial species from the marine environments. Marine microorganisms have evolved multifarious enzyme systems for the degradation of different kinds of polysaccharides; however, marine thermophiles capable of degrading polysaccharides have been studied to a lesser extent. In this chapter, we summarize the diversity of thermophiles and recent research on their application in biomass degradation.

Marine thermophiles in high-temperature habitats

The distribution of marine bacteria is influenced by various physical, chemical and biological factors such as temperature, pH, salinity, redox potential and nutrients. Temperature is an important environmental factor that influences the distribution and structure of microbial communities in a habitat, probably due to its direct impact on enzyme activities and metabolic processes in microbes. However, microbes have been found in marine habitats with high as well as low temperatures. Certain well-known marine habitats with high temperature include deep-sea hydrothermal vents, subsurface petroleum reservoirs

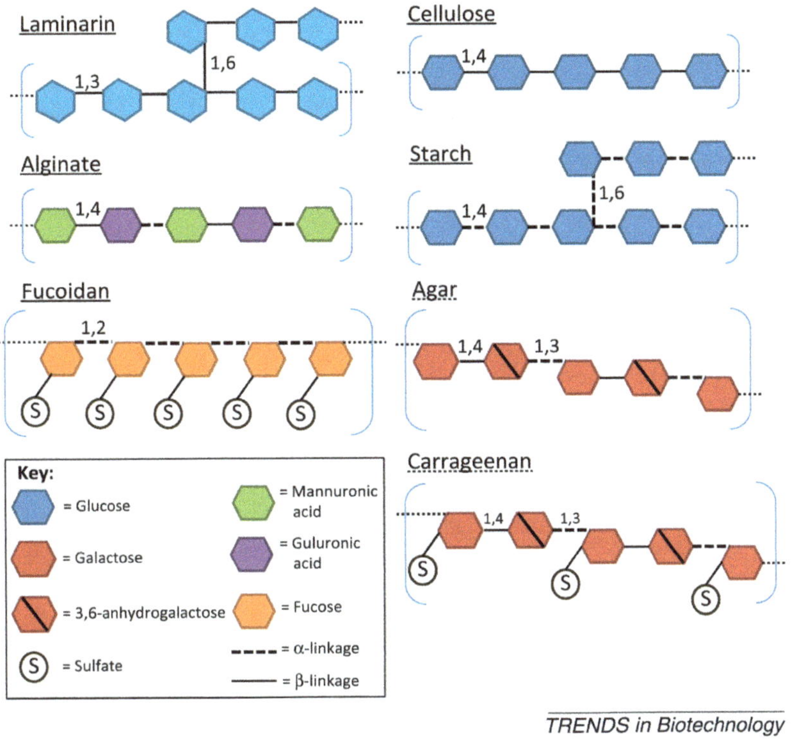

Figure 2.1 Structures of polysaccharides in seaweeds. Reprinted with permission from Wei *et al*. (2013). Copyright © 2012 Elsevier Ltd. With permission.

and marine hot springs, which are preferred by thermophilic microorganisms. The global distribution of hydrothermal vents is shown in Fig. 2.2 (Beaulieu *et al*., 2013), which reveals that they are most often located at the boundaries of tectonic plates. A detailed database of sites of more than 550 confirmed and inferred hydrothermal vents is available (Beaulieu *et al*., 2013). Hot seawater is ejected as black smoke from the cracks in the seafloor, prompting the name 'black smokers' (Fig. 2.3) (Connelly *et al*., 2012; Rogers *et al*., 2012). The water is acidic and contains high concentrations of salt and minerals.

A number of thermophilic bacteria have been isolated to date from marine geothermal habitats, particularly from hydrothermal vents. The isolated bacteria were widely distributed across various bacterial taxa; however, a few families from Aquificae, Thermotogae, Thermodesulfobacteria and Epsilonproteobacteria were represented to a much higher extent than other categories, including *Desulfurobacteriaceae*, *Aquificaceae*, *Hydrogenothermaceae*, *Hydrogenimonaceae*, *Thermotogaceae*, *Thermaceae*, *Thermodesulfobacteriaceae* and *Nautiliaceae* (Table 2.1). These highly represented families usually contain more than one genus that includes species identified as marine thermophiles. The common features shared by most species from these categories include their isolation from hydrothermal vent systems and chemolithoautotrophic characteristics, for instance, hydrogen-oxidizing, nitrate-reducing or sulfur-reducing. The bacterial species belonging to the family *Thermotogaceae* (including the genera *Kosmotoga*, *Marinitoga* and *Thermosipho*) are chiefly heterotrophic, and *Thermosipho affectus* was found to be capable of utilizing cellulose as a carbon source. Moreover, all the species of certain genera (*Marinithermus*, *Oceanithermus*, *Rhabdothermus* and *Vulcanithermus*) from the family *Thermaceae* were found to be marine thermophiles. These bacteria are capable of utilizing complex organic substrates; however, a significant difference from the families mentioned previously is that these bacteria grow under aerobic or microaerophilic conditions. As mentioned previously,

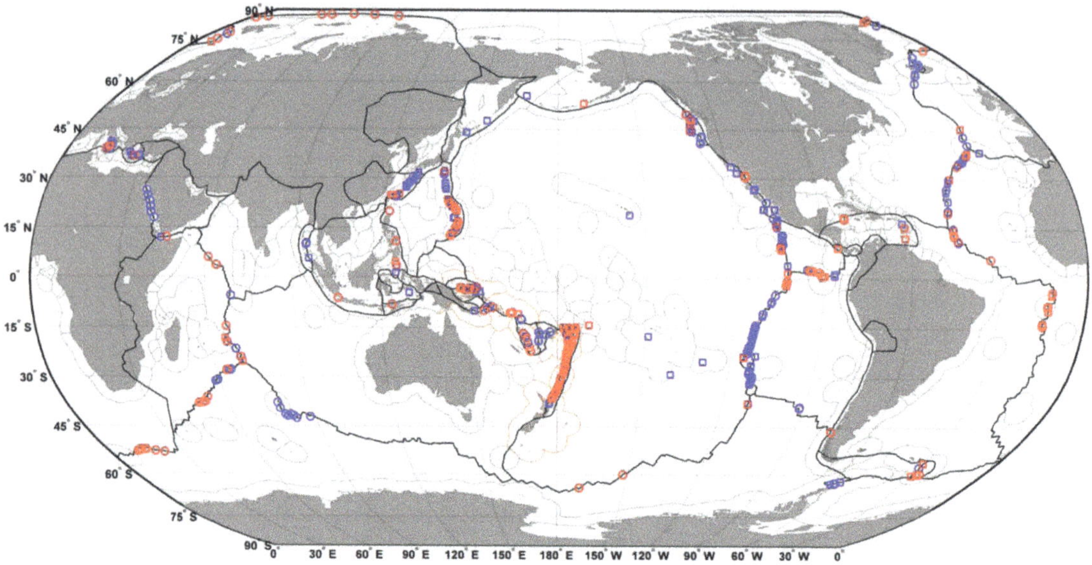

Figure 2.2 Global compilation of confirmed and inferred vent sites (confirmed (square) and inferred (circle), discovered and/or confirmed prior to (blue) or after (red) 2000). Reprinted with permission from Beaulieu *et al.* (2013). Copyright © 2013. American Geophysical Union. With permission.

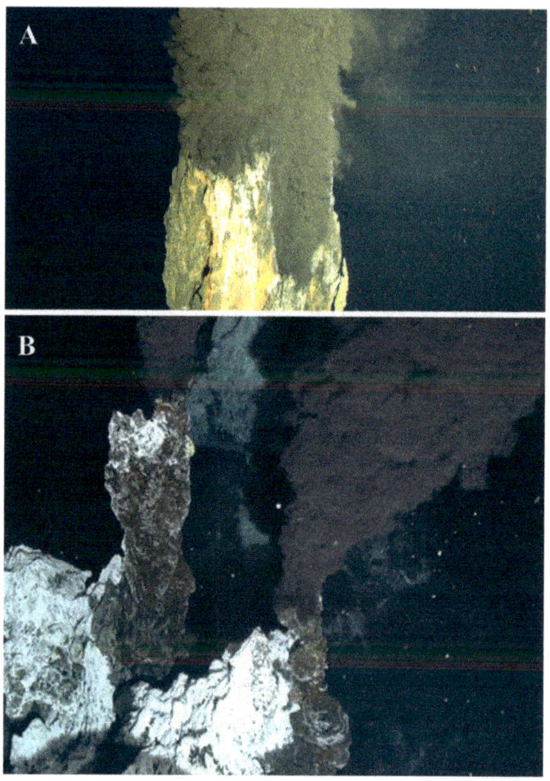

Figure 2.3 The black smokes ejected from the hydrothermal vents (A) The Beebe Vent, Piccard Hydrothermal Field on the Mid-Cayman Rise (4960 m depth) (Connelly *et al.*, 2012) (Figure 2.3A adapted with permission from Jon Copley, University of Southampton, UK); (B) Black smokers on the East Scotia Ridge south of the Polar Front (Rogers *et al.*, 2012).

Table 2.1 Taxonomic category mainly composed by marine thermophiles

Division or phylum	Family	Genera	Species	Reference
Aquificae	*Desulfurobacteriaceae*	*Balnearium* *Desulfurobacterium* *Phorcysia* *Thermovibrio*	All species from the four genera	Huber *et al*. (2002), L'Haridon *et al*. (2006), Perez-Rodriguez *et al*. (2012), Vetriani *et al*. (2004)
	Aquificaceae	*Aquifex* *Hydrogenivirga* *Hydrogenobaculum*	All species from the three genera	Huber *et al*. (1992), Nakagawa *et al*. (2004), Nunoura *et al*. (2008), Stohr *et al*. (2001)
	Hydrogenothermaceae	*Hydrogenothermus* *Persephonella*	All species from the three genera	Gotz *et al*. (2002), Nakagawa *et al*. (2003), Stohr *et al*. (2001)
	Hydrogenimonaceae	*Hydrogenimonas*	*Hydrogenimonas thermophila*	Takai *et al*. (2004)
Thermotogae	*Thermotogaceae*	*Kosmotoga Marinitoga* *Thermosipho*	All species from *Marinitoga*, and *K. arenicorallina*, *K. shengliensis*, *T. atlanticus*, *T. japonicas*, *T. melanesiensis*	Alain *et al*. (2002a), Antoine *et al*. (1997), DiPippo *et al*. (2009), Nunoura *et al*. (2007, 2010), Postec *et al*. (2005, 2010), Takai and Horikoshi (2000), Urios *et al*. (2004), Wery *et al*. (2001a)
Deinococcus-Thermus	*Thermaceae*	*Marinithermus* *Oceanithermus* *Rhabdothermus* *Vulcanithermus*	All species from the four genera	Miroshnichenko *et al*. (2003a), Miroshnichenko *et al*. (2003b), Mori *et al*. (2004), Sako *et al*. (2003), Steinsbu *et al*. (2011)
Thermodesulfobacteria	*Thermodesulfobacteriaceae*	*Thermodesulfatator* *Thermodesulfobacterium* *Thermosulfurimonas*	Most species from the four genera	Alain *et al*. (2010), Jeanthon *et al*. (2002), Moussard *et al*. (2004), Slobodkin *et al*. (2012)
Epsilonproteobacteria	*Nautiliaceae*	*Caminibacter* *Lebetimonas* *Nautilia*	All species from the three genera	Alain *et al*. (2002b, 2009), Miroshnichenko *et al*. (2002, 2004), Perez-Rodriguez *et al*. (2010), Smith *et al*. (2008), Takai *et al*. (2005), Voordeckers *et al*. (2005)

oxygen is typically limited in hydrothermal vent habitats, corresponding to the isolation of chiefly anaerobic or facultative anaerobic and few aerobic bacteria. Epsilonproteobacteria is increasingly recognized as an ecologically significant group of bacteria and includes large numbers of bacteria isolated from deep-sea hydrothermal habitats (Nakagawa *et al.*, 2005). Several of these bacteria have been characterized as thermophilic autotrophs utilizing hydrogen under anaerobic conditions (Sogin *et al.*, 2006). A study on diffuse flow hydrothermal vents using a massively parallel tag sequencing strategy revealed nearly 50% of the population in the sample as divergent Epsilonproteobacteria (Sogin *et al.*, 2006). Thus, thermophilic species of Epsilonproteobacteria likely play an important role in deep-sea high-temperature habitats.

Chemolithoautotrophic microorganisms such as hydrogen-oxidizing, nitrate-reducing or sulfur-reducing bacteria comprise a majority of the identified thermophiles from hydrothermal habitats. Moreover, a large number of sulfur-oxidizing species of bacteria from deep-sea vents belong to the genera *Thiomicrospira*, *Halothiobacillus* and *Beggiatoa* (Sievert *et al.*, 2008), which are chiefly mesophilic and therefore, beyond the scope of this chapter. The energy sources in hydrothermal habitats chiefly comprise chemicals released from high-temperature seawater–rock interactions. For instance, the sulfur-reducing bacterium *Nautilia abyssi* is capable of utilizing H_2 as an energy source, S° as an electron acceptor and CO_2 as a carbon source (Alain *et al.*, 2009).

As discussed previously, most of the isolated marine thermophiles belonged to a few genera with distribution across several families. Strains from other taxa isolated from typical geothermal areas under the ocean have also been reported as thermophiles. Examples are indicated in Table 2.2 and include marine thermophiles from the taxa Firmicutes, Chloroflexi, Bacteroidetes, Deferribacteres and Spirochaetes. The phylum Firmicutes, for instance, mainly comprises thermophilic bacteria from terrestrial environments and mesophiles; however, a few individual species were isolated from coastal or shallow hydrothermal vents or submarine sediments, including *Thermaerobacter marianensis* (and two other species belonging to this genus), *Bacillus vulcani*, *Calderihabitans maritimus*, *Caloranaerobacter azorensis* and *Carboxydobrachium pacificum*. These species are not distributed according to specific families or higher taxa, suggesting that the adaptation of these species to the marine environment is a result of the evolution of individual or small populations and not a systemic evolution of a large group. Moreover, the habitats of these species (shallow or submarine hydrothermal vents, Table 2.2) were closer to those of species from the main land compared to the deep-sea species (deep-sea hydrothermal vents, Table 2.1). We therefore hypothesize that a number of terrestrial thermophiles drifted to marine environments, evolved and adapted to the high-salinity marine environment. This hypothesis explains the distribution of these bacteria in habitats closer to the main land. Therefore, the species mentioned in Table 2.1 are more likely to be indigenous deep-sea microbes compared to those mentioned in Table 2.2.

To obtain a more comprehensive understanding of bacterial populations that thrive in shallow hydrothermal vents, Maugeri *et al.* (2009) investigated two hydrothermal vents off Panarea Island (Aeolian Islands, Italy) using PCR-DGGE fingerprinting and cultivation methods. Species of Epsilonproteobacteria were found to have very high representation among the bacteria inhabiting these two sites, with closely related bacteria isolated from various thermal or high-salt environments. Sulfur-oxidizing bacteria were detected by both culture-dependent and culture-independent methods. A total of 53 thermophilic, thermo-resistant and spore-forming strains were isolated using cultivation technique, and identified as species belonging to the genera *Bacillus* and *Geobacillus*, and the phylum Firmicutes (Maugeri *et al.*, 2009). Thus, submarine hydrothermal vents were concluded to be inhabited by a variety of microorganisms that tolerate environmental extremes, and possibly possess as yet uncharacterized industrial potential.

With the exception of thermophilic bacteria, most of the hyperthermophilic strains isolated from deep-sea hydrothermal vents belonged to the domain Archaea, including few genera from the families *Desulfurococcaceae*, *Pyrodictiaceae*, *Methanocaldococcaceae* and *Thermococcaceae*.

Table 2.2 Marine thermophilic and hyperthermophilic bacteria from other scattered taxons

Species name	Taxon	Optimum temperature (°C)	Origin	Reference
Thermaerobacter marianensisgen	Firmicutes	74–76	Deep sea	Takai *et al.* (1999)
Thermaerobacter nagasakiensis	Firmicutes	70	Shallow marine hydrothermal vent	Nunoura *et al.* (2002)
Thermomarinilinea lacunofontalis	Chloroflexi	60	Submarine hot spring	Nunoura *et al.* (2013)
Paenibacillus marinum	Firmicutes	55	Marine hot spring	Bouraoui *et al.* (2013)
Rhodothermus marinus Rhodothermus obamensis	Bacteroidetes	65, 80	Submarine alkaline hot springs, shallow hydrothermal vent	Alfredsson *et al.* (1988), Sako *et al.* (1996)
Bacillus vulcani	Firmicutes	60	Shallow hydrothermal vent	Caccamo *et al.* (2000)
Calderihabitans maritimus	Firmicutes	65	Marine sediment core of an undersea caldera	Yoneda *et al.* (2013)
Caldithrix palaeochoryensis	Deferribacteres	60	Geothermally heated sediment	Miroshnichenko *et al.* (2010)
Deferribacter autotrophicus	Deferribacteres	60	Deep-sea hydrothermal vent	Slobodkina *et al.* (2009)
Desulfacinum hydrothermale	Proteobacteria	60	Geothermally heated sediments	Sievert and Kuever (2000)
Caloranaerobacter azorensis	Firmicutes	65	Deep-sea hydrothermal chimney	Wery *et al.* (2001b)
Carboxydobrachium pacificum	Firmicutes	70	Submarine hot vent	Sokolova *et al.* (2001)
Spirochaeta thermophila	Spirochaetes	66–68	A marine hot spring	Aksenova *et al.* (1992)

Marine thermophiles in low-temperature habitats

In contrast to the optimum growth temperatures of thermophilic microbes, most parts of marine and sedimentary environments have low temperatures. Although several of these environments experience temperature fluctuations, the highest temperature encountered appears to be less than 30°C, which is much lower than the lower temperature boundary of thermophiles.

An enormous number of various types of microorganisms are found in marine habitats. In general, the mesophilic microbial population is considered to be highly active in these environments due to their relatively low temperature, while thermophilic bacteria are largely ignored due to low activities and smaller populations in such low-temperature environments. While the large diversity of thermophiles in marine geothermal habitats is well known, the origin and distribution of thermophiles in cold environments remains enigmatic. An increasing quantity of recent experimental evidence indicates diversity in the dormant spores of thermophilic bacteria found in low temperature environments; these spores become active at temperatures much higher than the *in situ* temperature. Existing evidence increasingly indicates that the geographic distribution of thermophiles is not restricted to high-temperature environments.

A few strains of thermophilic bacteria have been isolated from low-temperature habitats. The genus *Desulfotomaculum* is a large group comprising 31 species of bacteria; most of these bacterial species are sulfate-reducing thermophiles isolated from geothermal environments, capable of forming spores to survive unfavourable environmental conditions. Thus, strains of *Desulfotomaculum* have been isolated from environments that do not support their growth and reproduction (Vandieken *et al.*, 2006). For instance, *Desulfotomaculum arcticum*, a spore-forming, moderately thermophilic sulfate-reducing bacterium, was isolated from a permanently cold fjord sediment in Svalbard (Vandieken *et al.*, 2006). During research on sulfate reduction and microbial fuel cells (MFC), large numbers of thermophiles were found in low-temperature environments (Mai Faurschou Isaksen, 1994; Mathis *et al.*, 2008). Sulfate reduction is an extremely important process in the anaerobic mineralization of organic matter in marine shelf sediments (Leloup *et al.*, 2009). The consumption of organic matter by sulfate-reducing bacteria (SRB) occurs through the pathways depicted in Fig. 2.4. A research on SRB from sediment slurry in Aarhus Bay, Denmark, where seasonal temperatures range from 0°C to 15°C, revealed sulfate reduction activity at 60°C after a lag period of 1 day (Mai Faurschou Isaksen, 1994). Armed with the knowledge that environmental temperatures supporting the growth of thermophiles are not found in these habitats, the researchers discovered viable spores of thermophilic species at a density of 2.8×10^4 spore/cm^3 in the surface sediment (Mai Faurschou Isaksen, 1994). The isolation and characterization of another species of the genus *Desulfotomaculum*, named *Desulfotomaculum kuznetsovii*, was also described, proving that at least a few thermophilic species with physiological advantages for adaptation to unfavourable environments were widely distributed in cold marine environments. Another report showed that marine sediments from temperate water bodies were good sources of thermophilic electrode-reducing bacteria and that electric current was ten times greater when sediment fuel cells were incubated at 60°C rather than 22°C (Mathis *et al.*, 2008). The 16S rRNA genes from the anodes of MFC were cloned and sequenced. The members of the electrode-reducing community identified are listed in Table 2.3, with minor modifications. A cursory glance at Table 2.3 reveals that the closest relatives of the clones obtained in the study are found among the phyla Deferribacteres and Firmicutes. Species

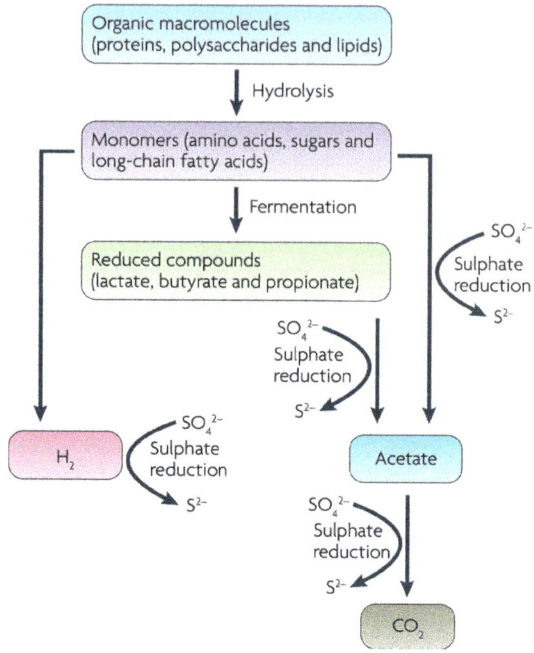

Figure 2.4 The pathways for sulfate-reducing bacteria in degradation of complex organic matter in anoxic environments in the presence of sulfate (Muyzer and Stams, 2008). Reprinted with permission from Muyzer and Stams (2008). Copyright © 2008, Rights Managed by Nature Publishing Group. With permission.

belonging to Deferribacteres are chiefly isolated from hydrothermal vents and thermophilic MFC, with few species of this phylum also distributed among mesophilic sediments. Species belonging to Firmicutes are isolated from various thermal environments, both marine and terrestrial (Tables 2.2 and 2.4). The distribution of Firmicutes-like species in marine sediments from low temperate water bodies was also confirmed in the same study. Several species of Firmicutes are capable of forming endospores, defined as a resistant differentiated cell type produced under hostile conditions.

Marine sediments constitute an important site of degradation of organic matter, which comprises a complex mixture of organisms from open sea as well as land. Anaerobic microbial communities play the major role in decomposition of organic matter in marine sediments. As described previously, several thermophiles were identified as sulfate-reducing bacteria, and high

Table 2.3 Blast results on bacterial 16S rRNA sequences derived from excised DGGE bands from samples collected from the work of Maugeri *et al.* (2009) with slight changes

Accession No.	BLASTn result	Phylogenetic affiliation	Similarity (%)	Found environment
FJ234151	Uncultured clone ATOS Iris 7 Rainbow 98	ε-Proteobacteria	97	Deep hydrothermal vent
FJ234152	Uncultured *Arcobacter* sp. clone IG-1-19	ε-Proteobacteria	97	Sea water
FJ234153	Uncult. clone MS12-2-B02	ε-Proteobacteria	94	Hydrothermal vent
FJ234154	Uncultured clone T6-Ph07-962	ε-Proteobacteria	96	Deep hydrothermal vent
FJ234163	Uncultured *Thiotrichales* clone SIMO-493	γ-Proteobacteria	92	Salt marsh
FJ234155	Uncultured. clone GoM HDB-15	ε-Proteobacteria	95	Gas hydrates
FJ234156	*Sulfurimonas denitrificans*	ε-Proteobacteria	97	Hydrothermal vent
FJ234157	*Sulfuricurvum kujiense*	ε-Proteobacteria	97	Underground crude oil
FJ234158	Uncultured clone PS-B28	ε-Proteobacteria	90	deep-sea vent
FJ234159	*Sulfurimonas paralvinellae*	ε-Proteobacteria	90	Deep-sea hydrothermal vent
FJ234160	Uncultured clone VC1.2-cl32	ε-Proteobacteria	82	Deep hydrothermal vent
FJ234172	*Chlorobium ferrooxidans*	Chlorobi	92	Sediment
FJ234161	Uncultured clone PYK04_18B_6	ε-Proteobacteria	95	Black smoker
FJ234164	*Caulobacter* sp. Tibet-S913	α-Proteobacteria	97	Permafrost region
FJ234165	Rhodomicrobium sp. ShRmc01	α-Proteobacteria	75	Lake Shira
FJ234169	Uncultured *Desulfobacteraceae* clone cLaki-JM30	δ-Proteobacteria	72	Lake Kinneret

Table 2.4 Clones and their closest relatives of a thermophilic MFC enriched from marine sediment from temperate waters (Mathis *et al.*, 2008)

Clones (number of clones)	Phylum (>90%)	Closest match (accession number) and origin	Similarity (%)
EU194828 (4)	Deferribacteres	Uncultured bacterium clone (DQ925879), hydrothermal vent chimneys	87
EU194830 (48)	Firmicutes	*Thermincola carboxydiphila* (AY603000), hot spring	99
EU194829 (1)	Deferribacteres	Uncultured bacterium clone (DQ424926), thermophilic MFC	96
EU194834 (6)	Deferribacteres	Uncultured bacterium clone 1A162 (DQ424915), thermophilic MFC	89
EU194831 (6)	Firmicutes	*Thermincola carboxydiphila* (AY603000), hot spring	99
EU194832 (4)	Firmicutes	*Thermincola carboxydiphila* (AY603000), hot spring	90
EU194835 (3)	Firmicutes	Uncultured bacterium clone (AY297976), thermophilic anaerobic hybrid reactor	98
EU194833 (1)	Firmicutes	*Thermincola carboxydiphila* (AY603000), hot spring	88
EU194837 (1)	Firmicutes	Uncultured low G+C Gram-positive bacterium clone (DQ234647), subsurface water	92
EU194836 (5)	Firmicutes	Uncultured soil bacterium clone (DQ248237), Carbon Tetrachloride Contaminated Soil	89
EU194827 (1)	Deferribacteres	Uncultured bacterium clone (DQ925879), hydrothermal vent chimneys	87

sulfate reduction rates (SRRs) were detected in anoxic sediments. Thus, studies on temperature adaptation could be carried out using SRR as a proxy for organic matter mineralization (Hubert *et al.*, 2010). Endospore germination experiments revealed a stable supply of thermophilic bacteria in the permanently cold Arctic marine sediments at densities exceeding 10^8 spores per m^2 per year (Hubert *et al.*, 2009). These Firmicutes are characterized by the lack of detectable activity at cold, *in situ* temperatures but rapid mineralization of organic matter by hydrolysis, fermentation and sulfate reduction upon induction at 50°C (Hubert *et al.*, 2009). The construction of clone libraries before (time zero) and after (56 hours) induction at 50°C revealed a complete replacement of the active microbial community in the sediment sample as the cold-adapted organisms were killed and the distinctly thermophilic bacterial community was activated. The thermophilic bacterial community chiefly included Firmicutes, with a large proportion of endospore-forming thermophilic Clostridiales (Table 2.5). The closest relatives of these thermophilic clones are found in warm subsurface petroleum reservoirs and ocean crust ecosystems, suggesting that seabed fluid flow from these environments delivered thermophiles to the cold ocean (Hubert *et al.*, 2009). This result is concomitant with the aforementioned study by Maugeri *et al.* (2009) on shallow hydrothermal vents, where they report the isolation of 53 strains under temperature conditions of 60°C, 70°C and 80°C, all of which belonged to Firmicutes. This result also suggests that thermophiles originating from geothermal areas supply large quantities of spore-forming bacteria to colder sites that are unsuitable for their growth; however, members of Epsilonproteobacteria and unique members of Gammaproteobacteria, which are also detected in hydrothermal environments, are not detected in cold environments probably because of their inability to form spores. In another work (Hubert *et al.*, 2010), thermophile diversity, abundance and distribution in Svalbard sediments were investigated. Volatile fatty acids (VFA) and sulfate reduction were monitored after incubation of the sediment slurry at 50°C for stimulation of thermophilic microbial activity. The production of abundant VFA was found to stimulate sulfate reduction, and SRR were found to increase exponentially corresponding to the consumption of VFA (Hubert *et al.*, 2010). Moreover, four different thermostable fluorescently labelled polysaccharides were employed for assessing extracellular thermotolerant hydrolytic enzymes. The observation that two of the four polysaccharides were hydrolysed upon incubation at 50°C indicated that these thermophiles possibly possess diverse extracellular hydrolytic enzymes. Another example that supports this conclusion is presented in the next section. The significant increase in SRR reflected activation and growth of thermophilic SRB. A clone library constructed for *Desulfotomaculum* revealed the existence of four distinct phylotypes (based on 16S rRNA gene sequence identity threshold of 99%). One of the closest relatives of the clone was *Desulfotomaculum halophilum*, isolated from formation waters in oilfields (Hubert *et al.*, 2010). Thus, species of the genus *Desulfotomaculum* are widespread in cold sediment habitats, and considerable diversity of thermophilic SRB is found in Svalbard marine sediments.

Table 2.5 Firmicutes enriched at 50°C and their closest relatives (Hubert *et al.*, 2009)

OUT, No. of clones	Next relative (accession number)	16S rDNA identity	Origin
A: 15 (18%)	*Clostridiisalibacter* sp. SOL3f37 (EU567322)	93.6%	Petroleum-contaminated soil
B: 10 (12%)	Gram-positive thermophile ODP15902 (AY704384)	94.1%	Crustal fluid
C: 4 (5%)	North Sea Troll oil reservoir clone (DQ647124)	96%	Formation water from a high-temperature oil reservoir
D: 7 (8%)	Gas field formation fluid clone (EU999019)	95.7%	Deep geothermal groundwater

Untapped thermophiles in coastal sediments

Several species of marine thermophiles have been introduced in the previous sections, which should provide an overall picture of the composition and distribution of these thermophiles. Marine thermophilic microbial communities are widely distributed in marine hydrothermal systems, hot springs, high-temperature petroleum reservoirs, fluids in the ocean crust and unexpectedly, cold marine environments. Thermophiles likely to be the original inhabitants have been isolated from high-temperature environments and include species belonging to Aquificae, Thermotogae, Thermodesulfobacteria and Epsilonproteobacteria. The diversity of known thermophilic anaerobic taxa that utilize various metabolic strategies is notable; for example, sulfur-reducing capability has been reported within the families *Aquificaceae* and *Thermotogaceae*, while marine thermophiles from *Nautiliaceae* are hydrogen-oxidizing or nitrate-reducing chemolithoautotrophs. The occurrence of thermophiles in low-temperature marine habitats has also been revealed by a chance finding. The use of non-culture methods revealed that thermophiles with a wide distribution in cold or other low-temperature environments are predominantly species of Firmicutes with spore-forming ability. The study of marine hydrothermal systems presents a few difficulties; sampling difficulty and the requirement of high pressure for certain deep-sea species constitute the main limitations for the culture of thermophiles by conventional techniques. Therefore, the distribution of thermophiles in the coastal sediments offers better opportunities for the isolation and study of marine thermophiles, particularly in the light of knowledge that these thermophiles likely travelled thousands of miles from deep-sea vents over hundreds of years. High sedimentation rates are found in coastal areas, and the age of a buried mussel bed at a depth of approximately 200 cm was estimated to be about 650 years (Kopke *et al.*, 2005). Moreover, 8500-year-old sapropels have been found at shallow depths of 17 to 35 cm below the sea floor (Suss *et al.*, 2004). The transport and dispersion of sediments are important for the distribution of microorganisms. Fig. 2.5 shows the processes that influence sediment transport. We know that the microbial species in marine sediments include thermophiles, moderate thermophiles and mesophiles, which are activated at different temperatures, and that thermophiles chiefly exist

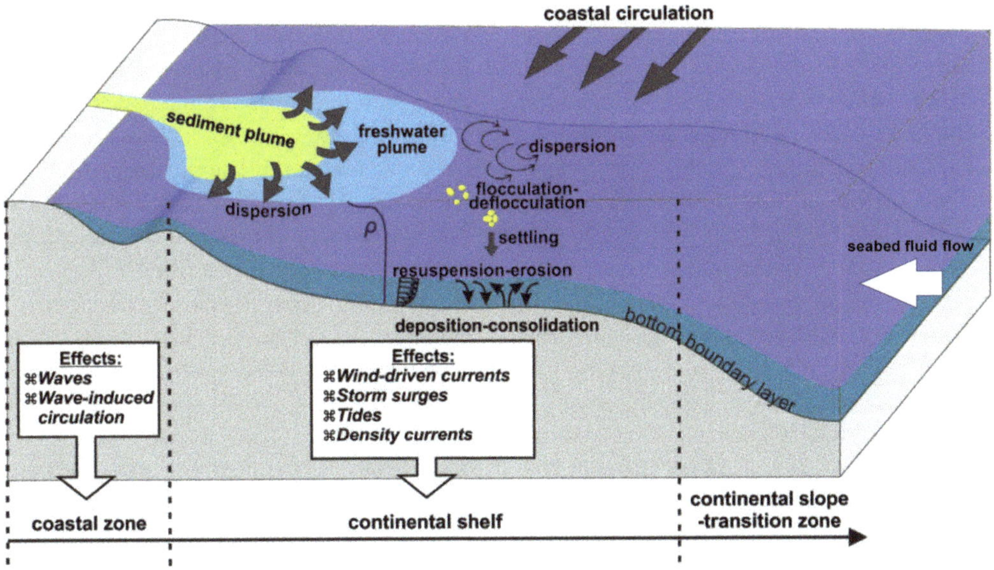

Figure 2.5 The main processes that control coastal sediment transport (Kombiadou and Krestenitis, 2013) (this diagram was cited from Kombiadou and Krestenitis with a modification of adding seabed fluid flow).

in the form of endospores. Sediment transport can be assumed to be, in reality, the transport of bacteria in the form of endospores, particularly with respect to the sediment from high-temperature environments.

In a recent study, microbial diversity in coastal subsurface sediments was assessed by a cultivation approach using various electron acceptors and substrate gradients; the study revealed that species belonging to Firmicutes constitute a large proportion of the microbial community. Further, the contribution of endospores to the viable community was evaluated, which revealed that absolute numbers of spores whose growth could be stimulated increased with depth, reaching a maximum at a depth of 50 cm (Kopke *et al.*, 2005). This observation suggests that spores from other parts of the ocean constantly accumulate on sediments. This characteristic of thermophiles has received little attention until now.

To obtain a clearer picture of the distribution of thermophiles, our group recently conducted a primary investigation of these microbial resources in the coastal sediments. Polysaccharides are important components of several biomasses; for example, cellulose in crop straw, alginate in brown algae and chitosan in shrimp shell. Coastal sediments involve the settling down of various polysaccharide-containing particles from both marine and terrestrial habitats on the sea floor. Hydrolysis of polysaccharides by anaerobic degradation pathways involves the secretion of extracellular enzymes, and results in the production of smaller monomeric products for cellular uptake and fermentation of the smaller organic matter. Therefore, we utilized polysaccharides for the enrichment of thermophiles from coastal sediments. Bacterial enrichment and isolation were carried out in the presence of individual polysaccharides of high molecular weight such as microcrystalline cellulose, alginate, chitosan, pectin and pullulan as sole carbon sources. Both culture-dependent and culture-independent methods were employed; culture-dependent methods such as plating were employed for the isolation of pure cultures. Although not all the tested polysaccharides were degraded by the thermophilic bacteria, certain interesting thermophilic bacterial communities were obtained. As an example, an anaerobic thermophilic bacterial community was enriched from coastal marine sediments, and a microbial community capable of efficiently degrading filter paper, the carbon source, was isolated under anaerobic conditions at 60°C (Ji *et al.*, 2012). For the identification of the community members, 16S rRNA gene library of the cellulolytic consortium was constructed. The most abundant OTUs accounted for 42% and 18% of the library, and shared similarity with the type strain *C. thermocellum* ATCC 27405, which is known for its high cellulolytic capacity attributable to the formation of cellulosomes (Ji *et al.*, 2012). In contrast to other cellulolytic enrichments, all sequences from the OTUs represented novel species. Moreover, most of the clones represented by the library shared 16S rRNA sequence similarities of lower than 90%, and all of them, of lower than 94% (Table 2.6). Their closest relatives are easily identifiable as Firmicutes, suggesting that the clones actually represent active spores under conditions of high temperature. These results indicate the existence of extensive unexplored bacterial resources in the ocean. Table 2.6 shows that their closest relatives are spread across various habitats, most of which have high temperatures, such as hot springs, petroleum fields, compost and deep-sea hydrothermal chimneys. The thermophilic members of the community could possibly have similar origins as their relatives, but due to exogenous forces, they drifted to certain novel environments and underwent relatively independent evolution.

Further, strain isolation was carried out for obtaining a clearer picture of the community, and the isolated pure cultures are listed in Table 2.7. The species that constituted a major proportion of the community as per 16S rRNA gene library analysis and which was closely related to *C. thermocellum* was not isolated. In contrast, species that accounted for minor proportions were more easily isolated, as were certain species not detected in the clone library. These pure cultures shared low 16S rRNA gene similarities with the identified type species, suggesting that the thermophiles in coastal sediments are extremely diverse and have remained largely undiscovered. In contrast to the study by Hubert *et al.* (2010) which revealed fluxing of diverse thermophilic bacteria into the cold arctic seabed from subsurface petroleum

Table 2.6 Represent species from the clone library of a thermophilic community enriched from coastal sediment and their closest relatives (Ji *et al.*, 2012)

Clone No.	Length (bp)	Nearest type strain	Origin	Similarity (%)	Accession
2	1499	*Acetomicrobium faecale*	Sewage sludge	88.0	FR749980
4	1496	*Planifilum yunnanense*	Hot spring	88.0	DQ119659
6	1512	*Desulfotomaculum alkaliphilum*	Mixed cow/pig manure	91.2	AF097024
8	1498	*Clostridium thermocellum*	Various terrestrial habitats	89.4	CP000568
9	1500	*Clostridium thermocellum*	Various terrestrial habitats	87.6	CP000568
11	1505	*Clostridium straminisolvens*	Cellulose-degrading bacterial community	88.8	AB125279
21	1624	*Caldicoprobacter oshimai*	Sheep faeces	89.8	AB450762
22	1505	*Desulfotomaculum halophilum*	Oil production facilities	90.6	U88891
23	1490	*Desulfotomaculum halophilum*	Oil production facilities	86.6	U88891
27	1494	*Sporosalibacterium faouarense*	Hydrocarbon-polluted soil surrounding a deep petroleum environment	92.4	EU567322
39	1501	*Clostridium purinilyticum*	Unclear	90.5	FR749894
41	1497	*Sporosalibacterium faouarense*	Hydrocarbon-polluted soil	86.7	EU567322
42	1492	*Planifilum fulgidum*	Compost	88.0	AB088362
46	1454	*Caloranaerobacter azorensis*	Deep-sea hydrothermal chimney	90.7	AJ272422
60	1487	*Bacillus thermolactis*	Dairy farms	93.4	AY397764

Table 2.7 Pure cultures isolated from the thermophilic community from our group

Pure culture	Closest type strain	Origin	Identity (%)	Accession
ASX2	*Desulfotomaculum halophilum*	Oil production facilities	90.4	U88891
ASW3	*Bacillus thermolactis*	Dairy farms	93.4	AY397764
ASX1	*Caldicoprobacter oshimai*	Sheep faeces	89.8	AB450762
ASX3	*Tepidimicrobium xylanilyticum*	Sludge of a thermophilic anaerobic Digester treating municipal solid waste and sewage	94.5	EF522948
ASX4	*Brassicibacter mesophilus*	Food industry wastewater	93.8	GU645013
CA	*Brassicibacter mesophilus*	Food industry wastewater	89.3	GU645013
S6	*Moorella humiferrea*	Terrestrial hydrothermal spring	86.6	GQ872425
H4	*Clostridium sporosphaeroides*	Unclear	92.7	X66002
DJ	*Thermosediminibacter litoriperuensis*	Coast of Peru	91.0	AY703479

reservoirs or ocean crust ecosystems, the origins of thermophiles from coastal sediments appears to be more complex. Further research is required for obtaining a clear picture of the origins of these thermophiles and their physiological and metabolic properties.

Diversity and enzymology of thermophilic cellulolytic microorganisms

The conversion of lignocellulosic biomass to fermentable sugars remains a big challenge due to the recalcitrance of the insoluble starting material. Therefore, the discovery of novel microorganisms that possess the ability to convert lignocellulosic biomass into sugars is very important. Moreover, employing active biomass-hydrolysing enzymes at extreme conditions of temperature and pH is likely to be advantageous for industrial scale production as these conditions are suitable for overcoming biomass recalcitrance and preventing the growth of contaminating microorganisms. To date, thermophilic cellulolytic microorganisms have been found among several genera, including *Thermotoga, Caldicellulosiruptor, Thermoanaerobacter, Thermoanaerobacterium* and *Clostridium.* The following section provides a brief summary of hyperthermophiles with polysaccharide-hydrolysing ability, and detailed discussion of two of the most studied cellulolytic genera *Caldicellulosiruptor* and *Clostridium.*

Hyperthermophiles

Hyperthermophiles that belong to the orders *Thermococcales, Sulfolobales* and *Thermotogales* have been reported to produce hydrolases. Archaeal species of the order *Thermococcales* have been reported to grow at temperatures above 100°C. These fermentative anaerobes are typically isolated from hydrothermal sea vents. The best-studied species, *Pyrococcus furiosus* (T_{opt} of 100°C), metabolizes α-linked glucosides such as starch and pullulan, and b-linked glucosides such as laminarin, barley glucan and chitin, producing hydrogen, carbon dioxide and acetate as primary fermentative products (Driskill *et al.*, 1999; Fiala and Stetter, 1986). A closely related species, *Pyrococcus horikoshii,* produces several glycoside hydrolases (Kaper *et al.*, 2002). The genome sequences of *P. furiosus, P. horikoshii* and two other members of *Thermococcales, P. abysii* and *Thermococcus kodakarensis* KOD1 (Fukui *et al.*, 2005; Kawarabayasi *et al.*, 1998; Lecompte *et al.*, 2001; Maeder *et al.*, 1999), reveal a variety of glucosidases and glucanases, some of which have been biochemically characterized (Ando *et al.*, 2002; Bauer *et al.*, 1999; Gueguen *et al.*, 1997; Kashima *et al.*, 2005). Unfortunately, none of them showed the ability to hydrolyse crystalline cellulose. The genus *Sulfolobus* of *Sulfolobales* has been well studied, and includes aerobic, extremely thermoacidophilic heterotrophs capable of growth on peptides, monosaccharides and α-linked polysaccharides (Sensen *et al.*, 1998). *Sulfolobus solfataricus* was reported to produce hemicellulolytic enzymes, including endoxylanase (Maurelli *et al.*, 2008) and a bifunctional β-xylosidase/α-L-arabinosidase (Kambourova *et al.*, 2007). *Thermotoga maritima* is the best-studied species of *Thermotogales* (T_{opt} of 80°C). About 7% of the predicted coding sequences in its genome are expected to be involved in the metabolism of monosaccharides and polysaccharides (Chhabra *et al.*, 2003; Rodionov *et al.*, 2013). *T. maritima* has been reported to possess the capacity for degrading β-glucan, xylan and pectin (Bronnenmeier *et al.*, 1995; Kluskens *et al.*, 2003; Xue *et al.*, 2006; Yang *et al.*, 2006). Other species of *Thermotoga,* including *Thermotoga lettingae, Thermotoga elfii, Thermotoga naphthophila* and *Thermotoga petrophila,* possess the capacity to degrade simple sugars (Ravot *et al.*, 1995; Takahata *et al.*, 2001).

Caldicellulosiruptor

In contrast to the hyperthermophilic microorganisms (T_{opt} of ≥ 80°C) described above, the order *Clostridiales* includes numerous species which are much less thermophilic and utilize crystalline cellulose and hemicellulose as growth substrates; these species play an important role in biomass deconstruction. Two of the most important genera in this order are *Caldicellulosiruptor* (T_{opt} of ≥ 70°C) and *Clostridium* (T_{opt} of ≥ 60°C).

Diversity of the genus *Caldicellulosiruptor*

The genus *Caldicellulosiruptor* utilizes carbohydrate components of plant cell walls, including cellulose and hemicellulose. This ability is crucial for the deconstruction of plant biomass into fermentable sugars. *Caldicellulosiruptor saccharolyticus* DSM 8903, the first strain of the genus, was isolated and identified in the 1980s, and seven other strains have been isolated since (Rainey *et al.*, 1994). The physiological and biochemical properties of lignocellulose degradation by these organisms have been intensively studied. *Caldicellulosiruptor* spp. carry out noncellulosomal deconstruction of plant biomass. The type species *C. saccharolyticus* of the genus, which was formerly known as Tp8T6331, was isolated from thermal springs in New Zealand,

and was described as an obligate anaerobe, and an extremely thermophilic, cellulolytic bacterium (Rainey *et al.*, 1994). Subsequently, *Caldicellulosiruptor lactoaceticus* was isolated from an alkaline hot spring in Iceland, and was found to share similar physiological properties with *C. saccharolyticus* (Mladenovska *et al.*, 1995). Several years later, *Caldicellulosiruptor owensensis* and *Caldicellulosiruptor kristjanssonii* were isolated (Bredholt *et al.*, 1999; Huang *et al.*, 1998) either from a sediment sample taken from Owens Lake, California, or from a biomat sample of a slightly alkaline hot spring (78°C) in Iceland. Several species were also isolated from terrestrial hot springs of Kamchatka (Far East, Russia), including *Caldicellulosiruptor kronotskyensis*, *Caldicellulosiruptor hydrothermalis* and *Caldicellulosiruptor bescii* strain DSM 6725 (Miroshnichenko *et al.*, 2008; Svetlichnyi *et al.*, 1990; Yang *et al.*, 2010). More recently, *Caldicellulosiruptor obsidiansis* was isolated from Obsidian Pool, Yellowstone National Park, through enrichment cultures containing dilute acid-pretreated switchgrass as the primary carbon and energy source for cultivation (Hamilton-Brehm *et al.*, 2010). In our group, a newly isolated strain *Caldicellulosiruptor* sp. F32 (GenBank Accession No: APGP00000000) exhibited strong lignocellulose degradation activity (Ying *et al.*, 2013). The phylogenetic tree of various species of *Caldicellulosiruptor* is shown in Fig. 2.6.

Caldicellulosiruptor spp. have a unique multi-functional cellulase system, which is similar to the non-cellulosomal system of cellulose-hydrolysing fungi such as *Trichoderma reesei*. The enzymatic system of *Caldicellulosiruptor* spp., however, is different, and could be considered midway between cellulosomes and free enzymes. The individual cellulases secreted in this system are multi-modular, and contain multiple binding (carbohydrate-binding modules, CBMs) and catalytic (glycoside hydrolase, GHs) domains. The efficiency of the cellulase system and its adaptation

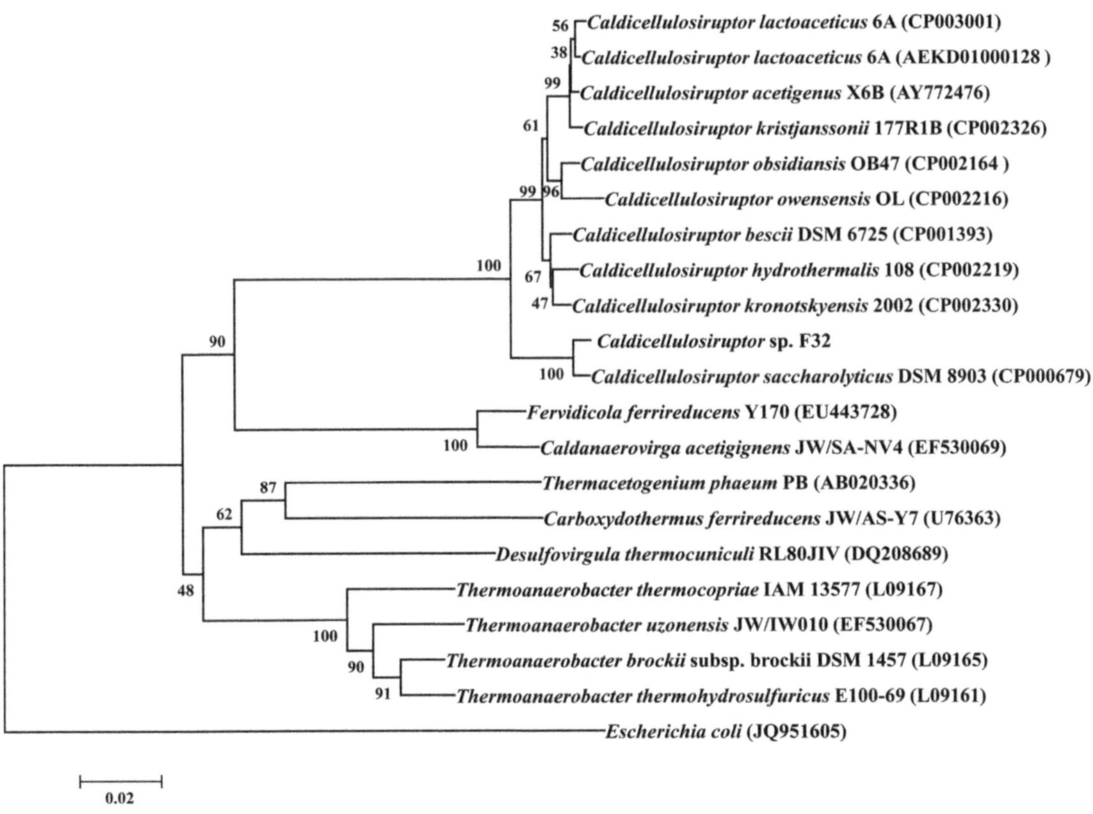

Figure 2.6 Phylogenetic tree using 16S rRNA sequences of *Caldicellulosiruptor* species plus related outliers.

to extremely high-temperature environments make *Caldicellulosiruptor* spp. strong candidates for lignocellulose degradation. Furthermore, several recent studies have promoted enthusiasm for the study of *Caldicellulosiruptor* strains. For example, the strain *C. saccharolyticus* is capable of simultaneously utilizing pentose and hexose sugars without glucose-induced carbon catabolite repression (CCR) (van de Werken *et al.*, 2008), while *C. bescii* can utilize lignocellulosic biomass without pretreatment (Yang *et al.*, 2009). The ability to simultaneously utilize a large range of sugars derived from biomass feedstocks is a highly desirable feature of these lignocellulose-utilizing and biofuel-producing bacteria.

Physiological and biochemical properties of the genus *Caldicellulosiruptor*

The members of the genus *Caldicellulosiruptor* are found in diverse habitats worldwide, including terrestrial hot springs and thermally heated mud flats. The lowest and highest optimum growth temperatures for *Caldicellulosiruptor* are 65°C and 78°C, respectively (Table 2.8). *Caldicellulosiruptor* spp. are typically associated with plant debris; consequently, all isolates characterized to date show the capacity to hydrolyse certain complex carbohydrates characteristic of plant cell walls, including cellulose and hemicellulose (Blumer-Schuette *et al.*, 2008). In addition, all the species of this genus show growth on xylan and acid-pretreated switchgrass, while growth on crystalline cellulose is variable. *Caldicellulosiruptor* spp. are capable of metabolizing various monosaccharides and oligosaccharides; glucans and xylans are utilized as carbon sources, and more importantly, lignocellulosic biomass is degraded without pretreatment. Therefore, the members of *Caldicellulosiruptor* are excellent genetic reservoirs of enzymes for plant biomass degradation (Blumer-Schuette *et al.*, 2012). Growth on various sugar substrates has been reported during the isolation of various species of *Caldicellulosiruptor* (Table 2.8). In a direct comparison of carbohydrate utilization by all the species of this genus (Blumer-Schuette *et al.*, 2010), seven *Caldicellulosiruptor* spp. were grown on various simple (C5 and C6) and complex sugars as well as dilute acid-pretreated biomass. Cell density was measured at 24 hours when grown on Avicel, pretreated switchgrass, glucose, xylose and yeast extract, which revealed 10-fold higher cell densities of *C. saccharolyticus* compared to *C. kristjanssonii* with the use of xylose as a carbon source. Growth of *C. lactoaceticus* was not observed on glucose, which is consistent with a previous study (Mladenovska *et al.*, 1995). The utilization of lignocellulosic material was also investigated; all the seven strains of *Caldicellulosiruptor* showed growth, with cell densities of higher than 10^8 cells/ml, upon the use of dilute acid-pretreated biomass (switchgrass) as growth substrate, indicating that the spectrum of polysaccharides present in the biomass (~50% glucan and 8% xylan) supported growth of all the species tested. Slight growth inhibition of the strains *C. saccharolyticus, C. bescii, C. kronotskyensis* and *C. lactoaceticus*, and stimulated growth of *C. kristjanssonii* and *C. owensensis* were observed on pretreated switchgrass compared to Avicel (Blumer-Schuette *et al.*, 2010). The results indicate that the most striking difference among the strains of *Caldicellulosiruptor* is their ability to grow on microcrystalline cellulose (Avicel). *C. saccharolyticus, C. bescii* and *C. kronotskyensis* showed vigorous growth on Avicel with final cell densities of over 10^9 cells/ml, while *C. kristjanssonii, C. hydrothermalis* and *C. owensensis* showed poorer growth, with cell densities of approximately 10^7–10^8 cells/ml. Moreover, different end products were obtained in different strains of the genus. All the species of *Caldicellulosiruptor,* with the exception of *C. obsidiansis* and *C. kronotskyensis*, produced H_2, a clean biofuel, as the main product. The high yield of H_2 (nearly 4 mol of H_2 per mol of glucose) obtained makes *Caldicellulosiruptor* spp. attractive candidates for lignocellulose-derived H_2 production. Of note, all species of *Caldicellulosiruptor* produced acetate and lactate; specifically, *C. hydrothermalis* and *C. kronotskyensis* produced these two organic acids as major products. In contrast, only a small amount of ethanol production was observed in five strains of this genus. Studying the pathways involved in the synthesis of acid and ethanol could help improve hydrogen and ethanol production.

Table 2.8 Characteristics of the genus *Caldicellulosiruptor*

Species	Optimal temperature (°C)	Carbon resources	Products	Reference
C. saccharolyticus DSM 8903	70	D-Glucose, D-xylose, maltose, starch, pullulan, amorphous cellulose and microcrystalline cellulose, xylan	H_2, acetate, lactate, small amounts of ethanol	van de Werken *et al.* (2008)
C. bescii DSM 6725	75	Able to use a variety of monosaccharides, microcrystalline cellulose, CMC, starch, glycogen, cellulose, xylan; unable to use xylose and pectin	H_2, acetate, lactate, CO_2 and small amounts of ethanol	Dam *et al.* (2011), Kataeva *et al.* (2009), Yang *et al.* (2010)
C. obsidiansis ATCC BAA2073	78	Glucose, xylose, maltose, starch, xylan, filter paper, pectin and dilute acid-pretreated switchgrass; unable to use mannitol, fucose, lignin, acetate, glycerol, ribose, sorbitol, carboxymethylcellulose	H_2, acetate, CO_2	Hamilton-Brehm *et al.* (2010)
C. hydrothermalis DSM 18901	65	Cellulose, CMC, filter paper, starch, xylan, pectin, cellobiose, glucose, xylose; unable to use pyruvate, chitin, raffinose, trehalose, ribose, peptone	Acetate, lactate	Elkins *et al.* (2010), Miroshnichenko *et al.* (2008)
C. kristjanssonii DSM 12137	78	Cellulose, xylan, starch, pectin; unable to use casein peptone, pyruvate, D-ribose, yeast extract	H_2, acetate, CO_2 and small amounts of ethanol and lactate	Blumer-Schuette *et al.* (2011), Bredholt *et al.* (1999)
C. kronotskyensis DSM 18902	70	Cellulose, CMC, filter paper, starch, xylan, dextran, pectin, cellobiose, glucose, xylose, maltose, lactose, arabinose, trehalose, yeast extract, peptone; unable to use pyruvate, chitin, raffinose, ribose	Acetate, lactate	Blumer-Schuette *et al.* (2011), Miroshnichenko *et al.* (2008)
C. lactoaceticus DSM 9545	68	Cellulose, xylan, starch, pectin, xylose, maltose; unable to use glucose, mannitose, pyruvate, ribose, sucrose	H_2, acetate, lactate and small amounts of ethanol and CO_2	Blumer-Schuette *et al.* (2011), Mladenovska *et al.* (1995)
C. owensensis DSM 13100	75	Glucose, xylose, pectin, xylan, starch, fructose, cellulose; unable to use glycerol lactate, trehalose	H_2, acetate, lactate, ethanol and CO_2	Blumer-Schuette *et al.* (2011), Huang *et al.* (1998)

Deconstruction of lignocellulose by *Caldicellulosiruptor* and cellulases

The characteristics of lignocellulosic biomass include cross-linking heterogeneity, complexity of constituents such as microfibrils and matrix polymers and high crystallinity, rendering the access of microorganisms or cellulases as well as degradation difficult. Strains of the genus *Caldicellulosiruptor* secrete large quantities of free multidomain thermophilic cellulases, which make them excellent candidates for lignocellulose deconstruction. Members of *Caldicellulosiruptor* have strong capabilities for cellulose and hemicellulose degradation, and are a huge repository of thermophilic cellulases; their glycoside hydrolases are distributed into 43 module families, and 26 GHs from 17 families constitute the core cellulase systems of the members of this genus (Blumer-Schuette *et al.*, 2012). Genome sequence data indicated the presence of certain gene clusters associated with lignocellulose degradation. Two GH-associated genomic loci of *C. saccharolyticus*

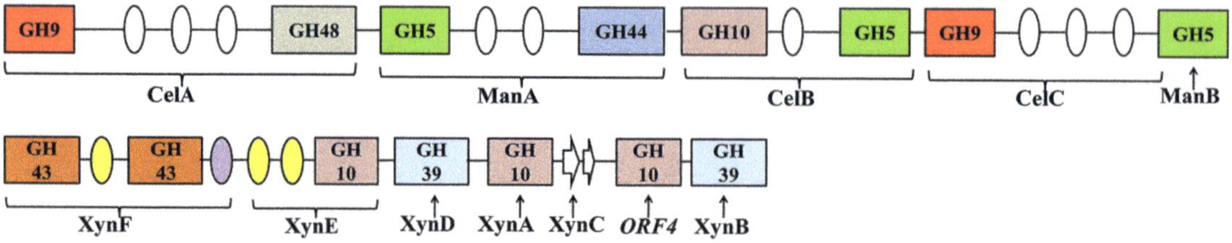

Figure 2.7 Biomass degrading enzyme genes located together in C. *saccharolyticus* (VanFossen *et al.*, 2011). A: Cellulase-hemicellulase gene cluster celA–manB (Csac_1076-1080) and xylanase gene cluster xynB–xynF (Csac_2404-2411). Boxes represent GH domains and the circles are CBM domains: solid white, CBM3; horizontal strips, CBM22; vertical strips, CBM6. Reprinted with permission from VanFossen *et al.* (2011). Copyright © 2011 Wiley Periodicals, Inc. With permission.

appear to be central to the hydrolysis of hemicellulosic and cellulosic substrates (Fig. 2.7). XynB–XynF (Csac_2404–Csac_2411) encodes intracellular and extracellular GHs with specificity for xylan and xylan side-chains as well as carboxymethyl cellulose (CMC), while CelA–ManB (Csac_1076–Csac_1080) encodes four multidomain extracellular cellulases (VanFossen *et al.*, 2011). In addition to GH-associated loci, several genes involved in lignocellulose degradation and metabolism also tend to aggregate into one cluster in certain species of this genus. Han *et al.* (2012) reported a hemicellulose utilization gene cluster, which encodes all the proteins required for hemicellulose utilization and metabolism. Such an organization greatly facilitates bacterial survival. Moreover, the lignocellulose-degrading enzymes exist not only in a single genomic loci or gene cluster, but also in the form of multi-functional enzymes. Several multidomain GHs have been characterized, such as CelA, CelB, Cel9B/Man5A and CbMan5B/Cel44A (Brunecky *et al.*, 2013; Park *et al.*, 2011; Su *et al.*, 2012; Ye *et al.*, 2012). In a recent work from our group, two xylanases were identified from *Caldicellulosiruptor* sp. F32, and found to exhibit distinctive xylan-hydrolysing properties (Ying *et al.*, 2013). The half-lives of xylanase JX030400 and JX030401 at 75°C were 4.1 hours and 4.6 hours, respectively; however, xylanase JX030400 showed 9-fold higher specific activity than JX030401.

Pretreatment of lignocellulose is an effective method to accelerate its degradation. Several pretreatment methods are carried out for enhancing the rates of subsequent enzymatic and microbial catalysis. Pretreatment methods typically include physical (e.g. milling, grinding), chemical (e.g. treatment with alkali, dilute acid), thermal (e.g. steam treatment) and biological (e.g. wood-degrading fungi) methods (Jorgensen *et al.*, 2007). For example, the use of dilute acid-pretreated biomass (switchgrass) as growth substrate allowed the growth of *C. bescii, C. hydrothermalis, C. kristjanssonii, C. kronotskyensis, C. lactoaceticus, C. owensensis* and *C. saccharolyticus* to cell densities of over 10^8 cells/ml, indicating that the spectrum of polysaccharides present in the biomass (~50% glucan and 8% xylan) supported growth of all the species tested (Blumer-Schuette *et al.*, 2010). In addition to these well-established technologies, a few novel approaches have also been reported. The use of ionic liquids (ILs) was first reported in 2002 during attempts to fractionate lignocellulose and decrystallize cellulose (Swatloski *et al.*, 2002). A phosphoric acid-acetone-water-based pretreatment conducted under mild conditions (50°C, 1 atm, 30–60 minutes) was also reported (Zhang *et al.*, 2007). In short, pretreatment facilitates the destruction of the chemical, physical and morphological characteristics of lignocellulose, which increases the accessibility of the material to hydrolysing enzymes. Although pretreatment typically results in vast improvements in the efficiency of lignocellulose conversion, several adverse effects have also been observed. The digestibility and recovery of cellulose and hemicellulose are likely to be lower than expected. A great number of lignin and hemicellulose-derived inhibitors are produced in the pretreatment process. Moreover, pretreatment is usually an energy-intensive process, which significantly increases the total cost of the degradation procedure.

The concept of effective bioconversion of lignocellulose without any requirement for pretreatment has increasingly gained attention. Only a few cultivated microorganisms such as *C. saccharolyticus* DSM 8903 and *C. bescii* DSM 6725 show the ability to degrade lignocellulosic biomass in the absence of pretreatment. These microbes efficiently utilize various types of un-pretreated plant biomass as well as crystalline cellulose and xylan. Insoluble substrates derived from switchgrass, poplar, Avicel and xylan were prepared by washing with water at 75°C after crushing. Each substrate was suspended in distilled water (1 g per 50 ml), stirred overnight at 75°C and washed twice with an equal volume of water at 75°C using a coarse glass filter (pore size of 40–60 μm). The substrate that was retained on the filter was dried overnight at 50°C, and used as un-pretreated insoluble substrate. *C. bescii* utilizing un-pretreated switchgrass and poplar as carbon sources reached stationary phase within approximately 20 hours, with a cell density of 1.5×10^8 cells/ml. In a recent work from our group, *Caldicellulosiruptor* sp. F32 capable of degrading cellulose and xylan was isolated from biocompost; the bacterium was closely related to *C. saccharolyticus* DSM 8903. Experiments revealed that the use of un-pretreated wheat straw as the carbon source allowed 3-fold higher cell numbers of strain F32 in the stationary phase compared to *C. saccharolyticus* DSM 8903 (Fig. 2.8A). Moreover, strain F32 was found to release greater quantities of the fermentation products acetate and hydrogen compared to *C. saccharolyticus* DSM 8903 (Fig. 2.8B). Analysis

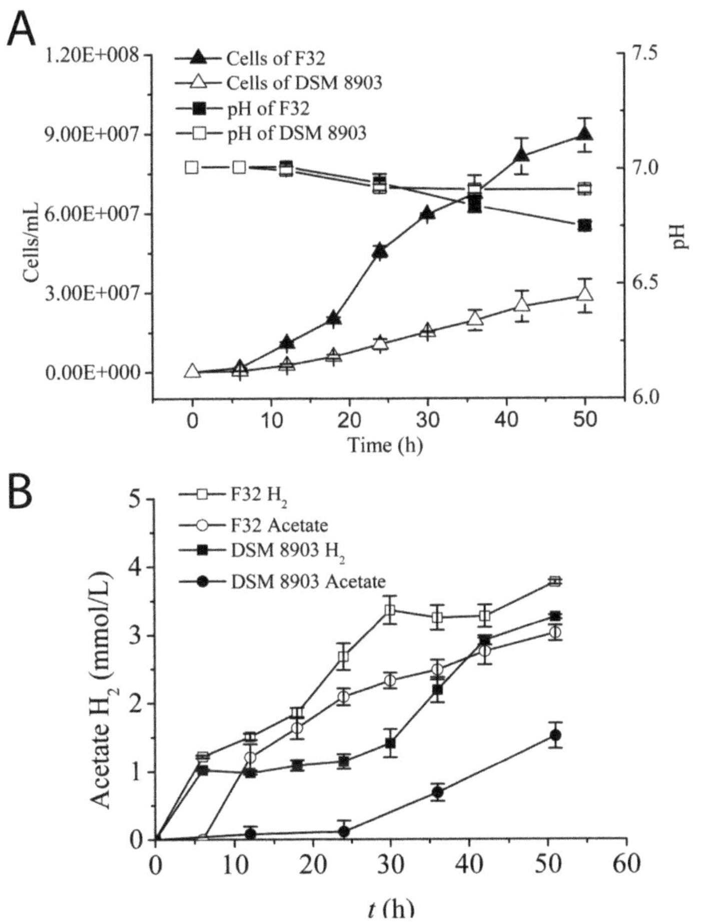

Figure 2.8 Growth, pH and products analysis of *C. saccharolyticus* F32 and *C. saccharolyticus* DSM 8903 with unpretreated wheat straw used as carbon source.

of lignocellulosic enzymes in the growth medium also revealed that strain F32 exhibits strong capacity for the utilization of un-pretreated wheat straw. As shown in Fig. 2.9, strain F32 displays high activities of both endoglucanase and xylanase.

Analysis of transcriptome, proteome and metabolic engineering of *Caldicellulosiruptor*

Genome sequencing data facilitated advances in transcriptome-sequencing studies in the genus *Caldicellulosiruptor*. The genome of *C. bescii* contains 2776 protein-coding sequences, of which 171 and 88 genes are associated with carbohydrate transport and utilization, respectively. The growth of *C. bescii* on crystalline cellulose (filter paper) resulted in the up-regulation of 32 CAZys, 61 transporters and 25 transcription factors as opposed to growth on glucose (Dam *et al.*, 2011). Transcriptome analysis of *C. saccharolyticus* grown on glucose versus galactose revealed >2-fold changes in 32 open reading frames (ORFs), and for glucose versus mannose, in 353 ORFs (VanFossen *et al.*, 2009). In addition, the mechanism of metabolic inhibition (carbon catabolite repression, CCR) was also investigated. Compiled data from transcriptome and genome analyses revealed the existence of at least 24 specific ABC transporters and a phosphotransferase system (PTS) in *C. saccharolyticus*. The absence of catabolite repression by glucose is an important characteristic of biohydrogen producers since it allows the simultaneous processing of an array of biomass-derived substrates. Transcriptome analysis conducted using a variety of monosaccharides (arabinose, fructose, galactose, glucose, mannose and xylose) indicated that most transporter genes responded to individual monosaccharides and polysaccharides;

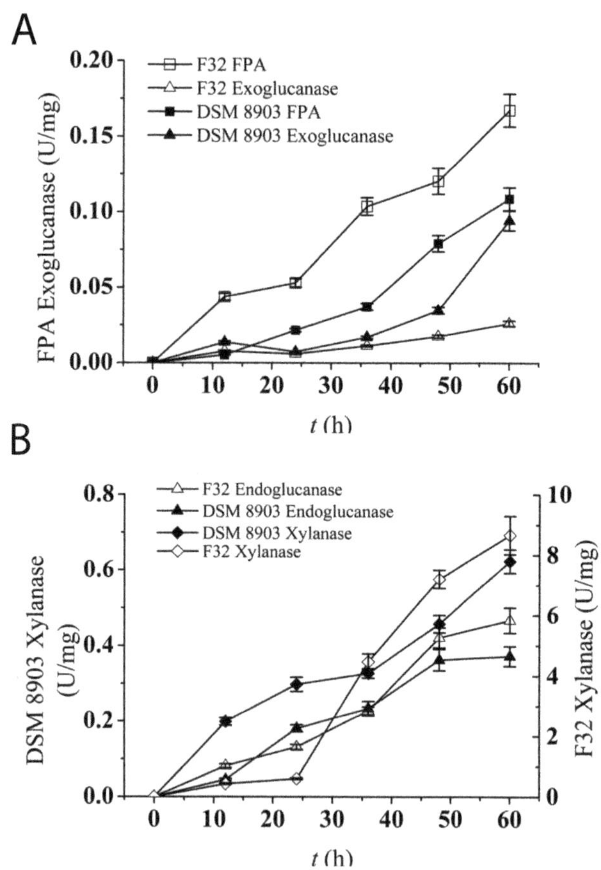

Figure 2.9 Enzyme activity analysis of *C. saccharolyticus* F32 and *C. saccharolyticus* DSM 8903 secretome induced by unpretreated wheat straw.

however, the genes Csac_0692 to Csac_0694 showed up-regulation only in the monosaccharide mixture. Genome and transcriptome analyses showed that a large number of carbohydrate utilization and ABC transport proteins function in concert, such that the strain exhibits a very broad spectrum of substrate utilization.

Secretomes from both *C. bescii* and *C. obsidiansis* consist of more than 400 proteins, most of which are multidomain glycosidases, extracellular solute-binding proteins, flagellin, putative pectate lyases and uncharacterized proteins with predicted secretion signals (Lochner *et al.*, 2011a). Mass spectrometry analysis in concert with cellulase activity measurements revealed a substantial abundance of multidomain glycosidases composed of GH families 5, 9, 10, 44 or 48 and CBM family 3 in *C. obsidiansis*. When grown on various carbon sources, the abundance of multifunctional GHs was significantly increased on cellobiose compared to Avicel and filter paper, while highly abundant flagellar, chemotaxis and pilus proteins were detected during growth on insoluble substrates (Lochner *et al.*, 2011b).

Research on genetic transformation has been conducted by Chung *et al.* (2011) for several years. This group initially characterized a novel thermostable restriction enzyme CbeI (Athe_2438) from *C. bescii* DSM 6725. Further studies revealed the requirement of a unique α-class N4-cytosine methyltransferase M. CbeI (Athe_2437) for transformation of DNA in *C. bescii*. The construction of 5-FOA-resistant/uracil auxotrophic mutants allows the transformation of *C. bescii* with exogenous DNA (Chung *et al.*, 2012). The abovementioned studies facilitated the successful knockout of *cbe* I (Athe_2438), the gene encoding the restriction enzyme CbeI, and enabled the successful transformation of *C. bescii* with exogenous plasmid DNA derived from *C. hydrothermalis* (Chung *et al.*, 2013). An engineered strain of *C. bescii* has been reported to exhibit direct conversion of switchgrass, a non-food renewable feedstock, to ethanol without pretreatment. The engineering of the bacterium involved deletion of lactate dehydrogenase and heterologous expression of a bifunctional acetaldehyde/alcohol dehydrogenase (Chung *et al.*, 2014).

Clostridia

Most cellulolytic strains that show optimum growth at 60°C belong to the genus *Clostridium*. Two types of glycoside hydrolase systems, cellulosomal and free-enzyme, occur in *Clostridium*.

Cellulosome-producing clostridia

Several anaerobic bacteria secrete a multiprotein complex of high molecular mass called cellulosome, which was initially discovered in *C. thermocellum* (Doi, 2008; Gilbert, 2007). Cellulosomes are cell protuberances which show tight binding to crystalline cellulose (Lamed *et al.*, 1987; Mayer *et al.*, 1987). Cellulosomes are easily isolated from cultures grown exclusively on microcrystalline cellulose or cellobiose through the use of affinity chromatography and gel filtration. They are macromolecular machines specifically designed for the hydrolysis of insoluble polysaccharides, and enable close contact between the cell and substrates, thereby minimizing the loss of hydrolytic products through diffusion, which is considered a major advantage in attached cells. An extract of cellulosomes contains a number of different proteins, most of which show enzymatic activity (Lamed and Bayer, 1988); however, attempts for mild denaturation, purification of single components and reconstitution were only met with partial success (Beattie *et al.*, 1994). Therefore, the vast majority of investigations on cellulosomes were accomplished through cloning and genetic engineering. Nearly 30 cellulosome-associated genes have been *hitherto* isolated through screening of genomic libraries, including the scaffoldin protein CipA which contains nine type I cohesin modules for the specific docking of enzymes and other protein components by virtue of their type I dockerin modules.

The most complex and well-investigated cellulosome is that of *C. thermocellum*, a moderately thermophilic bacterium (55–65°C) that has been repeatedly isolated from hot springs and wet, rotting biomass. However, cellulosomes have never been identified in bacteria that grow at temperatures above 65°C, or in archaea. The cellulosomes differ between species in terms of size (2.0–6.5 MDa) and composition (Beguin and Lemaire, 1996). Most of cellulosome producing strains aggregate to form larger supercomplexes, called polycellulosomes, with molecular mass ranging up to 100 MDa. The assembly of polycellulosomes involves the interaction of a type I dockerin module in the catalytic entities with one of several type I cohesin modules in the non-catalytic scaffoldin protein. Type II cohesin–dockerin modules mediate the binding of CipA protein to cell wall anchoring proteins (OlpB or SdbA) and possibly other proteins. The presence of dockerin module in a protein could be employed as a marker sequence for cellulosomal genes (Zverlov *et al.*, 2005).

The cellulosome contains a primary non-catalytic scaffoldin unit (CipA) capable of accommodating as many as nine catalytic units (Kruus *et al.*, 1995a). The catalytic units are non-covalently attached to the scaffoldin *via* high affinity type I interactions between dockerin domains present in the catalytic units and cohesins on the scaffoldin (Bayer *et al.*, 1985; Bayer *et al.*, 1998). In turn, the entire scaffoldin with bound subunits is attached to the cell surface *via* high affinity type II interactions between the dockerin domain of CipA and the cohesin(s) present on the anchoring proteins (OlpB, SdbA or Orf2p) (Lemaire *et al.*, 1995). The scaffoldin and several of the catalytic units also have CBMs that aid the direct attachment of the cellulosome to growth substrates to form a cell–cellulosome–substrate tri-complex (Fig. 2.10). CipA, the backbone scaffoldin protein, contains 9 type I cohesins and can accommodate up to nine type I dockerin-bearing catalytic units. CipA also contains a type II dockerin for cell-surface attachment *via* anchor proteins, and a cellulose-binding domain for attachment to the growth substrate. The genome of *C. thermocellum* encodes five proteins with type II cohesins, four of which have S-layer homology domain (SdbA, OlpB, Orf2p and Cthe0735) for cell-surface anchoring of type II dockerin-bearing CipA, while the other one (Cthe0736) lacks the SLH domain (Raman *et al.*, 2009).

The catalytic activity of cellulosomes is dependent on the presence of lipids (Bolobova *et al.*, 1994), emphasizing the necessity of *in vivo* investigations for determination of their overall structure. Ca^{2+} ions have been shown to stimulate the activity of cellulosomes (Wu *et al.*, 1988), while their removal leads to partial disintegration of cellulosome structure. Ca^{2+} ions play a crucial role in the correct folding of dockerin, thereby enabling the tight binding of catalytic components to scaffoldin (Choi and Ljungdahl, 1996; Lytle *et al.*, 2000; Salamitou *et al.*, 1994). The cellulosomes of *C. thermocellum* are glycosylated and contain 6–13% carbohydrates, the highest

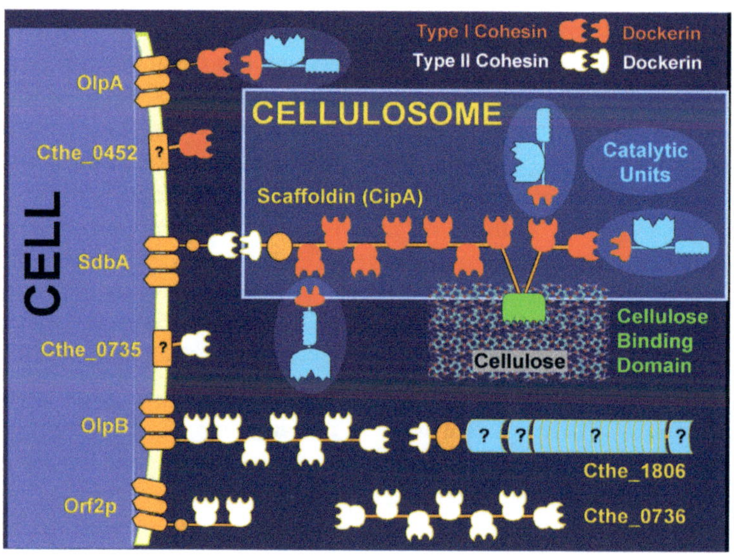

Figure 2.10 Simplified schematic representation of *Clostridium thermocellum* cellulosomal architecture (Raman *et al.*, 2009).

amounts of which are associated with scaffoldin. The role of cellulosome glycosylation is as yet undefined; however, protection against the action of proteases on the threonine-rich linker regions between the catalytic and binding modules, as well as specific functions of glycosylation in carbohydrate binding and eventually, in protein–protein recognition, can be assumed.

All the components of cellulosome in *C. thermocellum* contain dockerin repeat sequences, and dozens of components have been verified in the cellulosomal complex. CbhA, CelK, CelS and CelO are exoglucanases (Kruus *et al.*, 1995b; Zverlov *et al.*, 1999), while 27 genes code for endoglucanases (Brown *et al.*, 2012). In addition, a few cellulosomal genes encoding lichenase, chitinase, mannanase and xylanase have also been identified (Halstead *et al.*, 1999; Hayashi *et al.*, 1999; Zverlov *et al.*, 1994).

Free-acting glycoside hydrolases of clostridia

Several thermophilic, cellulolytic species of *Clostridium* produce free-acting glycoside hydrolases. *Clostridium stercorarium*, the most-studied species, showed excellent hemicellulose-utilizing activity (Adelsberger *et al.*, 2004). Another species, *Clostridium cellulosi*, rarely studied possibly due to the requirement of growth-supporting factors for its growth, was shown to contain several multidomain glycoside hydrolases in its genome (Zhang *et al.*, 2014). The species *Clostridium papyrosolvens* carried out hydrolysis of crystalline cellulose and xylan through the synergistic action of multiprotein complexes (Pohlschroder *et al.*, 1994). *Clostridium clariflavum*, a close relative of *C. thermocellum* due to the existence of cohesins in its cellulases, failed to form cellulosomes and showed higher capacity for xylan hydrolysis than *C. thermocellum* (Izquierdo *et al.*, 2012). *C. stercorarium* is a catabolically versatile bacterium that produces a wide range of hydrolases for the degradation of biomass, particularly for hemicellulose hydrolysis. It is moderately thermophilic with an optimum growth temperature of 65°C, and has been repeatedly isolated from self-heated compost (Kurose *et al.*, 1988; Zverlov *et al.*, 1998). The two-component cellulase system of *C. stercorarium* has been regarded as a model for simple cellulase systems capable of crystalline cellulose degradation (Riedel *et al.*, 1997). Cel9Z and Cel48Y were the only two cellulases isolated from the culture supernatant, and their synergistic action was found to be responsible for cellulose degradation (Riedel and Bronnenmeier, 1998). A great number of hemicellulases, glycosidases and esterases are produced by *C. stercorarium*, and have been characterized biochemically (Table 2.9).

The draft genome sequence of the strain *C. cellulosi* CS-4-4 showed the presence of 38 genes that encode glycoside hydrolases belonging to 15 families. The use of lignocellulose as a substrate corresponded to maximum abundance of the glycoside hydrolases encoded by ORF3880 and ORF3883 in the supernatant; both these hydrolases possess β-glucanase and xylanase but not glycosidase activities (Table 2.10). Moreover, ORF3880 contains a single GH9 module and five CBMs, as opposed to ORF3883 which contains one GH9 module and two CBMs (Fig. 2.11). Two genes belonging to the GH48 family annotated as exoglucanase were also present in its genome (Zhang *et al.*, 2014).

The cellulase system of *C. papyrosolvens* C7 contains at least seven multiprotein complexes of high molecular weight, each with different enzymatic and structural properties. The molecular weights of the complexes, as determined by gel filtration chromatography, ranged from 500 kDa to 660 kDa. Sodium dodecyl sulfate-polyacrylamide gel electrophoresis showed that each complex had a distinct polypeptide composition. The activity profiles of Avicelase, CMCase and xylanase differed between the protein complexes. Three of the complexes showed the ability to hydrolyse crystalline cellulose (Avicel). In-gel zymography (following electrophoresis under mildly denaturing conditions) revealed different CMCase-active proteins in all complexes, and xylanase-active proteins in only two of them. The xylanase-specific activity of these two complexes was more than 8-fold higher compared to that of the unfractionated cellulase preparation. A 125,000-M(r) glycoprotein with no apparent enzyme activity was the only polypeptide common to all seven complexes. Experiments involving the recombination of samples that were eluted from

Table 2.9 Temperature and pH optima and substrate specificity of the recombinant glycoside hydrolases

Characteristic	Recombinant protein							
	Xyn11A	Xyn10B	Xyn10C-cat	Arf43A	Arf51B (Schwarz *et al.*, 1990)	Bgl3Z	Bxl39A	Bxl3B
Temperature optimum (°C)	75	75	65	55	70	60	60	50–55
pH optimum	6.5–7.0	6.0–7.0	6.5	6.5	5.0	5.5–5.8	5.5–6.5	6.0
Substrate specificity (% activity) for								
pNP-β-glucoside	ND	ND	ND	ND	0.2	100	ND	0.9
pNP-β-cellobioside	ND	2.6	0.2	ND		++	3	ND
pNP-β-xyloside	ND	0.2	ND	65	0.4	7.3	100	100
pNP-α-arabinofuranoside	ND	ND	ND	100	100	ND	ND	16
pNP-β-fucoside	ND	0.1				19.8	3	
pNP-β-galactoside	ND	ND			0.3	ND	ND	
pNP-α-mannoside							22	ND
Xylobiose		+		–	–	–	+	++
Avicel	1.5	ND	ND			ND		
PASC	1.0	0.7				ND		
CMC	0.5	ND	ND	ND	ND		ND	ND
Arabinoxylan	100	100	100	ND	1.2	ND	ND	ND
Barley β-glucan	0.8	16.1	1.7	ND		0.5	ND	ND

Enzyme activity is given as a percentage of the maximum activity for the enzyme. The value of activity taken as 100% corresponds to 258 (Xyn11A), 1080 (Xyn10B), 549 (Xyn10C-cat), 0.15 (Arf43A), 0.8 (Arf51B), 0.4 (Bgl3Z), 0.4 (Bxl39A) and 0.75 (Bxl3B) U mg^{-1} protein, respectively.
ND, not detected; 2, not active; +, detectable; ++, highly active.

Table 2.10 Enzyme activities of glycoside hydrolases ORF3880 and ORF3883 on complex carbohydrates

Substrate	Specific activity (U/mg protein)	
	ORF3880	ORF3883
CMC-Na	0.41±0.12	0.85±0.05
Birchwood xylan	0.26±0.04	0.54±0.13
Avicel PH101	0.05±0.02	0.11±0.03
RAC	0.93±0.23	1.13±0.12
Whatman No.1 filter paper	0.17±0.08	0.14±0.05
Lichenan	2.75±0.34	2.51±0.56
Barley glucan	4.68±0.59	4.96±0.37
pNPG	ND[a]	ND
pNPC	ND	ND
pNPX	ND	ND

[a]ND, no detectable activity.

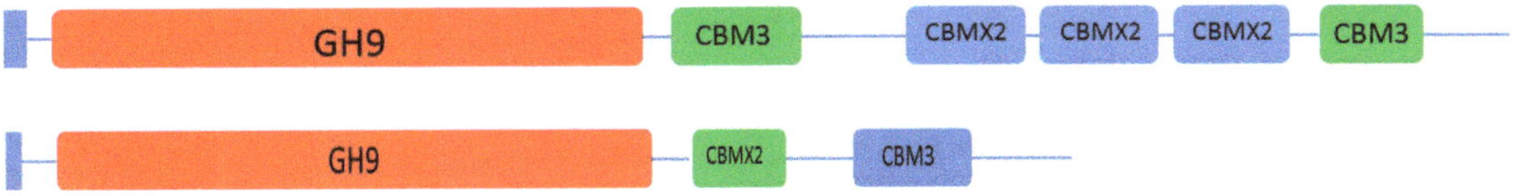

Figure 2.11 Schematic structures of ORF3880 and ORF3883.

ion-exchange chromatography column indicated that synergistic interactions played a role in the hydrolysis of crystalline cellulose by the cellulase system. This led to the hypothesis that the *C. papyrosolvens* enzyme system responsible for the hydrolysis of crystalline cellulose and xylan is a multi-complex system comprising at least seven diverse protein complexes (Pohlschroder *et al.*, 1994).

C. clariflavum was isolated from the anaerobic sludge of a cellulose-degrading methanogenic bioreactor (Shiratori *et al.*, 2009). The genome sequence of *C. clariflavum* DSM 19732 presented a total of 72 glycosyl hydrolases, representing 27 known GH families. Forty of these 72 enzymes were found to contain type-I dockerin domains, indicating their association with scaffoldin proteins containing complementary cohesin regions. Fifty of the glycosyl hydrolases (69.4%) of *C. clariflavum* DSM 19732 showed closest relation to those of *C. thermocellum* ATCC 27405 (Fig. 2.12). Of the remaining 22, a total of 15 are xylanases that belong to the GH families, i.e. GH10, GH11, GH39, GH43, GH51, GH67 and GH74. Cellulose-active GH families appear to show a similar distribution in both organisms, while a higher proportion and diversity of xylan-hydrolysing enzymes are found in *C. clariflavum*

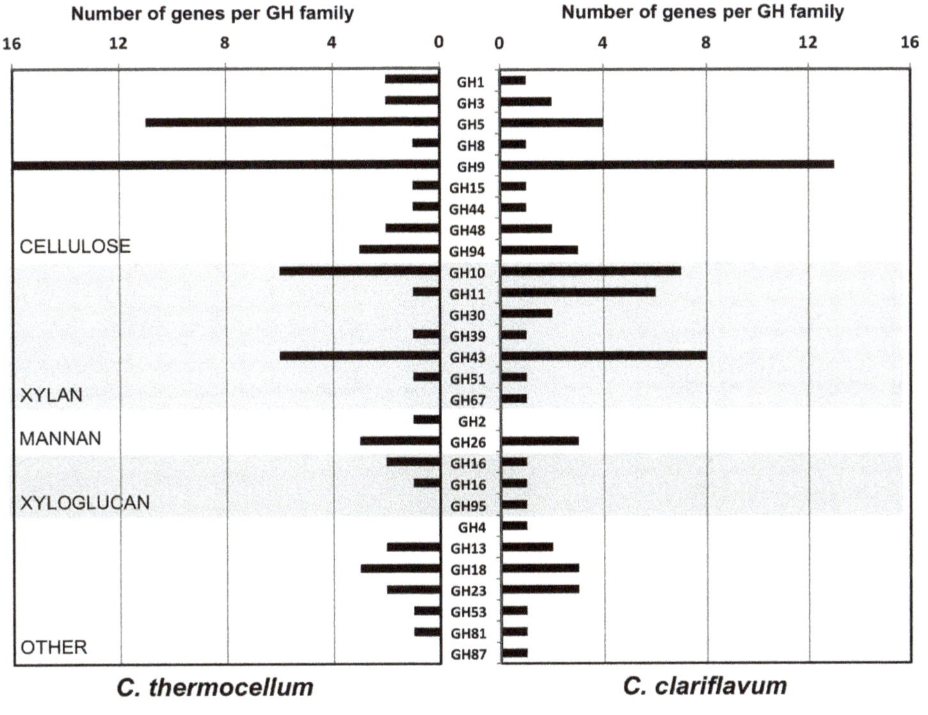

Figure 2.12 Comparison of glycosyl hydrolase inventory between *C. clariflavum* DSM 19732 and *C. thermocellum* ATCC 27405 (Izquierdo *et al.*, 2012). (The numbers of genes per GH family are shown, with GH families organized along the y-axis based on putative substrate specificity towards cellulose, xylan, mannan, xyloglucan and other.)

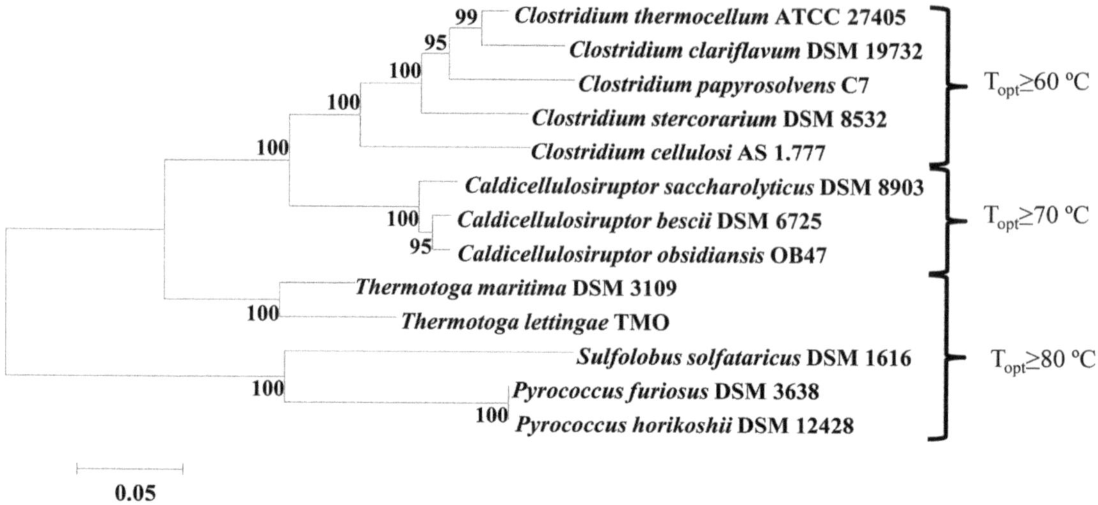

Figure 2.13 Phylogenetic tree of cellulolytic thermophiles.

compared with *C. thermocellum* (Izquierdo *et al.*, 2012). *Clostridium straminisolvens* CSK1, another thermophilic cellulolytic bacterium isolated from a cellulose-degrading bacterial community, has been rarely studied (Kato *et al.*, 2004).

A phylogenetic tree of the strains mentioned above revealed association between growth temperatures and phylogenetic position (Fig. 2.13). Hyperthermophiles (T_{opt} of ≥ 80°C) and moderate thermophiles (T_{opt} of ≥ 60°C) were separated into two branches, and greatest phylogenetic distance was observed between *Thermococcales*, a highly thermophilic order, and moderate thermophiles.

Adaptive mechanisms of thermophiles

Thermophiles encounter several challenges that accompany elevated temperatures, and considerable research interest has been elicited towards understanding the mechanisms of adaptation of these microorganisms. For example, bacterial and archaeal strains have shown the ability to adjust the permeability of their cytoplasmic membranes in accordance with growth temperatures to achieve high-temperature adaptation (vandeVossenberg *et al.*, 1995). In addition, the thermal stability of proteins in thermophiles necessitates the existence of mechanisms for preservation of protein structure.

In general, amino acid mutations, weak interactions and cofactors strongly promote thermal stability of proteins (Mesbah and Wiegel, 2008). Often, hydrophobic interactions and increase in density through salt-bridges or hydrogen bonds are consistent with thermophilic adaptation (Querol *et al.*, 1996; Robinson-Rechavi *et al.*, 2006; Szilagyi and Zavodszky, 2000). Increasing the rigidity of flexible fragments within the active site also aids in the improvement of enzyme thermostability (Xie *et al.*, 2014). The N- or C-terminal accessory domains influence thermal stability of proteins through properties of individual domains and interdomain interactions. The deletion of CBM domain was found to significantly increase thermostability in GH11 xylanase XynA from *Caldicellulosiruptor* sp. F32; however, opposing results were obtained with GH10 xylanase XynB. Through homology modelling and cross-linking experiments analysis, we demonstrated that for GH10 XynB, the resistance against thermoinactivation was generally contributed by both domain properties and interdomain interactions, whereas for GH11 XynA, no inter-domain interactions observed. Optimized intramolecular interactions contributed to improved thermostability (unpublished data). Fine-tuning of linker rigidity and flexibility leads to optimized intramolecular interactions between GH and CBM domains, which in turn determine thermal stability of proteins and provide microbes a powerful evolutionary strategy for adaptation to various ecological conditions.

High-temperature habitats are of special interest for microbiologist, in that they reveal the extremes to which evolution has been pushed. At present, we stand at the dawn of an exciting era in the study of evolution. Microorganisms, with their great exploitable diversity, will prove to be the highlights of the study, with those inhabiting extremely high-temperature environments expected to be of special relevance.

Acknowledgements

This work was supported by grants from the National Basic Research Program of China (No. 2011CB707404), the National Key Technology R&D Research Program (No. 2011BAD22B02-0), the Qingdao Institute of Biomass Energy and Bioprocess Technology Director Innovation Foundation for Young Scientists (No. Y37207210B) and '135' Projects Fund of CAS-QIBEBT Director Innovation Foundation.

References

Adelsberger, H., Hertel, C., Glawischnig, E., Zverlov, V.V., and Schwarz, W.H. (2004). Enzyme system of *Clostridium stercorarium* for hydrolysis of arabinoxylan: reconstitution of the *in vivo* system from recombinant enzymes. Microbiol.-SGM. *150*, 2257–2266.

Aksenova, H.Y., Rainey, F.A., Janssen, P.H., Zzvarzin, G.A., and Morgan, H.W. (1992). *Spirochaeta thermophila* sp. nov., an obligately anaerobic, polysaccharolytic, extremely thermophilic bacterium. Int. J. Syst. Bacteriol. *42*, 175–177.

Alain, K., Marteinsson, V.T., Miroshnichenko, M.L., Bonch-Osmolovskaya, E.A., Prieur, D., and Birrien, J.L. (2002a). *Marinitoga piezophila* sp. nov., a rod-shaped, thermo-piezophilic bacterium isolated under high hydrostatic pressure from a deep-sea hydrothermal vent. Int. J. Syst. Evol. Microbiol. *52*, 1331–1339.

Alain, K., Querellou, J., Lesongeur, F., Pignet, P., Crassous, P., Raguenes, G., Cueff, V., and Cambon-Bonavita, M.A. (2002b). *Caminibacter hydrogeniphilus* gen. nov., sp nov., a novel thermophilic, hydrogen-oxidizing bacterium isolated from an East Pacific Rise hydrothermal vent. Int. J. Syst. Evol. Microbiol. *52*, 1317–1323.

Alain, K., Callac, N., Guegan, M., Lesongeur, F., Crassous, P., Cambon-Bonavita, M.A., Querellou, J., and Prieur, D. (2009). *Nautilia abyssi* sp nov., a thermophilic, chemolithoautotrophic, sulfur-reducing bacterium isolated from an East Pacific Rise hydrothermal vent. Int. J. Syst. Evol. Microbiol. *59*, 1310–1315.

Alain, K., Postec, A., Grinsard, E., Lesongeur, F., Prieur, D., and Godfroy, A. (2010). *Thermodesulfatator atlanticus* sp nov., a thermophilic, chemolithoautotrophic, sulfate-reducing bacterium isolated from a Mid-Atlantic Ridge hydrothermal vent. Int. J. Syst. Evol. Microbiol. *60*, 33–38.

Alfredsson, G.A., Kristjansson, J.K., Hjorleifsdottir, S., and Stetter, K.O. (1988). *Rhodothermus marinus*, gen. nov., sp. nov., a thermophilic, halophilic bacterium from submarine hot springs in Iceland. J. Gen. Microbiol. *134*, 299–306.

Ando, S., Ishida, H., Kosugi, Y., and Ishikawa, K. (2002). *Hyperthermostable endoglucanase* from *Pyrococcus horikoshii*. Appl. Environ. Microbiol. *68*, 430–433.

Antoine, E., Cilia, V., Meunier, J.R., Guezennec, J., Lesongeur, F., and Barbier, G. (1997). *Thermosipho melanesiensis* sp. nov., a new thermophilic anaerobic bacterium belonging to the order Thermotogales, isolated from deep-sea hydrothermal vents in the Southwestern Pacific Ocean. Int. J. Syst. Bacteriol. *47*, 1118–1123.

Bauer, M.W., Driskill, L.E., Callen, W., Snead, M.A., Mathur, E.J., and Kelly, R.M. (1999). An endoglucanase, eglA, from the hyperthermophilic archaeon *Pyrococcus furiosus* hydrolyzes beta-1,4 bonds in mixed-linkage (1->3),(1->4)-beta-D-glucans and cellulose. J. Bacteriol. *181*, 284–290.

Bayer, E.A., Setter, E., and Lamed, R. (1985). Organization and distribution of the cellulosome in *Clostridium thermocellum*. J. Bacteriol. *163*, 552–559.

Bayer, E.A., Shimon, L.J., Shoham, Y., and Lamed, R. (1998). Cellulosomes-structure and ultrastructure. J. Struct. Biol. *124*, 221–234.

Beattie, L., Bhat, K.M., and Wood, T.M. (1994). The effect of cations on reassociation of the components of the cellulosome cellulase complex synthesized by the bacterium *Clostridium thermocellum*. Appl. Microbiol. Biotechnol. *40*, 740–744.

Beaulieu, S.E., Baker, E.T., German, C.R., and Maffei, A. (2013). An authoritative global database for active submarine hydrothermal vent fields. Geochem. Geophy. Geosy. *14*, 4892–4905.

Beguin, P., and Lemaire, M. (1996). The cellulosome: an exocellular, multiprotein complex specialized in cellulose degradation. Crit. Rev. Biochem. Mol. *31*, 201–236.

Blochl, E., Burggraf, S., Fiala, G., Laurerer, G., Huber, G., Huber, R., Rachel, R., Segerer, A., Stetter, K.O., and Volkl, P. (1995). Isolation, taxonomy and phylogeny of hyperthermophilic microorganisms. World J. Microb. Biot. *11*, 9–16.

Blumer-Schuette, S.E., Kataeva, I., Westpheling, J., Adams, M.W.W., and Kelly, R.M. (2008). Extremely thermophilic microorganisms for biomass conversion: status and prospects. Curr. Opin. Biotech. *19*, 210–217.

Blumer-Schuette, S.E., Lewis, D.L., and Kelly, R.M. (2010). Phylogenetic, microbiological, and glycoside hydrolase diversities within the extremely thermophilic, plant biomass-degrading genus *Caldicellulosiruptor*. Appl. Environ. Microbiol. *76*, 8084–8092.

Blumer-Schuette, S.E., Ozdemir, I., Mistry, D., Lucas, S., Lapidus, A., Cheng, J.F., Goodwin, L.A., Pitluck, S., Land, M.L., Hauser, L.J., *et al.* (2011). Complete genome sequences for the anaerobic, extremely thermophilic plant biomass-degrading bacteria *Caldicellulosiruptor hydrothermalis, Caldicellulosiruptor kristjanssonii, Caldicellulosiruptor kronotskyensis, Caldicellulosiruptor owensensis,* and *Caldicellulosiruptor lactoaceticus.* J. Bacteriol. *193,* 1483–1484.

Blumer-Schuette, S.E., Giannone, R.J., Zurawski, J.V., Ozdemir, I., Ma, Q., Yin, Y.B., Xu, Y., Kataeva, I., Poole, F.L., Adams, M.W.W., *et al.* (2012). *Caldicellulosiruptor* core and pangenomes reveal determinants for noncellulosomal thermophilic deconstruction of plant biomass. J. Bacteriol. *194,* 4015–4028.

Bolobova, A.V., Zhukov, A.V., and Klyosov, A.A. (1994). Lipids and Fatty-Acids in Cellulosomes of *Clostridium thermocellum.* Appl. Microbiol. Biot. *42,* 128–133.

Bouraoui, H., Rebib, H., Ben Aissa, M., Touzel, J.P., O'donohue, M., and Manai, M. (2013). *Paenibacillus marinum* sp nov., a thermophilic xylanolytic bacterium isolated from a marine hot spring in Tunisia. J. Basic Microbiol. *53,* 877–883.

Bredholt, S., Sonne-Hansen, J., Nielsen, P., Mathrani, I.M., and Ahring, B.K. (1999). *Caldicellulosiruptor kristjanssonii* sp nov., a cellulolytic extremely thermophilic, anaerobic bacterium. Int. J. Syst. Bacteriol. *49,* 991–996.

Brock, T.D. (1986). Introduction: an overview of the thermophiles. In Thermophiles: General Molecular and Applied Microbiology, T.D. Brock, ed. (John Wiley and Sons, New York), pp. 1–16.

Bronnenmeier, K., Kern, A., Liebl, W., and Staudenbauer, W.L. (1995). Purification of *Thermotoga maritima* enzymes for the degradation of cellulosic materials. Appl. Environ. Microbiol. *61,* 1399–1407.

Brown, S.D., Lamed, R., Morag, E., Borovok, I., Shoham, Y., Klingeman, D.M., Johnson, C.M., Yang, Z.M., Land, M.L., Utturkar, S.M., *et al.* (2012). Draft genome sequences for *Clostridium thermocellum* wild-type strain YS and derived cellulose adhesion-defective mutant strain AD2. J. Bacteriol. *194,* 3290–3291.

Brunecky, R., Alahuhta, M., Xu, Q., Donohoe, B.S., Crowley, M.F., Kataeva, I.A., Yang, S.J., Resch, M.G., Adams, M.W.W., Lunin, V.V., *et al.* (2013). Revealing nature's cellulase diversity: the digestion mechanism of *Caldicellulosiruptor bescii* CelA. Science *342,* 1513–1516.

Caccamo, D., Gugliandolo, C., Stackebrandt, E., and Maugeri, T.L. (2000). *Bacillus vulcani* sp nov., a novel thermophilic species isolated from a shallow marine hydrothermal vent. Int. J. Syst. Evol. Microbiol. *50,* 2009–2012.

Canganella, F., and Wiegel, J. (2014). Anaerobic thermophiles. Life *4,* 77–104.

Chhabra, S.R., Shockley, K.R., Conners, S.B., Scott, K.L., Wolfinger, R.D., and Kelly, R.M. (2003). Carbohydrate-induced differential gene expression patterns in the hyperthermophilic bacterium *Thermotoga maritima.* J. Biol. Chem. *278,* 7540–7552.

Choi, S.K., and Ljungdahl, L.G. (1996). Structural role of calcium for the organization of the cellulosome of *Clostridium thermocellum.* Biochemistry *35,* 4906–4910.

Chung, D., Farkas, J., Huddleston, J.R., Olivar, E., and Westpheling, J. (2012). Methylation by a unique alpha-class N4-cytosine methyltransferase is required for DNA transformation of *Caldicellulosiruptor bescii* DSM6725. PLoS One 7, e43844.

Chung, D., Cha, M., Guss, A.M., and Westpheling, J. (2014). Direct conversion of plant biomass to ethanol by engineered *Caldicellulosiruptor bescii.* Proc. Natl. Acad. Sci. U.S.A. *111,* 8931–8936.

Chung, D.H., Huddleston, J.R., Farkas, J., and Westpheling, J. (2011). Identification and characterization of CbeI, a novel thermostable restriction enzyme from Caldicellulosiruptor bescii DSM 6725 and a member of a new subfamily of HaeIII-like enzymes. J. Ind. Microbiol. Biot. *38,* 1867–1877.

Chung, D.W., Farkas, J., and Westpheling, J. (2013). Overcoming restriction as a barrier to DNA transformation in *Caldicellulosiruptor* species results in efficient marker replacement. Biotechnol. Biofuels *6,* 82.

Connelly, D.P., Copley, J.T., Murton, B.J., Stansfield, K., Tyler, P.A., German, C.R., Van Dover, C.L., Amon, D., Furlong, M., Grindlay N., *et al.* (2012). Hydrothermal vent fields and chemosynthetic biota on the world's deepest seafloor spreading centre. Nat. Commun. *3,* 620.

Dam, P., Kataeva, I., Yang, S.J., Zhou, F.F., Yin, Y.B., Chou, W.C., Poole, F.L., Westpheling, J., Hettich, R., Giannone, R., *et al.* (2011). Insights into plant biomass conversion from the genome of the anaerobic thermophilic bacterium Caldicellulosiruptor bescii DSM 6725. Nucleic Acids Res. *39,* 3240–3254.

DiPippo, J.L., Nesbo, C.L., Dahle, H., Doolittle, W.F., Birkland, N.K., and Noll, K.M. (2009). *Kosmotoga olearia* gen. nov., sp nov., a thermophilic, anaerobic heterotroph isolated from an oil production fluid. Int. J. Syst. Evol. Microbiol. *59,* 2991–3000.

Doi, R.H. (2008). Cellulases of mesophilic microorganisms: cellulosome and noncellulosome producers. Ann. N.Y. Acad. Sci. *1125,* 267–279.

Driskill, L.E., Bauer, M.W., and Kelly, R.M. (1999). Synergistic interactions among beta-laminarinase, beta-1,4-glucanase, and beta-glucosidase from the hyperthermophilic archaeon *Pyrococcus furiosus* during hydrolysis of beta-1,4-, beta-1,3-, and mixed-linked polysaccharides. Biotechnol. Bioeng. *66,* 51–60.

Elkins, J.G., Lochner, A., Hamilton-Brehm, S.D., Davenport, K.W., Podar, M., Brown, S.D., Land, M.L., Hauser, L.J., Klingeman, D.M., Raman, B., *et al.* (2010). Complete genome sequence of the cellulolytic thermophile *Caldicellulosiruptor obsidiansis* OB47T. J. Bacteriol. *192,* 6099–6100.

Fiala, G., and Stetter, K.O. (1986). *Pyrococcus furiosus* sp. nov. Represents a novel genus of marine heterotrophic Archaebacteria growing optimally at 100-degrees C. Arch. Microbiol. *145,* 56–61.

Fukui, T., Atomi, H., Kanai, T., Matsumi, R., Fujiwara, S., and Imanaka, T. (2005). Complete genome sequence of the hyperthermophilic archaeon *Thermococcus kodakaraensis* KOD1 and comparison with *Pyrococcus* genomes. Genome Res. *15,* 352–363.

Gilbert, H.J. (2007). Cellulosomes: microbial nanomachines that display plasticity in quaternary structure. Mol. Microbiol. *63,* 1568–1576.

Gotz, D., Banta, A., Beveridge, T.J., Rushdi, A.I., Simoneit, B.R.T., and Reysenbach, A. (2002). *Persephonella marina* gen. nov., sp nov and *Persephonella guaymasensis* sp nov., two novel, thermophilic, hydrogen-oxidizing microaerophiles from deep-sea hydrothermal vents. Int. J. Syst. Evol. Microbiol. *52*, 1349–1359.

Gueguen, Y., Voorhorst, W.G.B., van der Oost, J., and deVos, W.M. (1997). Molecular and biochemical characterization of an endo-beta-1,3-glucanase of the hyperthermophilic archaeon *Pyrococcus furiosus*. J. Biol. Chem. *272*, 31258–31264.

Halstead, J.R., Vercoe, P.E., Gilbert, H.J., Davidson, K., and Hazlewood, G.P. (1999). A family 26 mannanase produced by *Clostridium thermocellum* as a component of the cellulosome contains a domain which is conserved in mannanases from anaerobic fungi. Microbiol.-SGM *145*, 3101–3108.

Hamilton-Brehm, S.D., Mosher, J.J., Vishnivetskaya, T., Podar, M., Carroll, S., Allman, S., Phelps, T.J., Keller, M., and Elkins, J.G. (2010). *Caldicellulosiruptor obsidiansis* sp nov., an anaerobic, extremely thermophilic, cellulolytic bacterium isolated from obsidian pool, Yellowstone National Park. Appl. Environ. Microbiol. *76*, 1014–1020.

Han, Y.J., Agarwal, V., Dodd, D., Kim, J., Bae, B., Mackie, R.I., Nair, S.K., and Cann, I.K.O. (2012). Biochemical and structural insights into xylan utilization by the thermophilic bacterium *Caldanaerobius polysaccharolyticus*. J. Biol. Chem. *287*, 34946–34960.

Hayashi, H., Takehara, M., Hattori, T., Kimura, T., Karita, S., Sakka, K., and Ohmiya, K. (1999). Nucleotide sequences of two contiguous and highly homologous xylanase genes *xynA* and *xynB* and characterization of XynA from *Clostridium thermocellum*. Appl. Microbiol. Biot. *51*, 348–357.

Himmel, M.E., Ding, S.Y., Johnson, D.K., Adney, W.S., Nimlos, M.R., Brady, J.W., and Foust, T.D. (2007). Biomass recalcitrance: engineering plants and enzymes for biofuels production. Science *315*, 804–807.

Huang, C.Y., Patel, B.K., Mah, R.A., and Baresi, L. (1998). *Caldicellulosiruptor owensensis* sp. nov., an anaerobic, extremely thermophilic, xylanolytic bacterium. Int. J. Syst. Bacteriol. *48*, 91–97.

Huber, H., Diller, S., Horn, C., and Rachel, R. (2002). *Thermovibrio ruber* gen. nov., sp nov., an extremely thermophilic, chemolithoautotrophic, nitrate-reducing bacterium that forms a deep branch within the phylum Aquificae. Int. J. Syst. Evol. Microbiol. *52*, 1859–1865.

Huber, R., Wilharm, T., Huber, D., Trincone, A., Burggraf, S., Konig, H., Rachel, R., Rockinger, I., Fricke, H., and Stetter, K.O. (1992). *Aquifex pyrophilus* gen. nov. sp. nov. represents a novel group of marine hyperthermophilic hydrogen-oxidizing bacteria. Syst. Appl. Microbiol. *15*, 340–351.

Hubert, C., Loy, A., Nickel, M., Arnosti, C., Baranyi, C., Bruchert, V., Ferdelman, T., Finster, K., Christensen, F.M., de Rezende, J.R., *et al.* (2009). A constant flux of diverse thermophilic bacteria into the cold arctic seabed. Science *325*, 1541–1544.

Hubert, C., Arnosti, C., Bruchert, V., Loy, A., Vandieken, V., and Jorgensen, B.B. (2010). Thermophilic anaerobes in Arctic marine sediments induced to mineralize complex organic matter at high temperature. Environ. Microbiol. *12*, 1089–1104.

Isaksen, M.F., Bak, F., and Jørgense, B.B. (1994). Thermophilic sulfate-reducing bacteria in cold marine sediment. FEMS Microbiol. Ecol. *14*, 1–8.

Izquierdo, J.A., Goodwin, L., Davenport, K.W., Teshima, H., Bruce, D., Detter, C., Tapia, R., Han, S.S., Land, M., Hauser, L., *et al.* (2012). Complete genome sequence of *Clostridium clariflavum* DSM 19732. Stand. Genomic Sci. *6*, 104–115.

Jeanthon, C., L'Haridon, S., Cueff, V., Banta, A., Reysenbach, A.L., and Prieur, D. (2002). *Thermodesulfobacterium hydrogeniphilum* sp. nov., a thermophilic, chemolithoautotrophic, sulfate-reducing bacterium isolated from a deep-sea hydrothermal vent at Guaymas Basin, and emendation of the genus *Thermodesulfobacterium*. Int. J. Syst. Evol. Microbiol. *52*, 765–772.

Ji, S.Q., Wang, S.A., Tan, Y., Chen, X.H., Schwarz, W., and Li, F.L. (2012). An untapped bacterial cellulolytic community enriched from coastal marine sediment under anaerobic and thermophilic conditions. FEMS Microbiol. Lett. *335*, 39–46.

Jorgensen, H., Kristensen, J.B., and Felby, C. (2007). Enzymatic conversion of lignocellulose into fermentable sugars: challenges and opportunities. Biofuel. Bioprod. Bior. *1*, 119–134.

Kambourova, M., Mandeva, R., Fiume, I., Maurelli, L., Rossi, M., and Morana, A. (2007). Hydrolysis of xylan at high temperature by co-action of the xylanase from *Anoxybacillus flavithermus* BC and the beta-xylosidase/alpha-arabinosidase from *Sulfolobus solfataricus* O alpha. J. Appl. Microbiol. *102*, 1586–1593.

Kaper, T., van Heusden, H.H., van Loo, B., Vasella, A., van der Oost, J., and de Vos, W.M. (2002). Substrate specificity engineering of beta-mannosidase and beta-glucosidase from *Pyrococcus* by exchange of unique active site residues. Biochemistry *41*, 4147–4155.

Kashima, Y., Mori, K., Fukada, H., and Ishikawa, K. (2005). Analysis of the function of a hyperthermophilic endoglucanase from *Pyrococcus horikoshii* that hydrolyzes crystalline cellulose. Extremophiles *9*, 37–43.

Kataeva, I.A., Yang, S.J., Dam, P., Poole, F.L., Yin, Y., Zhou, F.F., Chou, W.C., Xu, Y., Goodwin, L., Sims, D.R., *et al.* (2009). Genome sequence of the anaerobic, thermophilic, and cellulolytic bacterium "*Anaerocellum thermophilum*" DSM 6725. J. Bacteriol. *191*, 3760–3761.

Kato, S., Haruta, S., Cui, Z.J., Ishii, M., Yokota, A., and Igarashi, Y. (2004). *Clostridium straminisolvens* sp nov., a moderately thermophilic, aerotolerant and cellulolytic bacterium isolated from a cellulose-degrading bacterial community. Int. J. Syst. Evol. Microbiol. *54*, 2043–2047.

Kawarabayasi, Y., Sawada, M., Horikawa, H., Haikawa, Y., Hino, Y., Yamamoto, S., Sekine, M., Baba, S., Kosugi, H., Hosoyama, A., Nagai, Y., *et al.* (1998). Complete sequence and gene organization of the genome of a hyper-thermophilic archaebacterium, *Pyrococcus horikoshii* OT3. DNA Res. *5*, 55–76.

Kluskens, L.D., van Alebeek, G.J.W.M., Voragen, A.G.J., de Vos, W.M., and van der Oost, J. (2003). Molecular and biochemical characterization of the thermoactive family 1 pectate lyase from the hyperthermophilic bacterium *Thermotoga maritima*. Biochem. J. *370*, 651–659.

Kopke, B., Wilms, R., Engelen, B., Cypionka, H., and Sass, H. (2005). Microbial diversity in coastal subsurface sediments: a cultivation approach using various electron acceptors and substrate gradients. Appl. Environ. Microbiol. *71*, 7819–7830.

Kombiadou, K, and Krestenitis, Y.N. (2013). Modelling cohesive sediment dynamics in the marine environment. In Sediment Transport Processes and their Modelling Applications, Manning, A.J., ed. (InTech, Rijeka, Croatia), pp. 213–246.

Kruus, K., Lua, A.C., Demain, A.L., and Wu, J.H.D. (1995a). The anchorage function of cipa (cell), a scaffolding protein of the *Clostridium thermocellum* cellulosome. Proc. Natl. Acad. Sci. U.S.A. *92*, 9254–9258.

Kruus, K., Wang, W.K., Ching, J.T., and Wu, J.H.D. (1995b). Exoglucanase activities of the recombinant *Clostridium Thermocellum* CelS, a major cellulosome component. J. Bacteriol. *177*, 1641–1644.

Kumar, R., Singh, S., and Singh, O.V. (2008). Bioconversion of lignocellulosic biomass: biochemical and molecular perspectives. J. Ind. Microbiol. Biotechnol. *35*, 377–391.

Kurose, N., Kinoshita, S., Yagyu, J., Uchida, M., Hanai, S., and Obayashi, A. (1988). Improvement of ethanol-production of *Thermophilic clostridium* Sp by mutation. J. Ferment. Bioeng. *66*, 467–472.

L'Haridon, S., Reysenbach, A.L., Tindall, B.J., Schonheit, P., Banta, A., Johnsen, U., Schumann, P., Gambacorta, A., Stackebrandt, E., and Jeanthon, C. (2006). *Desulfurobacterium atlanticum* sp. nov., *Desulfurobacterium pacificum* sp. nov. and *Thermovibrio guaymasensis* sp. nov., three thermophilic members of the *Desulfurobacteriaceae* fam. nov., a deep branching lineage within the Bacteria. Int. J. Syst. Evol. Microbiol. *56*, 2843–2852.

Lamed, R., and Bayer, E.A. (1988). Cellulosomes from *Clostridium thermocellum*. Method. Enzymol. *160*, 472–482.

Lamed, R., Naimark, J., Morgenstern, E., and Bayer, E.A. (1987). Specialized cell surface structures in cellulolytic bacteria. J. Bacteriol. *169*, 3792–3800.

Lecompte, O., Ripp, R., Puzos-Barbe, V., Duprat, S., Heilig, R., Dietrich, J., Thierry, J.C., and Poch, O. (2001). Genome evolution at the genus level: Comparison of three complete genomes of hyperthermophilic archaea. Genome Res. *11*, 981–993.

Leloup, J., Fossing, H., Kohls, K., Holmkvist, L., Borowski, C., and Jorgensen, B.B. (2009). Sulfate-reducing bacteria in marine sediment (Aarhus Bay, Denmark): abundance and diversity related to geochemical zonation. Environ. Microbiol. *11*, 1278–1291.

Lemaire, M., Ohayon, H., Gounon, P., Fujino, T., and Beguin, P. (1995). Olpb, a new outer layer protein of *Clostridium thermocellum*, and binding of its S-layer-like domains to components of the cell-envelope. J. Bacteriol. *177*, 2451–2459.

Lochner, A., Giannone, R.J., Keller, M., Antranikian, G., Graham, D.E., and Hettich, R.L. (2011a). Label-free quantitative proteomics for the extremely thermophilic bacterium *Caldicellulosiruptor obsidiansis* reveal distinct abundance patterns upon growth on cellobiose, crystalline cellulose, and switchgrass. J. Proteome Res. *10*, 5302–5314.

Lochner, A., Giannone, R.J., Rodriguez, M., Shah, M.B., Mielenz, J.R., Keller, M., Antranikian, G., Graham, D.E., and Hettich, R.L. (2011b). Use of label-free quantitative proteomics to distinguish the secreted cellulolytic systems of *Caldicellulosiruptor bescii* and *Caldicellulosiruptor obsidiansis*. Appl. Environ. Microbiol. *77*, 4042–4054.

Lytle, B.L., Volkman, B.F., Westler, W.M., and Wu, J.H.D. (2000). Secondary structure and calcium-induced folding of the *Clostridium thermocellum* dockerin domain determined by NMR spectroscopy. Arch. Biochem. Biophys. *379*, 237–244.

Madigan, M.T., and Martinko, J.M. (2006). Brock Biology of Microorganisms, 11th edn (Pearson Prentice Hall, Upper Saddle River).

Maeder, D.L., Weiss, R.B., Dunn, D.M., Cherry, J.L., Gonzalez, J.M., DiRuggiero, J., and Robb, F.T. (1999). Divergence of the hyperthermophilic archaea *Pyrococcus furiosus* and *P. horikoshii* inferred from complete genomic sequences. Genetics *152*, 1299–1305.

Mathis, B.J., Marshall, C.W., Milliken, C.E., Makkar, R.S., Creager, S.E., and May, H.D. (2008). Electricity generation by thermophilic microorganisms from marine sediment. Appl. Microbiol. Biot. *78*, 147–155.

Maugeri, T.L., Lentini, V., Gugliandolo, C., Italiano, F., Cousin, S., and Stackebrandt, E. (2009). Bacterial and archaeal populations at two shallow hydrothermal vents off Panarea Island (Eolian Islands, Italy). Extremophiles *13*, 199–212.

Maurelli, L., Giovane, A., Esposito, A., Moracci, M., Fiume, I., Rossi, M., and Morana, A. (2008). Evidence that the xylanase activity from *Sulfolobus solfataricus* O alpha is encoded by the endoglucanase precursor gene (sso1354) and characterization of the associated cellulase activity. Extremophiles *12*, 689–700.

Mayer, F., Coughlan, M.P., Mori, Y., and Ljungdahl, L.G. (1987). Macromolecular Organization of the Cellulolytic Enzyme Complex of *Clostridium Thermocellum* as Revealed by Electron Microscopy. Appl. Environ. Microbiol. *53*, 2785–2792.

Mesbah, N.M., and Wiegel, J. (2008). Life at extreme limits – the anaerobic halophilic alkalithermophiles. Ann. N.Y. Acad. Sci. *1125*, 44–57.

Miroshnichenko, M.L., Kostrikina, N.A., L'Haridon, S., Jeanthon, C., Hippe, H., Stackebrandt, E., and Bonch-Osmolovskaya, E.A. (2002). *Nautilia lithotrophica* gen. nov., sp. nov., a thermophilic sulfur-reducing epsilon-proteobacterium isolated from a deep-sea hydrothermal vent. Int. J. Syst. Evol. Microbiol. *52*, 1299–1304.

Miroshnichenko, M.L., L'Haridon, S., Jeanthon, C., Antipov, A.N., Kostrikina, N.A., Tindall, B.J.,

Schumann, P., Spring, S., Stackebrandt, E., and Bonch-Osmolovskaya, E.A. (2003a). *Oceanithermus profundus* gen. nov., sp. nov., a thermophilic, microaerophilic, facultatively chemolithoheterotrophic bacterium from a deep-sea hydrothermal vent. Int. J. Syst. Evol. Microbiol. *53*, 747–752.

Miroshnichenko, M.L., L'Haridon, S.L., Nercessian, O., Antipov, A.N., Kostrikina, N.A., Tindall, B.J., Schumann, P., Spring, S., Stackebrandt, E., Bonch-Osmolovskaya, E.A., and Jeanthon, C. (2003b). *Vulcanithermus mediatlanticus* gen. nov., sp. nov., a novel member of the family Thermaceae from a deep-sea hot vent. Int. J. Syst. Evol. Microbiol. *53*, 1143–1148.

Miroshnichenko, M.L., L'Haridon, S., Schumann, P., Spring, S., Bonch-Osmolovskaya, E.A., Jeanthon, C., and Stackebrandt, E. (2004). *Caminibacter profundus* sp. nov., a novel thermophile of *Nautiliales* ord. nov. within the class '*Epsilonproteobacteria*', isolated from a deep-sea hydrothermal vent. Int. J. Syst. Evol. Microbiol. *54*, 41–45.

Miroshnichenko, M.L., Kublanov, I.V., Kostrikina, N.A., Tourova, T.P., Kolganova, T.V., Birkeland, N.K., and Bonch-Osmolovskaya, E.A. (2008). *Caldicellulosiruptor kronotskyensis* sp. nov. and *Caldicellulosiruptor hydrothermalis* sp. nov., two extremely thermophilic, cellulolytic, anaerobic bacteria from Kamchatka thermal springs. Int. J. Syst. Evol. Microbiol. *58*, 1492–1496.

Miroshnichenko, M.L., Kolganova, T.V., Spring, S., Chernyh, N., and Bonch-Osmolovskaya, E.A. (2010). *Caldithrix palaeochoryensis* sp. nov., a thermophilic, anaerobic, chemo-organotrophic bacterium from a geothermally heated sediment, and emended description of the genus Caldithrix. Int. J. Syst. Evol. Microbiol. *60*, 2120–2123.

Mladenovska, Z., Mathrani, I.M., and Ahring, B.K. (1995). Isolation and characterization of *Caldicellulosiruptor Lactoaceticus* sp. nov., an extremely thermophilic, cellulolytic, anaerobic bacterium. Arch. Microbiol. *163*, 223–230.

Mori, K., Kakegawa, T., Higashi, Y., Nakamura, K., Maruyama, A., and Hanada, S. (2004). *Oceanithermus desulfurans* sp. nov., a novel thermophilic, sulfur-reducing bacterium isolated from a sulfide chimney in Suiyo Seamount. Int. J. Syst. Evol. Microbiol. *54*, 1561–1566.

Moussard, H., L'Haridon, S., Tindall, B.J., Banta, A., Schumann, P., Stackebrandt, E., Reysenbach, A.L., and Jeanthon, C. (2004). *Thermodesulfatator indicus* gen. nov., sp. nov., a novel thermophilic chemolithoautotrophic sulfate-reducing bacterium isolated from the Central Indian Ridge. Int. J. Syst. Evol. Microbiol. *54*, 227–233.

Muyzer, G., and Stams, A.J. (2008). The ecology and biotechnology of sulphate-reducing bacteria. Nat. Rev. Microbiol. *6*, 441–454.

Nakagawa, S., Takai, K., Horikoshi, K., and Sako, Y. (2003). *Persephonella hydrogeniphila* sp nov., a novel thermophilic, hydrogen-oxidizing bacterium from a deep-sea hydrothermal vent chimney. Int. J. Syst. Evol. Microbiol. *53*, 863–869.

Nakagawa, S., Nakamura, S., Inagaki, F., Takai, K., Shirai, N., and Sako, Y. (2004). *Hydrogenivirga caldilitoris* gen. nov., sp nov., a novel extremely thermophilic, hydrogen- and sulfur-oxidizing bacterium from a coastal hydrothermal field. Int. J. Syst. Evol. Microbiol. *54*, 2079–2084.

Nakagawa, S., Takai, K., Inagaki, F., Hirayama, H., Nunoura, T., Horikoshi, K., and Sako, Y. (2005). Distribution, phylogenetic diversity and physiological characteristics of epsilon-Proteobacteria in a deep-sea hydrothermal field. Environ. Microbiol. *7*, 1619–1632.

Nunoura, T., Akihara, S., Takai, K., and Sako, Y. (2002). *Thermaerobacter nagasakiensis* sp nov., a novel aerobic and extremely thermophilic marine bacterium. Arch. Microbiol. *177*, 339–344.

Nunoura, T., Oida, H., Miyazaki, M., Suzuki, Y., Takai, K., and Horikoshi, K. (2007). *Marinitoga okinawensis* sp nov., a novel thermophilic and anaerobic heterotroph isolated from a deep-sea hydrothermal field, Southern Okinawa Trough. Int. J. Syst. Evol. Microbiol. *57*, 467–471.

Nunoura, T., Miyazaki, M., Suzuki, Y., Takai, K., and Horikoshi, K. (2008). *Hydrogenivirga okinawensis* sp nov., a thermophilic sulfur-oxidizing chemolithoautotroph isolated from a deep-sea hydrothermal field, Southern Okinawa Trough. Int. J. Syst. Evol. Microbiol. *58*, 676–681.

Nunoura, T., Hirai, M., Imachi, H., Miyazaki, M., Makita, H., Hirayama, H., Furushima, Y., Yamamoto, H., and Takai, K. (2010). *Kosmotoga arenicorallina* sp. nov a thermophilic and obligately anaerobic heterotroph isolated from a shallow hydrothermal system occurring within a coral reef, southern part of the Yaeyama Archipelago, Japan, reclassification of *Thermococcoides shengliensis* as *Kosmotoga shengliensis* comb. nov., and emended description of the genus *Kosmotoga*. Arch. Microbiol. *192*, 811–819.

Nunoura, T., Hirai, M., Miyazaki, M., Kazama, H., Makita, H., Hirayama, H., Furushima, Y., Yamamoto, H., Imachi, H., and Takai, K. (2013). Isolation and Characterization of a Thermophilic, Obligately Anaerobic and Heterotrophic Marine Chloroflexi Bacterium from a Chloroflexi-dominated Microbial Community Associated with a Japanese Shallow Hydrothermal System, and Proposal for *Thermomarinilinea lacunofontalis* gen. nov., sp nov. Microbes Environ. *28*, 228–235.

Park, J.I., Kent, M.S., Datta, S., Holmes, B.M., Huang, Z.H., Simmons, B.A., Sale, K.L., and Sapra, R. (2011). Enzymatic hydrolysis of cellulose by the cellobiohydrolase domain of CelB from the hyperthermophilic bacterium *Caldicellulosiruptor saccharolyticus*. Bioresource Technol. *102*, 5988–5994.

Perez-Rodriguez, I., Ricci, J., Voordeckers, J.W., Starovoytov, V., and Vetriani, C. (2010). *Nautilia nitratireducens* sp nov., a thermophilic, anaerobic, chemosynthetic, nitrate-ammonifying bacterium isolated from a deep-sea hydrothermal vent. Int. J. Syst. Evol. Microbiol. *60*, 1182–1186.

Perez-Rodriguez, I., Grosche, A., Massenburg, L., Starovoytov, V., Lutz, R.A., and Vetriani, C. (2012). *Phorcysia thermohydrogeniphila* gen. nov., sp nov., a thermophilic, chemolithoautotrophic, nitrate-ammonifying bacterium from a deep-sea hydrothermal vent. Int. J. Syst. Evol. Microbiol. 62, 2388–2394.

Pohlschroder, M., Leschine, S.B., and Canaleparola, E. (1994). Multicomplex Cellulase Xylanase System of *Clostridium Papyrosolvens* C7. J. Bacteriol. *176*, 70–76.

Postec, A., Le Breton, C., Fardeau, M.L., Lesongeur, F.O., Pignet, P., Querellou, J., Ollivier, B., and Godfroy, A. (2005). *Marinitoga hydrogenitolerans* sp nov., a novel member of the order Thermotogales isolated from a black smoker chimney on the Mid-Atlantic Ridge. Int. J. Syst. Evol. Microbiol. *55*, 1217–1221.

Postec, A., Ciobanu, M., Birrien, J.L., Bienvenu, N., Prieur, D., and Le Romancer, M. (2010). *Marinitoga litoralis* sp nov., a thermophilic, heterotrophic bacterium isolated from a coastal thermal spring on Ile Saint-Paul, Southern Indian Ocean. Int. J. Syst. Evol. Microbiol. *60*, 1778–1782.

Prieur, D., Voytek, M., Jeanthon, C., and Reysenbach, A.-L. (2001). Deep-sea thermophilic prokaryotes. In Thermophiles: Biodiversity, Ecology, and Evolution, Reysenbach, A.-L., Voytek, M., and Mancinelli, R., eds. (Kluwer Academic/Plenum Publishers, New York), pp. 11–19.

Querol, E., PerezPons, J.A., and MozoVillarias, A. (1996). Analysis of protein conformational characteristics related to thermostability. Protein Eng. *9*, 265–271.

Rainey, F.A., Donnison, A.M., Janssen, P.H., Saul, D., Rodrigo, A., Bergquist, P.L., Daniel, R.M., Stackebrandt, E., and Morgan, H.W. (1994). Description of *Caldicellulosiruptor saccharolyticus* gen. nov., sp. nov. an obligately anaerobic, extremely thermophilic, cellulolytic bacterium. FEMS Microbiol. Lett. *120*, 263–266.

Raman, B., Pan, C., Hurst, G.B., Rodriguez, M. Jr., McKeown, C.K., Lankford, P.K., Samatova, N.F., and Mielenz, J.R. (2009). Impact of pretreated Switchgrass and biomass carbohydrates on Clostridium thermocellum ATCC 27405 cellulosome composition: a quantitative proteomic analysis. PLoS One *4*, e5271.

Ravot, G., Magot, M., Fardeau, M.L., Patel, B.K.C., Prensier, G., Egan, A., Garcia, J.L., and Ollivier, B. (1995). *Thermotoga Elfii* sp. nov., a novel thermophilic bacterium from an african oil-producing well. Int. J. Syst. Bacteriol. *45*, 308–314.

Riedel, K., and Bronnenmeier, K. (1998). Intramolecular synergism in an engineered exo-endo-1,4-beta-glucanase fusion protein. Mol. Microbiol. *28*, 767–775.

Riedel, K., Ritter, J., and Bronnenmeier, K. (1997). Synergistic interaction of the *Clostridium stercorarium* cellulases Avicelase I (CelZ) and Avicelase II (CelY) in the degradation of microcrystalline cellulose. FEMS Microbiol. Lett. *147*, 239–243.

Robinson-Rechavi, M., Alibes, A., and Godzik, A. (2006). Contribution of electrostatic interactions, compactness and quaternary structure to protein thermostability: Lessons from structural genomics of *Thermotoga maritima*. J. Mol. Biol. *356*, 547–557.

Rodionov, D.A., Rodionova, I.A., Li, X.Q., Ravcheev, D.A., Tarasova, Y., Portnoy, V.A., Zengler, K., and Osterman, A.L. (2013). Transcriptional regulation of the carbohydrate utilization network in *Thermotoga maritima*. Front. Microbiol. *4*.

Rogers, A.D., Tyler, P.A., Connelly, D.P., Copley, J.T., James, R., Larter, R.D., Linse, K., Mills, R.A., Naveira G., Alberto, P., *et al.* (2012) The discovery of new deep-sea hydrothermal vent communities in the southern ocean and implications for biogeography. PLoS Biol. *10*, e1001234.

Sako, Y., Takai, K., Ishida, Y., Uchida, A., and Katayama, Y. (1996). *Rhodothermus obamensis* sp nov, a modern lineage of extremely thermophilic marine bacteria. Int. J. Syst. Bacteriol. *46*, 1099–1104.

Sako, Y., Nakagawa, S., Takai, K., and Horikoshi, K. (2003). *Marinithermus hydrothermalis* gen. nov., sp nov., a strictly aerobic, thermophilic bacterium from a deep-sea hydrothermal vent chimney. Int. J. Syst. Evol. Microbiol. *53*, 59–65.

Salamitou, S., Raynaud, O., Lemaire, M., Coughlan, M., Beguin, P., and Aubert, J.P. (1994). Recognition specificity of the duplicated segments present in *Clostridium Thermocellum* endoglucanase celd and in the cellulosome-integrating protein cipa. J. Bacteriol. *176*, 2822–2827.

Schwarz, W.H., Adelsberger, H., Jauris, S., Hertel, C., Funk, B., and Staudenbauer, W.L. (1990). Xylan degradation by the thermophile *Clostridium stercorarium* – cloning and expression of xylanase, beta-D-xylosidase, and alpha-L-arabinofuranosidase genes in *Escherichia coli*. Biochem. Bioph. Res. Co. *170*, 368–374.

Sensen, C.W., Charlebois, R.L., Chow, C., Clausen, I.G., Curtis, B., Doolittle, W.F., Duguet, M., Erauso, G., Gaasterland, T., Garrett, R.A., *et al.* (1998). Completing the sequence of the *Sulfolobus solfataricus* P2 genome. Extremophiles *2*, 305–312.

Shiratori, H., Sasaya, K., Ohiwa, H., Ikeno, H., Ayame, S., Kataoka, N., Miya, A., Beppu, T., and Ueda, K. (2009). *Clostridium clariflavum* sp. nov. and *Clostridium caenicola* sp. nov., moderately thermophilic, cellulose-/cellobiose-digesting bacteria isolated from methanogenic sludge. Int. J. Syst. Evol. Microbiol. *59*, 1764–1770.

Sievert, S.M., and Kuever, J. (2000). *Desulfacinum hydrothermale* sp. nov., a thermophilic, sulfate-reducing bacterium from geothermally heated sediments near Milos Island (Greece). Int. J. Syst. Evol. Microbiol. *50 (Pt. 3)*, 1239–1246.

Sievert, S.M., Hügler, M., Taylor, C.D., and Wirsen, C.O. (2008). Sulfur oxidation at deep-sea hydrothermal vents. In Microbial Sulfur Metabolism, Dahl, C., and Friedrich, C.G., eds. (Springer, Berlin), pp. 238–258.

Slobodkin, A.I., Reysenbach, A.L., Slobodkina, G.B., Baslerov, R.V., Kostrikina, N.A., Wagner, I.D., and Bonch-Osmolovskaya, E.A. (2012). *Thermosulfurimonas dismutans* gen. nov., sp nov., an extremely thermophilic sulfur-disproportionating bacterium from a deep-sea hydrothermal vent. Int. J. Syst. Evol. Microbiol. *62*, 2565–2571.

Slobodkina, G.B., Kolganova, T.V., Chernyh, N.A., Querellou, J., Bonch-Osmolovskaya, E.A., and Slobodkin, A.I. (2009). *Deferribacter autotrophicus* sp nov., an iron(III)-reducing bacterium from a deep-sea hydrothermal vent. Int. J. Syst. Evol. Microbiol. *59*, 1508–1512.

Smith, J.L., Campbell, B.J., Hanson, T.E., Zhang, C.L., and Cary, S.C. (2008). *Nautilia profundicola* sp nov., a thermophilic, sulfur-reducing epsilonproteobacterium from deep-sea hydrothermal vents. Int. J. Syst. Evol. Microbiol. *58*, 1598–1602.

Sogin, M.L., Morrison, H.G., Huber, J.A., Mark Welch, D., Huse, S.M., Neal, P.R., Arrieta, J.M., and Herndl, G.J. (2006). Microbial diversity in the deep sea and the underexplored "rare biosphere". Proc. Natl. Acad. Sci. U.S.A. *103*, 12115–12120.

Sokolova, T.G., Gonzalez, J.M., Kostrikina, N.A., Chernyh, N.A., Tourova, T.P., Kato, C., Bonch-Osmolovskaya, E.A., and Robb, F.T. (2001). *Carboxydobrachium pacificum* gen. nov., sp nov., a new anaerobic, thermophilic, CO-utilizing marine bacterium from Okinawa Trough. Int. J. Syst. Evol. Microbiol. *51*, 141–149.

Steinsbu, B.O., Tindall, B.J., Torsvik, V.L., Thorseth, I.H., Daae, F.L., and Pedersen, R.B. (2011). *Rhabdothermus arcticus* gen. nov., sp nov., a member of the family Thermaceae isolated from a hydrothermal vent chimney in the Soria Mona vent field on the Arctic Mid-Ocean Ridge. Int. J. Syst. Evol. Microbiol. *61*, 2197–2204.

Stohr, R., Waberski, A., Volker, H., Tindall, B.J., and Thomm, M. (2001). *Hydrogenothermus marinus* gen. nov., sp. nov., a novel thermophilic hydrogen-oxidizing bacterium, recognition of *Calderobacterium hydrogenophilum* as a member of the genus *Hydrogenobacter* and proposal of the reclassification of *Hydrogenobacter acidophilus* as *Hydrogenobaculum acidophilum* gen. nov., comb. nov., in the phylum 'Hydrogenobacter/Aquifex'. Int. J. Syst. Evol. Microbiol. *51*, 1853–1862.

Su, X.Y., Mackie, R.I., and Cann, I.K.O. (2012). Biochemical and mutational analyses of a multidomain cellulase/mannanase from *Caldicellulosiruptor bescii*. Appl. Environ. Microbiol. *78*, 2230–2240.

Suss, J., Engelen, B., Cypionka, H., and Sass, H. (2004). Quantitative analysis of bacterial communities from *Mediterranean sapropels* based on cultivation-dependent methods. FEMS Microbiol. Ecol. *51*, 109–121.

Svetlichnyi, V.A., Svetlichnaya, T.P., Chernykh, N.A., and Zavarzin, G.A. (1990). *Anaerocellum thermophilum* gen. nov. sp. nov. – an extremely thermophilic cellulolytic eubacterium isolated from hot-springs in the Valley of Geysers. Microbiology 59, 598–604.

Swatloski, R.P., Spear, S.K., Holbrey, J.D., and Rogers, R.D. (2002). Dissolution of cellose with ionic liquids. J. Am. Chem. Soc. *124*, 4974–4975.

Szilagyi, A., and Zavodszky, P. (2000). Structural differences between mesophilic, moderately thermophilic and extremely thermophilic protein subunits: results of a comprehensive survey. Structure *8*, 493–504.

Takahata, Y., Nishijima, M., Hoaki, T., and Maruyama, T. (2001). *Thermotoga petrophila* sp. nov. and *Thermotoga naphthophila* sp. nov., two hyperthermophilic bacteria from the Kubiki oil reservoir in Niigata, Japan. Int. J. Syst. Evol. Microbiol. *51*, 1901–1909.

Takai, K., and Horikoshi, K. (2000). *Thermosipho japonicus* sp nov., an extremely thermophilic bacterium isolated from a deep-sea hydrothermal vent in Japan. Extremophiles *4*, 9–17.

Takai, K., Inoue, A., and Horikoshi, K. (1999). *Thermaerobacter marianensis* gen. nov., sp, nov., an aerobic extremely thermophilic marine bacterium from the 11000 m deep Mariana Trench. Int. J. Syst. Bacteriol. *49*, 619–628.

Takai, K., Nealson, K.H., and Horikoshi, K. (2004). *Hydrogenimonas thermophila* gen. nov., sp nov., a novel thermophilic, hydrogen-oxidizing chemolithoautotroph within the epsilon-Proteobacteria, isolated from a black smoker in a Central Indian Ridge hydrothermal field. Int. J. Syst. Evol. Microbiol. *54*, 25–32.

Takai, K., Hirayama, H., Nakagawa, T., Suzuki, Y., Nealson, K.H., and Horikoshi, K. (2005). *Lebetimonas acidiphila* gen. nov., sp nov., a novel thermophilic, acidophilic, hydrogen-oxidizing chemolithoautotroph within the Epsilonproteobacteria', isolated from a deep-sea hydrothermal fumarole in the Mariana Arc. Int. J. Syst. Evol. Microbiol. *55*, 183–189.

Urios, L., Cueff-Gauchard, V., Pignet, P., Postec, A., Fardeau, M.L., Ollivier, B., and Barbier, G. (2004). *Thermosipho atlanticus* sp. nov., a novel member of the Thermotogales isolated from a Mid-Atlantic Ridge hydrothermal vent. Int. J. Syst. Evol. Microbiol. *54*, 1953–1957.

Vandieken, V., Knoblauch, C., and Jorgensen, B.B. (2006). *Desulfotomaculum arcticum* sp nov., a novel spore-formin, moderately thermophilic, sulfate-reducing bacterium isolated from a permanently cold fjord sediment of Svalbard. Int. J. Syst. Evol. Microbiol. *56*, 687–690.

VanFossen, A.L., Lewis, D.L., Nichols, J.D., and Kelly, R.M. (2008). Polysaccharide degradation and synthesis by extremely thermophilic anaerobes. Ann. N.Y. Acad. Sci. *1125*, 322–337.

VanFossen, A.L., Verhaart, M.R.A., Kengen, S.M.W., and Kelly, R.M. (2009). Carbohydrate utilization patterns for the extremely thermophilic bacterium *Caldicellulosiruptor saccharolyticus* reveal broad growth substrate preferences. Appl. Environ. Microbiol. *75*, 7718–7724.

VanFossen, A.L., Ozdemir, I., Zelin, S.L., and Kelly, R.M. (2011). Glycoside hydrolase inventory drives plant polysaccharide deconstruction by the extremely thermophilic bacterium *Caldicellulosiruptor saccharolyticus*. Biotechnol. Bioeng. *108*, 1559–1569.

Vetriani, C., Speck, M.D., Ellor, S.V., Lutz, R.A., and Starovoytov, V. (2004). *Thermovibrio ammonificans* sp nov., a thermophilic, chemolithotrophic, nitrate-ammonifying bacterium from deep-sea hydrothermal vents. Int. J. Syst. Evol. Microbiol. *54*, 175–181.

Voordeckers, J.W., Starovoytov, V., and Vetriani, C. (2005). *Caminibacter mediatlanticus* sp nov., a thermophilic, chemolithoautotrophic, nitrate-ammonifying bacterium isolated from a deep-sea hydrothermal vent on the Mid-Atlantic Ridge. Int. J. Syst. Evol. Microbiol. *55*, 773–779.

van de Vossenberg, J.L.C.M., Ubbink-Kok, T., Elferink, M.G.L., Driessen, A.J.M., and Konings, W.N. (1995). Ion permeability of the cytoplasmic membrane limits the maximum growth temperature of bacteria and archaea. Mol. Microbiol. *18*, 925–932.

Wei, N., Quarterman, J., and Jin, Y.S. (2013). Marine macroalgae: an untapped resource for producing fuels and chemicals. Trends Biotechnol. *31*, 70–77.

van de Werken, H.J.G., Verhaart, M.R.A., VanFossen, A.L., Willquist, K., Lewis, D.L., Nichols, J.D., Goorissen, H.P., Mongodin, E.F., Nelson, K.E., van Niel, E.W.J., *et al.* (2008). Hydrogenomics of the extremely thermophilic bacterium *Caldicellulosiruptor saccharolyticus*. Appl. Environ. Microbiol. *74*, 6720–6729.

Wery, N., Lesongeur, F., Pignet, P., Derennes, V., Cambon-Bonavita, M.A., Godfroy, A., and Barbier, G. (2001a). *Marinitoga camini* gen, nov., sp nov., a rod-shaped bacterium belonging to the order Thermotogales, isolated from a deep-sea hydrothermal vent. Int. J. Syst. Evol. Microbiol. *51*, 495–504.

Wery, N., Moricet, J.M., Cueff, V., Jean, J., Pignet, P., Lesongeur, F., Cambon-Bonavita, M.A., and Barbier, G. (2001b). *Caloranaerobacter azorensis* gen. nov., sp nov., an anaerobic thermophilic bacterium isolated from a deep-sea hydrothermal vent. Int. J. Syst. Evol. Microbiol. *51*, 1789–1796.

Wu, J.H.D., Ormejohnson, W.H., and Demain, A.L. (1988). 2 Components of an extracellular protein aggregate of *Clostridium thermocellum* together degrade crystalline cellulose. Biochemistry *27*, 1703–1709.

Xie, Y., An, J., Yang, G.Y., Wu, G., Zhang, Y., Cui, L., and Feng, Y. (2014). Enhanced enzyme kinetic stability by increasing rigidity within the active site. J. Biol. Chem. *289*, 7994–8006.

Xue, Y.M., Wu, A.L., Zeng, H.Y., and Shao, W.L. (2006). High-level expression of an alpha-L-arabinofuranosidase from *Thermotoga maritima* in *Escherichia coli* for the production of xylobiose from xylan. Biotechnol. Lett. *28*, 351–356.

Yang, H., Ichinose, H., Yoshida, M., Nakajima, M., Kobayashi, H., and Kaneko, S. (2006). Characterization of a thermostable endo-beta-1,4-D-galactanase from the hyperthermophile *Thermotoga maritima*. Biosci. Biotech. Biochem. *70*, 538–541.

Yang, S.J., Kataeva, I., Hamilton-Brehm, S.D., Engle, N.L., Tschaplinski, T.J., Doeppke, C., Davis, M., Westpheling, J., and Adams, M.W.W. (2009). Efficient degradation of lignocellulosic plant biomass, without pretreatment, by the thermophilic anaerobe "*Anaerocellum thermophilum*" DSM 6725. Appl. Environ. Microbiol. *75*, 4762–4769.

Yang, S.J., Kataeva, I., Wiegel, J., Yin, Y.B., Dam, P., Xu, Y., Westpheling, J., and Adams, M.W.W. (2010). Classification of '*Anaerocellum thermophilum*' strain DSM 6725 as *Caldicellulosiruptor bescii* sp. nov. Int. J. Syst. Evol. Microbiol. *60*, 2011–2015.

Ye, L.B., Su, X.Y., Schmitz, G.E., Moon, Y.H., Zhang, J., Mackie, R.I., and Cann, I.K.O. (2012). Molecular and biochemical analyses of the GH44 module of CbMan5B/Cel44A, a bifunctional enzyme from the hyperthermophilic bacterium *Caldicellulosiruptor bescii*. Appl. Environ. Microbiol. *78*, 7048–7059.

Ying, Y., Meng, D., Chen, X., and Li, F. (2013). An extremely thermophilic anaerobic bacterium *Caldicellulosiruptor* sp. F32 exhibits distinctive properties in growth and xylanases during xylan hydrolysis. Enzyme. Microb. Technol. *53*, 194–199.

Yoneda, Y., Yoshida, T., Yasuda, H., Imada, C., and Sako, Y. (2013). A thermophilic, hydrogenogenic and carboxydotrophic bacterium, *Calderihabitans maritimus* gen. nov., sp. nov., from a marine sediment core of an undersea caldera. Int. J. Syst. Evol. Microbiol. *63*, 3602–3608.

Zhang, K., Chen, X., Schwarz, W.H., and Li, F. (2014). Synergism of glycoside hydrolase secretomes from two thermophilic bacteria cocultivated on lignocellulose. Appl. Environ. Microbiol. *80*, 2592–2601.

Zhang, Y.H.P., Ding, S.Y., Mielenz, J.R., Cui, J.B., Elander, R.T., Laser, M., Himmel, M.E., McMillan, J.R., and Lynd, L.R. (2007). Fractionating recalcitrant lignocellulose at modest reaction conditions. Biotechnol. Bioeng. *97*, 214–223.

Zverlov, V.V., Fuchs, K.P., Schwarz, W.H., and Velikodvorskaya, G.A. (1994). Purification and Cellulosomal Localization of Clostridium-Thermocellum Mixed Linkage Beta-Glucanase Licb (1,3–1,4-Beta-D-Glucanase). Biotechnol. Lett. *16*, 29–34.

Zverlov, V.V., Liebl, W., Bachleitner, M., and Schwarz, W.H. (1998). Nucleotide sequence of arfB of *Clostridium stercorarium*, and prediction of catalytic residues of alpha L-arabinofuranosidases based on local similarity with several families of glycosyl hydrolases. FEMS Microbiol. Lett. *164*, 337–343.

Zverlov, V.V., Velikodvorskaya, G.A., Schwarz, W.H., Kellermann, J., and Staudenbauer, W.L. (1999). Duplicated *Clostridium thermocellum* cellobiohydrolase gene encoding cellulosomal subunits S3 and S5. Appl. Microbiol. Biotechnol. *51*, 852–859.

Zverlov, V.V., Kellermann, J., and Schwarz, W.H. (2005). Functional subgenomics of *Clostridium thermocellum* cellulosomal genes: Identification of the major catalytic components in the extracellular complex and detection of three new enzymes. Proteomics *5*, 3646–3653.

Biochemical Properties and Applications of Heat-active Biocatalysts

3

Christian Schäfers, Skander Elleuche and Garabed Antranikian

Abstract

The discovery of extremophilic microorganisms (extremophiles) and their enzyme systems (extremozymes) has opened new opportunities for various industrial applications over the last decades. Depending on their specific properties such as high thermostability, tolerance at extreme pH values or their performance at high organic solvent concentrations, these microorganisms are suitable tools for a wide range of applications. Extremozymes are suitable for a vast number of industrial areas such as waste treatment, food and feed industries, paper industry, pharmaceutical and chemical industries. Thermoactive biocatalysts like proteases, lipases and carbohydrate-degrading enzymes are only a few examples that are of great industrial value and play a crucial role in high-temperature processes. This chapter focuses on different classes of heat-active enzymes that are currently used in various industrial applications. It highlights the value of these biocatalysts and their biochemical properties for the corresponding industry. Moreover, some light will be shed on additional enzymes that are less frequently applied, but are of potential use for new applications.

Introduction

In recent years, a number of biochemical processes has replaced traditional chemical industrial processes. Depending on the prevailing process conditions, like high temperatures or extreme pH values, there are further requirements that should be fulfilled by the new generations of enzymes. It is noteworthy to mention that most commonly used biocatalysts show low stability under the industrial harsh conditions (Zentgraf and Ahern, 1991). Fortunately, in the last four decades, extremophilic microorganisms have been described, unlike their non-extremophilic counterparts, to show a high adaptability to extraordinary milieus with extremes of temperature (0 to 120°C) or pH values (0 to 12), salt concentrations up to 35% and high pressure (Antranikian and Egorova, 2007; Demirjian *et al.*, 2001; Niehaus *et al.*, 1999). Enzymes derived from extremophiles (extremozymes) were isolated from different microorganisms that populate extreme environments such as hot springs, solfatara fields, alkaline lakes and deep-sea hot vents or volcanic eruptions (Fig. 3.1) (Antranikian and Egorova, 2007; Ladenstein and Antranikian, 1998; Madigan and Marrs, 1997; Niehaus *et al.*, 1999; Vieille and Zeikus, 2001). Thermophilic microorganisms, (thermophiles) with the highest reported growth temperature (hyperthermophiles) belong to the genera *Pyrococcus*, *Pyrobaculum*, *Pyrodictium* or *Methanopyrus* (Stetter, 1996). The hyperthermophilic archaeon *Methanopyrus kandleri* strain 116 grows up to 122°C and *Thermotoga maritima* and *Aquifex pyrophilus* represent the bacterial species with the highest reported growth temperature between 90°C and 95°C (Niehaus *et al.*, 1999; Takai *et al.*, 2008). Enzymes that are active under these conditions often display a high tolerance against organic solvents or detergents. Enzymes produced by extremophilic microorganisms are accordingly predestined as biocatalysts for industrial applications and outcompete traditionally mesophilic enzymes (Bruins *et al.*, 2001; Demirjian *et al.*, 2001; Kushner, 1978; Leuschner and

Figure 3.1 A hot spring representing a source for thermophilic microorganisms and DNA.

Antranikian, 1995; Sonnleitner, 1983; Zeikus *et al.*, 1998; Zentgraf and Ahern, 1991). High temperature ranges during industrial processes come along with several positive aspects like a reduced risk of microbial contamination, faster reaction rates and improved substrate solubility (Cowan *et al.*, 1985; Silva *et al.*, 2007; Sonnleitner, 1983; Zeikus *et al.*, 1998). Likewise, thermozymes enable an efficient remediation through an increased availability of hardly biodegradable and insoluble environmental pollutants. The corresponding outcomes are not only of economic nature with higher yields of product that go along with a reduction of production costs but is also of sustainable ecological significance. This shift towards a more ecological or 'greener' industry represents a trend that is nowadays discussed in many industrial fields and has steered interest from present enzymes to biodegradable alternatives such as biosurfactants. Moreover, thermozymes and their outstanding stability represent good candidates to close the gap between chemical and biological processes with environmental sustainability (Madigan and Marrs, 1997). It has been proposed that the increased stability of thermoactive enzymes is dependent on a vast number of structural properties like additional salt bridges, hydrophobic interactions, hydrogen bonds, packing, secondary structure propensity and rigidity (Eijsink *et al.*, 2004; Kumar *et al.*, 2000; Li *et al.*, 2005; Matthews, 1993; Szilagyi and Zavodszky, 2000; Vieille and Zeikus, 2001; Zeikus *et al.*, 1998).

Different strategies are applied for the finding of novel enzymes from extremophiles. These include: (i) development of intelligent screening systems, (ii) expression of extremozymes encoding genes in mesophilic hosts, (iii) establishments of extremophilic expression systems or (iv) protein engineering approaches (Vieille and Zeikus, 2001; Yang *et al.*, 2013). Beside activity-based screening approaches using dye-coupled substrates, sequence-based screening became very popular in the postgenomic era (Jabbour *et al.*, 2013; Martinez-Martinez *et al.*, 2013; Pang *et al.*, 2009; Zheng *et al.*, 2013). In contrast to the screening of novel thermophilic organisms, several reports describe the screening of heterologously expressed gene banks that are derived from extremophiles in hosts like *Escherichia coli* or yeasts, including *Saccharomyces cerevisiae* and *Pichia pastoris* (Latiffi *et al.*, 2013; Maciver *et al.*, 1994; Morana *et al.*, 1995; Sørensen *et al.*, 2003; Yang *et al.*, 2010). Production rates of recombinant enzymes in well-established mesophilic hosts are usually still superior over the thermophilic expression systems. There are multiple examples of thermozymes that have been produced in an active form in a mesophilic host without affecting catalytic properties and stability of the protein (Morana *et al.*, 1995). Moreover, easy purification is often achieved by an initial heat denaturation step. The alternative strategy represents the establishment of thermophilic expression hosts including *Thermus thermophilus* and *Sulfolobus solfataricus* (Cava *et al.*, 2009). The drawbacks include slow growth rates and low cell yields.

The focus of this chapter will primarily be on thermostable enzymes, such as starch-degrading enzymes, lipases, proteases, chitinases, pectinases and phytases, their biochemical properties and industrial applications.

Starch-degrading enzymes

Starch is a heterogeneous polysaccharide composed of the α-D-glucose high-molecular-weight components amylose and amylopectin. The linear and water-insoluble compound amylose exclusively exhibits α-1,4-linked glucopyranose residues, while the soluble amylopectin is a branched polymer, which contains α-1,4-linkages and α-1,6-linked branch points (Elleuche and Antranikian, 2013). The complex structure of starch necessitates an appropriate portfolio of hydrolysing enzymes to depolymerize the molecule to small sugars, such as glucose and maltose (Fig. 3.2). According to their primary structure and their mode of action, enzymes involved in carbohydrate metabolism can be assigned to the large family of glycosyl hydrolases (Henrissat, 1991). Amylolytic enzyme systems have been identified and characterized in several prokaryotes. The majority of starch-hydrolysing extremozymes belong to the groups of α-amylases, glucoamylases, α-glucosidases and pullulanases. These enzymes are attractive candidates for various applications, e.g. sugar syrup production, textile, brewing and pharmaceutical industries.

Amylases

α-Amylases (α-1,4-glucan-glucanohydrolase, EC 3.2.1.1) randomly attack α-1,4-linkages at the inner part of the starch molecule, thereby producing oligosaccharides and glucose in α-anomeric configuration as final products. These enzymes have been solely grouped into GH families 13

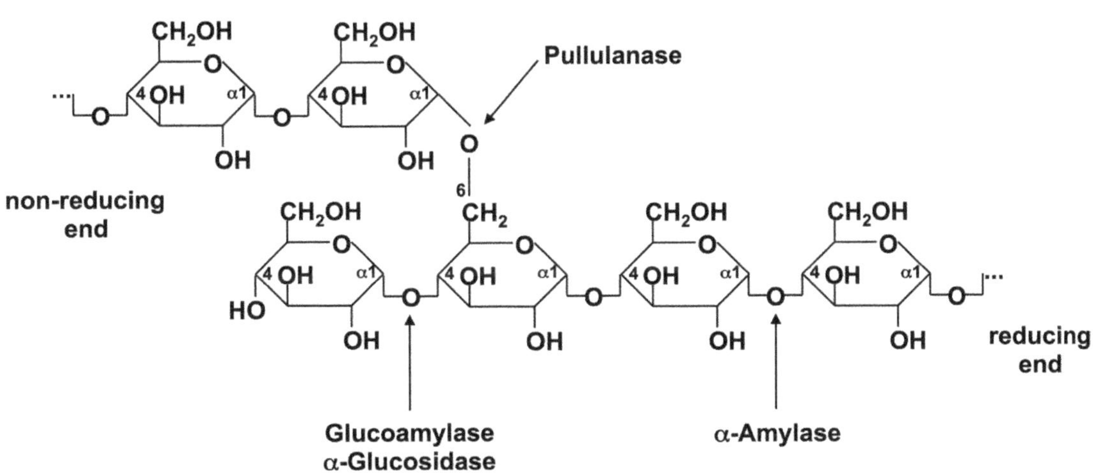

Figure 3.2 Starch molecule including attacking sites of starch-hydrolysing enzymes.

(α-amylase family) and 57 (www.cazy.org). Moreover, a single α-amylase from *Bacillus circulans* has been characterized and assigned to GH family 119 (Watanabe *et al.*, 2006).

The polypeptide chain is folded into three main domains (A, B and C) with A forming a characteristic catalytic domain in a $(\beta/\alpha)_8$-barrel conformation at the N-terminus of the protein (Prakash and Jaiswal, 2010). This motif is called the TIM barrel, which was first investigated in chicken muscle triose phosphate isomerase (Banner *et al.*, 1975). In all α-amlyases only eight amino acid residues were proposed to be invariant, seven of which were shown to be in close proximity to the active site of the protein (Antranikian and Egorova, 2007). Domain B contains several irregular β-strands and is highly diverse among α-amylases from various sources while domain C forms a β-sandwich. Domains B and C are located on opposite sites of the $(\beta/\alpha)_8$-barrel (Prakash and Jaiswal, 2010). However, domain B was proposed to be important for substrate specificity. Furthermore, it was shown that a deletion of an extra loop in domain B of the α-amylase Amy US100 from *Bacillus stearothermophilus* reinforces the active site conformation and the thermostability of the enzyme. In addition to the increased thermostability, the mutant also displayed a decreased calcium requirement (Khemakhem *et al.*, 2009). Most mesophilic α-amlyases contain a conserved calcium ion that is embedded at the interface between domain A and domain B. In case of the removal of the ion; the enzyme completely loses its activity.

The first example of a heat and acid stable protein that has been characterized in detail is an extracellular α-amylase from the bacterium *Alicyclobacillus acidocaldarius*. This 160 kDa enzyme exhibits temperature and pH optima at 75°C and 3, respectively (Schwermann *et al.*, 1994). This acidophilic amylase contains approximately 30% fewer charged amino acid residues than its mesophilic counterparts. Most of the amino acids are replaced by neutral polar residues and located on the protein surface, indicating a putative adaptation to the environment by reducing the amount of both negative and positive charges. This effect avoids electrostatic repulsion at low pH, and increases the acidostability. Thermostable α-amylases were not only found in Bacteria and Archaea, but also in plants, fungi and animals, where they are involved in carbohydrate metabolism (Prakash and Jaiswal, 2010). To date, the most thermoactive α-amylases were found in members of the archaeal genera *Methanocaldococcus* and *Pyrococcus* (Kim *et al.*, 2001; Koch *et al.*, 1991, 1990). The extracellular *Pyrococcus woesei* α-amylase has been extensively investigated and was shown to be active in a temperature range between 40 and 130°C with a temperature optimum at 100°C and with a half-life of 4 hours at 110°C (Table 3.1). This enzyme is capable of cleaving α-1,4 linkages in soluble and native starch, amylose, amylopectin and glycogen thereby catalysing the formation of different sized oligosaccharides with maltoheptaose being the smallest product (Koch *et al.*, 1991). Activity of the *Pyrococcus furiosus* α-amylase predominantly yielded glucose and maltose as final products (Koch *et al.*, 1990). Furthermore, the GH family 13 extracellular α-amylase and an intracellular enzyme from *P. furiosus* have been characterized after heterologous expression in *Bacillus subtilis* and *Escherichia coli* (Jorgensen *et al.*, 1997; Laderman *et al.*, 1993). The gene encoding the extracellular α-amylase from *P. woesei* was functionally expressed in heterologous hosts in *E. coli* and *Halomonas elongata*. The recombinant protein was extensively studied and crystallized, which led to the identification of several parameters supporting the heat adaptation properties. In addition to the identification of a notable number of salt bridges, a calcium ion was identified at the interface between domains A and B and was hypothesized to be important for the structural organization of the polypeptide (Frillingos *et al.*, 2000; Linden *et al.*, 2003; Linden and Wilmanns, 2004). It is widely accepted that the thermostability of α-amylases could be enhanced in the presence of divalent metal ions, although these ions were more often found in amylolytic enzymes from mesophiles (Antranikian and Egorova, 2007).

Another example is the extracellular protein AmyA from the thermoalkaliphilic bacterium *Anaerobranca gottschalkii*, which is an α-amylase with β-cyclodextrin forming activity, while the intracellular isozyme AmyB does not show comparable activities (Ballschmiter *et al.*, 2005). The

Table 3.1 Heat-active starch-hydrolysing enzymes from prokaryotic microorganisms

Enzyme	Host organism	T_{opt} (°C)	$T_{1/2}$	pH	Applications[a]/properties	Reference
α-Amylase						
–	*Alicyclobacillus acidocaldarius*	75	–	3	Thermoacidophilicity	Schwermann *et al.* (1994)
ASKA	*Anoxybacillus* sp.	60	65°C, 48 h	8.0	Wide industrial applications	Chai *et al.* (2012)
ADTA	*Anoxybacillus* sp.	60	65°C, 3 h	8.0	Wide industrial applications	Chai *et al.* (2012)
–	*Geobacillus thermoleovorans*	100	100°C, >5 h	8	Saccharification of raw pearl millet starch	Uma Maheswar Rao and Satyanarayana (2007)
–	*Halorubrum xinjiangense*	70	80–90°C, 1 h	8.5	Industrial desirable properties, salt- and solvent-tolerance	Moshfegh *et al.* (2013)
MJ1611	*Methanocaldococcus jannaschii*	120	100°C, 50 h	5.0–8.0	Heat-resistance	Kim *et al.* (2001)
Amy13A	Pilot-plant biogas reactor	80	70°, 3 h	7	Salt-tolerance	Jabbour *et al.* (2013)
PFA	*Pyrococcus furiosus*	100	98°C, 13 h	5.5–6.0	Liquefying enzyme	Dong *et al.* (1997a)
PWA	*Pyrococcus woesei*	100	–	5.5	Inactive after autoclaving for 8 h at 120°C	Koch *et al.* (1991)
THJA	*Thermococcus* sp.	95	90°C, 5 h	5	Starch-conversion biotechnologies	Wang *et al.* (2008)
AmyC	*Thermotoga maritima*	90	100°C, >1 h	8.5	Heat-resistance	Ballschmiter *et al.* (2006)
Glucoamylase						
MGA	*Methanocaldococcus jannaschii*	80	–	6.5	–	Uotsu-Tomita *et al.* (2001)
–	*Picrophilus oshimae*	90	90°C, 20 h	2	Thermoacidophilicity	Serour and Antranikian (2002)
–	*Picrophilus torridus*	90	90°C, 24 h	2	Thermoacidophilicity	Serour and Antranikian (2002)
SSG	*Sulfolobus solfataricus*	90	–	5.5–6.0	Industrial starch processing	Kim *et al.* (2004)
TtcGA	*Thermoanaerobacter tengcongensis*	75	80°C, 2.6 h	5.0	Sugar processing	Zheng *et al.* (2010)
TaGA	*Thermoplasma acidophilum*	75	80°C, 2 h	5.0	Industrial processes at high temperatures	Dock *et al.* (2008)
–	*Thermoplasma acidophilum*	90	90°C, 24 h	2	Thermoacidophilicity	Serour and Antranikian (2002)
α-Glucosidase						
–	*Pyrococcus furiosus*	105–115	98°C, 48 h	5.5	Starch-conversion technology	Chang *et al.* (2001)
AglA	*Picrophilus torridus*	80	–	5.0	Stable at 80°C for >120 min	Angelov *et al.* (2006)
–	*Pyrococcus woesei*	105	–	5.0–5.5	Starch-conversion technology	Chang *et al.* (2001)
MalA	*Sulfolobus solfataricus*	85	85°C, 39 h	4.5	Heat-resistance	Rolfsmeier *et al.* (1998)

Table 3.1 continued

Enzyme	Host organism	T_{opt} (°C)	$T_{1/2}$	pH	Applications[a]/properties	Reference
–	*Thermococcus* sp. AN1	98	98°C, 215 min	7	Immobilized enzyme is stable at 130°C	Piller *et al.* (1996)
Pullulanase type I						
–	*Bacillus flavocaldarius*	75–85	–	6.3	Debranching enzyme in starch-processing industry	Suzuki *et al.* (1991)
PulUS105	*Bacillus thermoleovorans*	75	–	5.6–6.0	–	Ben Messaoud *et al.* (2002)
PulA	*Caldicellulosiruptor saccharolyticus*	85	–	–	Saccharification of starch	Albertson *et al.* (1997)
PulA	*Fervidobacterium pennivorans*	80	80°C, 2 h	6	Starch-processing industry	Bertoldo *et al.* (1999)
–	*Thermotoga maritima*	90	90°C, 3.5 h	5.9	Starch-processing industry	Kriegshäuser and Liebl (2000)
PulA	*Thermotoga neapolitana*	80	80°C, 88 min	5.0–7.0	Production of glycoconjugate compounds	Kang *et al.* (2011)
–	*Thermus caldophilus*	75	78°C, 72 h	5.5	Detergent and starch industries	Kim *et al.* (1996)
Pullulanase type-II						
APU	*Pyrococcus furiosus*	105	106°C, 64 min	5.5	Heat-resistance	Dong *et al.* (1997b)
–	*Pyrococcus furiosus*	98	–	5.5	Production of maize sweetener	Brown and Kelly (1993)
–	*Pyrococcus woesei*	100	110°C, 20 min	6.0	Heat-resistance	Rüdiger *et al.* (1995)
Tk0977	*Thermococcus kodakarensis KOD1*	100	100°C, 2 h	5.5–6.0	Maltose syrup	Han *et al.* (2013)
–	*Thermococcus litoralis*	98	–	5.5	Production of maize sweetener	Brown and Kelly (1993)
TsiApu	*Thermococcus siculi*	95	95°C, >4 h	6.0	Industrial saccharification of starch	Jiao *et al.* (2011)

[a]Indicates potential applications that were proposed by the authors of the respective studies.

open reading frame *MJ1611* derived from *Methanocaldococcus jannaschii* even encodes a structural hybrid of the two distinct glycosyl hydrolase families 13 and 57. The respective recombinant enzyme MJA1 is one of the most thermoactive α-amylase investigated so far. It was shown to be optimally active at 120°C in a pH range between 5.0 and 8.0. Moreover, it is very active towards soluble starch, amylose, amylopectin and pullulan (Kim *et al.*, 2001). The latter is the preferred substrate for pullulanases. There are only a few amylolytic enzymes that were shown to be active at high salt concentrations in combination with elevated temperatures. Such biocatalysts were predominantly isolated from halophilic prokaryotes. An α-amylase from the archaeon *Halorubrum xinjiangense* was shown to display maximum activity at 70°C in the presence of 4 M NaCl or 4.5 M KCl, respectively (Moshfegh *et al.*, 2013). Another α-amylase with extraordinary catalytic properties has been recently isolated from a pilot-plant biogas reactor. Amy13A is not only thermoactive, it is also highly salt-tolerant exhibiting around 54% remaining activity in the presence of high salt concentrations (4 M NaCl). Low concentrations of $CaCl_2$ were shown to be advantageous for the catalytic performance and the thermostability of Amy13A, while concentration of 5 mM or higher resulted in slight inactivation (Jabbour *et al.*, 2013).

Glucoamylases

Glucoamylases hydrolyse successive terminal α-1,4-linked glucose units from the non-reducing end of polysaccharides, thereby releasing β-D-glucose. Most of the glucoamylases (glucan 1,4-α-glucosidase, amyloglucosidases, γ-amylase, lysosymal α-glucosidase, acid maltase, exo-1,4-α-glucosidase, EC 3.2.1.3) that have been characterized so far are of fungal origin and they have not been detected in animals or plants (Doman-Pytka and Bardowski, 2004). The distribution of glucoamylases (EC 3.2.1.2) in Archaea and Bacteria is rare and few enzymes have been detected in thermophilic species. These enzymes follow an inverting mechanism of reaction, which starts with a proton transfer from the general acid catalyst to the glycosidic oxygen. In a second reaction step the general base catalyst mediates the nucleophilic attack of a deprotonated water molecule. Glucoamylases have been structurally exclusively assigned to glycosyl hydrolase family 15 (Kumar and Satyanarayana, 2009). The typical structure of glucoamylases is composed of three independent domains. The catalytic domain displays a typical $(\alpha/\alpha)_6$-motif and is usually located at the N-terminus of the protein, interrupted by a linker region from the C-terminal starch-binding module (SBM) in eukaryotes. The SBM is responsible for the connection between the enzyme and starch or cell walls. However, SBM is completely absent in prokaryotic glucoamylases. Moreover, the solved crystal structure from the thermophile *Thermoanaerobacterium thermosaccharolyticum* revealed an additional super-β-sandwich domain at the N-terminus, typical for prokaryotic enzymes making them different from eukaryotic enzymes within the glycosyl hydrolase family 15 group (Marin-Navarro and Polaina, 2010). Additionally, glucoamylases also contain four conserved regions (I–IV) which are characteristic for the family of amylolytic enzymes and two glutamate residues that are most probably essential for catalytic activity (Doman-Pytka and Bardowski, 2004).

Extracellular glucoamylases from archaeal thermoacidophiles have also been purified. These extracellular enzymes are remarkably robust; enzymatic activity was highest at 90°C and pH 2, but even detectable at 100°C and pH 0.5 (Serour and Antranikian, 2002). The representatives from *Picrophilus torridus*, *Picrophilus oshimae* and *Thermoplasma acidophilum* displayed a half-life of more than 20 hours at 90°C. These enzymes are the most thermostable glucoamylases known so far. Catalytic properties were not affected by a variety of ions, chelator or disulfide bond disturbing reagents (Serour and Antranikian, 2002). Since the intracellular pH of thermoacidophiles is only slightly acidic (pH 4.6), therefore the intracellular enzymes are expected to be less acid tolerant. The glucoamlyase of *M. jannaschii* has been expressed in a heterologous host in *E. coli* and the enzyme in the crude insoluble extract showed a pH and temperature optimum of 6.6 and 80°C, respectively (Uotsu-Tomita *et al.*, 2001). Kim *et al.* described the detailed characterization of a recombinant glucoamylase from another thermoacidophilic host. Originally encoded by *S. solfataricus*, the putative glucoamylase encoding gene *ssg* was successfully expressed in *E. coli* (Kim *et al.*, 2004). This tetrameric enzyme is thermostable and prefers maltotriose as substrate under optimal conditions at 90°C and pH 5.5 to 6.0.

While α-amylases seem to be widely distributed, only few glucoamylases were found in thermophilic Archaea. Most of the glucoamylases that have been characterized so far are derived from various fungal species, but these enzymes are unstable at elevated temperatures (Kumar and Satyanarayana, 2009). There is great interest in finding heat-active enzymes from thermophilic fungi. To date, however, only few glucoamylase genes from thermophilic fungi have been sequenced and cloned.

The most thermostable glucoamylase from Bacteria, reported to date, was isolated from *Thermoanaerobacter tengcongensis* (Table 3.1). A signal peptide of 21 amino acids was predicted to be located at the N-terminus of the enzyme TtcGA and the expression of the full-length gene in *E. coli* revealed only poor expression levels, while the deletion of the signal sequence resulted in an increased production of the target protein. TtcGA displayed a temperature optimum at 75°C and no detectable loss of activity was observed during incubation at 75°C for 6 hours. Furthermore, the pH optimum is around pH 5.0 (Zheng *et al.*, 2010).

α-Glucosidases

Mammalian α-glucosidases are often referred to as glucoamylases, because they attack the same substrates (malto-oligosaccharides and starch), but they differ in their structure and their mode of action from microbial enzymes (Marin-Navarro and Polaina, 2010). α-Glucosidases (α-D-glucoside glucohydrolase, EC 3.2.1.20) also break α-1,4 linkages from the terminal non-reducing end, but they usually prefer smaller substrates than glucoamylases, which are produced by the action of other polysaccharide degrading enzymes. Glucose with an α-anomeric configuration is mainly liberated from maltose and maltotriose (Antranikian, 2008). Heat-active α-glucosidases are highly diverse in their amino acid composition and molecular weights (Leveque *et al.*, 2000). According to their primary structure these enzymes have been assigned to glycosyl hydrolase families 4, 13, 31, 63 and 122 (Henrissat, 1991).

The gene *malA* was the first open reading frame encoding an α-glucosidase (maltase) from the archaeal domain, which has been characterized in detail. Interestingly, this gene is closely related to mammalian homologues. MalA is the major soluble α-glucosidase from *S. solfataricus* and has been successfully expressed in *E. coli*. Recombinant MalA was purified and exhibited a temperature optimum at 85°C and a pH optimum at 4.5 for maltose (Table 3.1). The heterologously produced protein was more thermostable than the native variant (Rolfsmeier *et al.*, 1998). Another α-glucosidase (AglA) has been investigated from the thermoacidophile *P. torridus*. Although the optimal growth temperature of this archaeon is only 60°C, the enzyme displayed highest activity at 87°C and pH 5.0 (Angelov *et al.*, 2006).

Pullulanases

Pullulanases (EC 3.2.1.41) mainly cleave the α-1,6-bonds in the water-soluble linear α-glucan exopolysaccharide pullulan in an endo-fashion, yielding maltotriose with a molecular weight between 1500 and 810,000 kDa (Doman-Pytka and Bardowski, 2004) (Fig. 3.3). Pullulan is synthesized by the ascomycetous fungus *Aureobasidium pullulans*, when this organism is grown on medium containing glucose or sucrose as carbon-source. It is industrially produced mainly for the food industry (Wu *et al.*, 2010).

Classical pullulanases are divided into type I pullulanase (true pullulanase, α-dextrin 6-glucanohydrolase, debranching enzyme) and type II (amylopullulanase or α-amylase pullulanase). Enzymes of type I exclusively attack α-1,6-linkages in pullulan and branched oligo- and polysaccharides yielding linear oligosaccharides or maltotriose, respectively, while type II pullulanases exhibit a dual specificity attacking α-1,4-bonds in starch and α-1,6-glycosidic linkages in pullulan to give glucose, maltose and maltotriose as major products (Antranikian and Egorova, 2007). Amylopullulanases (EC.3.2.1.1/41) are unique because they combine the catalytic properties of both, α-amylases and pullulanases. However, a type I pullulanase from the extreme thermophile *Thermotoga neapolitana* was recently described to exhibit an additional α-1,6-transglycosylation activity (Kang *et al.*, 2011). Furthermore, four additional types of enzymes also cleave the polysaccharide pullulan. Pullulan hydrolases of types I and II within the third class of pullulan-degrading enzymes are active towards α-1,4-bonds in pullulan, amylose and starch, but are rarely effective in hydrolysing α-1,6 linkages. Pullulan hydrolase type II (isopullulanase, pullulan 4-glucanohydrolase, EC 3.2.1.57) exclusively cleaves the α-1,4-glycosidic linkages of pullulan to produce isopanose (6-α-maltosylglucose), while panose is produced by the action of pullulan hydrolase type I (neopullulanase, pullulan 4-D-glucanohydrolase, EC 3.2.1.135). Pullulan hydrolase type III is the exception within the subgroup of pullulan hydrolases that is capable of hydrolysing both α-1,4 and α-1,6-linkages in pullulan. Finally, some known glucoamylases are cleaving single glucose residues from the non-reducing end of pullulan (Doman-Pytka and Bardowski, 2004).

Pullulanases are widely distributed in nature and were predominantly characterized from mesophilic and thermophilic Bacteria (Doman-Pytka and Bardowski, 2004). Type I Pullulanases

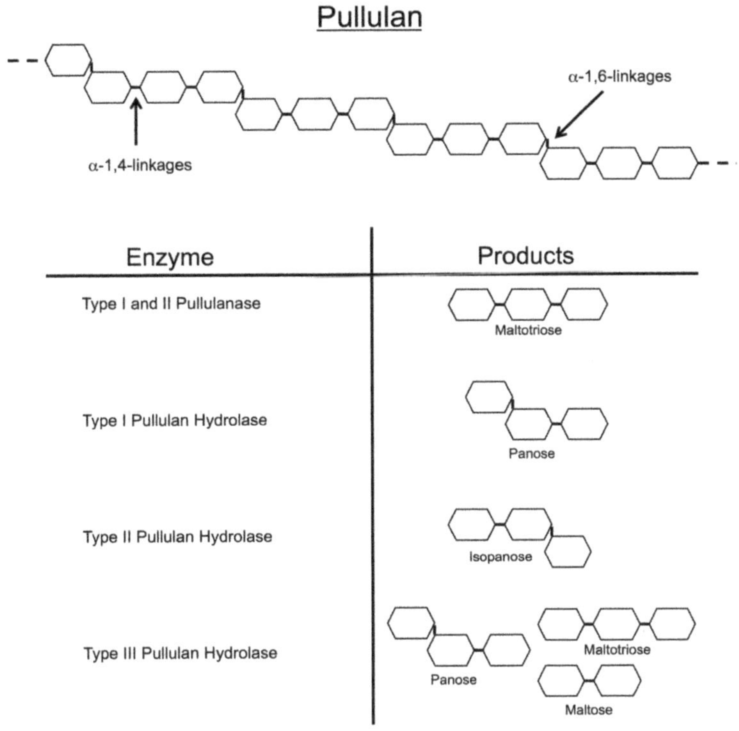

Figure 3.3 Schematic presentation of the action of different types of pullulanases.

seem to be not widely distributed among extreme thermophilic Bacteria and hyperthermophilic Archaea, while type II amylopullulanases are widely distributed among aerobic and anaerobic prokaryotes. However, pullulanase type I has been isolated and characterized from a number of moderate thermophilic bacterial species and was demonstrated to be intra- or extracellular. The first heat-adapted debranching enzymes (pullulanase type I) have been found in the aerobic thermophile *Bacillus flavocaldarius*, in the anaerobic thermophile *Fervidobacterium pennivorans* and in the anaerobic extreme thermophile *T. maritima* (Table 3.1) (Bertoldo *et al.*, 1999; Kriegshäuser and Liebl, 2000; Suzuki *et al.*, 1991). *B. flavocaldarius* encodes a cell-associated pullulanase that has been purified to homogeneity. This enzyme exhibits maximal activity between 75 and 80°C using pullulan as substrate and between 80 and 85°C with soluble starch (Suzuki *et al.*, 1991). Using this enzyme as a model, it has been speculated that an increased amount of proline residues is accompanied by an increase of the thermal stability in a linear fashion (Doman-Pytka and Bardowski, 2004). Another extremely thermostable pullulanase from *T. maritima* was purified using a two-step protocol with an affinity chromatography step based on the enzyme's capacity to bind to β-cyclodextrins (Kriegshäuser and Liebl, 2000). The pullulanase-encoding gene *pulA* from *F. pennivorans* was cloned without its signal sequence. After expression in *E. coli* using the *trc*-promoter system, the recombinant enzyme displayed 112 times higher activity in comparison with the enzyme produced in the parental wild type strain (Bertoldo *et al.*, 1999). This enzyme is specifically inhibited in the presence of divalent cations, such as Fe^{2+}, Cu^{2+} and Zn^{2+}, while other pullulanases were activated by divalent cations like Ca^{2+}, Mn^{2+} or Co^{2+} (Antranikian and Egorova, 2007; Antranikian *et al.*, 1987; Bertoldo *et al.*, 1999; Doman-Pytka and Bardowski, 2004). This was shown for the activity of a heat-stable type II pullulanase from *P. woesei*, which is dramatically increased (up to 370%) by the addition of calcium (Rüdiger *et al.*, 1995; Schwerdtfeger *et al.*, 1999).

The most thermoactive type II pullulanases have been isolated from archaeal species of the genera *Desulfurococcus, Pyrococcus* and *Thermococcus* (Antranikian and Egorova, 2007). Most of these enzymes exhibit remarkable thermostability profiles that are often strictly dependent on the addition of calcium to the reaction sample. Brown and Kelly demonstrated in 1993 that pullulanase activity was detectable at temperatures up to 140°C in the presence of calcium (Brown and Kelly, 1993). Pyrococcal type II pullulanases have been identified in both *P. furiosus* and *P. woesei* (Table 3.1). Half-life at 90°C of the *P. furiosus* amylopullulanase was drastically increased by incubation with 0.5 mM of calcium ions (Doman-Pytka and Bardowski, 2004). The enzyme from *P. woesei* is composed of a single polypeptide chain with a molecular mass of 90 kDa and has been shown to be optimally active at 100°C and pH 6.0. The pullulanase from *P. furiosus* is 110 kDa and optimally active at 98°C (Brown and Kelly, 1993). Another type II pullulanase has been recently identified in *Thermococcus kodakarensis* and was characterized after heterologous production in *E. coli*. This enzyme was active on both substrates, pullulan and starch, and showed highest activity at 100°C so far (Han *et al.*, 2013).

Cyclodextrin glycosyltransferases

Cyclodextrin glycosyltransferase (CGTase, EC 2.4.1.19) is an enzyme capable of catalysing the conversion of starch or oligodextrins into circular α-1,4-linked glucose molecules. Based on the number of linked glucose sugars, cyclodextrins are defined as α- (six glucose molecules), β- (seven glucose units) or γ-cyclodextrins (eight glucose molecules) displaying different properties with β-cyclodextrin being less water-soluble than α- and γ-types (Fig. 3.4). β-Cyclodextrin is mostly formed by enzymatic starch conversion allowing for its cheap industrial production compared to other cyclodextrins. CGTases were shown to be present in some thermophilic microorganisms of prokaryotic origin with several enzymes isolated from species of the genus *Bacillus* to be readily used in industrial applications (Antranikian, 2008). Most of known CGTases were derived from Bacteria, but there are some CGTases that have been discovered in Archaea as well (Table 3.2).

Among the first heat-active CGTase that has been characterized in detail is an enzyme from strain *Bacillus stearothermophilus* ET1 with an optimal temperature of 80°C and a half-life of 10 minutes when incubated at 80°C. Due to its high affinity to β-cyclodextrin, this enzyme has been purified to homogeneity by using a cyclodextrin affinity chromatography approach (Chung *et al.*, 1998). The most stable CGTase with the highest temperature optimum has been isolated from the anaerobic archaeon *Thermoanaerobacter* sp. B1001. The α-cyclodextrin forming enzyme retains 50% of its activity after incubation at 110°C for 40 minutes and displays an optimum at 90–100°C (Tachibana *et al.*, 1999). An α-amylase with cyclodextrin forming activity has been described from the thermoalkaliphilic bacterium

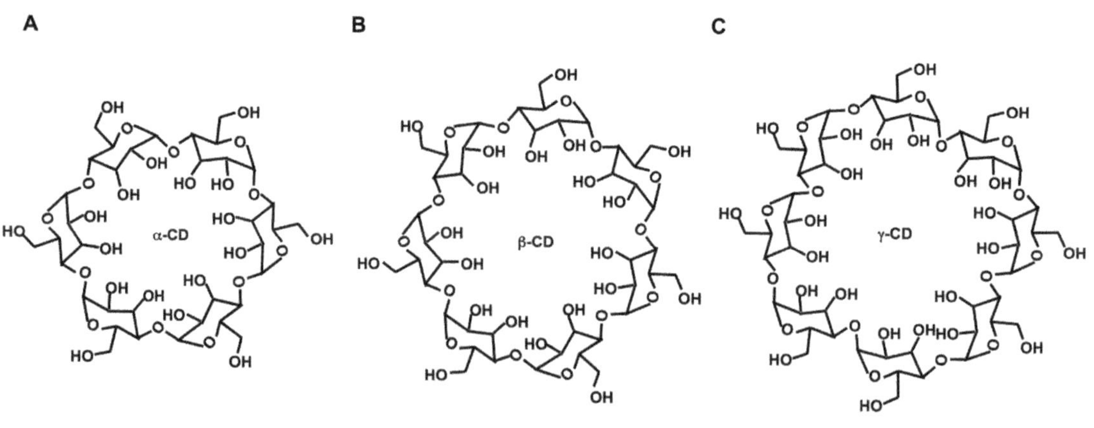

Figure 3.4 Structure of cyclodextrins. (A) α-cyclodextrin, (B) β-cyclodextrin and (C) γ-cyclodextrin. Cyclodextrins contain 6 (α), 7 (β) or 8 (γ) glucose units, which are linked via α-1,4-linkages.

Table 3.2 CGTases from Archaea and Bacteria

Enzyme	Host organism	T_{opt} (°C)	$T_{1/2}$	pH	Applications[a]/properties	Reference
–	*Amphibacillus* sp. NPST-10	50	70°C, >60 min	8.0	Cyclodextrin production industry	Ibrahim *et al.* (2012)
AmyA	*Anaerobranca gottschalkii*	70	70°C, 48 h	8.0	Production of β-cyclodextrin	Ballschmiter *et al.* (2005)
–	*Bacillus stearothermophilus* ET1	80	80°C, 10 min	8.0	Maize starch conversion	Chung *et al.* (1998)
PFCGT	*Pyrococcus furiosus*	95	90°C, <10 min	5.0	Production of β-cyclodextrin	Lee *et al.* (2007)
–	*Thermoanaerobacter* sp.	85	–	5.5	Production of β-cyclodextrin	Alcalde *et al.* (1998)
CGT_{TK}	*Thermococcus kodakaraensis* KOD1	80	100°C, 20 min	5.5–6.0	–	Rashid *et al.* (2002)
–	*Thermococcus* sp. B1001	90–100	110°C, 40 min	7.0	Production of α-cyclodextrin	Tachibana *et al.* (1999)

Anaerobranca gottschalkii. AmyA has been proposed to be an extracellular lipoprotein and has been shown to withstand alkaline conditions present in *Anaerobranca gottschalkii*'s natural habitats. It functions at alkaline conditions (> 50% residual activity at pH 9.5) and is optimally active at 70°C with a half-life of 48 hours at 70°C (Ballschmiter *et al.*, 2005). A novel CGTase that is capable of predominantly producing β-cyclodextrin at high yield from raw starch has recently been identified in an alkaliphilic *Amphibacillus* species (Ibrahim *et al.*, 2012).

Applications of starch-degrading enzymes

Diastase, the first starch-hydrolysing enzyme, has been isolated from germinating barley by Anselme Payen in 1833 and was shown to convert starch into sugars (Buchholz and Collins, 2013). Since then, many valuable discoveries have been made making starch-degrading enzymes among the most used biocatalysts in industrial applications (Kumari *et al.*, 2012). Owing to their unique attributes, amylolytic enzymes from thermophiles are of tremendous importance for biotechnological applications and often superior over their mesophilic counterparts. They are utilized in several industries including brewing, textile, food and pharmaceutical industry.

α-Amylases are applied as anti-staling agents in the baking industry. However, staling is a consequence of the continuous degradation of the amylopectin moiety in the starch molecule resulting in a loss of crispness and a loss of bread flavour. Thermostable amylopullulanases were considered to be superior over traditional amylases due to the hydrolysis of branched maltodextrins, which is the major factor for the increased stickiness of enzymatic treated bread. Moreover, these enzymes are suitable for application in other processes in industry including beverage industries, as well as selected production of different sugar syrups containing maltose and maltooligosaccharides for the production of fine chemicals and in pharmaceutical industries as substitutes for glucose in intravenous feeding (Nisha and Satyanarayana, 2013).

The production of fermentable sugars and syrups is mainly achieved by the application of α-amylase and glucoamylase. Starch that is derived from maize, wheat, rice, barley and potatoes, is used for the production of sweeteners. A traditional conversion process involves two major steps, namely liquefaction and saccharification (Fig. 3.5). Starch is gelatinised using heat in a first reaction step, which leads to the production of a viscous starch slurry. Subsequently, a starch-hydrolysing bacterial α-amylase is added to the reaction mixture to produce linear and branched dextrins. A fungal glucoamylase converts the oligo- and polysaccharides during the saccharification step to glucose (Elleuche and Antranikian,

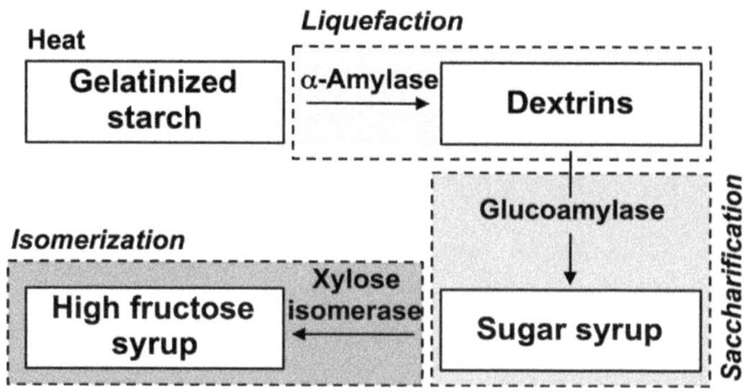

Figure 3.5 Production of high fructose syrup from starch. Enzymatic steps of liquefaction, saccharification and isomerization are indicated in grey. Process conditions: liquefaction (95°C, pH 6), saccharification (60°C, pH 4.5), isomerization (60°C, pH 7–8).

2013). Depending on the substrate and the process, pullulanases are used to debranch the remaining α-1,6-bonds in amylopectin moieties. Amylopullulanases are useful for industrial starch conversion processes, due to their debranching activity in combination with their common bifunctionality and metal-ion independence. This enzyme can be applied in a single-step process combining liquefaction and saccharification steps (Nisha and Satyanarayana, 2013). By the action of xylose-isomerase, glucose is converted into fructose in an isomerization reaction yielding HFS (high-fructose syrup) (Fig. 3.5).

Of special interest are cyclodextrins that are produced by CGTases. Cyclodextrins are useful circular oligosaccharides that are important in pharmaceutical, agricultural and cosmetic industries as stabilizing agents or emulsifiers (Biwer *et al.*, 2002). Due to the hydrophobic inner part of the circular compound, cyclodextrins are mainly used to encapsulate hydrophobic molecules.

Lipolytic enzymes

Lipolytic enzymes are widely distributed in nature and mainly used for the synthesis, hydrolysis or transesterification of simple esters and short chain fatty acids (esterases) or water-insoluble long-chain triglycerides (lipases) (Fig. 3.6). While lipases prefer long chain acyl esters (> 10 carbon atoms), esterases catalyse the synthesis and hydrolysis of short chain acetyl esters (< 10 carbon atoms) (Bornscheuer, 2002; Gupta *et al.*, 2004).

Lipolytic enzymes are serine hydrolases displaying conserved structures, namely an α/β-hydrolase fold containing six α-helices and eight β-strands. The so-called nucleophilic elbow includes a Gly-X-Ser-X-Gly-pentapeptide motif sequence with a conserved serine as part of the catalytically active site, and X as any amino acid residue (Jaeger and Reetz, 1998). The role of the central serine residue is mirrored by the drastic susceptibility of these enzymes towards specific serine inhibitors, such as PMSF and Pefabloc, which irreversibly abolish the catalytic activity (Cieslinski *et al.*, 2007; Royter *et al.*, 2009; Salameh and Wiegel, 2007). In addition to the nucleophilic serine, this triad comprises an acidic residue (aspartic or glutamic acid) and a histidine (Arpigny and Jaeger, 1999). The catalytic mechanism is mostly cofactor-independent and identical in esterases and lipases and is accomplished in the following fashion: (i) the catalytic Ser binds the substrate forming a tetrahedral intermediate and is stabilized by conserved triad residues His and Asp; (ii) an acyl complex is formed releasing alcohol; (iii) a nucleophilic moiety (water in hydrolysis and ester or alcohol in esterification) attacks the complex forming another tetrahedral intermediate; (iv) an acid or an ester is produced releasing the enzyme (Bornscheuer, 2002).

Based on catalytic activity and structural properties, lipolytic enzymes were originally assigned to eight different families including a variety of subfamilies, which were continuously updated and expanded (Arpigny and Jaeger, 1999). Due to sequence comparison, true lipases

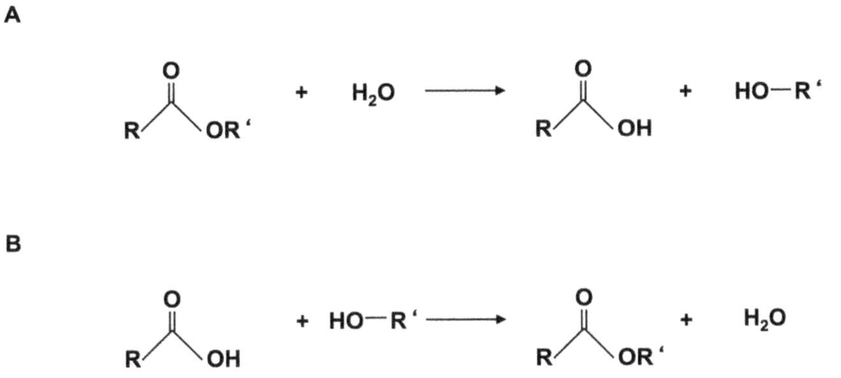

Figure 3.6 Reaction scheme depicting cleavage and formation of ester bonds catalysed by lipolytic enzymes. (A) Hydrolysis. (B) Esterification.

are assigned to the largest family, which is named family I and contains several sub-families indicated as families I.1, I.2 etc. Family I lipolytic enzymes contain two aspartic residues, which are mandatory for Ca^{2+}-binding. In addition, family I lipases from Gram-positive bacteria were shown to exhibit the former conserved consensus sequence, with an Ala-residue replacing the first Gly (Ala-X-Ser-X-Gly). Several other properties are exclusive in families II to VIII. Esterases and lipases containing a characteristic Gly-Asp-Ser-Leu consensus motif (GDSL group), instead of the conventional Gly-X-Ser-X-Gly-motif, belong to family II of lipolytic enzymes (Arpigny and Jaeger, 1999). A Ser-His dyad motif has been identified in some esterases instead of the highly conserved Ser-His-Asp(Glu)-triad. Furthermore, structure elucidation also revealed an α/β-tertiary fold that substantially differs from the common α/β-hydrolase fold (Wei *et al.*, 1995). Members of family III display the conserved fold of α/β-hydrolases, but the catalytic triad aspartic acid is located at a different position to family I isozymes. Bacterial family IV lipolytic enzymes share sequence similarities with mammalian hormone-sensitive lipases (Arpigny and Jaeger, 1999). One of the rare family IV heat-active esterases has been isolated from a metagenomic library consisting of DNA-fragments that were isolated from a thermal environmental sample (Rhee *et al.*, 2005). Family V members were shown to often resemble non-lipolytic enzymes, while the family VI esterases and lipases are of characteristically small-sized as compared to the rather large family VII enzymes (Bornscheuer, 2002). Family VIII lipolytic enzymes contain a non-functional Gly-X-Ser-X-Gly-motif in addition to an operating Ser-X-X-Lys consensus sequence (Petersen *et al.*, 2001).

Lipases and esterases have been both characterized from psychrophilic and thermophilic microbial sources (Al Khudary *et al.*, 2010; Ikeda and Clark, 1998; Li *et al.*, 2013; Royter *et al.*, 2009). They are usually identified in activity-based screening approaches using tributyrin-containing agar plates. Further activity tests with short- and long-chain fatty acid esters are used to distinguish esterases from lipases. Many possibilities for industrial applications favour the screening and identification of lipid-modifying enzymes operating at high temperatures (Levisson *et al.*, 2009b).

Lipases

Lipases (EC 3.1.1.3, triacylglycerol acylhydrolases) are among the most widely used biocatalysts in industrial applications (Gupta *et al.*, 2004). These enzymes catalyse the reversible hydrolysis of long-chain triglycerides to give di- and monoacylglycerol, glycerol and fatty acids. The reverse reaction results in the formation of ester bonds that attach fatty acids to a glycerol backbone. Lipases are chemo-, regio- and enantio-selective and are shown to be highly active at the water–lipid interface, while a hydrophobic α-helical lid-like structure covers the catalytic region in a non-aqueous environment (Sharma *et al.*, 2013b). Lipases are in need of a certain extent of substrate before exhibiting high activities, since the structural lid domain covers the active site and is only released in the presence of a minimum amount of substrate. Therefore these enzymes do not obey classical Michaelis–Menten kinetics (Bornscheuer, 2002).

Mostly extracellular lipases from mesophiles generally display high activity in a wide range of pH, but are unstable at temperatures above 60°C. Heat-stable microbial lipases are mainly characterized from prokaryotes and lower eukaryotes of fungal origin, while lipases of plant or animal origin are usually not relevant for biotechnological applications at harsh conditions. Several thermostable lipases have been isolated from moderate thermophiles mainly from fungi or diverse genera of the family *Bacillaceae* (*Bacillus*, *Geobacillus* and *Anoxybacillus*) including *B. thermoleovorans*, *B. stearothermophilus*, *B. thermocatenulatus*, *B. licheniformis*, *G. thermoleovorans* and *G. thermodenitrificans* (Balan *et al.*, 2012; Bordes *et al.*, 2011; Kim *et al.*, 1998a; Lee *et al.*, 1999; Schmidt-Dannert *et al.*, 1994). Extremophiles were shown to be a valuable resource for industrial relevant lipolytic enzymes. While the first heat-stable esterase from a hyperthermophilic microorganism had been investigated in detail around 25 years ago, lipases from Archaea or extreme thermophilic Bacteria have been only recently described (Chow *et al.*, 2012; Levisson *et al.*, 2009b; Royter *et al.*, 2009; Sobek and Gorisch, 1988). These promising lipases were mostly isolated from microorganisms thriving in hydrothermal vents and hot springs (Table 3.3) (Akanbi *et al.*, 2010; Johri *et al.*, 2012; Royter *et al.*, 2009; Sharma *et al.*, 2013b).

Two very stable and heat-active alkaline lipases have been isolated from the anaerobic thermophilic bacterium *Thermosyntropha lipolytica*. Both show an optimal activity at 96°C and retain 50% of activity after storage at 100°C for at least 2 hours. These two lipases are among the most thermostable and heat-active isozymes that have been reported (Salameh and Wiegel, 2007). Another two heat-active lipases were recently identified by a metagenomic screening approach. LipS and LipT were optimally active at 70°C and 75°C, and displayed a half-life of 48 hours and 3 hours, respectively (Chow *et al.*, 2012). Moreover, both enzymes preferred long chained fatty acids with more than eight carbon atoms, but they could not be assigned to any of the known lipase and carboxylesterase families (Arpigny and Jaeger, 1999). They are probably the prototype members of two novel lipase families. Furthermore, the first heat-active and solvent-tolerant family V lipases were described from the thermophilic anaerobic bacteria *Thermoanaerobacter thermohydrosulfuricus* and *Caldanaerobacter subterraneus* subsp. *tengcongensis*. These enzymes were active over a broad pH and temperature range and displayed a half-life of 48 hours at 75°C (Royter *et al.*, 2009).

Although the functionality of most lipases is not dependent on the presence of a co-factor, divalent metal ions were discussed to be of relevance for structural stabilization of lipases at high temperatures or even as catalytic activators. The depletion of calcium was shown to cause inactivation and aggregation of a lipase from *Burkholderia glumae* (Invernizzi *et al.*, 2009). It was hypothesized that parallel calcium and zinc ion coordination in lipases from thermophilic microorganisms favours the stability of these enzymes at elevated temperatures. A large comparison of sequence data revealed several striking features, such as long β-strands, an increase in neutrally charged hydrogen bonding pairs and less β-branched residues in helices, to be important for thermostability of lipases from extremophiles (Chakravorty *et al.*, 2011). Moreover, several approaches were developed to improve the catalytic activity of lipases. For instance, immobilization strategies were long shown to be useful for the stabilization of the fungal model enzymes from *Candida antarctica* and *Thermomyces lanuginosus* (Barbosa *et al.*, 2012; Cowan and Fernandez-Lafuente, 2011; Galvis *et al.*, 2012). Immobilization of lipolytic enzymes improved catalytic properties and stability using different techniques such as physical adsorption on different matrices and adsorption by ionic and covalent binding procedures (Cowan and Fernandez-Lafuente, 2011; Kumar *et al.*, 2006; Sanchez-Otero *et al.*, 2008; Sharma *et al.*, 2013b; Song *et al.*, 2013). Adsorption of a heat-active lipase from *Bacillus thermoleovorans* was achieved using porous polypropylene and the enzyme from *Bacillus coagulans* was immobilized on silica, which significantly improved its thermostability (Kumar *et al.*, 2006; Sanchez-Otero *et al.*, 2008). A lipase

Table 3.3 Heat-active lipases and esterases from prokaryotes

Enzyme	Host organism	T_{opt} (°C)	$T_{1/2}$	pH	Applications[a]/properties	Reference
Lipases						
–	*Archaeoglobus fulgidus*	95	80°C, 10h	9.0	Heat-resistance	Levisson *et al.* (2009b)
LipCst	*Caldanaerobacter subterraneus* subsp. *tengcongensis*	75	75°C, 48h	8.0	Biotransformation in water-free medium	Royter *et al.* (2009)
–	*Geobacillus* sp. Iso5	70	85°C, 3h	8.0	Significant potential for industrial purposes	Mahadevan and Neelagund (2013)
–	*Geobacillus thermodenitrificans* IBRL-nra	65	65–75°C, 16h	7.5	Food, dairy, pharmaceutical, detergent, organic synthesis, biodiesel	Balan *et al.* (2012)
LipS	Metagenomic library	70	90°C, >4h	8.0	Production of chiral pharmaceuticals	Chow *et al.* (2012)
LipT	Metagenomic library	75	70°C, 3h	8.0	Various biotechnological applications	Chow *et al.* (2012)
LipTth	*Thermoanaerobacter thermohydrosulfuricus*	75	75°C, 48h	8.0	(*S*)-stereoselective towards esters of secondary alcohols	Royter *et al.* (2009)
LipA	*Thermosyntropha lipolytica*	96	100°C, 6h	9.4	Great interest for specific industrial applications	Salameh and Wiegel (2007)
LipB	*Thermosyntropha lipolytica*	96	100°C, 2h	9.6	Great interest for specific industrial applications	Salameh and Wiegel (2007)
Esterases						
Est-AF	*Archaeoglobus fulgidus*	80	90°C, >80min	6.0–7.0	Slow reaction rate, low enantioselectivity	Kim *et al.* (2008)
EstE1	Metagenomic library	95	90°C, 20min	6.0	Heat-resistance	Rhee *et al.* (2005)
EstA	*Picrophilus torridus*	70	90°C, 21h	6.5	Production and/or optimization of ibuprofen and naproxen	Hess *et al.* (2008)
EstB	*Picrophilus torridus*	55	90°C, 10h	7.0	Production and/or optimization of naproxen	Hess *et al.* (2008)
Est_{Pc}	*Pyrobaculum calidifontis*	90	110°C, 56min	7.0	High potential in future applications	Hotta *et al.* (2002)
–	*Pyrococcus furiosus*	100	126°C, 50min	7.6	Heat-resistance	Ikeda and Clark (1998)
–	*Sulfolobus shibatae*	90	120°C, 20min	6.0	Heat-resistance	Huddleston *et al.* (1995)
Est3	*Sulfolobus solfataricus*	80	80°C, 40min	7.4	Synthesis of chiral compounds	Park *et al.* (2006)
EstA	*Thermotoga maritima*	95	100°C, 90min	8.5	Heat-resistance	Levisson *et al.* (2009a)
EstD	*Thermotoga maritima*	95	100°C, 60min	7.0	Heat-resistance	Levisson *et al.* (2007)
EstTs1	*Thermus scotoductus* SA-01	80	70°C, 48h	7.0	Reporter for gene expression	du Plessis *et al.* (2010)

[a]Indicates potential applications that were proposed by the authors of the respective studies.

from the thermophilic bacterium *Thermus thermophilus* was produced in recombinant form in *E. coli* and immobilized onto supermagnetic nanoparticles for efficient purification and reusability. Moreover, this enzyme displayed better resistance to temperature, pH and several additives (Song *et al.*, 2013). Thus, the simplified recovery of immobilized over soluble enzymes makes this method attractive for versatile industrial applications. Different strategies are generally used for the high-yield production of lipases. The thermostability of enzymes from mesophiles can be improved by evolutionary or rational design approaches based on sequence and structure-derived knowledge of their heat-stable counterparts (Jaeger and Reetz, 1998; Sharma *et al.*, 2013b). An extracellular, thermostable lipase from bacterial strain *B. circulans* IIIB153 was compared to a related heat-labile enzyme from *Chromobacterium viscosum* at the structural level, indicating the presence of a large amount of aromatic amino acids and salt bridges, which were proposed to be relevant for thermal stability (Johri *et al.*, 2012).

Esterases

Esterases (EC 3.1.1.1, carboxyl ester hydrolases) catalyse either the hydrolytic cleavage of ester bonds to give an acid and an alcohol moiety or the formation of ester bonds in the reverse reaction (Bornscheuer, 2002). Furthermore, these enzymes are capable of catalysing the transesterification of ester bonds. Esterases prefer short-chain acyl esters that are shorter than 10 carbon atoms and they are not active on micelles-forming substrates (Levisson *et al.*, 2009b). Several esterases were described to exhibit side activities towards tertiary alcohol esters or phospholipase activity (Antranikian, 2008). Although their physiological role is often not well understood, it was proposed that wide substrate tolerances might be a hint that these enzymes are paramount during evolution to access new carbon sources.

This enzyme family is highly diverse and several members were identified in pro- and eukaryotic microorganisms including (hyper-)thermophilic Bacteria and Archaea. Most of the heat-stable carboxylic ester hydrolases that were isolated from thermophiles belong to this group of esterases (Levisson *et al.*, 2009b). The classical screening in wild type strains led to the identification of most enzymes known to date. Nowadays, genome mining and (meta-)genomic libraries constantly replace activity-based screenings approaches. Most of esterases, that have been characterized, display optimal activities that are close or even above the optimal growth temperature of the host organism (Table 3.3).

Thermoacidophilic species (growth at high temperatures and low pH) have long been investigated with regard to unique lipolytic enzymes. The first esterase from a thermoacidophilic microorganism was isolated from *S. acidocaldarius*. Purification of this enzyme was achieved in a four-step purification approach and it displayed a half-life of around one hour when stored at 100°C (Sobek and Gorisch, 1988). A highly thermoactive esterase with optimal performance at a temperature of 90°C was found in another *Sulfolobus* species, namely *S. shibatae*. The respective enzyme displayed a transient activation at high temperatures and retained 50% of its activity after incubation at 120°C for 20 minutes (Huddleston *et al.*, 1995). Meanwhile, a heat-dependent activation of esterases and lipases has been described for a couple of members (Kim *et al.*, 2008; Royter *et al.*, 2009). Recently, the first esterase from a thermophilic *Thermus* species was identified and characterized. The genome of *Thermus scotoductus* strain SA-01 encodes an esterase, designated EstTs1, which showed only little identity when compared with known esterases of families IV–VI (<20%). Nevertheless, putative homologues were found in the genomes of *Thermus aquaticus* and *T. thermophilus*, indicating the presence of another non-related lipolytic enzyme family that has been exclusively formed in *Thermus* species during evolution (du Plessis *et al.*, 2010). This enzyme exhibited a half-life of 48 hours at 70°C, while an esterase from the archaeon *S. acidocaldarius* DSM639 was only 50% stable for 45 minutes when incubated at 80°C (Arpigny *et al.*, 1998). Among the most heat-stable esterases that were characterized so far, are two isozymes, EstA and EstB, encoded within the genome of the thermoacidophile *P. torridus*, which grows optimally at pH 0.7. The determined half-lifes at 90°C were 21 hours for EstA and 10 hours for EstB. Interestingly, their temperature profiles revealed optimal activities at lower temperatures, namely at 70°C and 55°C, respectively (Hess *et al.*, 2008). The enzyme from the archaeal species *P. furiosus* (growth at 100°C) was described to be the most thermostable and heat-active esterase with a temperature optimum of 100°C and half life of 50 minutes at 126°C (Ikeda and Clark, 1998).

The complex relationship between stability and structure, however, is far from being understood. The structure of a heat-stable esterase, which prefers *p*-nitrophenols with C2- and C4-acyl chains, has been recently solved. EstA from the thermophilic bacterium *T. maritima* is a member of the α/β-hydrolase family with an unusual hexameric structure. It forms a dimer of trimers, with the two dimers facing each other. Interestingly, the enzyme structure is composed of an N-terminal Ig-like domain in addition to the α/β-hydrolase domain. The Ig-like domain is involved in the dimerization of the trimers (Levisson *et al.*, 2009a). In contrast to this, several trimeric and tetrameric esterases from thermophiles have been postulated to not contain an additional Ig-like domain (Ejima *et al.*, 2004; Morana *et al.*, 2002). Although the physiological function of this enzyme is not clear, the role of Ig-like domains in other bacterial enzymes were often speculated to be involved in binding the substrate or directing the substrate to the catalytic region. However, deletion of the Ig-like domain in *T. maritima* EstA interfered with dimerization and resulted in almost complete loss of activity (Levisson *et al.*, 2009a). Another physiological function of an atypical part of an esterase was postulated for an OsmC-like domain in EstO of the psychrophilic bacterium *Pseudoalteromonas arctica* (growth at 10°C) (Al Khudary *et al.*, 2010). The function of OsmC-like proteins in other microorganisms remains unknown, but the expression of the corresponding genes is usually induced as a response to osmotic stress (Jenkins *et al.*, 2007; Rehse *et al.*, 2004). It has been shown that the deletion of OsmC-like domain in EstO resulted in a lower tolerance towards increasing salt concentrations. This domain might be involved in enzymatic activation under physiological salt conditions (up to 9% NaCl). Moreover, EstO is highly unstable at elevated temperatures (half-life at 40°C is 5 hours), while the OsmC-truncation variant displays a half-life of 2.5 hours at 90°C. This observation is typical for enzymes from (hyper-) thermophiles and might indicate a role of OsmC-like domain for higher flexibility at low temperatures and has been strengthened by destabilizing a cold-active xylanase from *P. arctica* at increasing temperatures fused to the OsmC-like domain (Al Khudary *et al.*, 2010; Elleuche *et al.*, 2011).

Industrial application of lipases and esterases

Biotransformations using lipases and esterases are getting continuous increasing attention due to their wide substrate variety, precision and selectivity. These hydrolases exhibit enantio- and regioselectivity, which makes them highly desired in the industrial production of drugs. Lipases and esterases represent a predominant family of technical enzymes used in organic chemistry (Gupta *et al.*, 2004; Jaeger and Eggert, 2002; Jaeger and Reetz, 1998; Sharma *et al.*, 2013b). Their biotechnological utilization has several advantages over traditional chemical processes, because more pure products and fewer wastes are generated by lower energy consumption. Most of these enzymes are either of prokaryotic or fungal origin with enzymes isolated from species of the genera *Pseudomonas* and *Bacillus* dominating the group of bacterial lipolytic enzymes. Lipases are also used for the degradation of natural materials and pollutants, such as cereal wastes and plastics. However, most esterases and lipases known so far are derived from mesophilic microorganisms (Levisson *et al.*, 2009b). Most industrial processes in which lipolytic biocatalysts can be applied run at temperatures higher than 45°C, indicating the relevance of these heat-stable enzymes (Sharma *et al.*, 2013b).

Esterases and lipases are already widely applied in organic synthesis for the resolution of racemates to give optically pure compounds. Therapeutics and further valuable compounds are usually highly complex molecules that are often hard to synthesize by pure chemical methods. In general only a single enantiomer is usually functional and applicable as a drug in medical approaches, while the second one is not active (Jaeger and Eggert, 2002). In this context, esterases can be used for the production of enantiomerically pure

S-2-chloropropionic acid from racemic mixtures of p-nitrophenyl-2-chlorpropionate to subsequently synthesize optimally pure molecules for the generation of valuable end products, such as the herbicide agrochemical compound Mecoprop. Research efforts were undertaken to improve the catalytic performance of esterases to produce S-2-chloropropionic acid by random mutagenesis and molecular modelling (Manco *et al.*, 2002). The two esterases from the thermoacidophilic archaeon *P. torridus* displayed a remarkable substrate specificity including the ability to hydrolyse naproxen, ketoprofen and ibuprofen. Therefore, these enzymes might be potential candidates for the production and/or optimization of non-steroidal anti-inflammatory drugs (Hess *et al.*, 2008). The metagenome-derived lipase LipS was able to hydrolyse complex substrates with aromatic side chains. Moreover, this enzyme is highly (*R*)-selective for naproxen, ibuprofen and other pharmaceutical relevant substrates and might be applicable for the production of enantiomere pure products (Chow *et al.*, 2012). The lipase from the thermophilic bacterium *T. thermohydrosulfuricus* was shown to be resistant towards extremely high concentrations (up to 99%) of organic solvents, which promises versatile industrial applications including the kinetic resolution of useful compounds under harsh conditions. Equally, this enzyme is also highly (*S*)-stereoselective towards esters of secondary alcohols (Royter *et al.*, 2009). The substrate spectrum and temperature properties including thermal enzyme activation could be changed by the incorporation of non-canonical amino acids, which might even widen the range of possible industrial applications of this enzyme (Hoesl *et al.*, 2011).

In addition, thermostable lipases and esterases are of interest for the food industry, since they can be used to enhance flavours through the production and modification of fats and oils. In the oleochemical industry, lipases are routinely used to produce lipids of defined structures such as cocoa butter equivalents, substitutes for human milk fats or margarine ingredients (Sharma *et al.*, 2013b). Higher temperatures will favour better solubility and reaction rates. Feruloyl esterases are used to release ferulic acid from hemicellulose in plant cell walls. This phenolic acid can be used as a precursor in the enzymatic production of vanillin (Bornscheuer, 2002). Furthermore, the production of enantiomerically pure (−)-menthol contains a transesterification step with (+)-menthol that was shown to be catalysed by a lipase of bacterial origin (Athawale *et al.*, 2001).

In the laundry industry, alkaline lipases are applied in the degradation of long-chain, water-insoluble triacylglycerols to give mono- and diacylglycerols and free fatty acids, thereby improving the washing performance of the detergent. Lipolase™ was the first commercially produced lipase in detergent industry. This enzyme has been isolated from the fungus *T. lanuginosus* and was initially produced in 1994 by Novo Nordisk (Jaeger and Reetz, 1998).

Proteolytic enzymes

Proteases catalyse the hydrolysis of proteins to produce amino acids and small peptides (Fig. 3.7). Since proteases support the catabolic metabolism in a cell, the cell is obliged to fine-tune the expression and catalytic activity of these enzymes. Since their introduction as detergent additives in 1914 by Rohm and Haas, proteases are widely distributed and intensively used in many industrial areas. Moreover, their industrial significance is also represented by the amount of commercially produced proteases that outmatch any other industrially produced enzyme (Maurer, 2004). Intracellular proteases have a big impact on a set of metabolic and cellular processes including differentiation and enzyme maturation, while extracellular proteases play an important role in hydrolysis of proteins as well as in the uptake and utilization of hydrolytic products. A distinctive cleavage pattern between two particular amino acid residues is specific to each kind of protease, allowing controlled hydrolysis of proteins at certain regions. Reported proteolytic cleavage sites on oxidized insulin B chain as model substrate include Leu^{15} and Tyr^{16}, Tyr^{16} and Leu^{17} or Glu^{13}-Ala^{14} amongst others (Morihara, 1974; Peek *et al.*, 1993).

The appearance of proteolytic enzymes distinguished between simple monomeric enzymes and complex multi-subunit structures with high molecular masses (MEROPS database http://

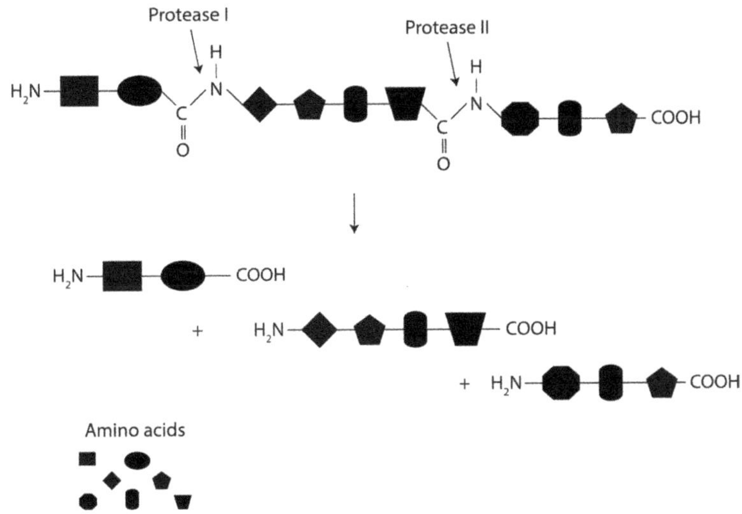

Figure 3.7 Reaction scheme illustrating the cleavage pattern of proteolytic enzymes and the resulting release of smaller peptide chains.

merops.sanger.ac.uk) (Rawlings *et al.*, 2014). Microbial proteases are classified based on the characteristics of their active site and pH optima into serine, cysteine, aspartic, threonine, glutamic acid or metalloproteases, and acidic, neutral or alkaline (Barrett *et al.*, 2004; Hartley, 1960). Based on this versatility, proteases that are active under extreme conditions are of great value for widely distributed industrial areas.

Serine proteases

Serine proteases (EC 3.4.21) represent the majority of thermostable proteases that are currently used for industrial purposes (Table 3.4). As mentioned above, its specific cleavage reaction leads to the production of amino acids and small peptides from polypeptide chains.

Serine proteases contain a catalytic triad typical for hydrolase and transferase enzymes, including lipases and esterases (see above). This catalytic triad is formed by a conserved motif including a serine, a histidine and an aspartate. During the proteolytic reaction the histidine acts both as the general base and general acid, while the aspartate supports the orientation of the histidine and neutralizes its charges during transition states. Finally, the serine serves as the important nucleophile of serine proteases (Buller and Townsend, 2013; Hedstrom, 2002). Serine proteases can be further subdivided into chymotrypsin-like and subtilisin-like proteases according to the first reported serine proteases, the mammalian chymotrypsin and subtilisin BPN as well as subtilisin Carlsberg from the *Bacilli* (DeLange and Smith, 1968; Markland and Smith, 1967). Prototype subtilisin and chymotrypsin are often used in detergents, but due to their thermolabile nature they will not be discussed further in this chapter. *Bacillus* strains also possess thermostable serine proteases that are widely used in industrial applications, predominantly in laundry detergent industries, due to their alkalophilic properties (Table 3.4) (Jaouadi *et al.*, 2009; Jayakumar *et al.*, 2012; Nadeem *et al.*, 2013). One of the most thermostable *Bacillus* serine proteases is a 36.9 kDa enzyme from *Bacillus* sp. AK.1 that showed a half-life of 13 hours at 80°C in the presence of divalent calcium ions. Similar to other reported serine proteases (subtilisin BPN, aqualysin I and thermitase), the cleavage sites of oxidized insulin B chain are Leu^{15}-Tyr^{16}, Gln^{4}-His^{5} and Glu^{13}-Ala^{14} (Morihara, 1974; Peek *et al.*, 1993).

Another source of thermostable serine proteases is represented by the genus *Thermus*. Caldolysin™ and PreTaq™ are two heat-active serine proteases that are prominently used in industrial applications. Depending on their thermophilic hosts, the 21 kDa enzyme

Table 3.4 Prokaryotic heat-active proteolytic enzymes

Enzyme	Host organism	T_{opt} (°C)	$T_{1/2}$	pH	Applications[a]/properties	Reference
Serine proteases						
–	*Bacillus licheniformis* UV-9	60	70°C, 30 min	11.0	Additive in Laundry detergent	Nadeem *et al.* (2013)
SAPB	*Bacillus pumilus* CBS	65	60°C, 80 min	10.6	Converting feather-rich waste, dehairing	Jaouadi *et al.* (2008, 2009)
–	*Bacillus pumilus* MCAS8	60	70°C, 30 min	9.0	Additive in laundry detergent, dehairing of animal skins	Jayakumar *et al.* (2012)
–	*Bacillus* sp. AK1	70	80°C, 13 h	7.5	Additive in laundry detergent	Peek *et al.* (1993)
Proteolysin	*Coprothermobacter proteolyticus*	85	80°C, 2 h	10.0	Processes with harsh conditions	Toplak *et al.* (2013)
Fervidolysin	*Fervidobacterium pennivorans*	80	90°C, 60 min	10.0	Dehairing, leather industry	Friedrich and Antranikian (1996)
–	*Pyrococcus abyssi* 549	95	115°C, 2 min	9.0	Heat-resistance	Dib *et al.* (1998)
Pyrolysin	*Pyrococcus furiosus*	115	100°C, 4 h	6.5–10.5	Heat-resistance	Eggen *et al.* (1990)
Protease I	*Sulfolobus solfataricus*	90	92°C, 5 h	6.5–8.0	Industrial processes at high temperatures	Burlini *et al.* (1992), Eggen *et al.* (1990)
Caldolysin	*Thermus aquaticus* T-351	65–85	90°C, 1 h	8.0	Industrial processes at high temperatures	Cowan and Daniel (1982)
Aqualysin I	*Thermus aquaticus* YT-1	80	80°C, 30 min	10.0	Additive in laundry detergent	Matsuzawa *et al.* (1988)
PreTaq™	*Thermus* sp. Rt41A	90	90°C, 20 min	8.0	DNA clean-up prior PCR	Life Technologies Bruins *et al.* (2001)
Cysteine proteases						
PfPcp	*Pyrococcus furiosus*	90	75°C, 150 min	6–9	Deblocking of pGlu prior sequencing	Tsunasawa *et al.* (1998)
TT protease	*Pyrococcus furiosus* KOD1	110	100°C, 60 min	7.0	Heat-resistance	Morikawa *et al.* (1994)
Protease III	*Sulfolobus solfataricus*		90°C, 15 min	–	Industrial processes at high temperatures	Fusi *et al.* (1991)
TlPcp	*Thermococcus litoralis*	70	70°C, 1 h	–	Deblocking of pGlu prior sequencing	Singleton *et al.* (2000)
Aspartic proteases						
Kumamolysin	*Bacillus* sp. MN-32	70	80°C, 10 min	3.0	Thermoacidophilicity	Murao *et al.* (1993)
Wai21.A1	*Bacillus* sp. Wai21.A1	60	80°C, 2 min	3.5	Thermoacidophilicity	Prescott *et al.* (1995)
Wp22.A1	*Bacillus* sp. Wp22.A1	60	80°C, 21 min	3.5	Thermoacidophilicity	Toogood *et al.* (1995)
Thermopsin	*Sulfolobus acidocaldarius*	90	80°C, 48 h	2.0	Thermoacidophilicity	Lin and Tang (1990)
Metalloproteases						
NprM	*Bacillus stearothermophilus*	70	90°C, 25 min	7.5	Peptide synthesis	Kubo *et al.* (1988)
API	*Bacillus stearothermophilus*	80	80°C, 15 h	7.5–9.4	Peptide synthesis	Roncari *et al.* (1976)

Table 3.4 continued

Enzyme	Host organism	T_{opt} (°C)	$T_{1/2}$	pH	Applications[a]/properties	Reference
Thermolysin	*Bacillus thermoproteolyticus*	70	87°C, 30 min	8.0	Peptide synthesis/ production of aspartame precursor	Eijsink *et al.* (1995), Endo (1962), Ohta *et al.* (1974)
CPase Taq	*Thermus aquaticus* YT-1	80	80°C, 5 h	8.0	C-terminal sequencing	Lee *et al.* (1992)
Amino-peptidase T	*Thermus aquaticus* YT-1	75–80	80°C, 20 h	8.5–9.0	C-terminal sequencing	Minagawa *et al.* (1988)
Other proteases						
–	*Methanocaldococcus jannaschii*	116	116°C, 45 min	–	Thermo-barophil, high-pressure applications, organic synthesis	Michels and Clark (1997)
–	*Pyrococcus horikoshi*	95	90°C, 2 h	7.5	N-terminal sequencing	Ando *et al.* (1999), Onoe *et al.* (2002)
Maritimacin	*Thermotoga maritima*	90	95°C, 36 min	7.1	Antibacterial activity	Hicks *et al.* (1998)

[a]Indicates potential applications that were proposed by the authors of the respective studies.

Caldolysin™ from *Thermus aquaticus* strain T-351 ($T_{1/2}$ = 90°C, 1 hour) and Pretaq™ ($T_{1/2}$ = 90°C, 20 minutes) from *Thermus* sp. strain Rt41A are extremely thermostable exhibiting activities at high temperatures between 65°C to 85°C and 90°C, respectively (Bruins *et al.*, 2001; Cowan and Daniel, 1982). Moreover, *T. aquaticus* produces aqualysin I, another alkaline thermostable protease, that showed highest activity at pH 10 and 80°C (Matsuzawa *et al.*, 1988). Aqualysin I was intensively studied by Lee and colleagues, who produced this enzyme in several expression hosts like *E. coli* and *T. thermophilus* (Lee *et al.*, 1994, 1991). They determined the importance of the amino-terminal prosequence for the synthesis of the mature active enzyme and the necessity of the carboxyterminal pro-sequence for the extracellular location of the enzyme. Likewise, structural details from aqualysin I were successfully transferred to the mesophilic protease subtilisin E and have led to an enhanced thermostable derivative (Takagi *et al.*, 1990). Another thermostable serine protease with a promising potential for applications at harsh process conditions was recently produced from the thermophilic bacterium *Coprothermobacter proteolyticus*. In addition to its thermostability ($T_{1/2}$: 80°C, 2 hours) proteolysin is tolerant against high concentrations of organic solvents and detergents such as dithiothreitol and guanidinium hydrochloride (Toplak *et al.*, 2013).

The most thermostable serine proteases are from hyperthermophilic Archaea and show optimal activities at temperatures higher than 100°C (Burlini *et al.*, 1992; Cowan *et al.*, 1987; Dib *et al.*, 1998; Eggen *et al.*, 1990; Koga *et al.*, 2013). Pyrolysin, a serine protease from *P. furiosus* with an optimal temperature of 115°C and a half-life of 4 hours at 100°C as well as protease I from the extreme thermoacidophilic archaeon *S. solfatarius* with a reported half-life of 5 hours at 92°C and an optimal activity at 90°C are only two examples of extraordinary enzymes (Burlini *et al.*, 1992; Eggen *et al.*, 1990; Voorhorst *et al.*, 1996). Additional serine type proteases with comparable properties have been isolated from several hyperthermophilic organisms including *Thermococcus* species, *Staphylothermus marinus* and the extreme thermophilic bacterium *Coprothermobacter proteolyticus*. All of these serine proteases showed optimal activities at temperatures between 80°C and 95°C and were mostly stable for at least 60 minutes at 100°C (Klingeberg *et al.*, 1991). Another industrial promising enzyme is fervidolysin from the thermophile *F. pennivorans* that belongs to the serine proteases and shows keratinolytic activity. Keratin is a very insoluble structure element that gives feather and wool high stability. Since feathers are considered as waste material in the poultry industry, their degradation into useable compounds is of great interest. Although all described keratinases belong to the class of serine proteases not all serine proteases exhibit keratinolytic activity. Due to its heat-stability and keratinolytic activity fervidolysin is a good candidate for hydrolysis of keratinous material. The enzyme showed optimal activity at 80°C and pH 10 with a reported half-life of 60 minutes at 90°C (Friedrich and Antranikian, 1996; Kluskens *et al.*, 2002).

Cysteine proteases

The distinctive feature of cysteine proteases (thiol proteases) (EC 3.4.22) is a crucial cysteine residue within their active site. Characteristically, their catalytic activity is strongly reduced by a vast number of inhibitors including iodoacetic acid, thiolstatin and *p*-chloromercuribenzoate, since all of them modify the thiol-group in the crucial nucleophilic cysteine (Fusi *et al.*, 1991; Morikawa *et al.*, 1994; Murao *et al.*, 1993). The hyperthermophilic archaeon *P. furiosus* strain KOD1 produces a very thermoactive cysteine protease (TT protease) with a molecular weight of 44 kDa and optimal activity at 110°C. It has a half-life of 60 minutes at 100°C, and is inhibited in the presence of *p*-chloromercuribenzoate and stimulated by cysteine (Morikawa *et al.*, 1994). Another example of a hyperthermostable cysteine protease is Protease III from the thermoacidophilic Archaeon *S. solfataricus* that retains most of its activity for 15 minutes at 90°C (Fusi *et al.*, 1991).

A distinct group within the cysteine proteases is called pyrrolidone carboxyl peptidases (Pcp, EC 3.4.19.3). Pcps are characterized by a typical Cys-His-Glu motif within the catalytic triad and they hydrolyse amino-terminal L-pyroglutamic acids from proteins (Doolittle and Armentrout, 1968; Singleton *et al.*, 2000). They are classified into two different types distributed among prokaryotic and eukaryotic organisms (Awade *et al.*, 1994; Szewczuk and Kwiatkowska, 1970; Szewczuk and Mulczyk, 1969). The eukaryotic members are mostly mammalian enzymes that represent two types. Type I Pcps are soluble proteins with molecular masses smaller than 24 kDa, also described as cysteine proteases because of an essential sulfhydryl group. Brain-tissue membrane-associated proteins that degrade neuropeptides have been assigned to type II-Pcps. These enzymes belong to the metalloproteases and are strongly inhibited by chelating agents (Mantle *et al.*, 1991; Singleton *et al.*, 2000; Tsunasawa *et al.*, 1998). Thermostable type I bacterial Pcps were described in hyperthermophilic Archaea like *P. furiosus* (PfPcp) and *T. litoralis* (TlPcp) (Table 3.4) (Singleton and Littlechild, 2001; Singleton *et al.*, 2000; Tsunasawa, 1998; Tsunasawa *et al.*, 1998). PfPcp was produced in recombinant form in *E. coli* and classified as a cysteine protease. A substitution of a cysteine by a serine at position 142 led to complete inactivation. The enzyme exhibited optimal activity at 90°C and pH 6–9 with a half-life of 150 minutes at 75°C (Tsunasawa *et al.*, 1998). Likewise, the pyrrolidone carboxyl peptidase TlPcp from *T. litoralis* was recombinantly produced in *E. coli*. Its tetrameric conformation consists of four 24 kDa-subunits with an optimal activity at 70°C and a corresponding half-life of 1 hour at 70°C (Doolittle and Armentrout, 1968; Singleton *et al.*, 2000; Tsunasawa *et al.*, 1998).

Aspartic proteases

Aspartic proteases (EC 3.4.23) possess two distinctive carboxy aspartic acid residues within their active site sequence Asp-Thr-Gly. These amino acids are crucial for catalytic activity. Aspartic proteases, alternatively named 'acid proteases' are extremely acidophilic, preferring environments with pH values between 1.5 and 5.0. Their preferred cleavage sites of oxidized insulin B chain are Leu^{15}-Tyr^{16} and Phe^{25}-Tyr^{26} (Murao *et al.*, 1993). In general, aspartic proteases can be inhibited by treatment with diazoacetyl norleucine methyl ester, 1,2-epoxy-p-nitrophenoxy propane or pepstatin (Davies, 1990; Rao *et al.*, 1998; Rich, 1985). Even though most of the described aspartic proteases are thermolabile, few examples of thermostable enzymes have been reported (Table 3.4). Thermopsin from *S. acidocaldarius* exhibits a half-life of 48 hours at 80°C and a maximal activity at pH 2 and 90°C (Lin and Tang, 1990). Interestingly, this acid protease showed no sequence similarity to proteases of the pepsin-like family or the pepstatin-insensitive acid proteases.

It probably belongs to the group of glycoproteins exhibiting 11 potential N-glycosylation Asn-X-Thr/Ser signals. However, the distinctive amino acid sequence Asp-Thr-Gly motif for aspartic proteases within its active site is lacking (Lin and Tang, 1990). Accordingly, thermopsin was described as a prototype member of a new subgroup of aspartic proteases, which are significantly different when compared to pepsin-like and pepstatin-insensitive types (Lin and Tang, 1990).

Although most of the described aspartic proteases are inhibited by pepstatin, a group of enzymes that showed insensitivity towards pepstatin is also reported (Murao *et al.*, 1993; Oda *et al.*, 1987a,b). Most of the published pepstatin-insensitive and thermostable aspartic proteases are produced by thermophilic *Bacillus* strains (Murao *et al.*, 1993; Prescott *et al.*, 1995; Toogood *et al.*, 1995). Kumamolysin, a 41 kDa-enzyme from *Bacillus* novosp. MN32 was isolated from hotspring water of the Aso Volcano region in Japan. Its optimal activity was detected at 70°C and pH 3.0 with a reported half-life of 10 minutes at 80°C (Murao *et al.*, 1993). More examples of thermostable and pepstatin-insensitive *Bacillus* enzymes include Wp22.A1 and Wai21.A1 with an optimal activity at 60°C and pH 3.5. Both enzymes differ from other described aspartic proteases in that they cleave oxidized insulin B chains primarily between Val^2 and Asn^3 as compared to Leu^{15} and Tyr^{16} (Prescott *et al.*, 1995; Toogood *et al.*, 1995).

Metalloproteases

Metalloproteases exhibit one or two metal ions within their active site, which are crucial for catalytic activity but not for stability (Daniel *et al.*, 1996). Metalloproteases can be assigned to two different groups: (i) thermolysin-like or metalloendoproteases (EC 3.4.24) and (ii) metalloexoproteases (EC 3.4.17) including carboxypeptidase A-like metalloproteases. Most of the described metalloproteases are characterized by a zinc ion within their catalytic region. However, some enzymes exhibit other ions such as cobalt, magnesium or nickel. A water molecule is activated by the metal ion to act as a nucleophile that attacks the carbonyl bond, leading to a gem-diol intermediate and the conclusive cleavage of the peptide bond. Since metal ions play a crucial role during their catalytic activity chelating agents like EDTA, talopeptin or o-phenanthroline inhibit metalloproteases (Coolbear *et al.*, 1991; Fontana, 1988; Murao *et al.*, 1980; Rawlings *et al.*, 2014; Takii *et al.*, 1987). The complete catalytic mechanism of metalloproteases and the importance of their metal ion were portrayed more than 50 years ago using the metalloendoprotease thermolysin (EC 3.4.24.27) from *Bacillus thermoproteolyticus* as a model protein (Endo, 1962). In 1972, the structure of thermolysin was resolved (Matthews *et al.*, 1972a,b). 'Thermolysin-like peptidases' (TLP) contain two domains; an amino-terminal domain predominantly consisting of β-sheets and a single α-helix separated from the carboxy-terminal domain, which contains a vast number of alpha helices. A deep flexible cleft, containing a zinc-binding active site is present between the two domains that are connected by a central α-helix (Colman *et al.*, 1972; Eijsink *et al.*, 1995; Holmes and Matthews, 1982; Matthews *et al.*, 1972b, 1974). Three highly conserved amino acid ligands are responsible for the coordination of the zinc ion. In thermolysin, two histidine residues within a conserved 'HEXXH' motif and a glutamate within a 'GXBNEXBSD' motif were described to hold the zinc ion in position (Hooper, 1994; Jiang and Bond, 1992). The essential function of the zinc ion was demonstrated by treatment with chelating agents resulting in a complete inhibition of activity (Roche and Voordouw, 1978). A small surface region in the amino-terminal half of the protein was described to be crucial for thermostability, since a less thermostable TLP from *B. stearothermophilus* CU21 (73.4°C) displayed an increased activity (93.3°C) after modification of this region by site directed mutagenesis (Eijsink *et al.*, 1995). Moreover, influence of four bounded calcium ions (Ca1-Ca4) on thermostability was described, by showing that a thermolabile TLP variant only binds Ca1 and Ca2. Furthermore, the binding site of Ca3 is located within the conserved N-terminal region and amino acid changes led to decreased thermostability (Eijsink *et al.*, 2011, 1995).

There are other reports of thermostable *Bacillus* metalloproteases in addition to thermolysin like NprM and API from *B. stearothermophilus*

(Table 3.4) (Kubo and Imanaka, 1988; Roncari *et al.*, 1976). However, thermolysin and NprM share a nearly identical amino acid sequence. The only difference is a two amino acid substitution and the formation of an additional hydrogen bond that seems responsible for the 10% higher stability of NprM ($T_{1/2}$ 90°C, 25 minutes) (Kubo and Imanaka, 1988; Kubo *et al.*, 1988). Another thermostable metalloprotease from *B. stearothermophilus* is the multi-subunit enzyme API (EC 3.4.11.22) that was described to exhibit an even higher thermostability and no loss of activity was observed at 80°C after 15 hours. Zinc was functionally replaced by cobalt, but the presence of cobalt resulted in a reduced thermostability. Interestingly, API exhibited a very unspecific cleavage pattern that appeared to be next to any kind of amino acid (basic, acidic and neutral), thereby showing a distinctive broad substrate specificity that is usually described for aminopeptidases (Roncari *et al.*, 1976). Similar substrate specificities were also described for an aminopeptidase (aminopeptidase T) and a metallocarboxypeptidase (CPase Taq, EC 3.4.17.19) from the thermophilic bacterium *T. aquaticus*. Both showed catalytic activities that strongly depend on cobalt ions instead of zinc and cleave either N-terminal amino acids like Gly and Leu or all amino acids with the exception of a C-terminal located proline (Lee *et al.*, 1992; Minagawa *et al.*, 1988). Their unique cleavage pattern, heat-activity (80°C) and thermostability at 80°C (CPase Taq 5 hours and aminopeptidase T 20 hours) make them promising candidates for C-terminal sequencing approaches (Lee *et al.*, 1992; Minagawa *et al.*, 1988).

Other thermostable proteases

In addition to the previously described proteases, there are several descriptions of proteolytic enzymes that have not yet been classified or that show no similarity to existent groups at all (Table 3.4). With the help of genomic and metagenomic approaches, several proteasomes including putative protease encoding sequences have been reported. The homomultimeric protease maritimacin from the extreme thermophilic bacterium *T. maritima* consists of numerous 31 kDa subunits with a total molecular mass > 669 kDa. The enzyme displayed optimal activity on substrates like chymotrypsin and trypsin at 90°C and pH 7.1 in combination with a reported half-life of 36 minutes at 95°C (Hicks *et al.*, 1998). The name was given with regard to the sequence identity to linocin M18, a bacteriocin that is produced by the mesophilic bacterium *Brevibacterium linens* (Hicks *et al.*, 1998; Valdes-Stauber and Scherer, 1994). Such a homology to mesophilic enzymes with antimicrobial activity has led to speculations about microbial ecological strategies that take place in high temperature niches, making these enzymes good candidates for applications with thermostable antibiotics (Hicks *et al.*, 1998). A heat-stable and pressure-tolerant protease is produced by the extremely thermophilic Archaeon *M. jannaschii* that was isolated from a deep-sea hydrothermal vent (Bult *et al.*, 1996; Jones *et al.*, 1983; Michels and Clark, 1997). This enzyme contains a molecular mass of 29 kDa and prefers leucine residue at the P1 site of polypeptides for its catalytic activity. The reported half-life of this enzyme was 45 minutes at 116°C with an optimal temperature at 116°C. However, enzymatic activity was measured at temperatures up to 130°C. Moreover, its activity and thermostability increased with pressure up to 500 atm (Michels and Clark, 1997). These characteristics coupled with its unique performance under extreme conditions (high temperature and high pressure) make this enzyme suitable for industrial applications that take place at high pressure e.g. organic synthesis (Horikoshi, 1995; Michels and Clark, 1997; Van Eldik *et al.*, 1989). Another example is a cobalt-dependent aminopeptidase that is produced by *Pyrococcus horikoshii* showing the highest identity on protein sequence levels to both an endoglucanase from *Clostridium thermocellum* as well as to a deblocking aminopeptidase from *P. furiosus* (Ando *et al.*, 1999; Onoe *et al.*, 2002). Optimal activity values were detected at 95°C and pH 7.5 with a half-life of 2 hours at 90°C. This enzyme was classified as a prototype member of a novel aminopeptidase class, because it showed aminopeptidase activity towards N-terminal amino acids but no endoglucanase activity. Like other aminopeptidases, it is a promising candidate for N-terminal sequencing of short peptides (Ando *et al.*, 1999; Onoe *et al.*, 2002).

Industrial applications of proteolytic enzymes

The physiological function of proteases ranges from utilization of exogenous peptides as nutrients to degradation of intracellular proteins and peptides to protein maturation. Their hydrolysis of proteins and peptides make them suitable tools for several industrial fields including detergent, pharmaceutical, food, leather and textile industries (Maurer, 2004). Their additional resistance against organic solvents is beneficial in peptide synthesis and detergent applications. Conditions with characteristically low concentrations of water and environments without any water at all are possible to use in combination with these compatible thermostable proteases (Wilson *et al.*, 1994a,b). Thermolysin is used in the synthesis of the aspartate precursor N(Benzyloxycaarbonyl)-L-asparty-L-phenyl-alanine methyl ester (Isowa *et al.*, 1979; Nakanishi *et al.*, 1990). Since the immobilized enzyme is stable in ethyl acetate, its usage leads to a yield of approximately 95% of the synthetic sweetener after 500 hours in continuous synthesis (Nakanishi *et al.*, 1990). Likewise, a protease from *Pseudomonas aeruginosa* was described with an equal tolerance against organic solvents, that may also be a good candidate for similar fields of application (Gupta and Khare, 2006).

The use of alkaline proteases for detergent industry applications dates back to the approach of Haas and Rohm in 1914 that employed pancreatic proteases in the commercialized product Burnus. Thermostable proteases represent suitable tools for detergent applications not only because of their thermostability but also due to the fact that they show a high proteolytic activity even in very alkaline pH ranges of > pH 9. Since these properties perfectly match process conditions with temperatures higher than 60°C and pH 9, the use of thermostable alkaline proteases for the degradation of proteinaceous stains such as milk, eggs, blood, grass and sauces is highly demanded (Jaouadi *et al.*, 2009; Kanchana *et al.*, 2012; Maurer, 2004; Nadeem *et al.*, 2013; Niehaus *et al.*, 1999). Most of the currently used alkaline proteases in detergent industries are serine proteases that were derived from mesophilic organisms like e.g. Savinase™ (Subtilisin 309), Alcalase™ (Subtilisin Carlsberg) or Properase™ (Genecor Int., USA) (Jaouadi *et al.*, 2009; Maurer, 2004). Since these proteases are usually not very stable, there is a great demand of thermostable alkaline proteases with tolerance against detergents. Several reports of promising serine alkaline proteases from thermophilic *Bacillus* strains and hyperthermophilic Archaea with a wide substrate specificity, an optimal activity over 60°C, pH ranges of pH >7.5, and tolerance against high detergent concentrations, have been tested for applications in detergent industry (Dib *et al.*, 1998; Jaouadi *et al.*, 2009; Jayakumar *et al.*, 2012; Matsuzawa *et al.*, 1988; Nadeem *et al.*, 2013).

Alkaline serine proteases that exhibit an additional keratinolytic activity like fervidolysin from *Fervidobacterium pennivorans* are good candidates for application in the leather industry, e.g. soaking, bating and dehairing (Friedrich and Antranikian, 1996; Gupta *et al.*, 2002; Jaouadi *et al.*, 2009; Kluskens *et al.*, 2002). An enzymatic step of dehairing skins and hides will replace the chemical and environmentally unfriendly process with lime and sulfide. In feed and food industries, amino acid rich waste feathers from poultry industries would be degraded more efficiently by these enzymes instead of using steam pressure and chemical treatment (Gupta *et al.*, 2002; Ichida *et al.*, 2001).

Since thermostable proteases withstand degradation by moderate temperatures up to 60°C they are used in several molecular biology approaches like cleaning ultrafiltration membranes and clean-up steps of DNA solutions prior to polymerase chain reactions. An enzymatic approach will result in a reduced amount of strong alkali or acids and consequently prolonging the lifetime of membranes. A protease from *Thermus* strain Rt41A was described that showed promising results in the enzymatic cleaning of ultracentrifugation membranes (Coolbear *et al.*, 1992). Pretaq™ (Life Technologies) is a thermostable protease that is used for applications in DNA clean-up prior to polymerase chain reactions since it exhibited high proteolytic activity at 90°C. This enzyme breaks down cellular components and undergoes a rapid inhibition by EGTA, in which the proteolytic activity does not interfere with subsequent steps such as polymerase chain reactions.

Another molecular approach in which thermostable proteases play an important role is N- or C-terminal sequencing of peptides due to fragmentation of protein sequences. Several thermostable proteases from *P. furiosus* and *S. sulfactorius* were described as suitable tools since they showed specific cleavages of peptide sequences, broad substrate specificity, optimal activity above 85°C and long half-lives at elevated temperatures (Colombo *et al.*, 1992; Vieille and Zeikus, 2001).

In addition to the previously mentioned applications, proteases have a long history in meat tenderization applications due to their cleavage of peptide bonds and degradation of collagen, the main protein complex in meat. Depending on the elevated temperatures at occurring processes during cooking, the use of thermostable proteases was successfully tested for meat tenderize. Wilson and colleagues suggested that the most suitable enzyme should be active during the preparation step (cooking) and not during storage. Suitable candidates were produced from *Bacillus*- and *Thermus*-strains and showed optimal activity at high temperatures against meat powder and collagen (Wilson *et al.*, 1992).

Other industrial-relevant enzymes and their applications

In comparison with this well-established group of enzymes, described already, fewer applications have been described for enzymes including chitinases, pectinases and phytases.

Chitinases

Chitinolytic enzymes degrade chitin, a major component of the structural skeleton present in many fungal organisms and a few invertebrates such as crustacean and insects (Adams, 2004; Nwe and Stevens, 2002; Ravi Kumar, 2000). The glycopolymer chitin is an unbranched polysaccharide and consists of β-1,4-linked N-acetyl-glucosamine (GlcNAc) units (Fig. 3.8A). Chitin is one of the most abundant biopolymers in nature (80000 metric tons/a), especially in marine environments. It is used as a source for both nitrogen and carbon and causing its degradation and commercial exploitation to come into global focus (Chavan and Deshpande, 2013). Regarding their cleavage patterns, chitin-degrading enzymes are classified into two major groups, the endo- (i) and the exo-chitinases (ii). Both hydrolyse chitin by splitting the β-glycosidic bonds between each N-acetyl-glucosamine residue. The cleavage site for endo-chitin-hydrolases (EC 3.2.1.14) is randomly distributed over the whole molecule. Enzymatic activity results in the release of the dimeric di-acetylchitobiose or chitotetraose and chitotriose. Exo-chitinases can be divided into two subgroups: the chitobiosidases (EC 3.2.1.29) and the N-acetyl-D-glycosaminidases (EC 3.2.1.30). Chitobiosidases hydrolyse the chitin molecule itself to produce chitobiose by attacking the non-reducing ends, whereas N-acetyl-D-glycosaminidases cleave the oligomeric products of chitobiosidases and endo-chitinases leading to the production of monomeric N-acetyl-glucosamine (Fig. 3.8A) (Chavan and Deshpande, 2013; Cohen-Kupiec and Chet, 1998; Nagpure *et al.*, 2013). Chitinolytic enzymes are widely distributed and gain interest for many industrial applications. Since fungal diseases represent a huge problem in agriculture and the exploitation of chemical fungicides contributes to environmental pollution, environmentally friendly alternatives are highly demanded. Acting as biofungicides, the degradation of chitin by chitinases is thus preferable and leads to a commercial exploitation of chitinolytic enzymes for biological control of plant pathogenics (Chang *et al.*, 2003; Huang *et al.*, 2005; Wang *et al.*, 2002). Recently, several thermostable chitinases from mesophilic organisms like *Bacillus cereus* or *Brevibacillus laterosporus* were described as suitable candidates for biocontrol applications (Hammami *et al.*, 2013; Prasanna *et al.*, 2013) (Table 3.5). The industrial relevance of chitin or chitosan is represented by their usability in various industrial applications, e.g. wastewater clearing, papermaking, cosmetic and medical applications, textile finishing and for the production of photographic products and heavy metal chelating agents (Georgopapadakou and Tkacz, 1995; Ravi Kumar, 2000; Shahidi *et al.*, 1999; Spindler *et al.*, 1990).

Since chitin is insoluble in water and not easily accessible for chitinolytic enzymes due to its highly crystalline structure, high temperatures

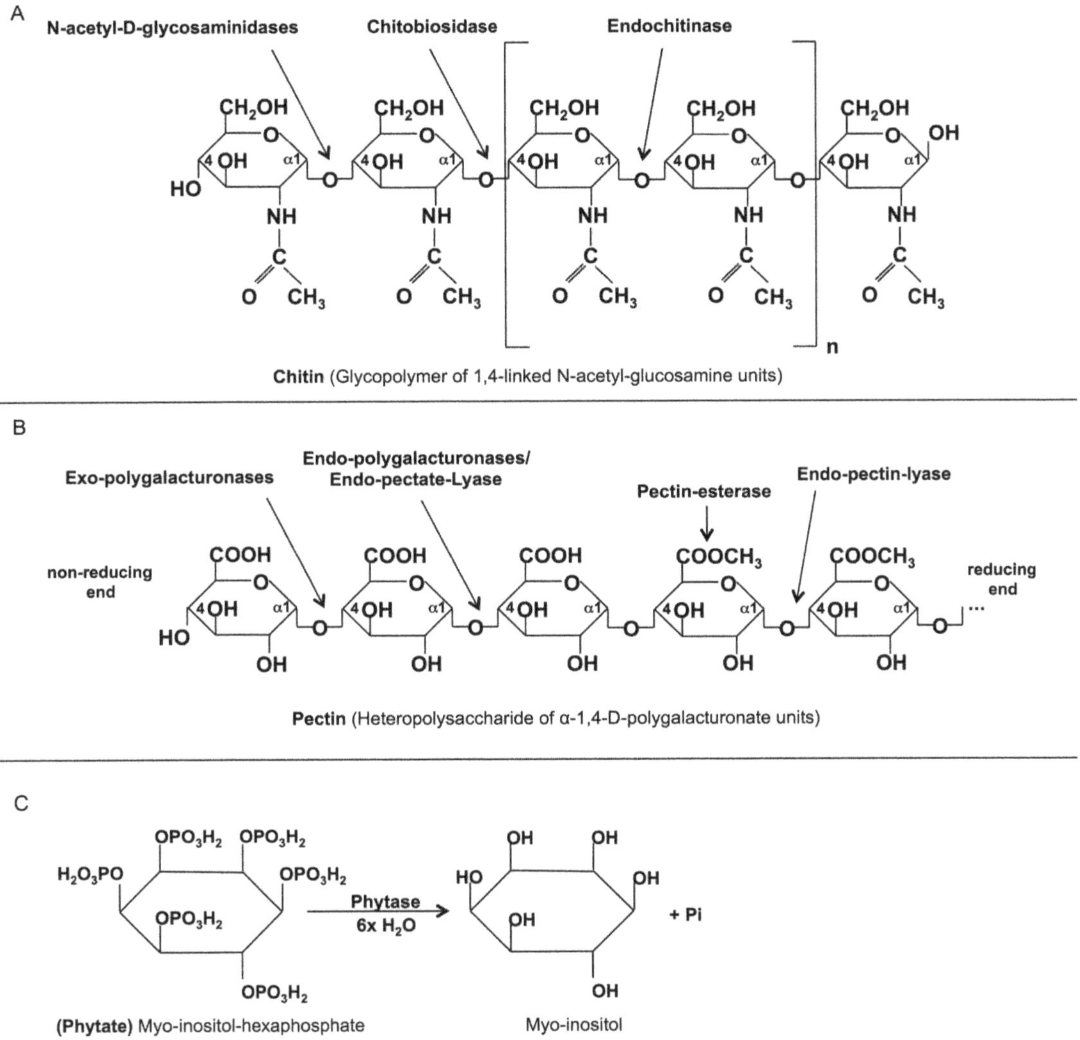

Figure 3.8 Enzymatic cleavage of chitin (A), pectin (B) and phytate (C) by appropriated enzymes.

are necessary and thus thermostable chitinases are needed (Haki and Rakshit, 2003; Roberts, 1992). Thermostable chitinases are found in both thermophilic and hyperthermophilic organisms. *Bacillus* species especially represent a source of thermostable chitinases (Table 3.5) (Bhushan and Hoondal, 1998; Hammami *et al.*, 2013; Sakai *et al.*, 1998; Takayanagi *et al.*, 1991; Yuli *et al.*, 2004). One report describes four thermostable chitinases, I, II, III and IV that were isolated from cell-free cultures from *Bacillus licheniformis* X-7u (Takayanagi *et al.*, 1991). All four investigated enzymes displaced optimal chitinolytic activity towards glycol chitin at 80°C and pH 10 (Takayanagi *et al.*, 1991). Recent reports of thermostable chitinases with antifungal properties and thus high potential as biofungicides were described from *B. cereus* strain IO8 and *B. laterosporus* (Hammami *et al.*, 2013; Prasanna *et al.*, 2013). The thermostable chitinase ChiIO8 from *B. cereus* was described to have a molecular mass of 30 kDa and an optimal activity at 65°C and pH 6.5. Its thermostability was represented by no loss of activity at 65°C after 100 minutes (Hammami *et al.*, 2013). A chitinolytic active mixture of one chitodextrinase and a chitinase from *B. laterosporus* exhibited antifungal activity against the phytopathogenic fungus *Fusarium equiseti* and

Table 3.5 Other industrial relevant enzymes from prokaryotic organisms

Enzyme	Host organism	T_{opt} (°C)	$T_{1/2}$	pH	Applications[a]/properties	Reference
Chitinases						
Chitinase I	*Bacillus licheniformis* X-7u	80	50°C, 30 min	10	Biocontrol	Takayanagi *et al.* (1991)
Chitinase II	*Bacillus licheniformis* X-7u	80	90°C, 30 min	10	Biocontrol	Takayanagi *et al.* (1991)
Chitinase III	*Bacillus licheniformis* X-7u	80	90°C, 30 min	10	Biocontrol	Takayanagi *et al.* (1991)
Chitinase IV	*Bacillus licheniformis* X-7u	80	90°C, 30 min	10	Biocontrol	Takayanagi *et al.* (1991)
ChiI08	*Bacillus cereus* IO8	65	65°C, 100 min	6.5	Antifungal activity for biocontrol	Hammami *et al.* (2013)
–	*Brevibacillus laterosporus*	70	80°C, 3 min	6–8	Antifungal activity for biocontrol	Prasanna *et al.* (2013)
PfChiA	*Pyrococcus furiosus*	90–95	–	6	Biocontrol	Gao *et al.* (2003)
PfChiB	*Pyrococcus furiosus*	90–95	–	6	Biocontrol	Gao *et al.* (2003)
–	*Sulfolobus tokodaii*	70	–	2.5	Thermoacidophilicity	Staufenberger *et al.* (2012)
Chi70	*Thermococcus chitinophagus*	70	120°C, 1 h	7	Heat-resistance	Andronopoulou and Vorgias (2004a,b)
TkChiA	*Thermococcus kodakaraensis* KOD1	85	–	5	High tolerance against organic solvents and detergents	Tanaka *et al.* (1999, 2004)
Pectinases						
Pectate, Lyase a	*Thermoanaerobacter italicus*	80	80°C, 60 min	9	Wide food industrial applications,	Kozianowski *et al.* (1997)
Pectate, Lyase b	*Thermoanaerobacter italicus*	80	80°C, 60 min	9	Wide food industrial applications	Kozianowski *et al.* (1997)
PelA	*Thermotoga maritima*	90	95°C, 2 h	9	Wide food industrial applications	Kluskens *et al.* (2003)
PelB	*Thermotoga maritima*	80	95°C, 15 min	6.4	Wide food industrial applications	Kluskens *et al.* (2005)
Phytases						
–	*Bacillus laevolacticus*	70	70°C, 3 h	7–8	Feed and food industrial applications	Gulati *et al.* (2007)
PhyL	*Bacillus licheniformis*	75	85°C, 6 h	6.5–7	Feed and food industrial applications	Borgi *et al.* (2013)
–	*Bacillus* sp. DS11	70	90°C, 10 min	7	Feed industry	Kim *et al.* (1998b)
–	*Bacillus* sp. MD2	67–73	–	6–7	Feed and food industrial applications	Tran *et al.* (2010)

[a]Indicates potential applications that were proposed by the authors of the respective studies.

showed optimal chitinolytic activity at 70°C and pH 6–8 with a half-life of 3 minutes at 80°C.

Thermostable chitinases were also isolated from hyperthermophilic Archaea like *Thermococcus chitinophagus* (Andronopoulou and Vorgias, 2004a,b), *T. kodakaraensis* KOD1 (Tanaka *et al.*, 1999, 2004), *P. furiosus* (Gao *et al.*, 2003; Oku and Ishikawa, 2006), and *Sulfolobus tokodaii* (Table 3.5) (Staufenberger *et al.*, 2012). *T. chitinophagus* produces a set of chitinolytic enzymes including a cell membrane associated endochitinase (Chi70), a periplasmic chitobiase (Chi90) and an extracellular exochitinase (Chi50). The expression of this multienzyme complex was mostly induced by chitin. The protein Chi70 exhibits a molecular mass of 70 kDa and showed optimal activity at 70°C and pH 7 with a half-life of 1 hour at 120°C. In addition, no inhibition was described after treatment with the chitinase inhibitor allosamidin and its substrate specificity was broad, because it showed activity towards several chitinous substrates and derivatives. Its classification into the group of endochitinases was determined due to its ability to hydrolyse colloidal chitin into chitobiose (Andronopoulou and Vorgias, 2004a,b). Another industrially relevant enzyme for use in biocontrol is the endochitinase TkChiA that was identified in the thermophilic Archaeon *T. kodakaraensis* KOD1. This extracellular chitinase was heterologously expressed and characterized in *E. coli* and showed a high tolerance towards organic solvents and detergents. Its optimal activity was determined at 85°C and pH 5. Characteristically, TkChiA exhibits a multidomain structure and consists of two active sites with different cleavage specificities in combination with three substrate-binding domains. It releases chitobiose as the major end product, and therefore was described as an endochitinase (Tanaka *et al.*, 1999, 2004). Another Archaeon that was described with the ability to grow on chitin was the hyperthermophile *P. furiosus* (Gao *et al.*, 2003). The corresponding thermostable chitinases PfChiA and PfChiB were produced and characterized in a recombinant form in *E. coli*. The catalytic regions of both enzymes showed highest similarity to members of family 18 bacterial chitinases. Their optimum activity was detected at pH 6 between 90°C and 95°C. The individual melting temperatures of PfChiA and PfChiB were 101°C and 114°C, respectively. PfChiA showed no detectable activity towards chitooligomers that were smaller than chitotetraose. Therefore, PfChiA was classified as endochitinase, whereas PfChiB was described as a chitobiosidase. The latter enzyme cleaves off the chitobiose from the non-reducing end of chitinous substrates. Synergistic activity of these two enzymes allows *P. furiosus* to grow on chitin as a sole carbon source (Gao *et al.*, 2003). The crenacheal chitinase from *S. tokodaii* was described as a dimer with the highest similarity to members of the bacterial chitinase family 18 and showed a specific activity towards colloidal substrates. The enzyme exhibits the highest activity at pH 2.5 and 70°C (Staufenberger *et al.*, 2012).

Pectinases

Pectinolytic enzymes hydrolyse pectin, one of the key elements in plants, that occurs especially in the primary cell wall and the middle lamellae in plant material (Kashyap *et al.*, 2001). The structure of this branched heterosaccharide is mainly built up of methyl esterified chains of α-1,4-D-polygalacturonate. Neutral sugar side chains like L-rhamnose, arabinose, galactose or xylose are connected to the main chains by L-rhamnopyranose residues (Arunachalam and Asha, 2010; Mehraj Pasha *et al.*, 2013). Due to their reaction mechanism and catalytic targets, pectinolytic enzymes can be distinguished into two major groups. The first one includes Pectinesterases (PE, EC 3.1.1.11) that degrade pectin into pectic acid by removing its methoxy groups, whereas the second group consists of pectin depolymerases (Fig. 3.8B). Pectin depolymerases can be further divided into hydrolases like polygalacturonases (PG, EC 3.2.1.15) and lyases like pectate lyases (PL, EC 4.2.2.2) acting on polygalacturonic acid or pectin lyases (PNL, EC 4.2.2.10) that prefer pectin as substrate. During their reaction mechanism, pectin is degraded by the cleavage of 1,4-D-glycosydic bonds within the D-galacturonate (Sakai *et al.*, 1993; Sharma *et al.*, 2013a). Polygalacturonases hydrolyse the glycosidic bonds near free carboxyl groups while lyases perform a β-elimination to split corresponding linkages. Another distinctive feature of pectinolytic enzymes is the resulting product. Endo-polygalacturonases and endo-lyases do not produce a single product at the end because their catalytic attack is randomly distributed overall the pectin molecule. In contrast, exo-polygalacturonases produce monomers or dimers by attacking the non-reducing end of the chain while unsaturated dimers from the reducing end are the products of exo-lyases attacking pectin molecules (Fig. 3.8B) (Pedrolli *et al.*, 2009; Sakai *et al.*, 1993).

At the structural level, pectinolytic enzymes show a characteristical single domain of parallel β-strands formed in a right-handed cylinder that is prism shaped. In addition, three parallel β-strands called PB1, PB2 and PB3 are formed by strands of consecutive turns. PB1 and PB2 form an antiparallel β-sandwich, whereas PB3 is located approximately perpendicular to PB2. Despite the different mode of action for pectinesterases, hydrolases and lyases, the binding site of the substrate is overall similarly located in a cleft exterior to the parallel β-helix between PB1 and the loops. Differences in this loop seem to be related to the variations in the enzymatic properties of each group of pectinolytic enzymes (Sharma *et al.*, 2013a)

Alongside their physiological role of breaking down pectinous substrates for nutrition uptake and use in metabolic processes like cell elongation, growth processes or fruit ripening, pectinases are widely used in industrial applications, mostly in food associated areas like the extraction of oils and fruit juices (Arunachalam and Asha, 2010; Hoondal *et al.*, 2002; Kertesz, 1930; Mehraj Pasha *et al.*, 2013; Sharma *et al.*, 2013a). Application of pectinases in fruit juice production was reported for the first time in 1930 (Kertesz, 1930). Pectinases are used to extract juice from fruits, such that the spoilage and decay of the fruits will be avoided and productivity is enhanced (Sharma *et al.*, 2013a). Further applications of pectinolytic enzymes include textile processing, degumming of plant-based fibres, coffee and tea fermentations and paper and pulp industry. Examples of commercially available pectinases are Panzym™ (Boehringer, Germany), Pectolase™ (Grinsteelvaeket, Denmark) or Pectinol™ (Rohm GmbH, Germany) (Hoondal *et al.*, 2002; Kashyap *et al.*, 2001; Mehraj Pasha *et al.*, 2013). During textile processing, pectinases along with other enzymes like amylases, lipases, cellulases and hemicellulases are used to remove sizing agents (Arunachalam and Asha, 2010; Hoondal *et al.*, 2002; Kashyap *et al.*, 2001; Pedrolli *et al.*, 2009). Degumming of plant fibres takes place at a high pH, favouring the application of alkaline pectinases in milder and more economic processes (Cao *et al.*, 1992; Kapoor *et al.*, 2001). For the fermentation of coffee and tea products, pectinolytic enzymes remove the coat from the beans and are sometimes used to remove the pulpy bean layer (Sharma *et al.*, 2013a; Sørensen *et al.*, 2000). During the papermaker process, pectinases are used for the depolymerization of galacturonic acids to contribute to a lower cationic demand of pectinous solutions (Reid and Ricard, 2000; Viikari *et al.*, 2001).

Microorganisms that produce pectinolytic enzymes are widely distributed in pro- and eukaryotes (Arias and Burns, 2002; Cao *et al.*, 1992; Fayyaz *et al.*, 1993; Horikoshi, 1990; Huang and Mahoney, 1999; Stratilova *et al.*, 1996; Suneetha and Prathyusha, 2011; Warrilow *et al.*, 1994). According to the occurring high process temperatures in industrial processes, pectinolytic enzymes from thermophilic microorganisms are preferred. Thus, several reports of thermostable pectinolytic enzymes are available and most of them are of alkaliphilic nature with a high stability at alkaline pH (Table 3.5). By an activity based screening approach on citrus pectin as substrate, two citrus pectate lyases were identified and characterized from the thermophilic bacterium *Thermoanaerobacter italicus*, that was isolated from a thermal spa in Italy (Kozianowski *et al.*, 1997). Lyase 'a' was described to be a monomer, which exhibited a molecular mass of 135 kDa, whereas Lyase 'b' was a heterodimer consisting of a 93 kDa and a 138 kDa subunit. Both enzymes were thermoactive and showed optimal activity at pH 9 and 80°C. In addition, they were heat-stable regarding a half-life of 60 minutes at 80°C and an increase of activity was observed after calcium treatment (Kozianowski *et al.*, 1997). Two other heat-stable pectinolytic enzymes were produced in *T. maritima* (Kluskens *et al.*, 2003; Kluskens *et al.*, 2005). The thermostable pectate lyase, PelA, was heterologously produced in *E. coli*. The enzyme was characterized as a tetramer with a crucial dependence on calcium ions for both activity and stability. The enzyme preferred polygalacturonic acid over pectin and showed optimal activity at 90°C and pH 9 with a half-life of 2 hours at 95°C. Regarding its catalytic target, PelA exhibited a unique cleavage pattern towards polygalacturonic acid that led mainly to unsaturated trigalacturonate (Kluskens *et al.*, 2003). The second enzyme PelB was one of the few described bacterial exopolygalacturonases with the ability to reduce monogalacturonate units from the non-reducing end of the polygalacturonate and showed an optimal activity at 80°C, an optimal pH 6.4 and a half-life of 15 minutes at 95°C (Kluskens *et al.*, 2005).

Phytases

Phytases (myo-inositol hexakisphosphate phosphohydrolases) catalyse the stepwise splitting and release of phosphates from phytate or phytic acid into smaller myo-inositol intermediates. Phytate (Fig. 3.8C) is the key phosphorus molecule that is predominantly present in oilseeds, cereal grains and legumes (Reddy *et al.*, 1982). Depending on their attacking dephosphorylation site, phytate-degrading enzymes are classified into three main groups: (i) 3-phytases (myo-inositol hexakisphosphate 3-phosphohydrolases), (ii) 6-phytases and (iii) 5-phytases. 3-Phytases (EC 3.1.3.8) split the phosphate group of the inositol ring of phytate by attacking the carbon at the 3-position while 5-phytases (EC 3.1.3.72) and 6-phytases (EC 3.1.3.26) attack the fifth and sixth carbon, respectively. A further classification of phytases relies on the catalytic mechanism and organization of the active site. Accordingly, phytases can be grouped into (i) histidine acid phosphatases (HAPhy), (ii) β-propeller phytases (BPPhy, alkaline phytases), (iii) β-purple acid phosphatases (PAPhy, metaloenzymes) and (iv) protein tyrosine phosphatases (PTPhy, dual-specificity phosphatases) that differ in their structural characteristics (Lei *et al.*, 2013; Mullaney and Ullah, 2003). In general, all phytases consist of two similar domains, an alpha and an alpha/beta domain with the active site localized between them, but they differ specifically in the number of included α-helices and β-sheets (Kostrewa *et al.*, 1999; Lei *et al.*, 2013; Lim *et al.*, 2000; Ullah and Dischinger, 1993). Despite this common features, they show characteristical differences like the highly conserved N-terminal sequence motif RHGXRXP in combination with the C-terminal HD consensus sequence for histidine acid phosphatases that were described in representative enzymes of *Aspergillus niger* and *E. coli* (Kostrewa *et al.*, 1999; Lim *et al.*, 2000). In contrast β-propeller phytases mainly consist of β-propeller and form a six-bladed molecule (Ha *et al.*, 2000; Shin *et al.*, 2001). Furthermore, β-purple acid phosphatases and protein tyrosine phosphatases show specific structural differences because they exhibit either seven crucial conserved metal-liganding residues (D, D, Y, N, H, H, H) localized in conserved motifs, or a unique p-loop structure within the active site with a conserved HCXXGXXR(T/S) sequence motif (Chu *et al.*, 2004; Schenk *et al.*, 2000; Than *et al.*, 1999). All phytases interact with the substrate via the conserved RHGXRXP motif within the binding pocket that targets the phosphate group and subsequently forms an enzyme-substrate-complex. After that, the catalytic domain catalyses the release of the phosphate group with the help of the HD element that leads to dephosphorylation of the substrate and results in the end products of inositol and P_i (Loewus and Murthy, 2000; Mullaney *et al.*, 2000).

Since the description and commercialization of phytase as an animal feed additive, these enzymes and their producing organisms are greatly demanded (Vohra and Satyanarayana, 2003). Phytase-producing organisms include eukaryotic species like plants (Hamada, 1996; Hubel and Beck, 1996; Nagai and Funahashi, 1962), yeasts (Lambrechts *et al.*, 1992; Nakamura *et al.*, 2000; Vohra and Satyanarayana, 2001) and filamentous fungi (Pasamontes *et al.*, 1997; Ullah and Gibson, 1987; van Hartingsveldt *et al.*, 1993) as well as thermophilic prokaryotes (Table 3.5) (Gulati *et al.*, 2007; Mittal *et al.*, 2012). Although most of the heat-stable phytases have been described from thermophilic fungi. There are only few reports on heat-stable phytases from Bacteria and no reports on Archaea. One of the first heat-stable phytases of bacterial origin was isolated from thermophilic *Bacillus* sp. DS11 (Kim *et al.*, 1998b). The heterologously produced protein in *E. coli* has a molecular mass of 44 kDa and showed an optimal activity at 70°C and pH 7 with a half-life of 10 minutes at 90°C in the presence of calcium ions. The dependence of metal ions was determined by EDTA treatment that led to a significant reduction of catalytic activity. It showed highest activity on phytate, while phosphate esters were only processed with low efficiency (Kim *et al.*, 1998b). Another thermostable alkaline *Bacillus* phytase has been described from *Bacillus* strain MD2. The 47.5 kDa protein showed high activity at temperatures over 100°C. Its optimal activity was between

67°C and 73°C and at pH 6 and pH 7. Moreover, it was activated in the presence of calcium ions. In addition to its thermostability, this enzyme also processed other phosphate substrates such as ADP, ATP and α-fructose-1,6-diphosphate. Therefore, it was suggested to be a suitable candidate for applications for feed and food industries (Tran *et al.*, 2010). *Bacillus laevolacticus* also produces a thermostable alkaline phytase with an optimal activity at 70°C and pH values between pH 7 and pH 8 with a high specificity towards phytate. In addition, the partially purified enzyme exhibited high thermostability retaining 80% of its activity after 3 hours incubation at 70°C (Gulati *et al.*, 2007). Recently, Borgi and colleagues described a thermostable β-propeller phytase that was produced by *B. licheniformis*. The recombinant enzyme PhyL exhibited an optimal activity at pH 6.5 to pH 7 and 75°C. In presence of Ca^{2+} the reported half-life was 6 hours at 85°C (Borgi *et al.*, 2013).

Conclusions and future trends

Research in the field of extremophiles in the last four decades has led to an increased research focus on the biochemical characterization of extremozymes and their usability for widespread industrial applications. The discovery of the first thermostable DNA polymerase from *T. aquaticus* and its utilization in PCR has changed the whole technology of molecular research in all its aspects. Owing to their unique properties such as activity at elevated temperatures, tolerance against organic solvents and detergents, especially enzymes from thermophiles are expected to fill the gap between biological and chemical processes. The advantages of using heat-stable enzymes include faster reaction rates, lower viscosity, higher substrate solubility, higher selectivity and lower risks of contamination.

In recent years, the major drawback of utilizing extremozymes in industrial applications was the low productivity of enzymes and consequently the high production costs. The dramatic progress in various technologies such as metagenomics, genomics, directed evolution, synthetic biology and advanced fermentation technologies will help to identify new markets for extremozymes. These technologies will be used for the production of a new generation of tailor-made enzymes that will contribute to the development of future sustainable green technologies.

Acknowledgement

The authors thank Mazen Rizk for critical reading of the manuscript.

Web resources

MEROPS the peptidase database: http://merops.sanger.ac.uk

BRENDA the comprehensive enzyme information system: http://www.brenda-enzymes.org

CAZY the Carbohydrate-Active enZYmes Database: http://www.cazy.org/

References

Adams, D.J. (2004). Fungal cell wall chitinases and glucanases. Microbiology *150*, 2029–2035.

Akanbi, T., Kamaruzaman, A., Abu Bakar, F., Sheikh Abdul Hamid, N., Radu, S., Abdul Manap, M., and Saari, N. (2010). Highly thermostable extracellular lipase-producing *Bacillus* strain isolated from a Malaysian hotspring and identified using 16S rRNA gene sequencing. Int. Food Res. J. *17*, 45–53.

Al Khudary, R., Venkatachalam, R., Katzer, M., Elleuche, S., and Antranikian, G. (2010). A cold-adapted esterase of a novel marine isolate, *Pseudoalteromonas arctica*: gene cloning, enzyme purification and characterization. Extremophiles *14*, 273–285.

Albertson, G.D., McHale, R.H., Gibbs, M.D., and Bergquist, P.L. (1997). Cloning and sequence of a type I pullulanase from an extremely thermophilic anaerobic bacterium, *Caldicellulosiruptor saccharolyticus*. Biochim. Biophys. Acta *1354*, 35–39.

Alcalde, M., Plou, F.J., Pastor, E., and Ballesteros, A. (1998). Effect of chemical modification of cyclodextrin glycosyltransferase (CGTase) from *Thermoanaerobacter* sp. on its activity and product selectivity. Ann. N.Y. Acad. Sci. *864*, 183–187.

Ando, S., Ishikawa, K., Ishida, H., Kawarabayasi, Y., Kikuchi, H., and Kosugi, Y. (1999). Thermostable aminopeptidase from *Pyrococcus horikoshii*. FEBS Lett. *447*, 25–28.

Andronopoulou, E., and Vorgias, C.E. (2004a). Isolation, cloning, and overexpression of a chitinase gene fragment from the hyperthermophilic archaeon *Thermococcus chitonophagus*: semi-denaturing purification of the recombinant peptide and investigation of its relation with other chitinases. Protein Expr. Purif. 35, 264–271.

Andronopoulou, E., and Vorgias, C.E. (2004b). Multiple components and induction mechanism of the chitinolytic system of the hyperthermophilic archaeon *Thermococcus chitonophagus*. Appl. Microbiol. Biotechnol. 65, 694–702.

Angelov, A., Putyrski, M., and Liebl, W. (2006). Molecular and biochemical characterization of alpha-glucosidase and alpha-mannosidase and their clustered genes from the thermoacidophilic archaeon *Picrophilus torridus*. J. Bacteriol. *188*, 7123–7131.

Antranikian, G. (2008). Industrial relevance of thermophiles and their enzymes. In Thermophiles – Biology and Technology at High Temperatures, Robb, F., Antranikian, G., Grogan, D., and Driessen, A., eds. (CRC Press, Taylor & Francis Group, Boca Raton), pp. 113–160.

Antranikian, G., and Egorova, K. (2007). Extremophiles, a unique source of biocatalysts for industrial biotechnology. In Physiology and Biochemistry of Extremophiles, Gerday, C., and Glansdorff, N., eds. (ASM Press, Washington D.C.), pp. 361–406.

Antranikian, G., Zablowski, P., and Gottschalk, G. (1987). Conditions for the overproduction and excretion of thermostable α-amylase and pullulanase from *Clostridium thermohydrosulfuricum* DSM 567. Appl. Microbiol. Biotechnol. *27*, 75–81.

Arias, C.R., and Burns, J.K. (2002). A pectinmethylesterase gene associated with a heat-stable extract from citrus. J. Agr. Food Chem. *50*, 3465–3472.

Arpigny, J.L., and Jaeger, K.E. (1999). Bacterial lipolytic enzymes: classification and properties. Biochem. J. *343 (Pt. 1)*, 177–183.

Arpigny, J.L., Jendrossek, D., and Jaeger, K.E. (1998). A novel heat-stable lipolytic enzyme from *Sulfolobus acidocaldarius* DSM 639 displaying similarity to polyhydroxyalkanoate depolymerases. FEMS Microbiol. Lett. *167*, 69–73.

Arunachalam, C., and Asha, S. (2010). Pectinolytic enzyme – a review of new studies. Adv. Biotech. J. *9*, 1–5.

Athawale, V., Manjrekar, N., and Athawale, M. (2001). Enzymatic synthesis of chiral menthyl methacrylate monomer by *Pseudomonas cepacia* lipase catalysed resolution of (±)-menthol. J. Mol. Catal. B-Enzym. *16*, 169–173.

Awade, A.C., Cleuziat, P., Gonzales, T., and Robert-Baudouy, J. (1994). Pyrrolidone carboxyl peptidase (Pcp): an enzyme that removes pyroglutamic acid (pGlu) from pGlu-peptides and pGlu-proteins. Proteins *20*, 34–51.

Balan, A., Ibrahim, D., Abdul Rahim, R., and Ahmad Rashid, F.A. (2012). Purification and Characterization of a Thermostable Lipase from *Geobacillus thermodenitrificans* IBRL-nra. Enzyme Res. *2012*, 987523.

Ballschmiter, M., Armbrecht, M., Ivanova, K., Antranikian, G., and Liebl, W. (2005). AmyA, an alpha-amylase with beta-cyclodextrin-forming activity, and AmyB from the thermoalkaliphilic organism *Anaerobranca gottschalkii*: two alpha-amylases adapted to their different cellular localizations. Appl. Environ. Microbiol. *71*, 3709–3715.

Ballschmiter, M., Fütterer, O., and Liebl, W. (2006). Identification and characterization of a novel intracellular alkaline alpha-amylase from the hyperthermophilic bacterium *Thermotoga maritima* MSB8. Appl. Environ. Microbiol. 72, 2206–2211.

Banner, D.W., Bloomer, A.C., Petsko, G.A., Phillips, D.C., Pogson, C.I., Wilson, I.A., Corran, P.H., Furth, A.J., Milman, J.D., Offord, R.E., *et al.* (1975). Structure of chicken muscle triose phosphate isomerase determined crystallographically at 2.5 angstrom resolution using amino acid sequence data. Nature 255, 609–614.

Barbosa, O., Ruiz, M., Ortiz, C., Fernández, M., Torres, R., and Fernandez-Lafuente, R. (2012). Modulation of the properties of immobilized CALB by chemical modification with 2,3,4-trinitrobenzenesulfonate or ethylendiamine. Advantages of using adsorbed lipases on hydrophobic supports. Proc. Biochem. *47*.

Barrett, A.J., Woessner, J.F., and Rawlings, N.D. (2004). Handbook of Proteolytic Enzymes, Vol 1 (Access Online via Elsevier, London).

Ben Messaoud, E., Ben Ammar, L., Mellouli, S., and Bejar, S. (2002). Thermostable pullulanase type I from new isolated *Bacillus thermoleovorans* US105: cloning, sequencing and expression of the gene in *E. coli*. Enzyme. Microb. Technol. *31*, 827–832.

Bertoldo, C., Duffner, F., Jorgensen, P.L., and Antranikian, G. (1999). Pullulanase type I from *Fervidobacterium pennavorans* Ven5: cloning, sequencing, and expression of the gene and biochemical characterization of the recombinant enzyme. Appl. Environ. Microbiol. *65*, 2084–2091.

Bhushan, B., and Hoondal, G.S. (1998). Isolation, purification and properties of a thermostable chitinase from an alkalophilic *Bacillus* sp. BG-11. Biotechnol. Lett. *20*, 157–159.

Biwer, A., Antranikian, G., and Heinzle, E. (2002). Enzymatic production of cyclodextrins. Appl. Microbiol. Biotechnol. *59*, 609–617.

Bordes, F., Tarquis, L., Nicaud, J.M., and Marty, A. (2011). Isolation of a thermostable variant of Lip2 lipase from *Yarrowia lipolytica* by directed evolution and deeper insight into the denaturation mechanisms involved. J. Biotechnol. *156*, 117–124.

Borgi, M.A., Boudebbouze, S., Aghajari, N., Szukala, F., Pons, N., Maguin, E., and Rhimi, M. (2013). The attractive recombinant phytase from *Bacillus licheniformis*: biochemical and molecular characterization. Appl. Microbiol. Biotechnol. *98*, 5937–5947.

Bornscheuer, U.T. (2002). Microbial carboxyl esterases: classification, properties and application in biocatalysis. FEMS Microbiol. Rev. *26*, 73–81.

Brown, S.H., and Kelly, R.M. (1993). Characterization of amylolytic enzymes, having both a-1,4 and a-1,6 hydrolytic activity, from the Thermophilic Archaea *Pyrococcus furiosus* and *Thermococcus litoralis*. Appl. Environ. Microbiol. *59*, 2614–2621.

Bruins, M.E., Janssen, A.E., and Boom, R.M. (2001). Thermozymes and their applications: a review of recent literature and patents. Appl. Biochem. Biotechnol. *90*, 155–186.

Buchholz, K., and Collins, J. (2013). The roots—a short history of industrial microbiology and biotechnology. Appl. Microbiol. Biotechnol. 97, 3747–3762.

Buller, A.R., and Townsend, C.A. (2013). Intrinsic evolutionary constraints on protease structure, enzyme acylation, and the identity of the catalytic triad. Proc. Natl. Acad. Sci. U.S.A. *110*, E653–661.

Bult, C.J., White, O., Olsen, G.J., Zhou, L., Fleischmann, R.D., Sutton, G.G., Blake, J.A., FitzGerald, L.M., Clayton, R.A., Gocayne, J.D., *et al.* (1996). Complete genome sequence of the methanogenic archaeon, *Methanococcus jannaschii*. Science *273*, 1058–1073.

Burlini, N., Magnani, P., Villa, A., Macchi, F., Tortora, P., and Guerritore, A. (1992). A heat-stable serine proteinase from the extreme thermophilic archaebacterium *Sulfolobus solfataricus*. Biochim. Biophys. Acta *1122*, 283–292.

Cao, J., Zheng, L., and Chen, S. (1992). Screening of pectinase producer from alkalophilic bacteria and study on its potential application in degumming of ramie. Enzyme Microb. Tech. *14*, 1013–1016.

Cava, F., Hidalgo, A., and Berenguer, J. (2009). *Thermus thermophilus* as biological model. Extremophiles *13*, 213–231.

Chai, Y.Y., Rahman, R.N., Illias, R.M., and Goh, K.M. (2012). Cloning and characterization of two new thermostable and alkalitolerant alpha-amylases from the *Anoxybacillus* species that produce high levels of maltose. J. Ind. Microbiol. Biotechnol. *39*, 731–741.

Chakravorty, D., Parameswaran, S., Dubey, V.K., and Patra, S. (2011). In silico characterization of thermostable lipases. Extremophiles *15*, 89–103.

Chang, S.T., Parker, K.N., Bauer, M.W., and Kelly, R.M. (2001). alpha-Glucosidase from *Pyrococcus furiosus*. Methods Enzymol. *330*, 260–269.

Chang, W.T., Chen, C.S., and Wang, S.L. (2003). An antifungal chitinase produced by *Bacillus cereus* with shrimp and crab shell powder as a carbon source. Curr. Microbiol. *47*, 102–108.

Chavan, S., and Deshpande, M. (2013). Chitinolytic enzymes: an appraisal as a product of commercial potential. Biotechnol. Prog. *29*, 833–846.

Chow, J., Kovacic, F., Dall Antonia, Y., Krauss, U., Fersini, F., Schmeisser, C., Lauinger, B., Bongen, P., Pietruszka, J., Schmidt, M., *et al.* (2012). The metagenome-derived enzymes LipS and LipT increase the diversity of known lipases. PLoS One *7*, e47665.

Chu, H.-M., Guo, R.-T., Lin, T.-W., Chou, C.-C., Shr, H.-L., Lai, H.-L., Tang, T.-Y., Cheng, K.-J., Selinger, B.L., and Wang, A.H.J. (2004). Structures of *Selenomonas ruminantium* phytase in complex with persulfated phytate: DSP Phytase fold and mechanism for sequential substrate hydrolysis. Structure *12*, 2015–2024.

Chung, H.-J., Yoon, S.-H., Lee, M.-J., Kim, M.-J., Kweon, K.-S., Lee, I.-W., Kim, J.-W., Oh, B.-H., Lee, H.-S., Spiridonova, V.A., *et al.* (1998). Characterization of a Thermostable Cyclodextrin Glucanotransferase Isolated from *Bacillus stearothermophilus* ET1+. J. Agr. Food Chem. *46*, 952–959.

Cieslinski, H., Bialkowska, A.M., Dlugolecka, A., Daroch, M., Tkaczuk, K.L., Kalinowska, H., Kur, J., and Turkiewicz, M. (2007). A cold-adapted esterase from psychrotrophic *Pseudoalteromas* sp. strain 643A. Arch. Microbiol. *188*, 27–36.

Cohen-Kupiec, R., and Chet, I. (1998). The molecular biology of chitin digestion. Curr. Opin. Biotechnol. *9*, 270–277.

Colman, P.M., Jansonius, J.N., and Matthews, B.W. (1972). The structure of thermolysin: an electron density map at 2–3 A resolution. J. Mol. Biol. *70*, 701–724.

Colombo, S., D'Auria, S., Fusi, P., Zecca, L., Raia, C.A., and Tortora, P. (1992). Purification and characterization of a thermostable carboxypeptidase from the extreme thermophilic archaebacterium *Sulfolobus solfataricus*. Eur. J. Biochem. *206*, 349–357.

Coolbear, T., Eames, C.W., Casey, Y., Daniel, R.M., and Morgan, H.W. (1991). Screening of strains identified as extremely thermophilic bacilli for extracellular proteolytic activity and general properties of the proteinases from two of the strains. J. Appl. Bacteriol. *71*, 252–264.

Coolbear, T., Monk, C., Peek, K., Morgan, H.W., and Daniel, R.M. (1992). Laboratory-scale investigations into the use of extremely thermophilic proteinases for cleaning ultrafiltration membranes fouled during whey processing. J. Memb. Sci. *67*, 93–101.

Cowan, D.A., and Daniel, R.M. (1982). Purification and some properties of an extracellular protease (caldolysin) from an extreme thermophile. Biochim. Biophys. Acta *705*, 293–305.

Cowan, D., Daniel, R., and Morgan, H. (1985). Thermophilic proteases: Properties and potential applications. Trends Biotechnol. *3*, 68–72.

Cowan, D.A., Smolenski, K.A., Daniel, R.M., and Morgan, H.W. (1987). An extremely thermostable extracellular proteinase from a strain of the archaebacterium *Desulfurococcus* growing at 88 degrees C. Biochem. J. *247*, 121–133.

Cowan, D.A., and Fernandez-Lafuente, R. (2011). Enhancing the functional properties of thermophilic enzymes by chemical modification and immobilization. Enzyme. Microb. Technol. *49*, 326–346.

Daniel, R., Toogood, H., and Bergquist, P. (1996). Thermostable proteases. Biotechnol. Genet. Eng. Rev. *13*, 51–100.

Davies, D.R. (1990). The structure and function of the aspartic proteinases. Annu. Rev. Biophys. Biophys. Chem. *19*, 189–215.

DeLange, R.J., and Smith, E.L. (1968). Subtilisin Carlsberg. I. Amino acid composition; isolation and composition of peptides from the tryptic hydrolysate. J. Biol. Chem. *243*, 2134–2142.

Demirjian, D.C., Moris-Varas, F., and Cassidy, C.S. (2001). Enzymes from extremophiles. Curr. Opin. Chem. Biol. *5*, 144–151.

Dib, R., Chobert, J.M., Dalgalarrondo, M., Barbier, G., and Haertle, T. (1998). Purification, molecular properties and specificity of a thermoactive and thermostable proteinase from *Pyrococcus abyssi*, strain st 549, hyperthermophilic archaea from deep-sea hydrothermal ecosystem. FEBS Lett. *431*, 279–284.

Dock, C., Hess, M., and Antranikian, G. (2008). A thermoactive glucoamylase with biotechnological relevance from the thermoacidophilic Euryarchaeon *Thermoplasma acidophilum*. Appl. Microbiol. Biotechnol. *78*, 105–114.

Doman-Pytka, M., and Bardowski, J. (2004). Pullulan degrading enzymes of bacterial origin. Crit. Rev. Microbiol *30*, 107–121.

Dong, G., Vieille, C., Savchenko, A., and Zeikus, J.G. (1997a). Cloning, sequencing, and expression of the gene encoding extracellular alpha-amylase from *Pyrococcus furiosus* and biochemical characterization of the recombinant enzyme. Appl. Environ. Microbiol. *63*, 3569–3576.

Dong, G., Vieille, C., and Zeikus, J.G. (1997b). Cloning, sequencing, and expression of the gene encoding amylopullulanase from *Pyrococcus furiosus* and biochemical characterization of the recombinant enzyme. Appl. Environ. Microbiol. *63*, 3577–3584.

Doolittle, R.F., and Armentrout, R.W. (1968). Pyrrolidonyl peptidase. An enzyme for selective removal of pyrrolidonecarboxylic acid residues from polypeptides. Biochemistry *7*, 516–521.

Eggen, R., Geerling, A., Watts, J., and de Vos, W.M. (1990). Characterization of pyrolysin, a hyperthermoactive serine protease from the archaebacterium *Pyrococcus furiosus*. FEMS Microbiol. Lett. *71*, 17–20.

Eijsink, V.G., Veltman, O.R., Aukema, W., Vriend, G., and Venema, G. (1995). Structural determinants of the stability of thermolysin-like proteinases. Nat. Struct. Biol. *2*, 374–379.

Eijsink, V.G., Bjork, A., Gaseidnes, S., Sirevag, R., Synstad, B., van den Burg, B., and Vriend, G. (2004). Rational engineering of enzyme stability. J. Biotechnol. *113*, 105–120.

Eijsink, V.G., Matthews, B.W., and Vriend, G. (2011). The role of calcium ions in the stability and instability of a thermolysin-like protease. Protein Sci. *20*, 1346–1355.

Ejima, K., Liu, J., Oshima, Y., Hirooka, K., Shimanuki, S., Yokota, Y., Hemmi, H., Nakayama, T., and Nishino, T. (2004). Molecular cloning and characterization of a thermostable carboxylesterase from an archaeon, *Sulfolobus shibatae* DSM5389: non-linear kinetic behavior of a hormone-sensitive lipase family enzyme. J. Biosci. Bioeng. *98*, 445–451.

Elleuche, S., and Antranikian, G. (2013). Starch-hydrolyzing enzymes from thermophiles. In Thermophilic Microbes in Environmental and Industrial Biotechnology: Biotechnology of Thermophiles, Satyanarayana, T., Littlechild, J., and Kawarabayasi, Y., eds. (Dordrecht, Springer), pp. 509–533.

Elleuche, S., Piascheck, H., and Antranikian, G. (2011). Fusion of the OsmC domain from esterase confers thermolability to the cold active xylanase Xyn8 from *Pseudoalteromonas artica*. Extremophiles *15*, 311–317.

Endo, S. (1962). Study on protease by thermophilic bacteria. J. Ferment. Technol. *40*, 346–353.

Fayyaz, A., Asbi, B.A., Ghazali, H.M., Che Man, Y.B., and Jinap, S. (1993). Pectinesterase extraction from papaya. Food Chem. *47*, 183–185.

Fontana, A. (1988). Structure and stability of thermophilic enzymes. Studies on thermolysin. Biophys. Chem. *29*, 181–193.

Friedrich, A.B., and Antranikian, G. (1996). Keratin degradation by *Fervidobacterium pennavorans*, a novel thermophilic anaerobic species of the order Thermotogales. Appl. Environ. Microbiol. *62*, 2875–2882.

Frillingos, S., Linden, A., Niehaus, F., Vargas, C., Nieto, J.J., Ventosa, A., Antranikian, G., and Drainas, C. (2000). Cloning and expression of alpha-amylase from the hyperthermophilic archaeon *Pyrococcus woesei* in the moderately halophilic bacterium *Halomonas elongata*. J. Appl. Microbiol. *88*, 495–503.

Fusi, P., Villa, M., Burlini, N., Tortora, P., and Guerritore, A. (1991). Intracellular proteases from the extremely thermophilic archaebacterium *Sulfolobus solfataricus*. Experientia *47*, 1057–1060.

Galvis, M., Barbosa, O., Torres, R., Ortiz, C., and Fernandez-Lafuente, R. (2012). Effect of solid-phase chemical modification on the features of the lipase from *Thermomyces lanuginosus*. Proc. Biochem. *47*, 460–466.

Gao, J., Bauer, M.W., Shockley, K.R., Pysz, M.A., and Kelly, R.M. (2003). Growth of hyperthermophilic archaeon *Pyrococcus furiosus* on chitin involves two family 18 chitinases. Appl. Environ. Microbiol. *69*, 3119–3128.

Georgopapadakou, N.H., and Tkacz, J.S. (1995). The fungal cell wall as a drug target. Trends Microbiol. *3*, 98–104.

Gulati, H.K., Chadha, B.S., and Saini, H.S. (2007). Production and characterization of thermostable alkaline phytase from *Bacillus laevolacticus* isolated from rhizosphere soil. J. Ind. Microbiol. Biotechnol. *34*, 91–98.

Gupta, A., and Khare, S.K. (2006). A protease stable in organic solvents from solvent tolerant strain of *Pseudomonas aeruginosa*. Bioresour. Technol. *97*, 1788–1793.

Gupta, R., Beg, Q.K., and Lorenz, P. (2002). Bacterial alkaline proteases: molecular approaches and industrial applications. Appl. Microbiol. Biotechnol. *59*, 15–32.

Gupta, R., Gupta, N., and Rathi, P. (2004). Bacterial lipases: an overview of production, purification and biochemical properties. Appl. Microbiol. Biotechnol. *64*, 763–781.

Ha, N.-C., Oh, B.-C., Shin, S., Kim, H.-J., Oh, T.-K., Kim, Y.-O., Choi, K.Y., and Oh, B.-H. (2000). Crystal structures of a novel, thermostable phytase in partially and fully calcium-loaded states. Nat. Struct. Biol. *7*, 147–153.

Haki, G.D., and Rakshit, S.K. (2003). Developments in industrially important thermostable enzymes: a review. Bioresour. Technol. *89*, 17–34.

Hamada, J.S. (1996). Isolation and identification of the multiple forms of soybean phytases. J. Am. Oil Chem. Soc. *73*, 1143–1151.

Hammami, I., Siala, R., Jridi, M., Ktari, N., Nasri, M., and Triki, M.A. (2013). Partial purification and characterization of chiIO8, a novel antifungal chitinase produced by *Bacillus cereus* IO8. J. Appl. Microbiol. *115*, 358–366.

Han, T., Zeng, F., Li, Z., Liu, L., Wei, M., Guan, Q., Liang, X., Peng, Z., Liu, M., Qin, J., *et al.* (2013). Biochemical characterization of a recombinant pullulanase from *Thermococcus kodakarensis* KOD1. Lett. Appl. Microbiol. *57*, 336–343.

van Hartingsveldt, W., van Zeijl, C.M., Harteveld, G.M., Gouka, R.J., Suykerbuyk, M.E., Luiten, R.G., van Paridon, P.A., Selten, G.C., Veenstra, A.E., van Gorcom, R.F., *et al.* (1993). Cloning, characterization and overexpression of the phytase-encoding gene (phyA) of *Aspergillus niger*. Gene *127*, 87–94.

Hartley, B.S. (1960). Proteolytic enzymes. Annu. Rev. Biochem. *29*, 45–72.

Hedstrom, L. (2002). Serine protease mechanism and specificity. Chem. Rev. *102*, 4501–4524.

Henrissat, B. (1991). A classification of glycosyl hydrolases based on amino acid sequence similarities. Biochem. J. *280*, 309–316.

Hess, M., Katzer, M., and Antranikian, G. (2008). Extremely thermostable esterases from the thermoacidophilic euryarchaeon *Picrophilus torridus*. Extremophiles *12*, 351–364.

Hicks, P.M., Rinker, K.D., Baker, J.R., and Kelly, R.M. (1998). Homomultimeric protease in the hyperthermophilic bacterium *Thermotoga maritima* has structural and amino acid sequence homology to bacteriocins in mesophilic bacteria. FEBS Lett. *440*, 393–398.

Hoesl, M.G., Acevedo-Rocha, C.G., Nehring, S., Royter, M., Wolschner, C., Wiltschi, B., Budisa, N., and Antranikian, G. (2011). Lipase congeners designed by genetic code engineering. Chemcatchem 3, 213–221.

Holmes, M.A., and Matthews, B.W. (1982). Structure of thermolysin refined at 1.6A resolution. J. Mol. Biol. *160*, 623–639.

Hoondal, G.S., Tiwari, R.P., Tewari, R., Dahiya, N., and Beg, Q.K. (2002). Microbial alkaline pectinases and their industrial applications: a review. Appl. Microbiol. Biotechnol. *59*, 409–418.

Hooper, N.M. (1994). Families of zinc metalloproteases. FEBS Lett. *354*, 1–6.

Horikoshi, K. (1990). Enzymes of alkalophiles. In Microbial Enzymes and Biotechnology (Springer, Netherlands), pp. 275–294.

Horikoshi, K. (1995). Discovering novel bacteria, with an eye to biotechnological applications. Curr. Opin. Biotech. *6*, 292–297.

Hotta, Y., Ezaki, S., Atomi, H., and Imanaka, T. (2002). Extremely stable and versatile carboxylesterase from a hyperthermophilic archaeon. Appl. Environ. Microbiol. *68*, 3925–3931.

Huang, C.J., Wang, T.K., Chung, S.C., and Chen, C.Y. (2005). Identification of an antifungal chitinase from a potential biocontrol agent, *Bacillus cereus* 28-9. J. Biochem. Mol. Biol. *38*, 82–88.

Huang, L.K., and Mahoney, R. (1999). Purification and characterization of an endo-polygalacturonase from *Verticillium albo-atrum*. J. Appl. Microbiol. *86*, 145–156.

Hubel, F., and Beck, E. (1996). Maize Root Phytase (Purification, Characterization, and Localization of Enzyme Activity and Its Putative Substrate). Plant Physiol. *112*, 1429–1436.

Huddleston, S., Yallop, C.A., and Charalambous, B.M. (1995). The identification and partial characterisation of a novel inducible extracellular thermostable esterase from the archaeon *Sulfolobus shibatae*. Biochem. Biophys. Res. Commun. *216*, 495–500.

Ibrahim, A.S., Al-Salamah, A.A., El-Tayeb, M.A., El-Badawi, Y.B., and Antranikian, G. (2012). A Novel Cyclodextrin Glycosyltransferase from Alkaliphilic *Amphibacillus* sp. NPST-10: Purification and Properties. Int. J. Mol. Sci. *13*, 10505–10522.

Ichida, J.M., Krizova, L., LeFevre, C.A., Keener, H.M., Elwell, D.L., and Burtt, E.H., Jr. (2001). Bacterial inoculum enhances keratin degradation and biofilm formation in poultry compost. J. Microbiol. Methods *47*, 199–208.

Ikeda, M., and Clark, D.S. (1998). Molecular cloning of extremely thermostable esterase gene from hyperthermophilic archaeon *Pyrococcus furiosus* in *Escherichia coli*. Biotechnol. Bioeng. *57*, 624–629.

Invernizzi, G., Papaleo, E., Grandori, R., De Gioia, L., and Lotti, M. (2009). Relevance of metal ions for lipase stability: structural rearrangements induced in the *Burkholderia glumae* lipase by calcium depletion. J. Struct. Biol. *168*, 562–570.

Isowa, Y., Ohmori, M., Ichikawa, T., Mori, K., Nonaka, Y., Kihara, K.-i., Oyama, K., Satoh, H., and Nishimura, S. (1979). The thermolysin-catalyzed condensation reactions of n-substituted aspartic and glutamic acids with phenylalanine alkyl esters. Tetrahedron Letters *20*, 2611–2612.

Jabbour, D., Sorger, A., Sahm, K., and Antranikian, G. (2013). A highly thermoactive and salt-tolerant alpha-amylase isolated from a pilot-plant biogas reactor. Appl. Microbiol. Biotechnol. *97*, 2971–2978.

Jaeger, K.E., and Eggert, T. (2002). Lipases for biotechnology. Curr. Opin. Biotechnol. *13*, 390–397.

Jaeger, K.E., and Reetz, M.T. (1998). Microbial lipases form versatile tools for biotechnology. Trends Biotechnol. *16*, 396–403.

Jaouadi, B., Ellouz-Chaabouni, S., Rhimi, M., and Bejar, S. (2008). Biochemical and molecular characterization of a detergent-stable serine alkaline protease from *Bacillus pumilus* CBS with high catalytic efficiency. Biochimie *90*, 1291–1305.

Jaouadi, B., Ellouz-Chaabouni, S., Ali, M.B., Messaoud, E.B., Naili, B., Dhouib, A., and Bejar, S. (2009). Excellent laundry detergent compatibility and high dehairing ability of the *Bacillus pumilus* CBS alkaline proteinase (SAPB). Biotechnology and Bioprocess Engineering *14*, 503–512.

Jayakumar, R., Jayashree, S., Annapurna, B., and Seshadri, S. (2012). Characterization of thermostable serine alkaline protease from an alkaliphilic strain *Bacillus pumilus* MCAS8 and its applications. Appl. Biochem. Biotechnol. *168*, 1849–1866.

Jenkins, C., Geary, S.J., Gladd, M., and Djordjevic, S.P. (2007). The *Mycoplasma gallisepticum* OsmC-like protein MG1142 resides on the cell surface and binds heparin. Microbiology *153*, 1455–1463.

Jiang, W., and Bond, J.S. (1992). Families of metalloendopeptidases and their relationships. FEBS Lett. *312*, 110–114.

Jiao, Y.L., Wang, S.J., Lv, M.S., Xu, J.L., Fang, Y.W., and Liu, S. (2011). A GH57 family amylopullulanase from deep-sea *Thermococcus siculi*: expression of the gene and characterization of the recombinant enzyme. Curr. Microbiol. *62*, 222–228.

Johri, S., Bhat, A., Sayed, S., Nargotra, A., Jain, A., and Qazi, G.N. (2012). Novel thermostable lipase from *Bacillus circulans* IIIB153: comparison with the mesostable homologue at sequence and structure level. World. J. Microbiol. Biotechnol. *28*, 193–203.

Jones, W., Leigh, J., Mayer, F., Woese, C., and Wolfe, R. (1983). *Methanococcus jannaschii* sp. nov., an extremely thermophilic methanogen from a submarine hydrothermal vent. Arch. Microbiol. *136*, 254–261.

Jorgensen, S., Vorgias, C.E., and Antranikian, G. (1997). Cloning, sequencing, characterization, and expression of an extracellular alpha-amylase from the hyperthermophilic archaeon *Pyrococcus furiosus* in *Escherichia coli* and *Bacillus subtilis*. J. Biol. Chem. *272*, 16335–16342.

Kanchana, R., Jadhav, S., Goletikar, Y., and Manerekar, G. (2012). Production of alkaliphilic protease (s) by *Enterobacter* sp. for application in biodetergent formulation. Ind. Biotechnol. *8*, 31–35.

Kang, J., Park, K.M., Choi, K.H., Park, C.S., Kim, G.E., Kim, D., and Cha, J. (2011). Molecular cloning and biochemical characterization of a heat-stable type I pullulanase from *Thermotoga neapolitana*. Enzyme. Microb. Technol. *48*, 260–266.

Kapoor, M., Beg, Q.K., Bhushan, B., Singh, K., Dadhich, K., and Hoondal, G. (2001). Application of an alkaline and thermostable polygalacturonase from *Bacillus* sp. MG-cp-2 in degumming of ramie (*Boehmeria nivea*) and sunn hemp (*Crotalaria juncea*) bast fibres. Process Biochem. *36*, 803–807.

Kashyap, D., Vohra, P., Chopra, S., and Tewari, R. (2001). Applications of pectinases in the commercial sector: a review. Bioresour. Technol. *77*, 215–227.

Kertesz, Z. (1930). New method for enzymic clarification of unfermented apple juice. US patent no. 1.932.833, New York State Agricultural Experimentation Station (Geneva) Bull. No. 689, 1930.

Khemakhem, B., Ali, M.B., Aghajari, N., Juy, M., Haser, R., and Bejar, S. (2009). The importance of an extra loop in the B-domain of an alpha-amylase from *B. stearothermophilus* US100. Biochem. Biophys. Res. Commun. *385*, 78–83.

Kim, C.H., Nashiru, O., and Ko, J.H. (1996). Purification and biochemical characterization of pullulanase type I from *Thermus caldophilus* GK-24. FEMS Microbiol. Lett. *138*, 147–152.

Kim, H.K., Park, S.Y., Lee, J.K., and Oh, T.K. (1998a). Gene cloning and characterization of thermostable lipase from *Bacillus stearothermophilus* L1. Biosci. Biotechnol. Biochem. *62*, 66–71.

Kim, J.W., Flowers, L.O., Whiteley, M., and Peeples, T.L. (2001). Biochemical confirmation and characterization of the family-57-like alpha-amylase of *Methanococcus jannaschii*. Folia Microbiol. (Praha) *46*, 467–473.

Kim, M.S., Park, J.T., Kim, Y.W., Lee, H.S., Nyawira, R., Shin, H.S., Park, C.S., Yoo, S.H., Kim, Y.R., Moon, T.W., *et al.* (2004). Properties of a novel thermostable glucoamylase from the hyperthermophilic archaeon *Sulfolobus solfataricus* in relation to starch processing. Appl. Environ. Microbiol. *70*, 3933–3940.

Kim, S.B., Lee, W., and Ryu, Y.W. (2008). Cloning and characterization of thermostable esterase from *Archaeoglobus fulgidus*. J. Microbiol. *46*, 100–107.

Kim, Y.O., Lee, J.K., Kim, H.K., Yu, J.H., and Oh, T.K. (1998b). Cloning of the thermostable phytase gene (phy) from *Bacillus* sp. DS11 and its overexpression in *Escherichia coli*. FEMS Microbiol. Lett. *162*, 185–191.

Klingeberg, M., Hashwa, F., and Antranikian, G. (1991). Properties of extremely thermostable proteases from anaerobic hyperthermophilic bacteria. Appl. Microbiol. Biotechnol. *34*, 715–719.

Kluskens, L.D., Voorhorst, W.G., Siezen, R.J., Schwerdtfeger, R.M., Antranikian, G., van der Oost, J., and de Vos, W.M. (2002). Molecular characterization of fervidolysin, a subtilisin-like serine protease from the thermophilic bacterium *Fervidobacterium pennivorans*. Extremophiles *6*, 185–194.

Kluskens, L., Van Alebeek, G., Voragen, A., de VOS, W., and van der Oost, J. (2003). Molecular and biochemical characterization of the thermoactive family 1 pectate lyase from the hyperthermophilic bacterium *Thermotoga maritima*. Biochem. J. *370*, 651–659.

Kluskens, L.D., van Alebeek, G.J., Walther, J., Voragen, A.G., de Vos, W.M., and van der Oost, J. (2005). Characterization and mode of action of an exopolygalacturonase from the hyperthermophilic bacterium *Thermotoga maritima*. FEBS J. *272*, 5464–5473.

Koch, R., Zablowski, P., Spreinat, A., and Antranikian, G. (1990). Extremely thermostable amylolytic enzyme from the archaebacterium *Pyrococcus furiosus*. FEMS Microbiol. Lett. *71*, 21–26.

Koch, R., Spreinat, A., Lemke, K., and Antranikian, G. (1991). Purification and properties of a hyperthermoactive a-amylase from the archaeobacterium *Pyrococcus woesei*. Arch. Microbiol. *155*, 572–578.

Koga, Y., Tanaka, S.I., Sakudo, A., Tobiume, M., Aranishi, M., Hirata, A., Takano, K., Ikuta, K., and Kanaya, S. (2013). Proteolysis of abnormal prion protein with a thermostable protease from *Thermococcus kodakarensis* KOD1. Appl. Microbiol. Biotechnol. *98*, 2113–2120.

Kostrewa, D., Wyss, M., D'Arcy, A., and van Loon, A.P. (1999). Crystal structure of *Aspergillus niger* pH 2.5 acid phosphatase at 2.4 Å resolution. J. Mol. Biol. *288*, 965–974.

Kozianowski, G., Canganella, F., Rainey, F.A., Hippe, H., and Antranikian, G. (1997). Purification and characterization of thermostable pectate-lyases from a newly isolated thermophilic bacterium, *Thermoanaerobacter italicus* sp. nov. Extremophiles *1*, 171–182.

Kriegshäuser, G., and Liebl, W. (2000). Pullulanase from the hyperthermophilic bacterium *Thermotoga maritima*: purification by beta-cyclodextrin affinity chromatography. J. Chromatogr. B Biomed. Sci. Appl. *737*, 245–251.

Kubo, M., and Imanaka, T. (1988). Cloning and nucleotide sequence of the highly thermostable neutral protease gene from *Bacillus stearothermophilus*. J. Gen. Microbiol. *134*, 1883–1892.

Kubo, M., Murayama, K., Seto, K., and Imanaka, T. (1988). Highly thermostable neutral protease from *Bacillus stearothermophilus*. J. Ferment. Technol. *66*, 13–17.

Kumar, P., and Satyanarayana, T. (2009). Microbial glucoamylases: characteristics and applications. Crit. Rev. Biotechnol. *29*, 225–255.

Kumar, S., Tsai, C.J., and Nussinov, R. (2000). Factors enhancing protein thermostability. Protein Eng. *13*, 179–191.

Kumar, S., Pahujani, S., Ola, R.P., Kanwar, S.S., and Gupta, R. (2006). Enhanced thermostability of silica-immobilized lipase from Bacillus coagulans BTS-3 and synthesis of ethyl propionate. Acta Microbiol. Immunol. Hung. *53*, 219–231.

Kumari, A., Singh, K., and Kayastha, A.M. (2012). Amylase: general properties, mechanism and biotechnological applications – a review. Curr. Biotechnol. *1*, 98–107.

Kushner, D.J. (1978). Microbial life in extreme environments (Academic Press London, New York, San Francisco).

Ladenstein, R., and Antranikian, G. (1998). Proteins from hyperthermophiles: stability and enzymatic catalysis close to the boiling point of water. Adv. Biochem. Eng. Biotechnol. *61*, 37–85.

Laderman, K.A., Asada, K., Uemori, T., Mukai, H., Taguchi, Y., Kato, I., and Anfinsen, C.B. (1993). Alpha-amylase from the hyperthermophilic archaebacterium *Pyrococcus furiosus*. Cloning and sequencing of the gene and expression in *Escherichia coli*. J. Biol. Chem. *268*, 24402–24407.

Lambrechts, C., Boze, H., Moulin, G., and Galzy, P. (1992). Utilization of phytate by some yeasts. Biotechnol. Lett. *14*, 61–66.

Latiffi, A.A., Salleh, A.B., Rahman, R.N., Oslan, S.N., and Basri, M. (2013). Secretory expression of thermostable alkaline protease from *Bacillus stearothermophilus* FI by using native signal peptide and alpha-factor secretion signal in *Pichia pastoris*. Genes Genet. Syst. *88*, 85–91.

Lee, D., Koh, Y., Kim, K., Kim, B., Choi, H., Kim, D., Suhartono, M.T., and Pyun, Y. (1999). Isolation and characterization of a thermophilic lipase from bacillus thermoleovorans ID-1. FEMS Microbiol. Lett. *179*, 393–400.

Lee, M.H., Yang, S.J., Kim, J.W., Lee, H.S., Kim, J.W., and Park, K.H. (2007). Characterization of a thermostable cyclodextrin glucanotransferase from *Pyrococcus furiosus* DSM3638. Extremophiles *11*, 537–541.

Lee, S.H., Minagawa, E., Taguchi, H., Matsuzawa, H., Ohta, T., Kaminogawa, S., and Yamauchi, K. (1992). Purification and characterization of a thermostable carboxypeptidase (carboxypeptidase Taq) from *Thermus aquaticus* YT-1. Biosci. Biotechnol. Biochem. *56*, 1839–1844.

Lee, Y.C., Miyata, Y., Terada, I., Ohta, T., and Matsuzawa, H. (1991). Involvement of NH2-terminal pro-sequence in the production of active aqualysin I (a thermophilic serine protease) in *Escherichia coli*. Agric. Biol. Chem. *55*, 3027–3032.

Lee, Y.C., Koike, H., Taguchi, H., Ohta, T., and Matsuzawa, H. (1994). Requirement of a COOH-terminal pro-sequence for the extracellular secretion of aqualysin I (a thermophilic subtilisin-type protease) in *Thermus thermophilus*. FEMS Microbiol. Lett. *120*, 69–74.

Lei, X.G., Weaver, J.D., Mullaney, E., Ullah, A.H., and Azain, M.J. (2013). Phytase, a New Life for an "Old" Enzyme. Annu Rev Anim Biosci *1*, 283–309.

Leuschner, C., and Antranikian, G. (1995). Heat-stable enzymes from extremely thermophilic and hyperthermophilic microorganisms. World J. Microb. Biot. *11*, 95–114.

Leveque, E., Janecek, S., Haye, B., and Belarbi, A. (2000). Thermophilic archael amylolytic enzymes. Enzyme. Microb. Technol. *26*, 3–14.

Levisson, M., van der Oost, J., and Kengen, S.W. (2007). Characterization and structural modeling of a new type of thermostable esterase from *Thermotoga maritima*. FEBS J. *274*, 2832–2842.

Levisson, M., Sun, L., Hendriks, S., Swinkels, P., Akveld, T., Bultema, J.B., Barendregt, A., van den Heuvel, R.H., Dijkstra, B.W., van der Oost, J., *et al.* (2009a). Crystal structure and biochemical properties of a novel thermostable esterase containing an immunoglobulin-like domain. J. Mol. Biol. *385*, 949–962.

Levisson, M., van der Oost, J., and Kengen, S.W. (2009b). Carboxylic ester hydrolases from hyperthermophiles. Extremophiles *13*, 567–581.

Li, M., Yang, L.R., Xu, G., and Wu, J.P. (2013). Screening, purification and characterization of a novel cold-active and organic solvent-tolerant lipase from *Stenotrophomonas maltophilia* CGMCC 4254. Bioresour. Technol. *148*, 114–120.

Li, W.F., Zhou, X.X., and Lu, P. (2005). Structural features of thermozymes. Biotechnol. Adv. *23*, 271–281.

Lim, D., Golovan, S., Forsberg, C.W., and Jia, Z. (2000). Crystal structures of *Escherichia coli* phytase and its complex with phytate. Nat. Struct. Mol. Biol. *7*, 108–113.

Lin, X.-l., and Tang, J. (1990). Purification, characterization, and gene cloning of thermopsin, a thermostable acid protease from *Sulfolobus acidocaldarius*. J. Biol. Chem. *265*, 1490–1495.

Linden, A., Mayans, O., Meyer-Klaucke, W., Antranikian, G., and Wilmanns, M. (2003). Differential regulation of a hyperthermophilic alpha-amylase with a novel (Ca,Zn) two-metal center by zinc. J. Biol. Chem. *278*, 9875–9884.

Linden, A., and Wilmanns, M. (2004). Adaptation of class-13 alpha-amylases to diverse living conditions. Chembiochem *5*, 231–239.

Loewus, F.A., and Murthy, P.P. (2000). *myo*-Inositol metabolism in plants. Plant Sci. *150*, 1–19.

Maciver, B., McHale, R.H., Saul, D.J., and Bergquist, P.L. (1994). Cloning and sequencing of a serine proteinase gene from a thermophilic *Bacillus* species and its expression in *Escherichia coli*. Appl. Environ. Microbiol. *60*, 3981–3988.

Madigan, M.T., and Marrs, B.L. (1997). Extremophiles. Sci. Am. *276*, 82–87.

Mahadevan, G.D., and Neelagund, S.E. (2013). Thermostable lipase from *Geobacillus* sp. Iso5: Bioseparation, characterization and native structural studies. J. Basic Microbiol. *54*, 386–396.

Manco, G., Carrea, G., Giosue, E., Ottolina, G., Adamo, G., and Rossi, M. (2002). Modification of the enantioselectivity of two homologous thermophilic carboxylesterases from *Alicyclobacillus acidocaldarius* and *Archaeoglobus fulgidus* by random mutagenesis and screening. Extremophiles *6*, 325–331.

Mantle, D., Lauffart, B., and Gibson, A. (1991). Purification and characterization of leucyl aminopeptidase and pyroglutamyl aminopeptidase from human skeletal muscle. Clin. Chim. Acta *197*, 35–45.

Marin-Navarro, J., and Polaina, J. (2010). Glucoamylases: structural and biotechnological aspects. Appl. Microbiol. Biotechnol. *89*, 1267–1273.

Markland, F.S., and Smith, E.L. (1967). Subtilisin BPN. VII. Isolation of cyanogen bromide peptides and the complete amino acid sequence. J. Biol. Chem. *242*, 5198–5211.

Martinez-Martinez, M., Alcaide, M., Tchigvintsev, A., Reva, O., Polaina, J., Bargiela, R., Guazzaroni, M.E., Chicote, A., Canet, A., Valero, F., *et al.* (2013). Biochemical diversity of carboxyl esterases and lipases from Lake Arreo (Spain): a metagenomic approach. Appl. Environ. Microbiol. *79*, 3553–3562.

Matsuzawa, H., Tokugawa, K., Hamaoki, M., Mizoguchi, M., Taguchi, H., Terada, I., Kwon, S.T., and Ohta, T. (1988). Purification and characterization of aqualysin I (a thermophilic alkaline serine protease) produced by *Thermus aquaticus* YT-1. Eur. J. Biochem. *171*, 441–447.

Matthews, B.W. (1993). Structural and genetic analysis of protein stability. Annu. Rev. Biochem. *62*, 139–160.

Matthews, B.W., Colman, P.M., Jansonius, J.N., Titani, K., Walsh, K.A., and Neurath, H. (1972a). Structure of thermolysin. Nature: New Biol. *238*, 41–43.

Matthews, B.W., Jansonius, J.N., Colman, P.M., Schoenborn, B.P., and Dupourque, D. (1972b). Three-dimensional structure of thermolysin. Nature: New Biol. *238*, 37–41.

Matthews, B.W., Weaver, L.H., and Kester, W.R. (1974). The conformation of thermolysin. J. Biol. Chem. *249*, 8030–8044.

Maurer, K.H. (2004). Detergent proteases. Curr. Opin. Biotechnol. *15*, 330–334.

Mehraj Pasha, K., Anuradha, P., and Subbarao, D. (2013). Applications of pectinases in industrial sector. Int. J. Pure Appli. Sci. Technol. *16*, 89–95.

Michels, P.C., and Clark, D.S. (1997). Pressure-enhanced activity and stability of a hyperthermophilic protease from a deep-sea methanogen. Appl. Environ. Microbiol. *63*, 3985–3991.

Minagawa, E., Kaminogawa, S., Matsuzawa, H., Ohta, T., and Yamauchi, K. (1988). Isolation and characterization of a thermostable aminopeptidase (Aminopeptidase T) from *Thermus aquaticus* YT-1, an extremely thermophilic bacterium (Biological Chemistry). Agric. Biol. Chem. *52*, 1755–1763.

Mittal, A., Singh, G., Goyal, V., Yadav, A., and Aggarwal, N.K. (2012). Production of phytase by acido-thermophilic strain of *Klebsiella* sp. DB-3FJ711774. 1 using orange peel flour under submerged fermentation. Inno Roma Food Biotechnol. *10*, 18–27.

Morana, A., Moracci, M., Ottombrino, A., Ciaramella, M., Rossi, M., and De Rosa, M. (1995). Industrial-scale production and rapid purification of an archaeal beta-glycosidase expressed in *Saccharomyces cerevisiae*. Biotechnol. Appl. Biochem. *22*, 261–268.

Morana, A., Di Prizito, N., Aurilia, V., Rossi, M., and Cannio, R. (2002). A carboxylesterase from the hyperthermophilic archaeon *Sulfolobus solfataricus*: cloning of the gene, characterization of the protein. Gene *283*, 107–115.

Morihara, K. (1974). Comparative specificity of microbial proteinases. Adv. Enzymol. R.A.M.B. *41*, 179–243.

Morikawa, M., Izawa, Y., Rashid, N., Hoaki, T., and Imanaka, T. (1994). Purification and characterization of a thermostable thiol protease from a newly isolated hyperthermophilic *Pyrococcus* sp. Appl. Environ. Microbiol. *60*, 4559–4566.

Moshfegh, M., Shahverdi, A.R., Zarrini, G., and Faramarzi, M.A. (2013). Biochemical characterization of an extracellular polyextremophilic alpha-amylase from the halophilic archaeon *Halorubrum xinjiangense*. Extremophiles *17*, 677–687.

Mullaney, E.J., and Ullah, A.H. (2003). The term phytase comprises several different classes of enzymes. Biochem. Biophys. Res. Comm. *312*, 179–184.

Mullaney, E.J., Daly, C.B., Sethumadhavan, K., Rodriquez, E., Lei, X.G., and Ullah, A.H. (2000). Phytase activity in *Aspergillus fumigatus* isolates. Biochem. Biophys. Res. Commun. *275*, 759–763.

Murao, S., Katsura, M., Fukuhara, K.-I., and Oda, K. (1980). New metallo proteinase inhibitor (MK-I) produced by *Streptomyces mozunensis* MK-23. Agric. Biol. Chem. *44*.

Murao, S., Ohkuni, K., Nagao, M., Hirayama, K., Fukuhara, K., Oda, K., Oyama, H., and Shin, T. (1993). Purification and characterization of kumamolysin, a novel thermostable pepstatin-insensitive carboxyl proteinase from *Bacillus* novosp. MN-32. J. Biol. Chem. *268*, 349–355.

Nadeem, M., Qazi, J.I., Syed, Q., and Gulsher, M. (2013). Purification and characterization of an alkaline protease from *Bacillus licheniformis* UV-9 for detergent formulations. Songklanakarin J. Sci. Technol. *35*, 187–195.

Nagai, Y., and Funahashi, S. (1962). Phytase (myo-inositol-hexaphosphate phosphohydrolase) from wheat bran. Part I. Purification and substrate specificity. Agric. Biol. Chem. *26*, 794–803.

Nagpure, A., Choudhary, B., and Gupta, R.K. (2013). Chitinases: in agriculture and human healthcare. Crit. Rev. Biotechnol. *34*, 215–232.

Nakamura, Y., Fukuhara, H., and Sano, K. (2000). Secreted phytase activities of yeasts. Biosci. Biotechnol. Biochem. *64*, 841–844.

Nakanishi, K., Takeuchi, A., and Matsuno, R. (1990). Long-term continuous synthesis of aspartame precursor in a column reactor with an immobilized thermolysin. Appl. Microbiol. Biotechnol. *32*, 633–636.

Niehaus, F., Bertoldo, C., Kahler, M., and Antranikian, G. (1999). Extremophiles as a source of novel enzymes for industrial application. Appl. Microbiol. Biotechnol. *51*, 711–729.

Nisha, M., and Satyanarayana, T. (2013). Recombinant bacterial amylopullulanases: developments and perspectives. Bioengineered 4, 388–400.

Nwe, N., and Stevens, W.F. (2002). Production of fungal chitosan by solid substrate fermentation followed by enzymatic extraction. Biotechnol. Lett. *24*, 131–134.

Oda, K., Nakazima, T., Terashita, T., Suzuki, K.-i., and Murao, S. (1987a). Purification and Properties of an S-PI (Pepstatin Ac)-insensitive Carboxyl Proteinase from a *Xanthomonas* sp. Bacterium (Biological Chemistry). Agric. Biol. Chem. *51*, 3073–3080.

Oda, K., Sugitani, M., Fukuhara, K., and Murao, S. (1987b). Purification and properties of a pepstatin-insensitive carboxyl proteinase from a Gram-negative bacterium. Biochim. Biophys. Acta *923*, 463–469.

Ohta, Y., Tanabe, S., Hachimori, A., and Samejima, T. (1974). Changes of thermostability and activity of thermolysin on nitration with tetranitromethane. J. Biochem. *76*, 871–878.

Oku, T., and Ishikawa, K. (2006). Analysis of the hyperthermophilic chitinase from *Pyrococcus furiosus*: activity toward crystalline chitin. Biosci. Biotechnol. Biochem. *70*, 1696–1701.

Onoe, S., Ando, S., Ataka, M., and Ishikawa, K. (2002). Active site of deblocking aminopeptidase from *Pyrococcus horikoshii*. Biochem. Biophys. Res. Commun. *290*, 994–997.

Pang, H., Zhang, P., Duan, C.J., Mo, X.C., Tang, J.L., and Feng, J.X. (2009). Identification of cellulase genes from the metagenomes of compost soils and functional characterization of one novel endoglucanase. Curr. Microbiol. *58*, 404–408.

Park, Y.J., Choi, S.Y., and Lee, H.B. (2006). A carboxylesterase from the thermoacidophilic archaeon *Sulfolobus solfataricus* P1; purification, characterization, and expression. Biochim. Biophys. Acta *1760*, 820–828.

Pasamontes, L., Haiker, M., Wyss, M., Tessier, M., and Van Loon, A. (1997). Gene cloning, purification, and characterization of a heat-stable phytase from the fungus *Aspergillus fumigatus*. Appl. Environ. Microbiol. *63*, 1696–1700.

Pedrolli, D.B., Monteiro, A.C., Gomes, E., and Carmona, E.C. (2009). Pectin and pectinases: production, characterization and industrial application of microbial pectinolytic enzymes. Open Biotechnol. J. *3*, 9–18.

Peek, K., Veitch, D.P., Prescott, M., Daniel, R.M., MacIver, B., and Bergquist, P.L. (1993). Some characteristics of a proteinase from a thermophilic *Bacillus* sp. expressed in *Escherichia coli*: comparison with the native enzyme and its processing in *E. coli* and *in vitro*. Appl. Environ. Microbiol. *59*, 1168–1175.

Petersen, E.I., Valinger, G., Solkner, B., Stubenrauch, G., and Schwab, H. (2001). A novel esterase from *Burkholderia gladioli* which shows high deacetylation activity on cephalosporins is related to beta-lactamases and DD-peptidases. J. Biotechnol. *89*, 11–25.

Piller, K., Daniel, R.M., and Petach, H.H. (1996). Properties and stabilization of an extracellular alpha-glucosidase from the extremely thermophilic archaebacteria *Thermococcus* strain AN1: enzyme activity at 130 degrees C. Biochim. Biophys. Acta *1292*, 197–205.

du Plessis, E.M., Berger, E., Stark, T., Louw, M.E., and Visser, D. (2010). Characterization of a novel thermostable esterase from *Thermus scotoductus* SA-01: evidence of a new family of lipolytic esterases. Curr. Microbiol. *60*, 248–253.

Prakash, O., and Jaiswal, N. (2010). alpha-Amylase: an ideal representative of thermostable enzymes. Appl. Biochem. Biotechnol. *160*, 2401–2414.

Prasanna, L., Eijsink, V.H., Meadow, R., and Gåseidnes, S. (2013). A novel strain of *Brevibacillus laterosporus* produces chitinases that contribute to its biocontrol potential. Appl. Microbiol. Biotechnol. *97*, 1601–1611.

Prescott, M., Peek, K., and Daniel, R.M. (1995). Characterisation of a thermostable pepstatin-insensitive acid proteinase from a *Bacillus* sp. Int. J. Biochem. Cell Biol. *27*, 729–739.

Rao, M.B., Tanksale, A.M., Ghatge, M.S., and Deshpande, V.V. (1998). Molecular and biotechnological aspects of microbial proteases. Microbiol. Mol. Biol. Rev. *62*, 597–635.

Rashid, N., Cornista, J., Ezaki, S., Fukui, T., Atomi, H., and Imanaka, T. (2002). Characterization of an archaeal cyclodextrin glucanotransferase with a novel C-terminal domain. J. Bacteriol. *184*, 777–784.

Ravi Kumar, M.N. (2000). A review of chitin and chitosan applications. React. Funct. Polym. *46*, 1–27.

Rawlings, N.D., Waller, M., Barrett, A.J., and Bateman, A. (2014). MEROPS: the database of proteolytic enzymes, their substrates and inhibitors. Nucleic Acids Res. *42*, D503–D509.

Reddy, N.R., Sathe, S.K., and Salunkhe, D.K. (1982). Phytates in legumes and cereals. Adv. Food Res. *28*, 1–92.

Rehse, P.H., Ohshima, N., Nodake, Y., and Tahirov, T.H. (2004). Crystallographic structure and biochemical analysis of the *Thermus thermophilus* osmotically inducible protein C. J. Mol. Biol. *338*, 959–968.

Reid, I.I., and Ricard, M. (2000). Pectinase in papermaking: solving retention problems in mechanical pulps bleached with hydrogen peroxide. Enzyme. Microb. Technol. *26*, 115–123.

Rhee, J.K., Ahn, D.G., Kim, Y.G., and Oh, J.W. (2005). New thermophilic and thermostable esterase with sequence similarity to the hormone-sensitive lipase family, cloned from a metagenomic library. Appl. Environ. Microbiol. *71*, 817–825.

Rich, D.H. (1985). Pepstatin-derived inhibitors of aspartic proteinases. A close look at an apparent transition-state analog inhibitor. J. Med. Chem. *28*, 263–273.

Roberts, G. (1992). Physical structures. Chitin Chemistry, 1st edition. (Macmillan, London), pp. 20–35.

Roche, R.S., and Voordouw, G. (1978). The structural and functional roles of metal ions in thermolysin. CRC Crit. Rev. Biochem. *5*, 1–23.

Rolfsmeier, M., Haseltine, C., Bini, E., Clark, A., and Blum, P. (1998). Molecular characterization of the alpha-glucosidase gene (*malA*) from the hyperthermophilic archaeon *Sulfolobus solfataricus*. J. Bacteriol. *180*, 1287–1295.

Roncari, G., Stoll, E., and Zuber, H. (1976). Thermophilic aminopeptidase I. Methods Enzymol. *45*, 522–530.

Royter, M., Schmidt, M., Elend, C., Hobenreich, H., Schafer, T., Bornscheuer, U.T., and Antranikian, G. (2009). Thermostable lipases from the extreme thermophilic anaerobic bacteria *Thermoanaerobacter thermohydrosulfuricus* SOL1 and *Caldanaerobacter subterraneus* subsp. tengcongensis. Extremophiles *13*, 769–783.

Rüdiger, A., Jorgensen, P.L., and Antranikian, G. (1995). Isolation and characterization of a heat-stable pullulanase from the hyperthermophilic archaeon *Pyrococcus woesei* after cloning and expression of its gene in *Escherichia coli*. Appl. Environ. Microbiol. *61*, 567–575.

Sakai, K., Yokota, A., Kurokawa, H., Wakayama, M., and Moriguchi, M. (1998). Purification and characterization of three thermostable endochitinases of a noble *Bacillus* strain, MH-1, isolated from chitin-containing compost. Appl. Environ. Microbiol. *64*, 3397–3402.

Sakai, T., Sakamoto, T., Hallaert, J., and Vandamme, E.J. (1993). Pectin, pectinase and protopectinase: production, properties, and applications. Adv. Appl. Microbiol. *39*, 213–294.

Salameh, M.A., and Wiegel, J. (2007). Purification and characterization of two highly thermophilic alkaline lipases from *Thermosyntropha lipolytica*. Appl. Environ. Microbiol. *73*, 7725–7731.

Sanchez-Otero, M.G., Valerio-Alfaro, G., Garcia-Galindo, H.S., and Oliart-Ros, R.M. (2008). Immobilization in the presence of Triton X-100: modifications in activity and thermostability of *Geobacillus thermoleovorans* CCR11 lipase. J. Ind. Microbiol. Biotechnol. *35*, 1687–1693.

Schenk, G., Korsinczky, M.L., Hume, D.A., Hamilton, S., and DeJersey, J. (2000). Purple acid phosphatases from bacteria: similarities to mammalian and plant enzymes. Gene *255*, 419–424.

Schmidt-Dannert, C., Sztajer, H., Stocklein, W., Menge, U., and Schmid, R.D. (1994). Screening, purification and properties of a thermophilic lipase from *Bacillus thermocatenulatus*. Biochim. Biophys. Acta *1214*, 43–53.

Schwerdtfeger, R.M., Chiaraluce, R., Consalvi, V., Scandurra, R., and Antranikian, G. (1999). Stability, refolding and Ca2+ binding of pullulanase from the hyperthermophilic archaeon *Pyrococcus woesei*. Eur. J. Biochem. *264*, 479–487.

Schwermann, B., Pfau, K., Liliensiek, B., Schleyer, M., Fischer, T., and Bakker, E.P. (1994). Purification, properties and structural aspects of a thermoacidophilic alpha-amylase from *Alicyclobacillus acidocaldarius* atcc 27009. Insight into acidostability of proteins. Eur. J. Biochem. *226*, 981–991.

Serour, E., and Antranikian, G. (2002). Novel thermoactive glucoamylases from the thermoacidophilic Archaea *Thermoplasma acidophilum*, *Picrophilus torridus* and *Picrophilus oshimae*. Anton. Leeuw. *81*, 73–83.

Shahidi, F., Arachchi, J.K.V., and Jeon, Y.-J. (1999). Food applications of chitin and chitosans. Trends Food Sci. Technol. *10*, 37–51.

Sharma, N., Rathore, M., and Sharma, M. (2013a). Microbial pectinase: sources, characterization and applications. Rev. Environ. Sci. Biotechnol. *12*, 45–60.

Sharma, R., Thakur, V., Sharma, M., and Birkeland, N.-K. (2013b). Biocatalysis through thermostable lipases: adding flavor to chemistry. In Thermophilic Microbes in Environmental and Industrial Biotechnology: Biotechnology of Thermophiles, Satyanarayana, T., Littlechild, J., and Kawarabayasi, Y., eds. (Dordrecht, Springer), pp. 905–927.

Shin, S., Ha, N.-C., Oh, B.-C., Oh, T.-K., and Oh, B.-H. (2001). Enzyme mechanism and catalytic property of β propeller phytase. Structure *9*, 851–858.

Silva, C.R.d., Delatorre, A.B., and Martins, M.L.L. (2007). Effect of the culture conditions on the production of an extracellular protease by thermophilic *Bacillus* sp and some properties of the enzymatic activity. Braz. J. Microbiol. *38*, 253–258.

Singleton, M.R., and Littlechild, J.A. (2001). Pyrrolidone carboxylpeptidase from *Thermococcus litoralis*. Methods Enzymol. *330*, 394–403.

Singleton, M.R., Taylor, S.J., Parrat, J.S., and Littlechild, J.A. (2000). Cloning, expression, and characterization of pyrrolidone carboxyl peptidase from the archaeon *Thermococcus litoralis*. Extremophiles *4*, 297–303.

Sobek, H., and Gorisch, H. (1988). Purification and characterization of a heat-stable esterase from the thermoacidophilic archaebacterium *Sulfolobus acidocaldarius*. Biochem. J. *250*, 453–458.

Song, C., Sheng, L., and Zhang, X. (2013). Immobilization and characterization of a thermostable lipase. Mar. Biotechnol. (NY) *15*, 659–667.

Sonnleitner, B. (1983). Biotechnology of thermophilic bacteria—growth, products, and application. In Microbial Activities (Springer, Berlin, Heidelberg), pp. 69–138.

Sørensen, H.P., Sperling-Petersen, H.U., and Mortensen, K.K. (2003). Production of recombinant thermostable proteins expressed in *Escherichia coli*: completion of protein synthesis is the bottleneck. J. Chromatogr. B *786*, 207–214.

Sørensen, S.O., Pauly, M., Bush, M., Skjøt, M., McCann, M.C., Borkhardt, B., and Ulvskov, P. (2000). Pectin engineering: modification of potato pectin by *in vivo* expression of an endo-1, 4-β-D-galactanase. Proc. Natl. Acad. Sci. U.S.A. *97*, 7639–7644.

Spindler, K.D., Spindler-Barth, M., and Londershausen, M. (1990). Chitin metabolism: a target for drugs against parasites. Parasitol. Res. *76*, 283–288.

Staufenberger, T., Imhoff, J.F., and Labes, A. (2012). First crenarchaeal chitinase found in *Sulfolobus tokodaii*. Microbiol. Res. *167*, 262–269.

Stetter, K.O. (1996). Hyperthermophiles in the history of life. Ciba Foundation symposium *202*, 1–10; discussion 11–18.

Stratilova, E., Dzurova, M., Markovic, O., and Jornvall, H. (1996). An essential tyrosine residue of *Aspergillus polygalacturonase*. FEBS Lett. *382*, 164–166.

Suneetha, V., and Prathyusha, K. (2011). Bacterial pectinases and their potent biotechnological application in fruit processing/juice production industry: a review. J. Phytol. 3, 16–19.

Suzuki, Y., Hatagaki, K., and Oda, H. (1991). A hyperthermostable pullulanase produced by an extreme thermophile, *Bacillus flavocaldarius* KP 1228, and evidence for the proline theory of increasing protein thermostability. Appl. Microbiol. Biotechnol. *34*, 707–714.

Szewczuk, A., and Kwiatkowska, J. (1970). Pyrrolidonyl peptidase in animal, plant and human tissues. Occurrence and some properties of the enzyme. Eur. J. Biochem. *15*, 92–96.

Szewczuk, A., and Mulczyk, M. (1969). Pyrrolidonyl peptidase in bacteria. The enzyme from *Bacillus subtilis*. Eur. J. Biochem. *8*, 63–67.

Szilagyi, A., and Zavodszky, P. (2000). Structural differences between mesophilic, moderately thermophilic and extremely thermophilic protein subunits: results of a comprehensive survey. Structure *8*, 493–504.

Tachibana, Y., Kuramura, A., Shirasaka, N., Suzuki, Y., Yamamoto, T., Fujiwara, S., Takagi, M., and Imanaka, T. (1999). Purification and characterization of an extremely thermostable cyclomaltodextrin glucanotransferase from a newly isolated hyperthermophilic archaeon, a *Thermococcus* sp. Appl. Environ. Microbiol. *65*, 1991–1997.

Takagi, H., Takahashi, T., Momose, H., Inouye, M., Maeda, Y., Matsuzawa, H., and Ohta, T. (1990). Enhancement of the thermostability of subtilisin E by introduction of a disulfide bond engineered on the basis of structural comparison with a thermophilic serine protease. J. Biol. Chem. *265*, 6874–6878.

Takai, K., Nakamura, K., Toki, T., Tsunogai, U., Miyazaki, M., Miyazaki, J., Hirayama, H., Nakagawa, S., Nunoura, T., and Horikoshi, K. (2008). Cell proliferation at 122 degrees C and isotopically heavy CH_4 production by a hyperthermophilic methanogen under high-pressure cultivation. Proc. Natl. Acad. Sci. U.S.A. *105*, 10949–10954.

Takayanagi, T., Ajisaka, K., Takiguchi, Y., and Shimahara, K. (1991). Isolation and characterization of thermostable chitinases from *Bacillus licheniformis* X-7u. Biochim. Biophys. Acta *1078*, 404–410.

Takii, Y., Taguchi, H., Shimoto, H., and Suzuki, Y. (1987). *Bacillus stearothermophilus* KP 1236 neutral protease with strong thermostability comparable to thermolysin. Appl. Microbiol. Biotechnol. *27*, 186–191.

Tanaka, T., Fujiwara, S., Nishikori, S., Fukui, T., Takagi, M., and Imanaka, T. (1999). A unique chitinase with dual active sites and triple substrate binding sites from the hyperthermophilic archaeon *Pyrococcus kodakaraensis* KOD1. Appl. Environ. Microbiol. *65*, 5338–5344.

Tanaka, T., Fukui, T., Fujiwara, S., Atomi, H., and Imanaka, T. (2004). Concerted action of diacetylchitobiose deacetylase and exo-beta-D-glucosaminidase in a novel chitinolytic pathway in the hyperthermophilic archaeon *Thermococcus kodakaraensis* KOD1. J. Biol. Chem. *279*, 30021–30027.

Than, R., Feldmann, A.A., and Krebs, B. (1999). Structural and functional studies on model compounds of purple acid phosphatases and catechol oxidases. Coord. Chem. Rev. *182*, 211–241.

Toogood, H.S., Prescott, M., and Daniel, R.M. (1995). A pepstatin-insensitive aspartic proteinase from a thermophilic *Bacillus* sp. Biochem. J. *307 (Pt 3)*, 783–789.

Toplak, A., Wu, B., Fusetti, F., Quaedflieg, P.J., and Janssen, D.B. (2013). Proteolysin, a novel highly thermostable and cosolvent-compatible protease from the thermophilic bacterium *Coprothermobacter proteolyticus*. Appl. Environ. Microbiol. *79*, 5625–5632.

Tran, T.T., Mamo, G., Mattiasson, B., and Hatti-Kaul, R. (2010). A thermostable phytase from *Bacillus* sp. MD2: cloning, expression and high-level production in *Escherichia coli*. J. Ind. Microbiol. Biotechnol. *37*, 279–287.

Tsunasawa, S. (1998). Purification and application of a novel N-terminal deblocking aminopeptidase (DAP) from *Pyrococcus furiosus*. J. Protein Chem. *17*, 521–522.

Tsunasawa, S., Nakura, S., Tanigawa, T., and Kato, I. (1998). Pyrrolidone carboxyl peptidase from the hyperthermophilic Archaeon *Pyrococcus furiosus*: cloning and overexpression in Escherichia coli of the gene, and its application to protein sequence analysis. J. Biochem. *124*, 778–783.

Ullah, A., and Dischinger, H. (1993). *Aspergillus ficuum* Phytase: Complete Primary Structure Elucidation by Chemical Sequencing. Biochem. Biophys. Res. Comm. *192*, 747–753.

Ullah, A.H., and Gibson, D.M. (1987). Extracellular phytase (EC 3.1. 3.8) from *Aspergillus ficuum* NRRL 3135: purification and characterization. Prep. Biochem. *17*, 63–91.

Uma Maheswar Rao, J.L., and Satyanarayana, T. (2007). Purification and characterization of a hyperthermostable and high maltogenic alpha-amylase of an extreme thermophile *Geobacillus thermoleovorans*. Appl. Biochem. Biotechnol. *142*, 179–193.

Uotsu-Tomita, R., Tonozuka, T., Sakai, H., and Sakano, Y. (2001). Novel glucoamylase-type enzymes from *Thermoactinomyces vulgaris* and *Methanococcus jannaschii* whose genes are found in the flanking region of the alpha-amylase genes. Appl. Microbiol. Biotechnol. *56*, 465–473.

Valdes-Stauber, N., and Scherer, S. (1994). Isolation and characterization of Linocin M18, a bacteriocin produced by *Brevibacterium linens*. Appl. Environ. Microbiol. *60*, 3809–3814.

Van Eldik, R., Asano, T., and Le Noble, W. (1989). Activation and reaction volumes in solution. 2. Chem. Rev. *89*, 549–688.

Vieille, C., and Zeikus, G.J. (2001). Hyperthermophilic enzymes: sources, uses, and molecular mechanisms for thermostability. Microbiol. Mol. Biol. Rev. *65*, 1–43.

Viikari, L., Tenkanen, M., and Suurnäkki, A. (2001). Biotechnology in the pulp and paper industry. In Biotechnology Set, 2nd edn, Rehm, H-J., and Reed, G., eds. (Wiley-VCH Verlag GmbH, Weinheim, Germany), pp. 523–546.

Vohra, A., and Satyanarayana, T. (2001). Phytase production by the yeast, *Pichia anomala*. Biotechnol. Lett. *23*, 551–554.

Vohra, A., and Satyanarayana, T. (2003). Phytases: microbial sources, production, purification, and potential biotechnological applications. Crit. Rev. Biotechnol. *23*, 29–60.

Voorhorst, W.G., Eggen, R.I., Geerling, A.C., Platteeuw, C., Siezen, R.J., and de Vos, W.M. (1996). Isolation and characterization of the hyperthermostable serine protease, pyrolysin, and its gene from the hyperthermophilic archaeon *Pyrococcus furiosus*. J. Biol. Chem. *271*, 20426–20431.

Wang, S.L., Shih, I.L., Liang, T.W., and Wang, C.H. (2002). Purification and characterization of two antifungal chitinases extracellularly produced by *Bacillus amyloliquefaciens* V656 in a shrimp and crab shell powder medium. J. Agr. Food Chem. *50*, 2241–2248.

Wang, S., Lu, Z., Lu, M., Qin, S., Liu, H., Deng, X., Lin, Q., and Chen, J. (2008). Identification of archaeon-producing hyperthermophilic alpha-amylase and characterization of the alpha-amylase. Appl. Microbiol. Biotechnol. *80*, 605–614.

Warrilow, A.G., Turner, R.J., and Jones, M.G. (1994). A novel form of pectinesterase in tomato. Phytochemistry *35*, 863–868.

Watanabe, H., Nishimoto, T., Kubota, M., Chaen, H., and Fukuda, S. (2006). Cloning, sequencing, and expression of the genes encoding an isocyclomaltooligosaccharide glucanotransferase and an alpha-amylase from a *Bacillus circulans* strain. Biosci. Biotechnol. Biochem. *70*, 2690–2702.

Wei, Y., Schottel, J.L., Derewenda, U., Swenson, L., Patkar, S., and Derewenda, Z.S. (1995). A novel variant of the catalytic triad in the *Streptomyces scabies* esterase. Nat. Struct. Biol. *2*, 218–223.

Wilson, S.A., Young, O.A., Coolbear, T., and Daniel, R.M. (1992). The use of proteases from extreme thermophiles for meat tenderisation. Meat Sci. *32*, 93–103.

Wilson, S.A., Daniel, R.M., and Peek, K. (1994a). Peptide synthesis with a proteinase from the extremely thermophilic organism *Thermus* Rt41A. Biotechnol. Bioeng. *44*, 337–346.

Wilson, S.A., Peek, K., and Daniel, R.M. (1994b). Immobilization of a proteinase from the extremely thermophilic organism *Thermus* Rt41A. Biotechnol. Bioeng. *43*, 225–231.

Wu, S.J., Kim, J.M., Zhou, C., Jin, Z.Y., and Tong, Q.Y. (2010). Estimation of pullulan by hydrolysis with pullulanase. Biotechnol. Lett. *32*, 1143–1145.

Yang, C.H., Lin, K.I., Chen, G.H., Chen, Y.F., Chen, C.Y., Chen, W.L., and Huang, Y.C. (2010). Constitutive Expression of *Thermobifida fusca* Thermostable Acetylxylan Esterase Gene in *Pichia pastoris*. Int. J. Mol. Sci. *11*, 5143–5151.

Yang, H., Li, J., Shin, H.D., Du, G., Liu, L., and Chen, J. (2013). Molecular engineering of industrial enzymes: recent advances and future prospects. Appl. Microbiol. Biotechnol. *98*, 23–29.

Yuli, P.E., Suhartono, M.T., Rukayadi, Y., Hwang, J.K., and Pyun, Y.R. (2004). Characteristics of thermostable chitinase enzymes from the indonesian *Bacillus* sp.13.26. Enzyme Microb. Tech. *35*, 147–153.

Zeikus, J.G., Vieille, C., and Savchenko, A. (1998). Thermozymes: biotechnology and structure–function relationships. Extremophiles *2*, 179–183.

Zentgraf, B., and Ahern, T. (1991). Practical importance of enzyme stability. Pure Appl. Chem. *63*, 1527–1540.

Zheng, J., Liu, C., Liu, L., and Jin, Q. (2013). Characterisation of a thermo-alkali-stable lipase from oil-contaminated soil using a metagenomic approach. Syst. Appli. Microbiol. *36*, 197–204.

Zheng, Y., Xue, Y., Zhang, Y., Zhou, C., Schwaneberg, U., and Ma, Y. (2010). Cloning, expression, and characterization of a thermostable glucoamylase from *Thermoanaerobacter tengcongensis* MB4. Appl. Microbiol. Biotechnol. *87*, 225–233.

Lignocellulosic Biomass Deconstruction by the Extremely Thermophilic Genus *Caldicellulosiruptor*

4

Jonathan M. Conway, Jeffrey V. Zurawski, Laura L. Lee, Sara E. Blumer-Schuette and Robert M. Kelly

Abstract

Extremely thermophilic bacteria from the genus *Caldicellulosiruptor* produce a variety of large multidomain proteins for the attachment to and degradation of lignocellulosic biomass. The currently sequenced genomes from the genus *Caldicellulosiruptor* encode 144 homologous groups of carbohydrate active enzymes (CAZymes) containing 128 glycoside hydrolase (GH), 58 carbohydrate binding module (CBM), 9 polysaccharide lyase (PL), and 11 carbohydrate esterase (CE) domains. Extracellular CAZymes from these species have a distinct multidomain architecture, with multiple catalytic and binding domains in the same protein. This promotes synergy between catalytic domains and enhances the ability of these bacteria to degrade biomass. A number of these extracellular multidomain proteins are also cell surface associated via surface layer homology (SLH) domains. These SLH domain-containing proteins are thought to help these species attach to the biomass substrate, while catalytic domains in the same proteins degrade it. Because of their highly effective biomass degradation strategy, *Caldicellulosiruptor* species are of particular interest for development into consolidated bioprocessing microorganisms for the processing of lignocellulosic biomass to biofuels.

Introduction

Lignocellulosic biomass

Lignocellulosic biomass is abundant, inexpensive, and renewable, making it of interest as a feedstock for liquid biofuel production (Maki *et al.*, 2009). Lignocellulosic biomass can be obtained from a variety of sources, including agricultural wastes (maize, small grain, and sugarcane residues), industrial forestry product wastes (logging residues, tree thinning, sawdust, and paper pulp), and dedicated biofuel crops (switchgrass, miscanthus, sorghum, and poplar). Two recent studies project the total United States production capacity of lignocellulosic biomass priced less than $60 per dry ton to reach 550–600 million dry tons annually by the year 2020, without significant impact to the environment or food supply (National Research Council, 2009; US Department of Energy, 2011). This is projected to increase to about 770 million dry tons of production by the year 2030, supported largely by an annual 120 dry ton production increase in dedicated biofuel crops over those ten years (US Department of Energy, 2011). Similar analysis applies globally to regions where significant lignocellulose reserves are available.

Lignocellulosic biomass dry weight is comprised generally of about 35–50% cellulose, 20–35% hemicellulose, and 10–30% lignin (Pauly and Keegstra, 2008; Sun and Cheng, 2002). The composition of biomass varies significantly from species to species and within different tissues of a single species (Pauly and Keegstra, 2010). Fig. 4.1A shows the general makeup of the polysaccharides, cellulose and hemicellulose, in biomass. Cellulose is a homologous polymer comprised of long chains of β-1,4-linked glucose molecules. These long chains are able to pack parallel to one another into ordered arrays 3–5 nm thick, called crystalline microfibrils, which assemble into the

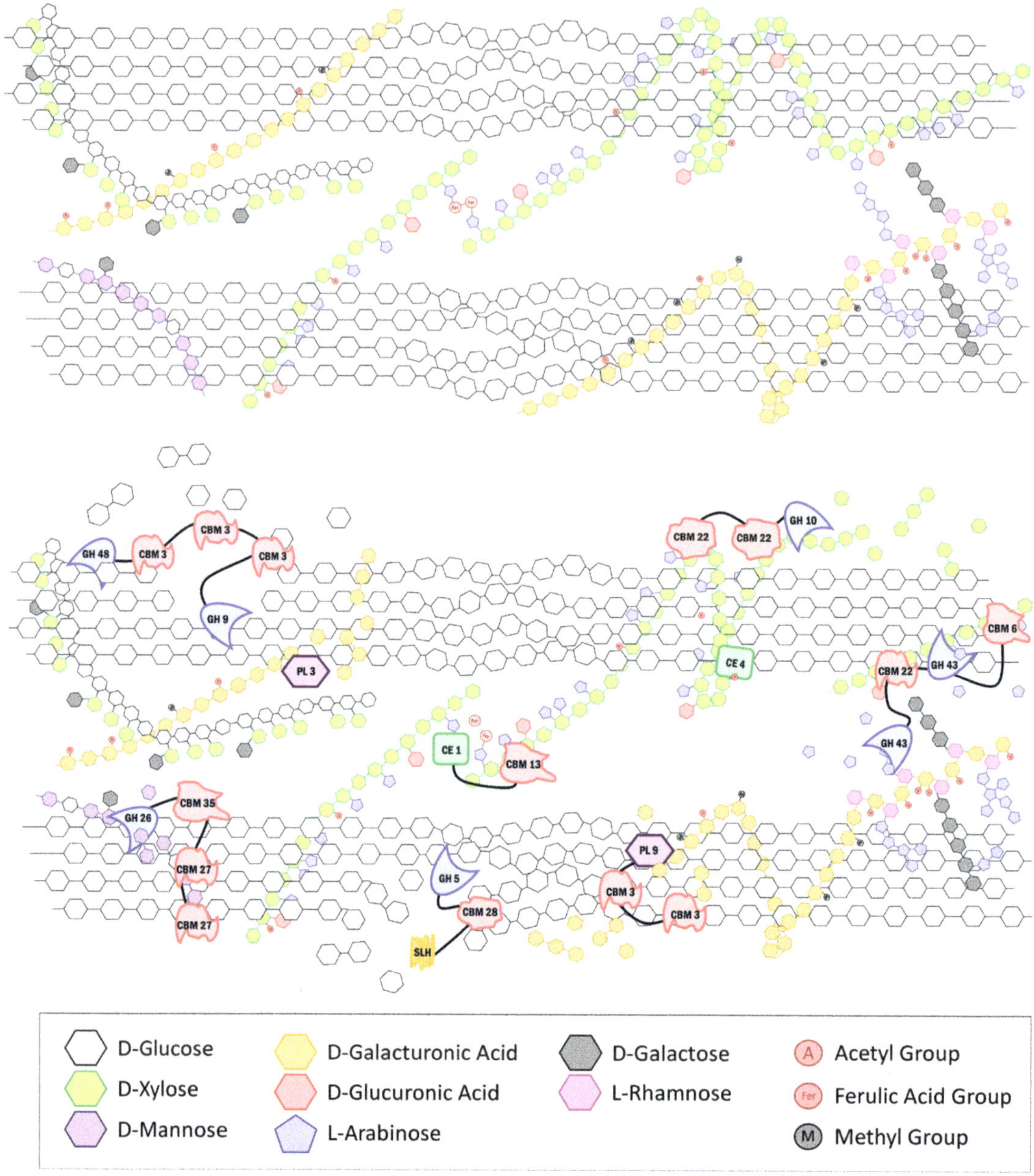

Figure 4.1 Biomass structure and degradation by selected *Caldicellulosiruptor* enzymes. (A) Cellulose microfibrils of β-1,4-linked D-glucose contain both crystalline (left and right) and amorphous regions (centre). Various forms of hemicellulose, xylan (top right, bottom centre), mannan (bottom left), and xyloglucan (top left), form a sheath around and crosslink the cellulose microfibrals. Pectin homogalacturonan (top left and bottom centre) and branched rhamnogalacturonan I (bottom right) also contribute to the sheathing of the cellulose microfibrals. (B) Selected *Caldicellulosiruptor* enzymes are shown degrading these biomass polysaccharides.

main structural element of plant cell walls (Cosgrove, 2005). The packing of these crystalline regions is sufficiently compact to preclude the penetration of enzymes and even small molecules, water in particular. However, irregularities in the packing of the polysaccharide chains create amorphous regions, pits, micropores, and capillaries, making the cellulose chains accessible to small molecules and, in some cases, cellulases, enzymes that break down cellulose (Lynd *et al.*, 2002).

Hemicellulose, on the other hand, is a highly variable heterologous polymer of pentoses (e.g. xylose, arabinose) and hexoses (e.g. mannose, glucose, galactose), integrated into a β-1,4- or mixed β-1,3–1,4-linked backbone of repeating disaccharide units, which are then substituted with various side chain sugars (Scheller and Ulvskov, 2010). Hemicellulose composition varies across different species, for example, hardwoods contain mainly xylans, while softwoods contain mainly mannans, and grasses contain primarily arabinoxylan (Pauly and Keegstra, 2008). This diverse composition necessitates an array of hemicellulases to recognize and cleave these linkages to completely degrade hemicellulose. Hemicellulosic glycans, particularly xyloglucan, are thought to non-covalently tether cellulose microfibrils in the cell wall, thereby increasing wall strength (Scheller and Ulvskov, 2010).

The non-carbohydrate fraction of plant biomass is lignin, a hydrophobic polymer synthesized by the oxidative coupling of the monomers p-hydroxyphenyl (H-), guaiacyl (G-), and syringyl (S-) lignin (Vanholme *et al.*, 2008). Lignin attaches covalently to hemicellulose, filling in gaps between hemicellulose and cellulose to increase the strength and rigidity of fibrous plant tissue, and its hydrophobicity improves water and solute conduction through plant vascular tissue (Grabber, 2005; Vanholme *et al.*, 2008). These characteristics of lignin interfere with the access of enzymes to the polysaccharides of biomass and enable lignin to adsorb enzymes, impeding their activity (Ju *et al.*, 2013).

Processing lignocellulose into biofuels

The main challenge in utilizing lignocellulosic biomass as an energy source is a deficiency of low-cost technologies for effectively accessing polysaccharides within the plant cell wall and breaking them down into fermentable sugars (Lynd *et al.*, 2005, 2002). Traditional approaches to bioprocessing of lignocellulosic biomass involve three process steps: pretreatment, enzymatic hydrolysis, and fermentation of released sugars (Margeot *et al.*, 2009). Thermal and/or chemical processing of raw biomass is done during pretreatment, with the goals of disrupting the plant cell wall and removing and separating lignin, and sometimes hemicellulose, from the substrate (Wu *et al.*, 2010). Cellulase and hemicellulase cocktails are then added to hydrolyse the biomass polysaccharides to mono- and oligo-saccharides, which are fermented by a microorganism to a desired product, typically ethanol. Many variations of these steps have been developed, either to keep hydrolysis and fermentation separate, or combine them in strategies called simultaneous saccharification and fermentation (SSF), where enzyme cocktails are mixed with ethanologens (e.g. yeast) to achieve biomass conversion (Ishola *et al.*, 2013; Margeot *et al.*, 2009). However, conditions within the SSF reaction must be amenable to the microorganism, as well as the enzyme mixtures, which is often far from the specific optimum conditions of either entity.

Ideally, a low-cost processing strategy would combine lignocellulose degradation and fermentation of the released sugars to fuel in a single process step performed by a single microbe, a strategy referred to as consolidated bioprocessing (CBP) (Lynd *et al.*, 2002). Currently, no naturally occurring microorganism performs the substrate utilization and product formation steps to the extent needed for CBP-based production of liquid fuels. However, efforts are under way to develop CBP microorganisms using two separate engineering strategies: improving product yield from cellulolytic microorganisms, or adding cellulases to non-cellulolytic microorganisms with high product yield to enable substrate utilization (Lynd *et al.*, 2005; Olson *et al.*, 2012). Both of these strategies require better understanding of the intrinsic mechanisms that cellulolytic microorganisms use to efficiently attack and break down cellulose. This will help to identify the most promising candidate microorganisms and cellulases for CBP development.

To date, many mesophilic recombinant microorganisms have been developed as potential CBP microorganisms (Yamada *et al.*, 2013; You *et al.*, 2012), but none can navigate the high variability of lignocellulosic feedstocks and have the robustness necessary to function under biologically unfavourable process conditions

(Blumer-Schuette *et al.*, 2008). Because of their ability to degrade cellulose at high temperatures and to utilize unpretreated biomass feedstocks, a number of cellulolytic, extremely thermophilic microorganisms show promise in overcoming the process barriers to CBP. The high temperatures at which extreme thermophiles grow optimally (i.e. 70°C or above) reduce the energy input for process cooling, facilitate volatile product recovery, and significantly limit the risk of microbial contamination by mesophilic methanogen and sulfate-reducing soil bacteria, likely to be present in unpretreated biomass feedstocks (Barnard *et al.*, 2010). In addition, the inventory of enzymes produced by these organisms provides a rich reservoir of highly thermostable, and often novel, enzymes for biomass processing.

Cellulolytic, thermophilic bacteria are rare, with many species clustered in the genera *Caldicellulosiruptor*, *Clostridium*, *Geobacillus*, *Acidothermus*, and *Thermoanaerobacter*. However, *Caldicellulosiruptor* is the only known genus with species that can fully deconstruct lignocellulose at temperatures above 70°C (Blumer-Schuette *et al.*, 2008). A number of non-cellulolytic thermophiles that produce carbohydrate-active enzymes related to biomass conversion are also of interest for developing CBP microorganisms, including *Thermotoga maritima*, *Pyrococcus furiosus*, *Sulfolobus solfataricus*, and *Dictyoglomus thermophilum* (Blumer-Schuette *et al.*, 2014; VanFossen *et al.*, 2008). Currently, genetic and metabolic engineering of cellulolytic thermophiles, now in the early stages, could enable fine-tuning the enzyme inventory and improving product yields from these organisms, looking towards CBP as the ultimate goal (Cai *et al.*, 2011; Cha *et al.*, 2013; Tripathi *et al.*, 2010). Environmental sampling and metagenomic approaches are also of interest to identify as yet undiscovered microorganisms and enzymes that could play a role in an engineered CBP biomass conversion system.

Carbohydrate active enzymes (CAZymes)

Microorganisms that degrade biomass substrates produce a large variety of extracellular cellulases, hemicellulases, and non-catalytic binding proteins for interacting with, and liberating sugars from, insoluble biomass substrates. These proteins are designated carbohydrate active enzymes (CAZymes), and have a modular architecture that may contain a single domain or multiple domains from various enzyme classes or binding motifs. These catalytic and binding domains that comprise CAZymes are classified into families according to amino acid sequence similarity, and cataloged in a database maintained at www.cazy.org (Cantarel *et al.*, 2009; Lombard *et al.*, 2014). The CAZy database is ever expanding by grouping proteins into new module families and incorporating putative CAZymes from an increasing number of sequenced genomes. The CAZy database is curated to include biochemical characterization and X-ray crystallographic structural information obtained for members of the various families (Lombard *et al.*, 2014).

Of the enzyme module classes catalogued in the CAZy database, the most abundant for biomass degradation are the catalytic glycoside hydrolases (GHs), which hydrolyse glycosidic bonds between various linked sugars, and the non-catalytic carbohydrate-binding modules (CBMs), which bind sugars both soluble and insoluble. The CAZy database also includes other carbohydrate-active domain classifications including: polysaccharide lyases (PLs), carbohydrate esterases (CEs), glycosyltransferases (GTs), and a newly designated 'Auxiliary Activities' (AAs) classification (Lombard *et al.*, 2014). PLs in the context of biomass degradation are typically involved in the degradation of pectin. CEs are involved in the removal of ester linked moieties (e.g. acetyl or methoxy groups) attached to the sugars in biomass. GTs are enzymes that form glycosidic bonds, rather than degrade them. While they are produced by cellulolytic microbes, GTs, because they are not involved in biomass deconstruction, will not be discussed here. The AAs class contains a variety of modules that do not fit into the other CAZy classes, but assist GHs, PLs, and CEs in accessing the polysaccharides of plant cell walls. Currently classified members of the AAs are predominantly from lignolytic and cellulolytic fungi, and the role of these modules in bacteria is currently not well characterized (Levasseur *et al.*, 2013). As such, the AAs class will not be discussed further here.

Currently, there are 133 GH, 71 CBM, 23 PL, and 16 CE families listed in the CAZy database. However, the constant curation of the database has deleted or reclassified some of the families as biochemical characterization data become available. For example, the family GH 61 was assigned to the GHs based on weak glucanase activity of one member. But as more members were characterized, the protein family has been reassigned as Auxiliary Activity family 9, rendering GH 61 obsolete. Owing to these types of changes, there are currently 126 GH, 69 CBM, 22 PL, and 15 CE families with classified members. A number of proteins that are known, or predicted, to have a carbohydrate active domain, but are not closely related to existing protein sequences, remain as 'non-classified sequences' or 'nc', meaning they have not been incorporated into a numbered family classification. As sequencing of novel organisms continues and new protein sequences are introduced into the database, these are grouped into new families when sufficient similarity is identified among non-classified sequences. In addition to the GH family designations, some families have been organized into 14 clans (GH-A through -N), based on tertiary protein structure similarities between GHs of different families (Naumoff, 2011). In the same way, selected CBM families are organized into seven 'fold' families, based on tertiary structure similarities (Boraston *et al.*, 2004; Guillen *et al.*, 2010). In the following sections, the role of each of these CAZy classified modules will be examined in relation to biomass attachment and deconstruction. A number of additional modules and proteins, also associated with CAZymes, will also be discussed.

Glycoside hydrolases (GHs)

The enormous diversity of saccharides in plant biomass necessitates the complementary action of a variety of GHs to dismantle them. Nearly all GHs exhibit specificity based on two structural elements of the carbohydrates they target: the configuration of the cleaved bond (α or β) and the ring size of the released sugar molecule (Naumoff, 2011). GHs hydrolyse glycosidic bonds, in almost all cases, by acid-base chemistry using two carboxylic acid moiety-containing amino acids in the active site, most often glutamic acid or aspartic acid (McCarter and Withers, 1994; Naumoff, 2011). Glycosidic bond hydrolysis proceeds by either an inverting mechanism, where the anomeric carbon is inverted by direct displacement of the leaving group with water, or a retaining mechanism where the anomeric carbon orientation is retained by double displacement through a glycosyl-enzyme intermediate (Maki *et al.*, 2009; McCarter and Withers, 1994). The transition state of GH-catalysed hydrolysis reactions involve the distortion of the pyranose ring into a puckered conformation to allow nucleophilic attack and bond hydrolysis (Vocadlo and Davies, 2008). These transition states have been examined for a number of GHs through various crystallographic and computational methods (Davies *et al.*, 2011; Knott *et al.*, 2013; Mayes *et al.*, 2013; Vocadlo and Davies, 2008).

GHs that break down polysaccharides are also classified as *endohydrolases* or *exohydrolases*, based on their ability to cleave glycosidic bonds either in the middle, or at the end of polysaccharide chains, respectively (Bayer *et al.*, 1998). In the case of cellulose hydrolysis, endo-β-1,4-glucanases cut the amorphous regions of cellulose to generate new cellulose chain ends, from which exo-β-1,4-glucanases work to remove small oligosaccharides, mainly cellobiose, resulting in the classification of many exo-β-1,4-glucanases enzymes as cellobiohydrolases. Exohydrolases can further be divided into two classes, based on their preference for acting at the reducing or non-reducing end of the polysaccharide chains they cleave (Barr *et al.*, 1996). Several GHs have been identified that display both endo- and exo- modes of action, so characterization based on enzymatic mechanism and structural properties result in more specific classifications (Bayer *et al.*, 1998; Maki *et al.*, 2009). Once soluble sugars are liberated from polysaccharides (e.g. cellodextrins from cellulose), another set of GHs, such as β-glucosidases or cellobiose phosphorylases, break these soluble carbohydrates into monomers for metabolism (Bianchetti *et al.*, 2011; Hong *et al.*, 2009; Lynd *et al.*, 2002).

The complex structure and diversity of sugars and bonds in biomass illustrate the need for a high degree of synergy between many GHs to efficiently hydrolyse the substrate. GHs that hydrolyse the β-1,4-glycosidic bonds found in cellulose and cellodextrins are most prevalent in families GH5, GH6, GH7, GH8, GH9, GH12, GH44, GH45, GH48, GH74, and GH124 (Lynd *et al.*, 2002). Most xylan-degrading enzymes are concentrated in GH5, GH8, GH10, and GH11, but bi-functional GHs have been identified in GH7, GH16, GH43, and GH62, families that are typically associated with the degradation of other polysaccharides (Pollet *et al.*, 2010). The α-1,4-glucosidic bonds, found in polysaccharides related to starch, are typically degraded by members of GH13, GH57, GH70, and GH77 (Janeček and Blesák, 2011; MacGregor *et al.*, 2001). Many other GH activities are relevant to biomass degradation, including: α-galactosidases in GH27 and GH36; β-mannanases in GH5 and GH26; β-mannosidases in GH2; α-arabinofuranosidases in GH43, GH51, GH54, and GH62; and α-glucuronidases in GH67 (Cantarel *et al.*, 2009; Gilbert *et al.*, 2008; Ryabova *et al.*, 2009; Sørensen *et al.*, 2006; Xu, 2011). Representative crystal structures have now been solved for 100 of the existing 126 GH families (Fushinobu *et al.*, 2013).

Carbohydrate-binding modules (CBMs)

CBMs are often components of catalytic proteins, which also contain catalytic GH, PL, or CE domains, or components of dedicated binding proteins that facilitate the attachment of microorganisms to biomass substrates. CBM domains enhance the function of the catalytic domains to which they are linked by increasing the proximity of the catalytic unit to the substrate, targeting specific areas of the substrate, and disrupting substrate interactions to enhance degradation (Boraston *et al.*, 2004). When CBMs are appended to catalytic protein units, GHs linked to these CBMs can have increased hydrolytic efficiency (Mahadevan *et al.*, 2008; Park *et al.*, 2011). This suggests that the CBM units increase the effective enzyme concentration at the substrate surface, thereby increasing efficiency. Other experimental data suggest that CBMs play a role in targeting GHs to specific areas of the substrate that differ in susceptibility to hydrolysis (Boraston *et al.*, 2003; McLean *et al.*, 2002). Microorganisms that grow on biomass attack a diversity of sugars. Congruency between CBM domain binding and GH domain catalytic specificity within the same protein is essential for efficient degradation by targeting particular sugars within the diverse plant cell wall (Guillen *et al.*, 2010). There is also significant evidence that CBMs play a role in disrupting or penetrating the structure of polysaccharides, enhancing the ability of the catalytic subunit to degrade the substrate (Arantes and Saddler, 2010; Din *et al.*, 1991; Reyes-Ortiz *et al.*, 2013). One model of cellulose microfibril disruption postulates CBM adsorption at a micro-defect in the cellulose chain packing, followed by penetration of the CBM and water molecules into the mircofibril, disrupts the hydrogen bonding between stacked cellulose chains, loosening individual chains for enzymatic degradation (Arantes and Saddler, 2010). To better understand the roles of CBMs, it is important to understand the types of sugars that they bind.

CBM binding is classified into types, based on the recognized ligands: Type A surface binding, Type B glycan chain binding, and Type C soluble sugar binding (Boraston *et al.*, 2004). Type A CBMs have planar binding sites, which reflect the flat surface of crystalline substrates, such as cellulose or chitin (Guillen *et al.*, 2010). Type B CBMs include the most CAZy classified CBM families, and bind to glycan chains via a cleft or groove binding site (Boraston *et al.*, 2004). The cleft or grove site limits the specificity of these CBMs to polysaccharide chains, for example, binding the free cellulose chains in amorphous microfibril regions, while no binding affinity is seen to crystalline regions of cellulose (Gourlay *et al.*, 2012). Similarly, the steric restrictions of the Type C CBM binding site only allow binding to soluble mono-, di-, or trisaccharides (Guillen *et al.*, 2010). Aromatic amino acids seem to play a universal role in carbohydrate binding in all CBM types, especially tryptophan and tyrosine, which stack against, in Type A CBMs, or sandwich, in Type B and C CBMs, the pyranose ring of carbohydrate targets (Boraston *et al.*, 2004; Lehtio *et al.*, 2003). In addition, site-directed mutations have shown that hydrogen bonds between CBM polar groups and sugar targets are key to CBM affinity in Type

B and C CBMs, but only play a minor role in Type A CBMs (Boraston *et al.*, 2004; Pell *et al.*, 2003).

Polysaccharide lyases (PLs)

PLs cleave uronic acid-containing polysaccharides and occur in a diversity of genomes from bacteria to higher eukaryotes. However, they are rarer in these genomes than GHs, presumably because only a small portion of polysaccharides contain uronic acids (Lombard *et al.*, 2010). The substrates for PLs can be divided generally into three groups, based on the uronic acid they contain at the PL cleavage site. Galacturonic acid containing polysaccharides including pectin and pectate; glucuronic/iduronic acid containing polysaccharides including heparin, chondroitin, and hyaluronan, among others; and mannuronate/guluronate containing polysaccharides including alginate are all degraded by PLs (Garron and Cygler, 2010). The first of these PL-degraded groups is only one that is significant in the context of biomass utilization, because pectin and pectate are significant components in the primary cell walls of all dicots and some monocots (Yadav *et al.*, 2009). Glucuronic acid containing glucuronoxylan and glucuronoarabinoxylan are significant components of the cell walls of dicots and grasses respectively (Scheller and Ulvskov, 2010). However, the single glucuronic acid moiety branching on the xylan backbone in these cases is not catalytically removed by PLs, but rather by GH4 or GH67 α-glucuronidases (Ryabova *et al.*, 2009; Xu, 2011).

Pectin consists of three different polysaccharide motifs: (1) homogalacturonan (HGA), the main backbone of α-(1,4)-linked D-galacturonic acid, (2) rhamnogalacturonan I (RGI), which has a [2)-L-rhamnose-α-(1,4)-D-galacturonic acid-α-(1,] linked backbone that is highly branched from rhamnose with L-arabinose and D-galactose and (3) rhamnogalacturonan II (RGII), which is a highly variable and complex branched region of the α-(1,4)-linked D-galacturonic acid backbone (Yadav *et al.*, 2009). Microorganisms that degrade plant biomass typically contain only PLs from the pectin/pectate lyase families PL1, PL2, PL3, PL9, PL10, and PL22 and from the rhamnogalacturonan lyase families PL4 and PL11 (Garron and Cygler, 2010).

Uronic acid-containing polysaccharides are cleaved by PLs via a β-elimination mechanism that generates a new reducing chain end and a hexenuronic acid moiety at the non-reducing chain end after cleavage (Lombard *et al.*, 2010). This occurs without the involvement of water, in contrast to the GH mechanism that does incorporate water to maintain an -OH group on the non-reducing chain end (Yip and Withers, 2004). Of the 22 PL families in the CAZy database, all but one, PL17, have a solved structure for a representative member and these structures group into eight 'fold' types (Lombard *et al.*, 2010). In pectate lyases with solved crystal structures, an Arg or Lys residue acts as the catalytic base (Alahuhta *et al.*, 2013; Yip and Withers, 2006). For the PL3 pectate lyase from *Caldicellulosiruptor bescii*, it was found that in digestion reactions of unpretreated switchgrass, where the PL3 was mixed with *C. bescii* cellobiohydrolase A (CelA), lower CelA loading was needed to achieve the same glucan conversion accomplished by higher loadings of CelA alone (Alahuhta *et al.*, 2013). This suggests that PLs play an important role in making areas of unpretreated biomass substrates more readily accessible to GHs of other biomass degrading CAZymes, thereby leading to more complete substrate utilization.

Carbohydrate esterases (CEs)

As a protection against microbial attack, many plant cell wall polysaccharides, particularly hemicelluloses and pectin, are highly substituted with a variety of side chains, including acetyl, feruloyl, and methoxy groups, linked to the sugar backbone by ester linkages (Williamson *et al.*, 1998). Esterification with phenolic acids, such as ferulic acid, can further react to form cross-linkages between adjacent cell wall polymers, which improve the mechanical strength of the plant cell wall, although less frequent than polysaccharide esterification with acetic acid (Biely, 2012). CEs remove these ester substitutions, allowing GHs and PLs access to the polysaccharide backbone for degradation. The synergistic relationship between CEs and GHs or PLs has been observed frequently (Biely *et al.*, 1986; Faulds and Williamson, 1995); for example, in maize stover degradation, the removal of acetyl substituting groups via the addition of an acetyl xylan esterase from CE family 5 improved enzymatic conversion of xylan by an endoxylanase from GH 11 (Selig *et al.*, 2009). CE domains can also be appended to GH or PL catalytic domains, resulting in bi-functional activity of a single enzyme, as observed in the CE1-GH11 containing XynS20E enzyme from *Neocallimastix patriciarum* (Pai *et al.*, 2010).

CEs can be grouped based on the substituted moiety that they remove: cinnamoyl esterases, including feruloyl esterases, are found in family CE1, although many have not yet been assigned a CE family (CEnc); acetylesterases, including acetylxylan esterases, belong to families CE1, CE2, CE3, CE4, CE5, CE6, CE7 and CE16; and, pectin methylesterases are in family CE8 (Brink and Vries, 2011; Cantarel *et al.*, 2009; Jolie *et al.*, 2010; Williamson *et al.*, 1998). The active centre of most CEs is based on a catalytic serine in the consensus sequence (Gly-X-Ser-X-Gly), which is part of a catalytic triad of Ser-His-Asp/Glu or catalytic dyad of Ser-His alone (Biely, 2012). In contrast, the catalytic centres of pectin methylesterases involve a catalytic Asp residue stabilized by Arg (Jolie *et al.*, 2010). The reaction mechanism for CEs is similar to that of lipases involved in the breakdown of triglyceride fats, where the substrate binds to the active serine, an alcohol is released, and an acyl-enzyme intermediate remains; subsequently, attack of a nucleophile like water and resolution results in product and the free enzyme (Aurilia *et al.*, 2008; Martínez-Martínez *et al.*, 2007).

Immunoglobulin-like superfamily

Two additional domain types, fibronectin type III (FN3) and bacterial immunoglobulin-like (BIg), both of which belong to the superfamily of immunoglobulin-like (Ig-like) proteins in Pfam clan CL0159, have been identified as members of the multidomain architecture of a number of *Caldicellulosiruptor* proteins (Chothia and Jones, 1997; Punta *et al.*, 2012). Members of this immunoglobulin-like superfamily typically only have 10–30% sequence identity, but share a similar structure of two face-to-face β-sheets with a hydrophobic core (Campbell and Spitzfaden, 1994; Chothia and Jones, 1997). FN3 domains are common components of eukaryotic proteins, but within bacterial species they have been identified almost exclusively in extracellular glycohydrolases (Kataeva *et al.*, 2002; Little *et al.*, 1994). There remains some disagreement as to how to classify FN3 domains and what their role is in bacterial enzymes. Some studies have characterized these domains as FN3 domains and 'X domains' (because of their unknown function) and have tried to differentiate phylogenetic groups of X domains from FN3 domains (Devillard *et al.*, 2004). Work with the cellulosomal cellobiohydrolase CbhA from *C. thermocellum*, containing domains from the N-terminus: CBM4-BIg-GH9-FN3-FN3-CBM3-Dockerin, showed increased hydrolysis of sugar from acid-swollen cellulose by protein truncations containing one or both FN3 domains and 'exfoliation' of cellulose chains by the FN3 domains alone (Kataeva *et al.*, 2002). Work by others to solve the crystal structures and characterize the same FN3 domains from *C. thermocellum* CbhA suggest that these domains play a role in improving the thermostability of the enzyme, and do not play a role in disrupting the substrate (Brunecky *et al.*, 2012). Other proposed roles for FN3 domains include promoting 'mechanical elasticity' of the enzyme, functioning as a CBM, and serving as a 'spacer' to maintain the optimal distance between the other components of multidomain enzymes (Campbell and Spitzfaden, 1994; Devillard *et al.*, 2004).

Bacterial immunoglobulin-like domains appear to be involved in cell-to-cell adhesion and extracellular glycohydrolysis, but a definitive role of these domains remains elusive (Fraser *et al.*, 2006). The structure of the BIg domain from *C. thermocellum* CbhA has been solved (Alahuhta *et al.*, 2010; Schubot *et al.*, 2004). The domain shows very close association with, and is required for, the catalytic ability of the GH9, suggesting a role in the structural stabilization of the catalytic domain (Schubot *et al.*, 2004). Similarly, an Ig-like domain in a thermostable esterase from *Thermotoga maritima* was found to play a role in the multimerization of the catalytic domain into a hexameric structure for increased activity and thermostability (Levisson *et al.*, 2009). Overall, very little characterization and few structures have been solved for members of the FN3 and BIg families, but the limited work that has been done suggest they play a role in both enzyme stabilization and substrate binding, both of which are essential for robust, thermostable, biomass-degrading enzymes.

Surface layer homology (SLH) domains

SLH domains are highly conserved, 50–60 amino acid sequences that are on the N- or C-terminus of proteins, and attach non-covalently to the cell surface within the surface layer (S-layer), a planar array of surface protein (Sara and Sleytr, 2000). S-layers are found in almost every taxonomic group of bacteria and archaea, and in most species are predominantly comprised of a single S-layer protein (SLP) that assembles into a protein lattice at the cell surface with oblique, square, or hexagonal symmetry (Sara and Sleytr, 2000). This structured protein array may serve many functions, including as a molecular sieve or protective coat, or an attachment site for extracellular enzymes and molecular moieties involved in cell adhesion or surface recognition (Pum *et al.*, 2013; Scott and Barnett, 2006). Proteins, other than the SLP, that contain SLH domains can be tethered to the cell surface within this array, while the remainder of the protein is free to serve its function (Engelhardt and Peters, 1998). CAZymes containing SLH domains can mediate the attachment of microorganisms to biomass substrates, by attaching to both the cell via SLH domains and to the substrate via CBM domains; they can also contribute to biomass deconstruction at the cell surface with catalytic domains (Ali *et al.*, 2001; Kosugi *et al.*, 2002; Ozdemir *et al.*, 2012). In Gram-positive bacteria, such as *Caldicellulosiruptor* and *Clostridium* species, SLH domains attach to the 15–30 nm thick peptidoglycan cell wall and its associated secondary cell wall polymers (e.g. teichoic and teichuronic acids) (Scott and Barnett, 2006).

Others (adhesin, expansin-like, and swollenin)

Expansins are proteins found in plants that participate in cell enlargement by loosening cell wall polysaccharides without breaking them, allowing plant cell walls to 'creep' and expand, which facilitates plant growth (Cosgrove, 2000; Cosgrove *et al.*, 2002; Sampedro and Cosgrove, 2005). Expansin-like gene sequences appear in the genomes of a small number of phylogenetically diverse bacteria and, from experimental evidence, display similar plant cell wall loosening characteristics to plant expansins (Georgelis *et al.*, 2014; Kerff *et al.*, 2008). Expansin-like proteins also include a set of fungal proteins, originally identified in *Trichoderma reesei*, called swollenins based on their ability to swell cotton fibres without degrading them (Kang *et al.*, 2013; Saloheimo *et al.*, 2002). These non-hydrolytic proteins are of particular interest as a biomass pretreatment in biofuels processes for their role in cellulose amorphogenesis, initial disruption of the crystalline microfibral structure prior to hydrolysis (Arantes and Saddler, 2010). Work is being done to better quantify their effect on cellulose and their ability to enhance the action of CAZymes (Jager *et al.*, 2011). Recently, *Caldicellulosiruptor* species have been shown to express two classes of novel adhesin proteins that are enriched in the crystalline cellulose-bound fraction of cultures analysed by proteomics analysis (Blumer-Schuette *et al.*, 2012). These adhesin proteins have no homology outside the *Caldicellulosiruptor* genus and are located downstream of a type IV pilus locus, suggesting association with type IV pili to attach to biomass substrates. This novel group of adhesin proteins have no hydrolytic activity, but may play a role in cellulose amorphogenesis, similar to expansin-like proteins (Blumer-Schuette *et al.*, 2015).

Biomass deconstruction by cellulolytic microorganisms

Cellulolytic microorganisms use three different degradation mechanisms, shown in Fig. 4.2, to coordinate their cellulase and hemicellulase inventory to break down biomass: free enzymes, cellulosome multienzyme complexes, and large multidomain enzymes (Vodovnik and Marinsek-Logar, 2010). Cellulolytic species use at least one, but often more than one, of these degradation mechanisms to effectively degrade lignocellulosic biomass. Aerobic cellulolytic fungi and bacteria typically produce large amounts of individual extracellular free enzymes, but nearly all cellulolytic species use free enzymes to some extent in

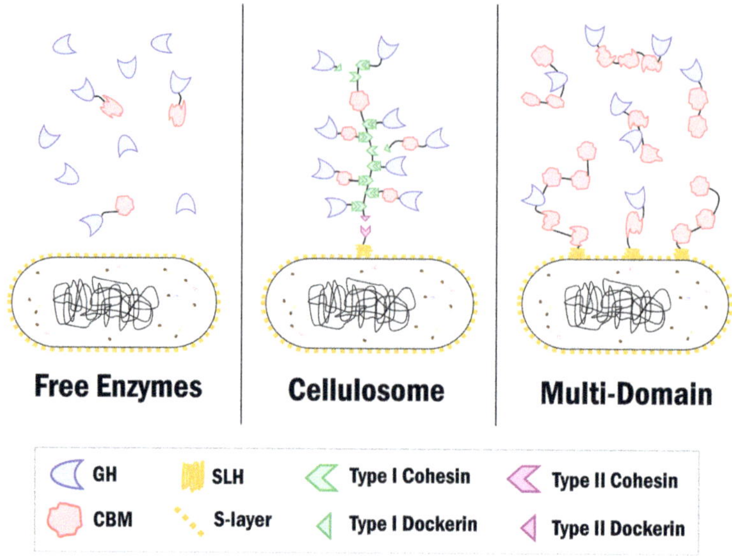

Figure 4.2 The three biomass degradation mechanisms used by cellulolytic microorganisms. Free enzymes contain either a single catalytic domain or one linked to a Carbohydrate Binding Module (CBM) domain. Free enzymes are secreted and disconnected from the cell. The cellulosome structure is shown structured around *Clostridium thermocellum* scaffoldin CipA, but a variety of other multienzyme cellulosome structures exist. Enzyme subunits attach to the scaffoldin by Type I cohesin–dockerin interactions. The scaffoldin attaches to an anchoring scaffoldin by a Type II cohesin–dockerin interaction. The anchoring scaffoldin is attached at the cell surface within the cell's surface layer (S-layer) by Surface Layer Homology (SLH) domains. Multi-domain proteins contain multiple catalytic and binding domains within the same protein. Some of these multidomain proteins are anchored to the cell surface within the S-layer by SLH domains. Microorganisms may use one or more of these mechanisms to effectively degrade biomass.

their biomass degradation strategy (Mazzoli *et al.*, 2012; Vodovnik and Marinsek-Logar, 2010). These free enzymes almost always contain a substrate-binding CBM domain attached to a catalytic GH domain by a small flexible linker (Wilson, 2011). While free enzymes are effective at degrading crystalline cellulose, multienzyme cellulosome complexes or large multidomain enzymes provide the opportunity for coordination of multiple catalytic domains for a more synergistic degradation strategy.

A number of anaerobic bacteria (e.g. *Clostridium* spp. and *Ruminococcus* spp.) and fungi (e.g. Chytridomycetes) produce large multi-enzyme complexes on the cell surface called cellulosomes, the most studied of which is from *Clostridium thermocellum* (Bayer et al., 2008). The cellulosome from C. *thermocellum* displays superior cellulose degradation activity to free enzyme cocktails and degrades biomass by a mechanism distinct from free enzymes in which the cellulosome physically separates cellulose microfibrils from the larger particle to improve digestion (Resch *et al.*, 2013). Cellulosomes are constructed around a large primary scaffoldin protein that acts as a scaffold for the enzyme units that degrade biomass. An example of a *C. thermocellum* cellulosome constructed around the main primary scaffoldin protein, CipA, is shown in Fig. 4.2. The primary scaffoldin protein contains many cohesin subunits and typically one CBM domain, which mediates attachment of the large cellulosome complex to the substrate. Enzymes containing GH, CBM, and terminal dockerin modules anchor to cohesin domains on the primary scaffoldin by high-affinity, calcium-dependent, type I cohesin–dockerin interactions (Vodovnik and Marinsek-Logar, 2010). The primary scaffoldin attaches to an anchoring scaffoldin protein via a type II cohesin–dockerin interaction that is divergent from the type I primary cohesin–dockerin interaction between the scaffoldin and enzyme subunit (Bayer *et al.*, 2008). The anchoring scaffoldin contains terminal surface layer homology

(SLH) domains that facilitate attachment of the entire cellulosome complex to the cell surface (Bayer *et al.*, 2008).

While the paradigm describing cellulosome assembly is modelled after the *C. thermocellum* version, further investigation into the cellulosomes of *C. thermocellum* and other species reveal a rich array of assembly motifs to increase the number of enzyme attachments or incorporate additional components to the cellulosome complex (Bayer *et al.*, 2008; Vodovnik and Marinsek-Logar, 2010). For example, in *C. thermocellum*, five cell surface attached scaffoldins, and one scaffoldin that is not cell associated, have been identified; these vary in assembly structures from attaching directly to a single enzyme subunit to attaching seven primary scaffoldins harbouring up to 63 enzymes (Fontes and Gilbert, 2010). In contrast, *Acetivibrio cellulolyticus* produces a 'three-tiered cellulosome', comprised of 83 enzymes attached in a complex to the cell surface via three interacting scaffoldins, one of which also contains a catalytic GH9 module (Bayer *et al.*, 2008). Cellulosomes provide a number of advantages to the microorganisms that produce them, including increased enzymatic synergism enhanced by favourable assembly and spacing of enzymatic and binding components, and proximity of the cell to the substrate for monitoring the microenvironment and uptake of released sugars for metabolism (Maki *et al.*, 2009).

Other anaerobic bacteria, particularly members of the genus *Caldicellulosiruptor*, do not coordinate enzymes into large multiprotein cellulosome complexes, but rather use large multidomain proteins, which can contain multiple catalytic and binding domains, to efficiently degrade biomass (Blumer-Schuette *et al.*, 2012). These multidomain proteins are characterized by their complex architecture of multiple GH, PL, CE, CBM, and/or other accessory domains in a single peptide. Several of these multidomain enzymes contain two distinct GH domains, creating a multi-functional protein, where it is possible for substrate intermediates to be mobile between the two GH module active sites resulting in increased synergism (Khandeparker and Numan, 2008; VanFossen *et al.*, 2011). The inclusion of SLH domains at either the N- or C-terminus of some of these proteins allows for attachment of the protein within the S-layer of the cell, while CBM domains in the same protein can mediate attachment to biomass to hold the cell in close proximity to the substrate (Ozdemir *et al.*, 2012). These complex multidomain proteins have been shown to be more active than the sum of the constituent domains as separate proteins, suggesting this multidomain enzyme strategy for degrading biomass provides an advantage to the microbes that produce them (Park *et al.*, 2011; Su *et al.*, 2012; VanFossen *et al.*, 2011).

Biomass degradation by the genus *Caldicellulosiruptor*

The genus *Caldicellulosiruptor*

Caldicellulosiruptor species are extremely thermophilic, Gram-positive, anaerobic, and asporogenous bacteria that are capable of breaking down a wide range of polysaccharides (Blumer-Schuette *et al.*, 2010). Currently, the closed genomes of eight *Caldicellulosiruptor* species, shown in Table 4.1, are available in the National Center for Biotechnology Information (NCBI) database (www.ncbi.nlm.nih.gov). In addition, two *Caldicellulosiruptor* genomes, *C. acetigenus* and *Caldicellulosiruptor* sp. F32, currently have permanent draft sequences and are listed as assemblies in the NCBI database. Other genomes that are currently unrelated include those for *Caldicellulosiruptor* species strain Rt8. B8 and strain Wai35.B1. These species are globally distributed through North America, Iceland, Russia, and New Zealand, however, while the genomes of species from similar locations share highly syntenous blocks, cellulolytic capacity does not correlate with the isolation location (Blumer-Schuette *et al.*, 2012).

Caldicellulosiruptor species have optimum growth temperatures between 70°C and 78°C, and certain species represent the most thermophilic, cellulolytic bacteria known (Blumer-Schuette *et al.*, 2010). One of the most extensively studied species in the genus is *C. saccharolyticus*, which has a large inventory of GHs in its genome that are differentially transcribed to adapt to varying substrates. These GHs metabolize a variety

Table 4.1 *Caldicellulosiruptor* genomes

Species	Accession number	Genome size (Mb)	Species nov. announcement	Genome announcement
C. acetigenus	ASM42172v1	2.59	Onyenwoke *et al.* (2006)	N/A
C. bescii	CP001393	2.93	Yang *et al.* (2010)	Kataeva *et al.* (2009)
C. hydrothermalis	CP002219	2.77	Miroshnichenko *et al.* (2008)	Blumer-Schuette *et al.* (2011)
C. kristjanssonii	CP002326	2.8	Bredholt *et al.* (1999)	Blumer-Schuette *et al.* (2011)
C. kronotskyensis	CP002330	2.84	Miroshnichenko *et al.* (2008)	Blumer-Schuette *et al.* (2011)
C. lactoaceticus	CP003001	2.62	Mladenovska *et al.* (1995)	Blumer-Schuette *et al.* (2011)
C. obsidiansis	CP002164	2.53	Hamilton-Brehm *et al.* (2010)	Elkins *et al.* (2010)
C. owensensis	CP002216	2.43	Huang *et al.* (1998)	Blumer-Schuette *et al.* (2011)
C. saccharolyticus	CP000679	2.97	Rainey *et al.* (1994)	van de Werken *et al.* (2008)

of monosaccharides, both pentose and hexose sugars, simultaneously without regulation by carbon catabolite repression (van de Werken *et al.*, 2008; VanFossen *et al.*, 2011, 2009). Another well studied species, *C. bescii*, has been shown to grow on industrially relevant loadings of unpretreated biomass and is able to digest up to 85% of insoluble unpretreated switchgrass, while simultaneously deconstructing lignin, cellulose, and hemicellulose (Basen *et al.*, 2014; Kataeva *et al.*, 2013). *Caldicellulosiruptor* species ferment the simple sugars liberated from the biomass substrate to molecular hydrogen, lactate, acetate, and a small amount of alcohols. A recent review of *Caldicellulosiruptor* metabolism discusses both the intermediary metabolism of these species for generating energy and reducing equivalents, as well as the metabolism for fuel production (Zurawski *et al.*, 2014). These characteristics make *Caldicellulosiruptor* species attractive candidates for development into consolidated bioprocessing (CBP) microorganisms.

Comparison of the eight closed *Caldicellulosiruptor* genomes revealed that the core genome, genes contained in the genomes of all eight species, includes 1543 ORFs, but is non-cellulolytic (Blumer-Schuette *et al.*, 2012). The pan genome, the collection of all genes in all species (4009 ORFs), was found to be open, indicating that sequencing of new species is expected to reveal novel genes not yet identified as part of the pan genome (Blumer-Schuette *et al.*, 2012). Within the genus, cellulolytic capability varies greatly with *C. bescii*, *C. kronotskyensis*, *C. obsidiansis*, *C. lactoaceticus*, and *C. saccharolyticus* having higher cellulolytic ability than the other species with completed genomes (Blumer-Schuette *et al.*, 2012). Comparison of these five highly cellulolytic species reveals three extracellular GH proteins present in all five genomes: GH9-CBM3-CBM3-CBM3-GH48 (often referred to as CelA), GH74-CBM3-CBM3-CBM3-GH48, and GH9-CBM3-CBM3-CBM3-GH5. These three CAZymes are implicated as being important for crystalline cellulose degradation, and in particular, the presence of GH48 and CBM3 domains appear to be essential for crystalline cellulose degradation, because the weakly cellulolytic species do contain GH5 and/or GH9 enzymes, but no GH48 domains (Blumer-Schuette *et al.*, 2012, 2010). In addition, proteomic analyses of *C. bescii* and *C. obsidiansis* identified the cellulolytic inventory of multi-functional, multidomain enzymes as primarily containing GH5, GH9, GH10, GH43, GH44, GH48, and GH74 domains, as well as highly conserved CBM3 domains (Lochner *et al.*, 2011b). The many multidomain enzymes, found in the genus *Caldicellulosiruptor*, that contain multiple catalytic and binding domains in a single protein are thought to have evolved by domain shuffling (Bergquist *et al.*, 1999; Gibbs *et al.*, 2000). These *Caldicellulosiruptor* multidomain enzymes are often co-located in areas of the genome, including one termed the 'glucan degradation locus' (GDL), which contains many of the CBM3-containing enzymes, and a separate xylanase locus (Blumer-Schuette *et al.*, 2014, 2012).

The *Caldicellulosiruptor* pan-CAZome

The nine *Caldicellulosiruptor* species genomes listed in Table 4.1 encode for over 600 individual CAZymes that can be binned into 144 groups of protein homologues in the pan-CAZome. These 144 proteins contain 128 GH domains representing 44 families and one non-classified sequence, 58 CBM domains representing 17 families, 11 CE domains representing 7 families, and 9 PL domains representing 4 families and 2 non-classified sequences. To separate these 144 proteins into those involved in extracellular biomass processing and those involved in intracellular sugar catabolism, SignalP4.1 (input settings: Gram-positive bacteria, sensitive D-cutoff = 0.42, no TM regions) was used to predict signal peptides at the protein N-terminus, which is indicative of protein secretion (Petersen *et al.*, 2011). Using this method, 60 of the 144 CAZymes were identified as extracellular, and the remaining 84 were identified as intracellular. A graphical representation of the domain structure of each of the proteins found in the *Caldicellulosiruptor* extracellular pan-CAZome is shown in Fig. 4.3. Similarly, the intracellular pan-CAZome is shown in Fig. 4.4. These figures show the hierarchical structure of the pan-CAZome, where the 7 extracellular and 26 intracellular core CAZymes, found in all nine species, are shown at the top, with the boxes of increasing size encompassing all the proteins found in five, three, two, and one or more species, respectively. Put another way, the proteins found inside the box for three or more species, but outside the box for five or more species, have homologous proteins in three or four of the nine *Caldicellulosiruptor* species.

Taking a closer look at the extracellular pan-CAZome reveals the multidomain enzyme strategy *Caldicellulosiruptor* species use to degrade biomass. These extracellular proteins contain between 1 and 9 domains, with the average protein having 2.8 domains. In addition, 13 of the 60 extracellular proteins contain more than one catalytic domain from the GH, PL, or CE families, which allows for bi-functional enzymes and an increased level of catalytic synergism from individual catalytic proteins. This multidomain extracellular CAZyme profile is in stark contrast to the CAZyme profile of the intracellular proteins, where all but two of the 84 proteins contain only one CAZy domain. This is not particularly surprising, as intracellular CAZymes are involved in the catabolism of mono- and oligo-saccharides that have already been liberated from the recalcitrant biomass substrate and transported into the cell.

Notable *Caldicellulosiruptor* CAZymes

A recent review (Blumer-Schuette *et al.*, 2014) catalogues the biochemically characterized CAZymes from *Caldicellulosiruptor* species. The role that a selected number of these proteins have in degrading a biomass substrate are shown in Fig. 4.1B, including: β-1,4-glucanase (CelA) GH9-CBM3-CBM3-CBM3-GH48 (Brunecky *et al.*, 2013; Yi *et al.*, 2013; Zverlov *et al.*, 1998), β-1,4-glucanase and β-1,4-xylanase GH5-CBM28-SLH (Ozdemir *et al.*, 2012), β-1,4-xylanase CBM22-CBM22-GH10 (Dwivedi *et al.*, 1996; Morris *et al.*, 1999; VanFossen *et al.*, 2011), β-1,4-mannanase CBM27-CBM27-CBM35-GH26 (Dwivedi *et al.*, 1996; Sunna, 2010), α-L-arabinofuranosidase GH43-CBM22-GH43-CBM6 (Bergquist *et al.*, 1999; VanFossen *et al.*, 2011), and pectin lyase PL3 (Alahuhta *et al.*, 2013). Also shown are two CE proteins and one PL protein whose properties have not been biochemically characterized, but are inferred from CAZy family classifications. These include the CE4 removing acetyl groups, a CE1-CBM13 protein disrupting the ferulic acid cross-linkage between xylan chains, and PL9-CBM3-CBM3 degrading the α-1,4-linked D-galacturonic acid backbone of pectin.

One of the most noteworthy and novel enzymes produced by the five highly cellulolytic *Caldicellulosiruptor* species is CelA (GH9-CBM3-CBM3-CBM3-GH48). This enzyme was first characterized from *C. bescii* (Athe_1867) as having a temperature optimum of 85°C and pH optimum of 5.0–6.0 on Avicel, a crystalline cellulose substrate, with additional weak activity on oat spelt xylan (Zverlov *et al.*, 1998). Proteomics work on the secretome of *C. bescii* revealed that CelA is the dominant cellulase produced by the species (Lochner *et al.*, 2011b). More recent work with various truncations of CelA have shown that

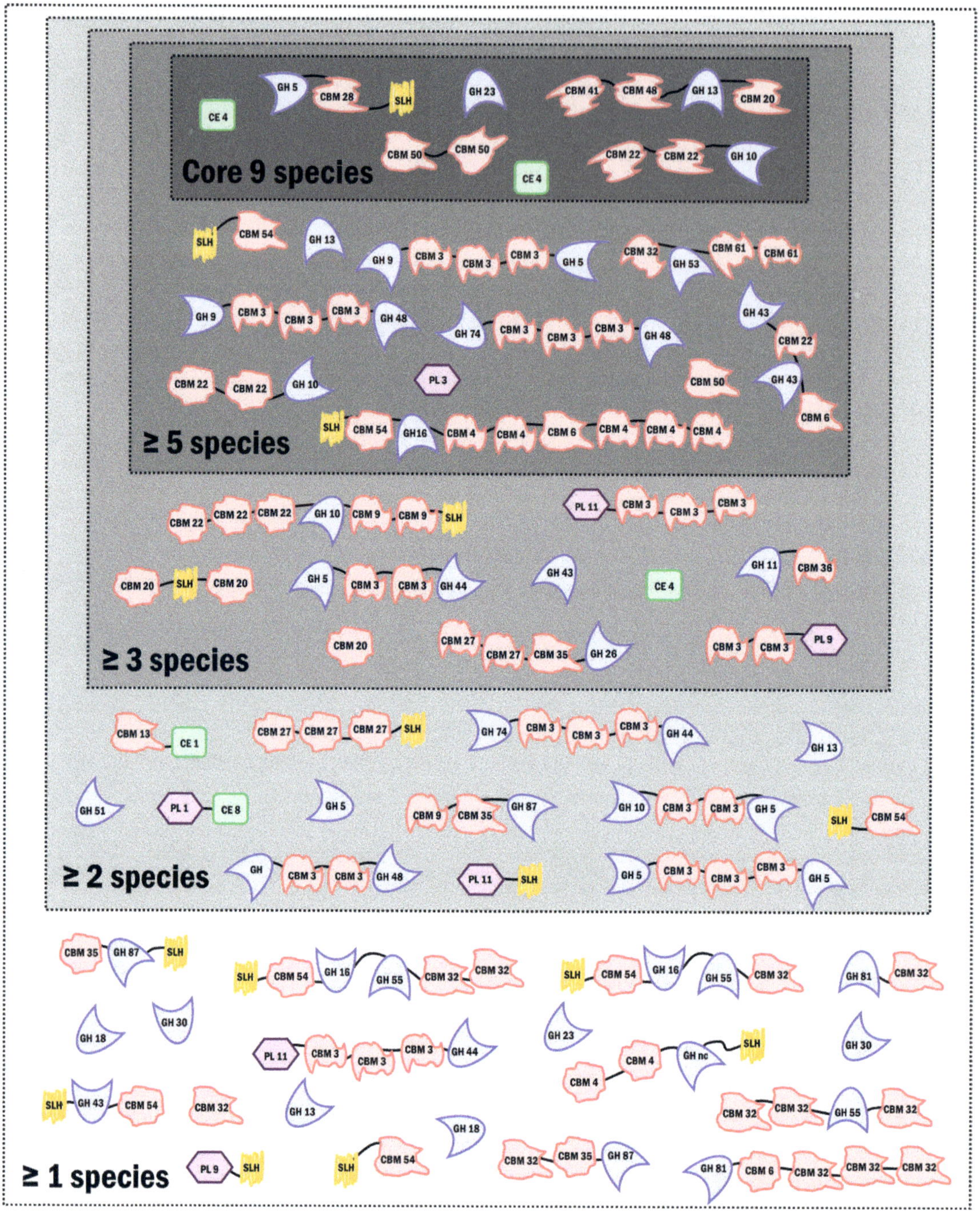

Figure 4.3 *Caldicellulosiruptor* extracellular pan-CAZome. Core extracellular CAZymes are contained in the genomes of all nine sequenced *Caldicellulosiruptor* species. The additional boxes encompass extracellular CAZymes contained in the genomes of five or more, three or more, two or more, and one or more species, respectively. CAZy domain family numbers are shown for Glycoside Hydrolase (GH), Carbohydrate Binding Module (CBM), Polysaccharide Lyase (PL), and Carbohydrate Esterase (CE) domains. Surface Layer Homology (SLH) domains are also shown, but additional predicted domains are omitted. Extracellular proteins were predicted to contain a signal peptide using SignalP4.1.

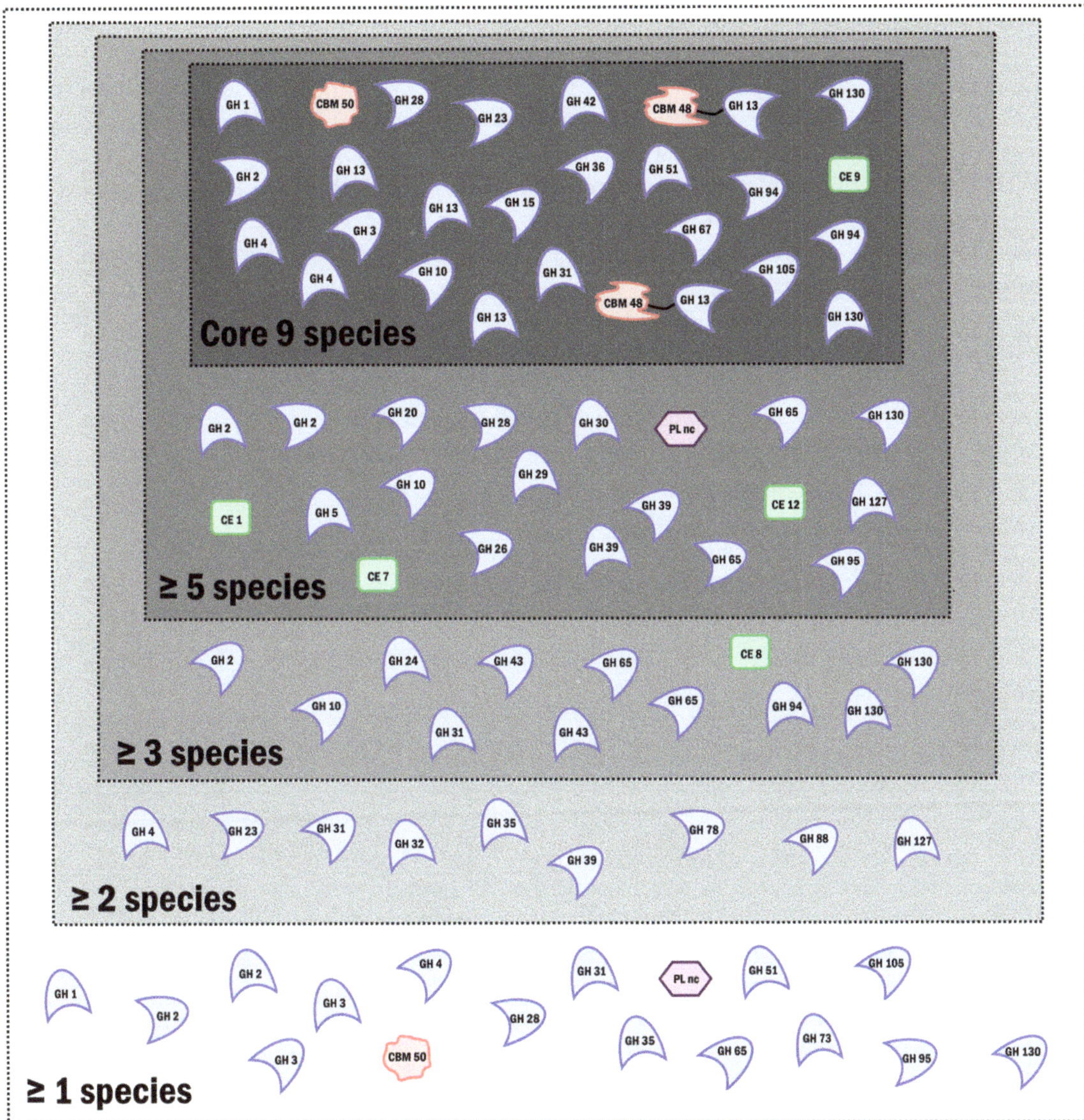

Figure 4.4 *Caldicellulosiruptor* intracellular pan-CAZome. Core intracellular CAZymes are contained in the genomes of all nine sequenced *Caldicellulosiruptor* species. The additional boxes encompass intracellular CAZymes contained in the genomes of five or more, three or more, two or more, and one or more species, respectively. CAZy domain family numbers are shown for Glycoside Hydrolase (GH), Carbohydrate Binding Module (CBM), Polysaccharide Lyase (PL), and Carbohydrate Esterase (CE) domains. The intracellular pan-CAZome contains no Surface Layer Homology (SLH) domain CAZymes. Additional predicted domains are omitted. Intracellular proteins are classified in this category if they are not predicted to contain a signal peptide using SignalP4.1.

the GH9 domain possesses the main catalytic activity on cellulose substrates, and the GH48 domain works synergistically with the GH9 to increase sugar release (Yi *et al.*, 2013). CelA has also been demonstrated to degrade the biomass substrates maize stover and switchgrass; while it appears to be inhibited by cellobiose, when a β-D-glucosidase was added with CelA, 100% glucan conversion from Avicel was achieved after 7 days at 75°C (Brunecky *et al.*, 2013). Additional work to elucidate the degradation mechanism of CelA showed that, unlike many cellulases which work from the ends of crystalline cellulose microfibrals, CelA burrows in the middle of the cellulose microfibrals creating cavities approximately 15–20 nm wide and 15–30 nm long (Brunecky *et al.*, 2013). This burrowing mechanism promotes the formation of new cellulose fibre chain ends that could be synergistically digested by other cellulases produced by *Caldicellulosiruptor* species. CelA is among the largest, most thermostable, and most active enzymes yet identified for the degradation of crystalline cellulose.

Linker regions in *Caldicellulosiruptor* CAZymes

CelA is located in the *Caldicellulosiruptor* 'glucan degrading locus' (GDL), along with a number of other CBM3-rich CAZymes (Blumer-Schuette *et al.*, 2012). Another feature of the genes in the GDL is that the linkers between the domains of these multidomain proteins are highly Pro/Thr-rich (VanFossen *et al.*, 2011). Linkers have been a noted feature in a number of CAZymes, often with high Gly, Ser, Pro, and Thr content, and are thought to be unfolded regions of disorder to allow flexibility between the folded ordered CAZy domains (Payne *et al.*, 2013). To probe the extracellular *Caldicellulosiruptor* CAZyme for these flexible linker sequences, the extracellular CAZymes from *C. kronotskyensis*, the species containing the most overall CAZymes, were submitted to the SEG segmentation algorithm, an online predictor of low complexity disordered regions (Wootton, 1994). SEG was used to identify the low complexity regions from these proteins, and further analysis was used to evaluate these regions for high Pro/Thr content. In *C. kronotskyensis*, all of the CAZymes located in the GDL from Calkro_0850 (CelA) to Calkro_0864 contain highly Pro/Thr-rich linkers, and, furthermore, these were the only CAZymes in the genome to exhibit this characteristic. Another potential linker motif was identified in Calkro_0402 in the region between the SLH domains and the reminder of the protein, where the SEG predicted region of disorder is highly Gln/Thr-rich. Flexible linkers that are Asn-rich or Gly-rich, instead of the typical Pro/Thr-rich, have been identified in CAZymes from the ruminal fungus, *Neocallimastix patriciarum* (Denman et al., 1996; Pai *et al.*, 2010). CAZyme linker length and flexibility has been probed by X-ray scattering, identifying a number of conformations, including those where the linker is extended, suggesting these regions are highly flexible (Receveur *et al.*, 2002; von Ossowski *et al.*, 2005). This characteristic of the linkers promotes synergy between the various linked domains, as each domain has the flexibility of movement to perform its role, binding or enzymatic, in the larger multidomain protein.

Another feature of these Pro/Thr-rich linkers is that that they are thought to be the sites of glycosylation of these *Caldicellulosiruptor* proteins. The Ser and Thr residues in linker regions of CAZymes are often O-glycosylated, which has been primarily studied in the CAZymes of *Trichoderma reesei* (Payne *et al.*, 2013; Stals *et al.*, 2004). Native purified CelA (Athe_1867) from *C. bescii* appears to be larger than the predicted molecular weight by gel electrophoresis (194 kDa predicted, 230 kDa observed), suggesting that the protein in glycosylated (Yi *et al.*, 2013; Zverlov *et al.*, 1998). Glycosylation has also been implicated for other CAZymes from the GDL: Athe_1857 (GH10-CBM3-CBM3-GH48), Athe_1865 (GH9-CBM3-CBM3-CBM3-GH5), and Athe_1859 (GH5-CBM3-CBM3-GH44), in a study of cell-free enzymes from *C. bescii* (Kanafusa-Shinkai *et al.*, 2013). In addition, glycosylation has been observed for components of the cellulosome in *C. thermocellum* and other cellulosomal microbes (Mazzoli *et al.*, 2012; Sabathe and Soucaille, 2003). This glycosylation may be a method to protect against proteases, a potential problem in the low complexity sequences of linker regions between domains in *Caldicellulosiruptor* enzymes (Upreti *et al.*, 2003). Alternatively, glycosylation may play a role in the dynamic binding to carbohydrate substrates and processive action of CAZymes, as was demonstrated by molecular dynamic simulation of *Trichoderma reesei* cellobiohydrolases Cel6A (GH6-CBM1) and Cel7A (GH7-CBM1) (Payne *et al.*, 2013). Glycosylation was also found to be an important factor in the rigidity of the linker in Cel45 from *Humicola insolens*, where a mutated linker with less glycosylation was found to be more rigid by X-ray scattering (Receveur *et al.*, 2002).

The *Caldicellulosiruptor* surface layer

A number of the multidomain CAZymes from the genus *Caldicellulosiruptor* also contain S-layer homology (SLH) domains to localize these proteins at the cell surface within the S-layer. This unique group of multidomain proteins includes both catalytic proteins containing GH5, GH10, GH16, GH43, GH55, GH87, GHnc, PL9, and PL11 domains and non-catalytic proteins containing CBM and immunoglobulin-like domains (Ozdemir *et al.*, 2012). A list of the *Caldicellulosiruptor* SLH domain containing proteins is shown in Table 4.2 and includes protein loci, amino acid length, and domain structure. These SLH multidomain proteins are present in all nine sequenced *Caldicellulosiruptor* species and a number of these proteins are unique to a single species. Because these proteins are localized to the cell surface, the interface between the cell and substrate, it is believed that these SLH multidomain proteins play a major role in the attachment to and degradation of lignocellulosic biomass by *Caldicellulosiruptor* species.

As stated previously, the S-layer in most microorganisms is made up of a single S-layer protein (SLP). Proteomic and transcriptomic experiments implicate Csac_2451, from *C. saccharolyticus,* and its homologues through the other *Caldicellulosiruptor* species, as the main SLP (Blumer-Schuette *et al.*, 2012, 2010; Dam *et al.*, 2011; Lochner *et al.*, 2011a). Nearly all of the SLH proteins in Table 4.2 are extracellular by the SignalP4.1 prediction, supporting the idea that these proteins are localized outside the cell at the cell surface. Those that do not have predicted signal peptides are truncated at the N-terminus, unlike homologues in the same group, which likely explains the lack of a predicted signal peptide. The SLH domains of these proteins are either at the N-terminus and C-terminus, with only one group of homologues having SLH domains between two CBM domains. The location of these domains at either terminus suggests that these proteins are tethered to the cell by the SLH domain containing end and then allowed to swing out away from the cell to interact with biomass with the remaining domains.

Csac_0678 (GH5-CBM28-SLH-SLH-SLH) is the only catalytic SLH protein to be conserved through all *Caldicellulosiruptor* species. Csac_0678 was found to be optimally active at 75°C and pH 5.0 on a variety of substrates containing β-1,3/4-glucan, β-1,4-xylan, and β-1,4-glucan linkages including cellulose (Ozdemir *et al.*, 2012). Other *Caldicellulosiruptor* SLH proteins are thought to play a role in helping the cell attach and tether itself to the biomass substrate. Csac_2722 is one such protein that has been found to bind Avicel, a crystalline cellulose substrate, and has been identified in the S-layer and cell membrane fraction of *C. saccharolyticus* grown on switchgrass (Ozdemir *et al.*, 2012). The proteins grouped as 'other' in Table 4.2 do not contain currently identified GH, PL, CE, or CBM domains, but a number of these proteins have been identified by proteomics as Avicel-bound (Blumer-Schuette *et al.*, 2012). These proteins presumably contain unidentified CAZy domains that facilitate their attachment to biomass. SLH proteins not involved in biomass attachment and degradation may play a role in the S-layer by helping to space the other SLH containing CAZymes, giving them adequate area to perform their enzymatic or binding roles.

FN3-containing *Caldicellulosiruptor* proteins

As previously discussed, little is known about the role of fibronectin type III (FN3)-like domains found in bacteria. A number of *Caldicellulosiruptor* proteins are predicted to contain these FN3 domains and are listed by homologous group in Table 4.3. These proteins are particularly large, ranging from 750 to 3027 amino acids in length. Of these, Athe_0012 is the largest protein found in the genomes of all currently sequenced *Caldicellulosiruptor* species. All of these proteins are predicted to be extracellular by SignalP4.1, except two GH3-FN3 groups, which are the smallest FN3-containing proteins. Three of these protein groups also contain SLH domains for localization at the surface of the bacterial cell.

At one time it was thought that in bacteria, FN3 domains were only found in extracellular glycoside hydrolase enzymes (Little *et al.*, 1994). However, four homologous groups are found

Table 4.2 Surface layer homology domain containing *Caldicellulosiruptor* proteins

	Locus	No. of amino acids	Domains
GH	Athe_0594, Calkro_2036, Calhy_2064, H557DRAFT_00803, Csac_0678, Calkr_2007, Calla_0352	755–756	GH5-CBM28[a]-SLH-SLH-SLH
	Calow_0459[a], COB47_0546[a]	567–568	
	Calkro_0072, Calhy_0060, COB47_0076	1732	SLH-SLH-SLH-CBM54-GH16-CBM4-CBM4-CBM6-CBM4-CBM4-CBM4
	Csac_2549[b]	526	
	H557DRAFT_02137	1411	
	Calkro_0402, Calow_1924	1672–1675	CBM22-CBM22-CBM22-GH10-CBM9-CBM9-(CE15[c])-SLH-SLH
	Calkr_2245[c]	2159	
	Calla_0206	1593	
	Calkro_0111	2435	SLH-SLH-SLH-CBM54-FN3-GH16-FLD-FN3-FN3-FLD-FN3-FN3-GH55-CBM32-CBM32
	Calkro_0121[b]	2229	
	Calhy_1629	1440	SLH-SLH-SLH-GH43-CBM54
	Calhy_2383	2007	CBM35-GH87-FN3-SLH-SLH-SLH
	Csac_2722	2593	CBM4-Big4-Coh-CBM4-Big4-GHnc-SLH-SLH-SLH
PL	Calkro_0550, Calow_1771	1789–1790	Big4-PL11-SLH-SLH
	Calow_2109	1711	Big3-Big3-Big3-PL9-SLH-SLH-SLH
CBM only	Calkr_1989[b,d]	326	CBM20-SLH-SLH-CBM20
	Calla_0367, Calow_0484, COB47_0564	1097	
	Calla_2176	1088	CBM27-CBM27-CBM27-SLH-SLH-SLH
	COB47_0167[b,d]	886	
	Calkr_2289	1495	SLH-SLH-SLH-CBM54
	Calla_0162[b]	480	
Other	Athe_0438, Calkro_2198, Calhy_2227, Calkr_2224, Calla_0227, Calow_0272, COB47_0390, Csac_0207, H557DRAFT_02082	1156–1157	SLH-SLH-Big5
	Athe_1943, Calkro_0749, Calhy_0832, Calkr_0758, Calla_1578, Calow_1666, COB47_1753, Csac_1009, H557DRAFT_02335	275–277	SLH-SLH
	Athe_2303, Calkro_0333, Calhy_0455, Calla_0125, Calkr_2323, COB47_2069, Csac_2451, H557DRAFT_01166	1013–1026	SLH
	Calow_1997	972	
	Athe_2341, Calkro_0210, Calhy_0185, Calkr_2370, Calla_0076, Calow_2070, COB47_2114, Csac_0591, H557DRAFT_01346	483–484	SLH-SLH
	Athe_2342, Calkro_0209, Calhy_0184, Calkr_2371, Calla_0075, Calow_2071, COB47_2115, Csac_0590, H557DRAFT_01347	1003–1010	SLH
	Athe_1839, Calkro_0875, Calkr_0834, Calla_1498, Calow_1583, COB47_1651, Csac_2381, H557DRAFT_00696	575–576	SLH-SLH-SLH
	Athe_2295, Calkro_0339, Calhy_0464, Calkr_2316, Calla_0133, COB47_2061, Csac_2445, H557DRAFT_01171	1074–1075	SLH-SLH

Table 4.2 continued

	Locus	No. of amino acids	Domains
Other	Calkr_2463, Calla_2324, Calow_0034, COB47_0063	1774–1779	SLH-SLH-SLH-Big1-Tglut
	Athe_0077	1710	
	Calhy_0047[d]	1626	
	Athe_0608, Calkro_2018, Calhy_2046, Calow_0482, COB47_0562	547	SLH-SLH-SLH
	Calkro_0360, Calkr_2450	793–798	SLH-SLH-SLH
	Calow_1967[d]	683	
	Calkro_0197, Calkr_2380	1052	SLH-SLH
	Calkro_0155, Calow_2111	549	MG2-SLH-SLH-SLH
	Calkro_0090, Calhy_0073	664	SLH-SLH-SLH
	Athe_0012	3027	SLH-SLH-FN3-vWFA-SH3-RHSrep-RHScore
	Calkro_0014	2994	
	Calkr_2514, Calla_2376	643	SLH-SLH-SLH
	Calkr_0022	254	SLH-SLH-SLH
	Calla_2302[b]	144	
	Athe_2223	591	SLH
	COB47_1200[bd]	477	
	Calkr_2453	1153	SLH-Big5
	H557DRAFT_02031[d]	128	SLH-SLH

[a]Calow_0459 and COB47_0546 have a truncated CBM28 domain.
[b]Fragment.
[c]Calkr_2245 contains an additional CE15 domain.
[d]No predicted signal peptide.
GH, glycoside hydrolase; PL, polysaccharide lyase; CE, carbohydrate esterase; CBM, carbohydrate binding module; SLH, surface layer homology (pfam00395); FN3, fibronectin type III-like (pfam00041); Big, bacterial immunoglobulin (CL0159); FLD, fascin-like domain (CL0066); Coh, cohesin-like (CL14606); Tglut, transglutaminase-like superfamily (pfam01841); MG2, macroglobulin2 (pfam01835); vWFA, Von Willebrand factor type A (pfam00092); SH3, bacterial SH3 (pfam08239); RHSrep, RHS repeat (pfam05593); RHScore, RHS associated core domain (TIGR03696).

in the *Caldicellulosiruptor* genus that contain no identified catalytic domains and, furthermore, no domains other than the predicted FN3 domains. An individual FN3 domain is only approximately 80 amino acids in length, leaving open the possibility that other, as yet unidentified, domains could be present in these proteins. Four of the homologous protein groups, found in Table 4.3, do contain catalytic glycoside hydrolase domains. For these proteins, the FN3 domain may contribute to catalytic ability, binding, thermostability, or domain spacing. However, the definitive role that these FN3 domains play in biomass degradation by the genus *Caldicellulosiruptor*, and more broadly in bacteria, remains unknown.

Comparison to *Clostridium thermocellum*

As mentioned, one of the best studied cellulolytic thermophilic microorganisms other than the *Caldicellulosiruptor* species is *C. thermocellum*, which primarily uses a cellulosomal mechanism to coordinate its inventory of CAZymes to degrade biomass. Both *Caldicellulosiruptor* species and *C. thermocellum* are of particular interest to engineer as metabolic engineering platforms for consolidated bioprocessing (CBP) of lignocellulosic biomass to biofuels. Table 4.4 compares the extracellular catalytic CAZyme domain inventory of *C. thermocellum* to that of *Caldicellulosiruptor kronotskyensis*, the *Caldicellulosiruptor*

Table 4.3 Fibronectin type III-like domain containing *Caldicellulosiruptor* proteins

Locus	# AA	SignalP	Domains
Athe_1944, Calhy_0831, Calkro_0748, COB47_1754, Calow_1667, Csac_1008, H557DRAFT_02334	1201–1207	Y	FN3-FN3
Calla_1579, Calkr_0757	798		
Athe_1945, Calhy_0830, Calkro_0747, COB47_1755, Calow_1668, Csac_1007, H557DRAFT_02333	1265	Y	FN3-FN3
Athe_0012	3027	Y	SLH-SLH-FN3-vWFA-SH3-RHSrep-RHScore
Calkro_0014	2994		
Calkro_0111	2435	Y	SLH-SLH-SLH-CBM54-FN3-GH16-FLD-FN3-FN3-FLD-FN3-FN3-GH55-CBM32-CBM32
Calkro_0121[a]	2229		
Calkro_0153	983	Y	FN3-FN3
Calkro_0113	1240	Y	CBM32-CBM32-FN3-GH55-CBM32
Calhy_2383	2007	Y	CBM35-GH87-FN3-SLH-SLH-SLH
Calow_2113	983	Y	FN3
Athe_2354, Calhy_0170, Calkr_2396, Calkro_0181, Calla_0066, COB47_2124, Calow_2083, Csac_0586, H557DRAFT_01368	770–771	N	GH3-FN3
Csac_1102	750	N	GH3-FN3

[a]Fragment.
GH, glycoside hydrolase; CBM, carbohydrate binding module; SLH, surface layer homology (pfam00395); FN3, fibronectin type III-like (pfam00041); FLD, fascin-like domain (CL0066); vWFA, Von Willebrand factor type A (pfam00092); SH3, bacterial SH3 (pfam08239); RHSrep, RHS repeat (pfam05593); RHScore, RHS associated core domain (TIGR03696).

species with the largest CAZyme inventory. Each catalytic domain found in multidomain proteins with multiple catalytic domains is cataloged separately. A number of *C. thermocellum* strains have sequenced genomes, but most work has focused on two of these strains: *C. thermocellum* ATCC 27405 and *C. thermocellum* DSM 1313. All extracellular CAZymes from the ATCC 27405 strain are found in DSM 1313 and only 3 proteins (containing GH11, GH43, and GH18) are unique to DSM 1313. Because of these very few differences between the two strains, they were collapsed into a single CAZyme domain profile for comparison to *C. kronotskyensis* in Table 4.4. Again, extracellular proteins for both *C. thermocellum* and *C. kronotskyensis* were predicted by identifying signal peptides using SignalP4.1 (Petersen *et al.*, 2011).

One of the major differences between the catalytic domain inventory of these microbes shown in Table 4.4 is that *C. thermocellum* has almost double the catalytic CAZymes domains of *C. kronotskyensis* (70 vs. 40). However the number of domain families is fairly consistent between the organisms: 17 vs. 15 GH families, 4 vs. 3 PL families, 7 vs. 2 CE families for *C. thermocellum* vs. *C. kronotskyensis*, respectively. Interestingly, almost all (61 out of 70) of the catalytic CAZyme domains from *C. thermocellum* are associated with a dockerin domain that would mediate their attachment to the cellulosome. The remaining 9 catalytic CAZyme domains, which are each on separate proteins, are likely free-enzymes. Some of these *C. thermocellum* free enzymes have high similarity to *C. kronotskyensis* CAZymes, particularly Cthe_0040 (GH9-CBM3-CBM3) and Cthe_0071 (GH48-CBM3), which mirror the structure of the CBM3-rich cellulases of the *Caldicellulosiruptor* glucan degradation locus (GDL), and Cthe_2809 (SLH-SLH-SLH-CBM54-GH16-CBM4-CBM4-CBM4-CBM4), which has high similarity to Calkro_0072. In addition to the catalytic CAZy domain containing proteins, *C. thermocellum* contains 9 extracellular proteins that contain only CBM domains compared to three exclusively CBM-containing proteins in

Table 4.4 *C. thermocellum* and *C. kronotskyensis* extracellular catalytic CAZy domain inventory comparison

Extracellular CAZy domains	*C. thermocellum*	*C. kronotskyensis*
GH3	1	0
GH5	8	4
GH8	1	0
GH9	15	2
GH10	4	4
GH11	2	1
GH13	0	3
GH16	2	3
GH18	3	0
GH23	1	1
GH26	3	0
GH30	2	0
GH39	0	0
GH43	6	2
GH44	2	1
GH48	2	3
GH53	1	1
GH55	0	3
GH74	1	1
GH81	0	2
GH87	0	1
GH94	1	0
GH124	0	0
PL1	2	1
PL3	1	1
PL9	1	0
PL11	1	2
CE1	3	0
CE2	1	0
CE3	1	0
CE4	2	3
CE8	1	1
CE12	1	0
CEnc	1	0
Total	**70**	**40**

C. kronotskyensis. These proteins are presumably involved in biomass attachment.

To compare the cellulosomal and multidomain mechanisms that *C. thermocellum* and *C. kronotskyensis*, respectively, use to degrade biomass, the number of CAZy domains (GH, PL, CE, CBM) per extracellular protein was calculated. *C. thermocellum* contains 136 CAZy domains in 72 total extracellular proteins, while *C. kronotskyensis* contains 103 CAZy domains in 34 extracellular proteins. *C. kronotskyensis* has about one more CAZy domain per extracellular protein

(3.0 domains/protein) than *C. thermocellum* (1.8 domains/protein) on average. This confirms the multidomain strategy that *C. kronotskyensis* uses to degrade biomass and highlights one of the major differences between these cellulolytic thermophiles.

Conclusions

Members of the genus *Caldicellulosiruptor* are the most thermophilic, cellulolytic bacteria known. *Caldicellulosiruptor* species accomplish degradation of unpretreated lignocellulosic biomass by using a collection of unique large multidomain CAZymes. The multidomain architecture of these proteins promotes synergy between the various catalytic and binding domains within a single protein to efficiently degrade biomass. An important subset of the CAZymes found in the genus *Caldicellulosiruptor* includes SLH domains to localize these proteins to the cell surface. These SLH proteins also contain CBM domains for mediating attachment of the cell to the biomass substrate, while various catalytic GH, PL, and CE domains in the same protein digest it. Additionally, a number of accessory domains, such as FN3 domains, are a part of a number of *Caldicellulosiruptor* CAZymes. The definitive functions of these domains are largely unknown in the context of biomass degradation, and remain an area of ongoing investigation. The *Caldicellulosiruptor* multidomain mechanism for biomass degradation is a distinct alternative to the cellulosome mechanism of other cellulolytic thermophiles, such as *Clostridium thermocellum*. The ability of *Caldicellulosiruptor* to degrade unpretreated biomass, using a unique inventory of multidomain enzymes, at temperatures above 70°C, makes these species leading candidates for development into CBP microorganisms for the conversion of biomass to biofuels.

Future trends

Metagenomics: identifying novel CAZymes

Identification of novel proteins related to biomass degradation is becoming more possible with the decreasing cost of DNA sequencing and the development of metagenomics. This sequence-based analysis takes advantage of the great diversity in microbial communities, particularly of microorganisms that would not be culturable using standard microbiological techniques (Handelsman, 2004; Sleator *et al.*, 2008). A standard metagenomic experiment would take an environmental sample, potentially enrich the natural microbiota in a particular subset of microbes by culturing under selective condition (e.g. high temperature for thermophiles, media with crystalline cellulose for cellulolytic microbes), isolate DNA from the microbial community in the sample, and sequence all of the isolated DNA for analysis (Handelsman, 2004). This technique has been used to identify novel CAZymes from the human gut microbiota (El Kaoutari *et al.*, 2013), decaying wood (Li *et al.*, 2011), buffalo rumen (Duan *et al.*, 2009), and German grassland soil samples (Nacke *et al.*, 2012). Once a CAZyme gene of interest is identified, it can be cloned for biochemical characterization, as was done in the German grassland soil study to identify a novel CBM9-GH9 cellulase optimally active at 45–50°C and two novel GH11 xylanases optimally active at 60°C (Nacke *et al.*, 2012). Novel enzymes of particular interest can then be inserted into the genome of a CBP microorganism to improve lignocellulosic biomass degradation.

Caldicellulosiruptor bescii genetic system

Ultimately, the development of a CBP microorganism depends on the ability to genetically manipulate a platform organism to adjust the cellulolytic enzyme inventory and metabolic pathways for optimized biomass degradation and fuel molecule production. A genetic system for *Caldicellulosiruptor bescii* has only recently been developed, and work to make it tractable and extend genetic tools into other *Caldicellulosiruptor* species is continuing (Cha *et al.*, 2013; Chung *et al.*, 2013a–c, 2012, 2011; Farkas *et al.*, 2013). This genetic system in *C. bescii* is constructed around a uracil auxotroph ΔpyrFA (Cbes_1377) disruption mutant. The gene *pyrF* encodes for orotidine monophosphate decarboxylase, an enzyme essential for uracil biosynthesis. This allows for selection of uracil auxotroph mutants because uracil prototrophs with functioning PyrF will ultimately convert 5-fluoroorotic acid (5-FOA) into the toxic product fluorodeoxyuridine (Chung *et al.*, 2013a). The uracil auxotroph mutant can then be complemented with a shuttle vector containing the *pyrF* gene under the control of the 30S ribosomal subunit (Cbes_2105) regulatory region (Chung *et al.*, 2013a). This ΔpyrFA strain was used to successfully knockout the gene Cbes_1918 encoding lactate dehydrogenase (*ldh*) to metabolically engineer *C. bescii* to divert metabolic flux away from lactate and increase acetate and hydrogen production (Cha *et al.*, 2013).

Metabolic engineering of a CBP thermophile

Even with a tractable genetic system, the genetic manipulations necessary to engineering a highly productive consolidating bioprocessing thermophile are non-trivial. Genetic manipulations will be needed to engineer both the front-end CAZymes inventory to promote complete degradation of the biomass substrate, and the back-end metabolism to divert flux to products of interest (i.e. alcohols or other fuel molecules). This front-end engineering will likely involve knocking genes into the CBP species. As seen in the *Caldicellulosiruptor* genus, a number of novel and potentially very productive biomass degrading enzymes are found only in a single species and these would need to be added together into the engineered strain. Likewise, metagenomic studies in cellulosic environments will no doubt reveal novel thermophilic CAZymes from a diversity of species that could complement the *Caldicellulosiruptor* pan-CAZome in an engineered strain. The two main challenges of the back-end metabolic engineering are a relative dearth of precise information about the metabolism of these cellulolytic thermophiles and an absence of a drop-in thermophilic liquid fuel production pathway. While we have a significant understanding of the metabolism of *Caldicellulosiruptor* species, there are still holes in our knowledge that are actively being pursued. A more precise understanding of *Caldicellulosiruptor* metabolism is necessary to target gene knockouts to direct metabolic flux away from byproducts to products of interest. While small amounts of alcohols are produced by *Caldicellulosiruptor* species, it is unclear at this early stage whether metabolism can be significantly diverted to fuels using only the *Caldicellulosiruptor* metabolism. In the case that it is not, another metabolic engineering strategy is to add all of the components of an alcohol production pathway from another source into the CBP host. However, currently such a thermophilic pathway does not exist at the growth temperatures of *Caldicellulosiruptor* species (>70°C). Work continues to develop our understanding and manipulate both the front-end biomass degradation and back-end metabolic flux of the *Caldicellulosiruptor* genus.

Acknowledgements

This work was funded in part by the BioEnergy Science Center, a US Department of Energy Bioenergy Research Center supported by the Office of Biological and Environmental Research in the DOE Office of Science.

Web resources

Carbohydrate-Active enZYmes Database: www.cazy.org
National Center for Biotechnology Information (NCBI): http://www.ncbi.nlm.nih.gov/
NCBI Conserved Domains Search: http://www.ncbi.nlm.nih.gov/Structure/cdd/wrpsb.cgi
Pfam 27.0 Database: http://pfam.sanger.ac.uk/
SignalP4.1 Server: http://www.cbs.dtu.dk/services/SignalP/
SEG Prediction of Low Complexity Regions: http://mendel.imp.ac.at/METHODS/seg.server.html

References

Alahuhta, M., Xu, Q., Bomble, Y.J., Brunecky, R., Adney, W.S., Ding, S.Y., Himmel, M.E., and Lunin, V.V. (2010). The unique binding mode of cellulosomal CBM4 from *Clostridium thermocellum* cellobiohydrolase A. J. Mol. Biol. *402*, 374–387.

Alahuhta, M., Brunecky, R., Chandrayan, P., Kataeva, I., Adams, M.W., Himmel, M.E., and Lunin, V.V. (2013). The structure and mode of action of *Caldicellulosiruptor bescii* family 3 pectate lyase in biomass deconstruction. Acta Crystallogr. D Biol. Crystallogr. *69*, 534–539.

Ali, M.K., Kimura, T., Sakka, K., and Ohmiya, K. (2001). The multidomain xylanase Xyn10B as a cellulose-binding protein in *Clostridium stercorarium*. FEMS Microbiol. Lett. *198*, 79–83.

Arantes, V., and Saddler, J.N. (2010). Access to cellulose limits the efficiency of enzymatic hydrolysis: the role of amorphogenesis. Biotechnol. Biofuels 3, 4.

Aurilia, V., Parracino, A., and D'Auria, S. (2008). Microbial carbohydrate esterases in cold adapted environments. Gene *410*, 234–240.

Barnard, D., Casanueva, A., Tuffin, M., and Cowan, D. (2010). Extremophiles in biofuel synthesis. Environ. Technol. *31*, 871–888.

Barr, B.K., Hsieh, Y.-L., Ganem, B., and Wilson, D.B. (1996). Identification of two functionally different classes of exocellulases. Biochemistry *35*, 586–592.

Basen, M., Rhaesa, A.M., Kataeva, I., Prybol, C.J., Scott, I.M., Poole, F.L., and Adams, M.W.W. (2014). Degradation of high loads of crystalline cellulose and of unpretreated plant biomass by the thermophilic bacterium *Caldicellulosiruptor bescii*. Bioresour. Technol. *152*, 384–392.

Bayer, E.A., Chanzy, H., Lamed, R., and Shoham, Y. (1998). Cellulose, cellulases and cellulosomes. Curr. Opin. Struct. Biol. *8*, 548–557.

Bayer, E.A., Lamed, R., White, B.A., and Flint, H.J. (2008). From cellulosomes to cellulosomics. Chem. Rec. *8*, 364–377.

Bergquist, P.L., Gibbs, M.D., Morris, D.D., Te'o, V.S.J., Saul, D.J., and Morgan, H.W. (1999). Molecular diversity of thermophilic cellulolytic and hemicellulolytic bacteria. FEMS Microbiol. Ecol. *28*, 99–110.

Bianchetti, C.M., Elsen, N.L., Fox, B.G., and Phillips, G.N., Jr. (2011). Structure of cellobiose phosphorylase from *Clostridium thermocellum* in complex with phosphate. Acta Crystallogr. Sect. F Struct. Biol. Cryst. Commun. *67*, 1345–1349.

Biely, P. (2012). Microbial carbohydrate esterases deacetylating plant polysaccharides. Biotechnol. Adv. *30*, 1575–1588.

Biely, P., Mackenzie, C.R., Puls, J., and Schneider, H. (1986). Cooperativity of esterases and xylanases in the enzymatic degradation of acetyl xylan. Bio-Technology *4*, 731–733.

Blumer-Schuette, S.E., Kataeva, I., Westpheling, J., Adams, M.W., and Kelly, R.M. (2008). Extremely thermophilic microorganisms for biomass conversion: status and prospects. Curr. Opin. Biotechnol. *19*, 210–217.

Blumer-Schuette, S.E., Lewis, D.L., and Kelly, R.M. (2010). Phylogenetic, microbiological, and glycoside hydrolase diversities within the extremely thermophilic, plant biomass-degrading genus *Caldicellulosiruptor*. Appl. Environ. Microbiol. *76*, 8084–8092.

Blumer-Schuette, S.E., Ozdemir, I., Mistry, D., Lucas, S., Lapidus, A., Cheng, J.-F., Goodwin, L.A., Pitluck, S., Land, M.L., Hauser, L.J., *et al.* (2011). Complete Genome Sequences for the Anaerobic, Extremely Thermophilic Plant Biomass-Degrading Bacteria *Caldicellulosiruptor hydrothermalis, Caldicellulosiruptor kristjanssonii, Caldicellulosiruptor kronotskyensis, Caldicellulosiruptor owensensis*, and *Caldicellulosiruptor lactoaceticus*. J. Bacteriol. *193*, 1483–1484.

Blumer-Schuette, S.E., Giannone, R.J., Zurawski, J.V., Ozdemir, I., Ma, Q., Yin, Y., Xu, Y., Kataeva, I., Poole, F.L. 2nd, Adams, M.W., *et al.* (2012). Caldicellulosiruptor core and pangenomes reveal determinants for noncellulosomal thermophilic deconstruction of plant biomass. J. Bacteriol. *194*, 4015–4028.

Blumer-Schuette, S.E., Brown, S.D., Sander, K.B., Bayer, E.A., Kataeva, I., Zurawski, J.V., Conway, J.M., Adams, M.W.W., and Kelly, R.M. (2014). Thermophilic lignocellulose deconstruction. FEMS Microbiol. Rev. *38*, 393–448.

Blumer-Schuette, S.E., Alahuhta, M., Conway, J.M., Lee, L.L., Zurawski, J.V., Giannone, R.J., Hettich, R.L., Lunin, V.V., Himmel, M.E., and Kelly, R.M. (2015). Discrete and structurally unique proteins (Tāpirins) mediate attachment of extremely thermophilic *Caldicellulosiruptor* species to cellulose. J. Biol. Chem. *290*, 10645–10656.

Boraston, A.B., Kwan, E., Chiu, P., Warren, R.A., and Kilburn, D.G. (2003). Recognition and hydrolysis of noncrystalline cellulose. J. Biol. Chem. *278*, 6120–6127.

Boraston, A.B., Bolam, D.N., Gilbert, H.J., and Davies, G.J. (2004). Carbohydrate-binding modules: fine-tuning polysaccharide recognition. Biochem. J. *382*, 769–781.

Bredholt, S., Sonne-Hansen, J., Nielsen, P., Mathrani, I.M., and Ahring, B.K. (1999). *Caldicellulosiruptor kristjanssonii* sp. nov., a cellulolytic, extremely thermophilic, anaerobic bacterium. Int. J. Syst. Bacteriol. *49 Pt 3*, 991–996.

Brink, J., and Vries, R. (2011). Fungal enzyme sets for plant polysaccharide degradation. Appl. Microbiol. Biotechnol. *91*, 1477–1492.

Brunecky, R., Alahuhta, M., Bomble, Y.J., Xu, Q., Baker, J.O., Ding, S.Y., Himmel, M.E., and Lunin, V.V. (2012). Structure and function of the *Clostridium thermocellum* cellobiohydrolase A X1-module repeat: enhancement through stabilization of the CbhA complex. Acta Crystallogr. D Biol. Crystallogr. *68*, 292–299.

Brunecky, R., Alahuhta, M., Xu, Q., Donohoe, B.S., Crowley, M.F., Kataeva, I.A., Yang, S.-J., Resch, M.G., Adams, M.W.W., Lunin, V.V., *et al.* (2013). Revealing nature's cellulase diversity: the digestion mechanism of *Caldicellulosiruptor bescii* CelA. Science *342*, 1513–1516.

Cai, Y., Lai, C., Li, S., Liang, Z., Zhu, M., Liang, S., and Wang, J. (2011). Disruption of lactate dehydrogenase through homologous recombination to improve bioethanol production in *Thermoanaerobacterium aotearoense*. Enzyme Microb. Tech. *48*, 155–161.

Campbell, I.D., and Spitzfaden, C. (1994). Building proteins with fibronectin type III modules. Structure *2*, 333–337.

Cantarel, B.L., Coutinho, P.M., Rancurel, C., Bernard, T., Lombard, V., and Henrissat, B. (2009). The Carbohydrate-Active EnZymes database (CAZy): an expert resource for glycogenomics. Nucleic Acids Res. *37*, D233–238.

Cha, M., Chung, D., Elkins, J.G., Guss, A.M., and Westpheling, J. (2013). Metabolic engineering of *Caldicellulosiruptor bescii* yields increased hydrogen production from lignocellulosic biomass. Biotechnol. Biofuels *6*, 85.

Chothia, C., and Jones, E.Y. (1997). The molecular structure of cell adhesion molecules. Annu. Rev. Biochem. *66*, 823–862.

Chung, D.H., Huddleston, J.R., Farkas, J., and Westpheling, J. (2011). Identification and characterization of CbeI, a novel thermostable restriction enzyme from *Caldicellulosiruptor bescii* DSM 6725 and a member of a new subfamily of HaeIII-like enzymes. J. Ind. Microbiol. Biotechnol. *38*, 1867–1877.

Chung, D., Farkas, J., Huddleston, J.R., Olivar, E., and Westpheling, J. (2012). Methylation by a unique alpha-class N4-cytosine methyltransferase is required for DNA transformation of *Caldicellulosiruptor bescii* DSM6725. PLoS One 7, e43844.

Chung, D., Cha, M., Farkas, J., and Westpheling, J. (2013a). Construction of a stable replicating shuttle vector for *Caldicellulosiruptor* species: use for extending genetic methodologies to other members of this genus. PLoS One *8*, e62881.

Chung, D., Farkas, J., and Westpheling, J. (2013b). Detection of a novel active transposable element in *Caldicellulosiruptor hydrothermalis* and a new search for elements in this genus. J. Ind. Microbiol. Biotechnol. *40*, 517–521.

Chung, D., Farkas, J., and Westpheling, J. (2013c). Overcoming restriction as a barrier to DNA transformation in Caldicellulosiruptor species results in efficient marker replacement. Biotechnol. Biofuels *6*, 82.

Cosgrove, D.J. (2000). Loosening of plant cell walls by expansins. Nature *407*, 321–326.

Cosgrove, D.J. (2005). Growth of the plant cell wall. Nat. Rev. Mol. Cell Biol. *6*, 850–861.

Cosgrove, D.J., Li, L.C., Cho, H.-T., Hoffmann-Benning, S., Moore, R.C., and Blecker, D. (2002). The growing world of expansins. Plant Cell Physiol. *43*, 1436–1444.

Dam, P., Kataeva, I., Yang, S.J., Zhou, F., Yin, Y., Chou, W., Poole, F.L. 2nd, Westpheling, J., Hettich, R., Giannone, R., *et al.* (2011). Insights into plant biomass conversion from the genome of the anaerobic thermophilic bacterium *Caldicellulosiruptor bescii* DSM 6725. Nucleic Acids Res. *39*, 3240–3254.

Davies, G.J., Planas, A., and Rovira, C. (2011). Conformational analyses of the reaction coordinate of glycosidases. Accounts Chem. Res. *45*, 308–316.

Denman, S., Xue, G.P., and Patel, B. (1996). Characterization of a *Neocallimastix patriciarum* cellulase cDNA (*celA*) homologous to *Trichoderma reesei* cellobiohydrolase II. Appl. Environ. Microbiol. *62*, 1889–1896.

Devillard, E., Goodheart, D.B., Karnati, S.K., Bayer, E.A., Lamed, R., Miron, J., Nelson, K.E., and Morrison, M. (2004). Ruminococcus albus 8 mutants defective in cellulose degradation are deficient in two processive endocellulases, Cel48A and Cel9B, both of which possess a novel modular architecture. J. Bacteriol. *186*, 136–145.

Din, N., Gilkes, N.R., Tekant, B., Miller, R.C., Warren, R.A.J., and Kilburn, D.G. (1991). Non-hydrolytic disruption of cellulose gibres by the binding domain of a bacterial cellulase. Nat. Biotech. *9*, 1096–1099.

Duan, C.J., Xian, L., Zhao, G.C., Feng, Y., Pang, H., Bai, X.L., Tang, J.L., Ma, Q.S., and Feng, J.X. (2009). Isolation and partial characterization of novel genes encoding acidic cellulases from metagenomes of buffalo rumens. J. Appl. Microbiol. *107*, 245–256.

Dwivedi, P.P., Gibbs, M.D., Saul, D.J., and Bergquist, P.L. (1996). Cloning, sequencing and overexpression in *Escherichia coli* of a xylanase gene, xynA from the thermophilic bacterium Rt8B.4 genus *Caldicellulosiruptor*. Appl. Microbiol. Biotechnol. *45*, 86–93.

El Kaoutari, A., Armougom, F., Gordon, J.I., Raoult, D., and Henrissat, B. (2013). The abundance and variety of carbohydrate-active enzymes in the human gut microbiota. Nat. Rev. Microbiol. *11*, 497–504.

Elkins, J.G., Lochner, A., Hamilton-Brehm, S.D., Davenport, K.W., Podar, M., Brown, S.D., Land, M.L., Hauser, L.J., Klingeman, D.M., Raman, B., *et al.* (2010). Complete Genome Sequence of the Cellulolytic Thermophile *Caldicellulosiruptor obsidiansis* OB47T. J. Bacteriol. *192*, 6099–6100.

Engelhardt, H., and Peters, J. (1998). Structural research on surface layers: a focus on stability, surface layer homology domains, and surface layer-cell wall interactions. J. Struct. Biol. *124*, 276–302.

Farkas, J., Chung, D., Cha, M., Copeland, J., Grayeski, P., and Westpheling, J. (2013). Improved growth media and culture techniques for genetic analysis and assessment of biomass utilization by *Caldicellulosiruptor bescii*. J. Ind. Microbiol. Biotechnol. *40*, 41–49.

Faulds, C.B., and Williamson, G. (1995). Release of ferulic acid from wheat bran by a ferulic acid esterase (FAE-III) from *Aspergillus niger*. Appl. Microbiol. Biotechnol. *43*, 1082–1087.

Fontes, C.M., and Gilbert, H.J. (2010). Cellulosomes: highly efficient nanomachines designed to deconstruct plant cell wall complex carbohydrates. Annu. Rev. Biochem. *79*, 655–681.

Fraser, J.S., Yu, Z., Maxwell, K.L., and Davidson, A.R. (2006). Ig-like domains on bacteriophages: a tale of promiscuity and deceit. J. Mol. Biol. *359*, 496–507.

Fushinobu, S., Alves, V.D., and Coutinho, P.M. (2013). Multiple rewards from a treasure trove of novel glycoside hydrolase and polysaccharide lyase structures: new folds, mechanistic details, and evolutionary relationships. Curr. Opin. Struct. Biol. *23*, 652–659.

Garron, M.L., and Cygler, M. (2010). Structural and mechanistic classification of uronic acid-containing polysaccharide lyases. Glycobiology *20*, 1547–1573.

Georgelis, N., Nikolaidis, N., and Cosgrove, D.J. (2014). Biochemical analysis of expansin-like proteins from microbes. Carbohyd. Polym. *100*, 17–23.

Gibbs, M.D., Reeves, R.A., Farrington, G.K., Anderson, P., Williams, D.P., and Bergquist, P.L. (2000). Multidomain and multifunctional glycosyl hydrolases from the extreme thermophile Caldicellulosiruptor isolate Tok7B.1. Curr. Microbiol. *40*, 333–340.

Gilbert, H.J., Stålbrand, H., and Brumer, H. (2008). How the walls come crumbling down: recent structural biochemistry of plant polysaccharide degradation. Curr. Opin. Plant Biol. *11*, 338–348.

Gourlay, K., Arantes, V., and Saddler, J.N. (2012). Use of substructure-specific carbohydrate binding modules to track changes in cellulose accessibility and surface morphology during the amorphogenesis step of enzymatic hydrolysis. Biotechnol. Biofuels *5*, 51.

Grabber, J. (2005). How Do Lignin Composition, Structure, and Cross-Linking Affect Degradability?

A Review of Cell Wall Model Studies. Crop Sci *45*, 820–831.

Guillen, D., Sanchez, S., and Rodriguez-Sanoja, R. (2010). Carbohydrate-binding domains: multiplicity of biological roles. Appl. Microbiol. Biotechnol. *85*, 1241–1249.

Hamilton-Brehm, S.D., Mosher, J.J., Vishnivetskaya, T., Podar, M., Carroll, S., Allman, S., Phelps, T.J., Keller, M., and Elkins, J.G. (2010). *Caldicellulosiruptor obsidiansis* sp. nov., an Anaerobic, Extremely Thermophilic, Cellulolytic Bacterium Isolated from Obsidian Pool, Yellowstone National Park. Appl. Environ. Microbiol. *76*, 1014–1020.

Handelsman, J. (2004). Metagenomics: application of genomics to uncultured microorganisms. Microbiol. Mol. Biol. Rev. *68*, 669–685.

Hong, M.-R., Kim, Y.-S., Park, C.-S., Lee, J.-K., Kim, Y.-S., and Oh, D.-K. (2009). Characterization of a recombinant β-glucosidase from the thermophilic bacterium *Caldicellulosiruptor saccharolyticus*. J. Biosci. Bioeng. *108*, 36–40.

Huang, C.Y., Patel, B.K., Mah, R.A., and Baresi, L. (1998). *Caldicellulosiruptor owensensis* sp. nov., an anaerobic, extremely thermophilic, xylanolytic bacterium. Int. J. Syst. Bacteriol. *48 (Pt 1)*, 91–97.

Ishola, M.M., Jahandideh, A., Haidarian, B., Brandberg, T., and Taherzadeh, M.J. (2013). Simultaneous saccharification, filtration and fermentation (SSFF): a novel method for bioethanol production from lignocellulosic biomass. Bioresour. Technol. *133*, 68–73.

Jager, G., Girfoglio, M., Dollo, F., Rinaldi, R., Bongard, H., Commandeur, U., Fischer, R., Spiess, A.C., and Buchs, J. (2011). How recombinant swollenin from *Kluyveromyces lactis* affects cellulosic substrates and accelerates their hydrolysis. Biotechnol. Biofuels *4*, 33.

Janeček, Š., and Blesák, K. (2011). Sequence-structural features and evolutionary relationships of family GH57 α-amylases and their putative α-amylase-like homologues. Protein J. *30*, 429–435.

Jolie, R.P., Duvetter, T., Van Loey, A.M., and Hendrickx, M.E. (2010). Pectin methylesterase and its proteinaceous inhibitor: a review. Carbohyd. Res. *345*, 2583–2595.

Ju, X., Engelhard, M., and Zhang, X. (2013). An advanced understanding of the specific effects of xylan and surface lignin contents on enzymatic hydrolysis of lignocellulosic biomass. Bioresour. Technol. *132*, 137–145.

Kanafusa-Shinkai, S., Wakayama, J.i., Tsukamoto, K., Hayashi, N., Miyazaki, Y., Ohmori, H., Tajima, K., and Yokoyama, H. (2013). Degradation of microcrystalline cellulose and non-pretreated plant biomass by a cell-free extracellular cellulase/hemicellulase system from the extreme thermophilic bacterium *Caldicellulosiruptor bescii*. J. Biosci. Bioeng. *115*, 64–70.

Kang, K., Wang, S., Lai, G., Liu, G., and Xing, M. (2013). Characterization of a novel swollenin from *Penicillium oxalicum* in facilitating enzymatic saccharification of cellulose. BMC Biotechnol. *13*, 42.

Kataeva, I.A., Seidel, R.D. 3rd, Shah, A., West, L.T., Li, X.L., and Ljungdahl, L.G. (2002). The fibronectin type 3-like repeat from the *Clostridium thermocellum* cellobiohydrolase CbhA promotes hydrolysis of cellulose by modifying its surface. Appl. Environ. Microbiol. *68*, 4292–4300.

Kataeva, I.A., Yang, S.J., Dam, P., Poole, F.L. 2nd, Yin, Y., Zhou, F., Chou, W.C., Xu, Y., Goodwin, L., Sims, D.R., *et al.* (2009). Genome sequence of the anaerobic, thermophilic, and cellulolytic bacterium *"Anaerocellum thermophilum"* DSM 6725. J. Bacteriol. *191*, 3760–3761.

Kataeva, I., Foston, M.B., Yang, S.-J., Pattathil, S., Biswal, A.K., Poole Ii, F.L., Basen, M., Rhaesa, A.M., Thomas, T.P., Azadi, P., *et al.* (2013). Carbohydrate and lignin are simultaneously solubilized from unpretreated switchgrass by microbial action at high temperature. Energ. Environ. Sci. *6*, 2186–2195.

Kerff, F., Amoroso, A., Herman, R., Sauvage, E., Petrella, S., Filee, P., Charlier, P., Joris, B., Tabuchi, A., Nikolaidis, N., *et al.* (2008). Crystal structure and activity of *Bacillus subtilis* YoaJ (EXLX1), a bacterial expansin that promotes root colonization. Proc. Natl. Acad. Sci. U.S.A. *105*, 16876–16881.

Khandeparker, R., and Numan, M.T. (2008). Bifunctional xylanases and their potential use in biotechnology. J. Ind. Microbiol. Biotechnol. *35*, 635–644.

Knott, B.C., Haddad Momeni, M., Crowley, M.F., Mackenzie, L.F., Götz, A.W., Sandgren, M., Withers, S.G., Ståhlberg, J., and Beckham, G.T. (2013). The mechanism of cellulose hydrolysis by a two-step, retaining cellobiohydrolase elucidated by structural and transition path sampling studies. J. Am. Chem. Soc. *136*, 321–329.

Kosugi, A., Murashima, K., Tamaru, Y., and Doi, R.H. (2002). Cell-surface-anchoring role of N-terminal surface layer homology domains of *Clostridium cellulovorans* EngE. J. Bacteriol. *184*, 884–888.

Lehtio, J., Sugiyama, J., Gustavsson, M., Fransson, L., Linder, M., and Teeri, T.T. (2003). The binding specificity and affinity determinants of family 1 and family 3 cellulose binding modules. Proc. Natl. Acad. Sci. U.S.A. *100*, 484–489.

Levasseur, A., Drula, E., Lombard, V., Coutinho, P.M., and Henrissat, B. (2013). Expansion of the enzymatic repertoire of the CAZy database to integrate auxiliary redox enzymes. Biotechnol. Biofuels *6*, 41.

Levisson, M., Sun, L., Hendriks, S., Swinkels, P., Akveld, T., Bultema, J.B., Barendregt, A., van den Heuvel, R.H., Dijkstra, B.W., van der Oost, J., *et al.* (2009). Crystal structure and biochemical properties of a novel thermostable esterase containing an immunoglobulin-like domain. J. Mol. Biol. *385*, 949–962.

Li, L.L., Taghavi, S., McCorkle, S.M., Zhang, Y.B., Blewitt, M.G., Brunecky, R., Adney, W.S., Himmel, M.E., Brumm, P., Drinkwater, C., *et al.* (2011). Bioprospecting metagenomics of decaying wood: mining for new glycoside hydrolases. Biotechnol. Biofuels *4*, 23.

Little, E., Bork, P., and Doolittle, R.F. (1994). Tracing the spread of fibronectin type III domains in bacterial glycohydrolases. J. Mol. Evol. *39*, 631–643.

Lochner, A., Giannone, R.J., Keller, M., Antranikian, G., Graham, D.E., and Hettich, R.L. (2011a).

Label-free quantitative proteomics for the extremely thermophilic bacterium *Caldicellulosiruptor obsidiansis* reveal distinct abundance patterns upon growth on cellobiose, crystalline cellulose, and switchgrass. J. Proteome Res. *10*, 5302–5314.

Lochner, A., Giannone, R.J., Rodriguez, M., Jr., Shah, M.B., Mielenz, J.R., Keller, M., Antranikian, G., Graham, D.E., and Hettich, R.L. (2011b). Use of label-free quantitative proteomics to distinguish the secreted cellulolytic systems of *Caldicellulosiruptor bescii* and *Caldicellulosiruptor obsidiansis*. Appl. Environ. Microbiol. *77*, 4042–4054.

Lombard, V., Bernard, T., Rancurel, C., Brumer, H., Coutinho, P.M., and Henrissat, B. (2010). A hierarchical classification of polysaccharide lyases for glycogenomics. Biochem. J. *432*, 437–444.

Lombard, V., Golaconda Ramulu, H., Drula, E., Coutinho, P.M., and Henrissat, B. (2014). The carbohydrate-active enzymes database (CAZy) in 2013. Nucleic Acids Res. *42*, D490–495.

Lynd, L.R., Weimer, P.J., van Zyl, W.H., and Pretorius, I.S. (2002). Microbial cellulose utilization: fundamentals and biotechnology. Microbiol. Mol. Biol. Rev. *66*, 506–577.

Lynd, L.R., van Zyl, W.H., McBride, J.E., and Laser, M. (2005). Consolidated bioprocessing of cellulosic biomass: an update. Curr. Opin. Biotechnol. *16*, 577–583.

McCarter, J.D., and Withers, S.G. (1994). Mechanisms of enzymatic glycoside hydrolysis. Curr. Opin. Struct. Biol. *4*, 885–892.

MacGregor, E.A., Janeček, Š., and Svensson, B. (2001). Relationship of sequence and structure to specificity in the α-amylase family of enzymes. Biochim. Biophys. Acta *1546*, 1–20.

McLean, B.W., Boraston, A.B., Brouwer, D., Sanaie, N., Fyfe, C.A., Warren, R.A., Kilburn, D.G., and Haynes, C.A. (2002). Carbohydrate-binding modules recognize fine substructures of cellulose. J. Biol. Chem. *277*, 50245–50254.

Mahadevan, S.A., Wi, S.G., Lee, D.-S., and Bae, H.-J. (2008). Site-directed mutagenesis and CBM engineering of Cel5A (*Thermotoga maritima*). FEMS Microbiol. Lett. *287*, 205–211.

Maki, M., Leung, K.T., and Qin, W. (2009). The prospects of cellulase-producing bacteria for the bioconversion of lignocellulosic biomass. Int J Biol. Sci *5*, 500–516.

Margeot, A., Hahn-Hagerdal, B., Edlund, M., Slade, R., and Monot, F. (2009). New improvements for lignocellulosic ethanol. Curr. Opin. Biotechnol. *20*, 372–380.

Martínez-Martínez, I., Montoro-García, S., Lozada-Ramírez, J.D., Sánchez-Ferrer, Á., and García-Carmona, F. (2007). A colorimetric assay for the determination of acetyl xylan esterase or cephalosporin C acetyl esterase activities using 7-amino cephalosporanic acid, cephalosporin C, or acetylated xylan as substrate. Anal. Biochem. *369*, 210–217.

Mayes, H.B., Broadbelt, L.J., and Beckham, G.T. (2014). How sugars pucker: electronic structure calculations map the kinetic landscape of five biologically paramount monosaccharides and their implications for enzymatic catalysis. J. Am. Chem. Soc. *136*, 1008–1022.

Mazzoli, R., Lamberti, C., and Pessione, E. (2012). Engineering new metabolic capabilities in bacteria: lessons from recombinant cellulolytic strategies. Trends Biotechnol. *30*, 111–119.

Miroshnichenko, M.L., Kublanov, I.V., Kostrikina, N.A., Tourova, T.P., Kolganova, T.V., Birkeland, N.-K., and Bonch-Osmolovskaya, E.A. (2008). *Caldicellulosiruptor kronotskyensis* sp. nov. and *Caldicellulosiruptor hydrothermalis* sp. nov., two extremely thermophilic, cellulolytic, anaerobic bacteria from Kamchatka thermal springs. Int. J. Syst. Evol. Microbiol. *58*, 1492–1496.

Mladenovska, Z., Mathrani, I., and Ahring, B. (1995). Isolation and characterization of *Caldicellulosiruptor lactoaceticus* sp. nov., an extremely thermophilic, cellulolytic, anaerobic bacterium. Arch. Microbiol. *163*, 223–230.

Morris, D.D., Gibbs, M.D., Ford, M., Thomas, J., and Bergquist, P.L. (1999). Family 10 and 11 xylanase genes from *Caldicellulosiruptor* sp. strain Rt69B.1. Extremophiles *3*, 103–111.

Nacke, H., Engelhaupt, M., Brady, S., Fischer, C., Tautzt, J., and Daniel, R. (2012). Identification and characterization of novel cellulolytic and hemicellulolytic genes and enzymes derived from German grassland soil metagenomes. Biotechnol. Lett. *34*, 663–675.

National Research Council (2009). Liquid Transportation Fuels from Coal and Biomass: Technological Status, Costs, and Environmental Impacts (Washington, DC, The National Academies Press).

Naumoff, D.G. (2011). Hierarchical classification of glycoside hydrolases. Biochemistry (Mosc) *76*, 622–635.

Olson, D.G., McBride, J.E., Shaw, A.J., and Lynd, L.R. (2012). Recent progress in consolidated bioprocessing. Curr. Opin. Biotechnol. *23*, 396–405.

Onyenwoke, R.U., Lee, Y.-J., Dabrowski, S., Ahring, B.K., and Wiegel, J. (2006). Reclassification of *Thermoanaerobium acetigenum* as *Caldicellulosiruptor acetigenus* comb. nov. and emendation of the genus description. Int. J. Syst. Evol. Microbiol. *56*, 1391–1395.

von Ossowski, I., Eaton, J.T., Czjzek, M., Perkins, S.J., Frandsen, T.P., Schulein, M., Panine, P., Henrissat, B., and Receveur-Brechot, V. (2005). Protein disorder: conformational distribution of the flexible linker in a chimeric double cellulase. Biophys. J. *88*, 2823–2832.

Ozdemir, I., Blumer-Schuette, S.E., and Kelly, R.M. (2012). S-layer homology domain proteins Csac_0678 and Csac_2722 are implicated in plant polysaccharide deconstruction by the extremely thermophilic bacterium *Caldicellulosiruptor saccharolyticus*. Appl. Environ. Microbiol. *78*, 768–777.

Pai, C.-K., Wu, Z.-Y., Chen, M.-J., Zeng, Y.-F., Chen, J.-W., Duan, C.-H., Li, M.-L., and Liu, J.-R. (2010). Molecular cloning and characterization of a bifunctional xylanolytic enzyme from *Neocallimastix patriciarum*. Appl. Microbiol. Biotechnol. *85*, 1451–1462.

Park, J.I., Kent, M.S., Datta, S., Holmes, B.M., Huang, Z., Simmons, B.A., Sale, K.L., and Sapra, R. (2011). Enzymatic hydrolysis of cellulose by the cellobiohydrolase domain of CelB from the hyperthermophilic bacterium *Caldicellulosiruptor saccharolyticus*. Bioresour. Technol. *102*, 5988–5994.

Pauly, M., and Keegstra, K. (2008). Cell-wall carbohydrates and their modification as a resource for biofuels. Plant J. *54*, 559–568.

Pauly, M., and Keegstra, K. (2010). Plant cell wall polymers as precursors for biofuels. Curr. Opin. Plant Biol. *13*, 304–311.

Payne, C.M., Resch, M.G., Chen, L., Crowley, M.F., Himmel, M.E., Taylor, L.E. 2nd, Sandgren, M., Stahlberg, J., Stals, I., Tan, Z., *et al.* (2013). Glycosylated linkers in multimodular lignocellulose-degrading enzymes dynamically bind to cellulose. Proc. Natl. Acad. Sci. U.S.A. *110*, 14646–14651.

Pell, G., Williamson, M.P., Walters, C., Du, H., Gilbert, H.J., and Bolam, D.N. (2003). Importance of hydrophobic and polar residues in ligand binding in the family 15 carbohydrate-binding module from *Cellvibrio japonicus* Xyn10C. Biochemistry *42*, 9316–9323.

Petersen, T.N., Brunak, S., von Heijne, G., and Nielsen, H. (2011). SignalP 4.0: discriminating signal peptides from transmembrane regions. Nat. Methods *8*, 785–786.

Pollet, A., Delcour, J.A., and Courtin, C.M. (2010). Structural determinants of the substrate specificities of xylanases from different glycoside hydrolase families. Crit. Rev. Biotechnol. *30*, 176–191.

Pum, D., Toca-Herrera, J.L., and Sleytr, U.B. (2013). S-layer protein self-assembly. Int. J. Mol. Sci. *14*, 2484–2501.

Punta, M., Coggill, P.C., Eberhardt, R.Y., Mistry, J., Tate, J., Boursnell, C., Pang, N., Forslund, K., Ceric, G., Clements, J., *et al.* (2012). The Pfam protein families database. Nucleic Acids Res. *40*, D290–301.

Rainey, F.A., Donnison, A.M., Janssen, P.H., Saul, D., Rodrigo, A., Bergquist, P.L., Daniel, R.M., Stackebrandt, E., and Morgan, H.W. (1994). Description of *Caldicellulosiruptor saccharolyticus* gen. nov., sp. nov: An obligately anaerobic, extremely thermophilic, cellulolytic bacterium. FEMS Microbiol. Lett. *120*, 263–266.

Receveur, V., Czjzek, M., Schülein, M., Panine, P., and Henrissat, B. (2002). Dimension, shape, and conformational flexibility of a two domain fungal cellulase in solution probed by small angle X-ray scattering. J. Biol. Chem. *277*, 40887–40892.

Resch, M.G., Donohoe, B.S., Baker, J.O., Decker, S.R., Bayer, E.A., Beckham, G.T., and Himmel, M.E. (2013). Fungal cellulases and complexed cellulosomal enzymes exhibit synergistic mechanisms in cellulose deconstruction. Energ. Environ. Sci. *6*, 1858–1867.

Reyes-Ortiz, V., Heins, R.A., Cheng, G., Kim, E.Y., Vernon, B.C., Elandt, R.B., Adams, P.D., Sale, K.L., Hadi, M.Z., Simmons, B.A., *et al.* (2013). Addition of a carbohydrate-binding module enhances cellulase penetration into cellulose substrates. Biotechnol. Biofuels *6*, 93.

Ryabova, O., Vršanská, M., Kaneko, S., van Zyl, W.H., and Biely, P. (2009). A novel family of hemicellulolytic α-glucuronidase. FEBS Lett. *583*, 1457–1462.

Sabathe, F., and Soucaille, P. (2003). Characterization of the CipA scaffolding protein and *in vivo* production of a minicellulosome in *Clostridium acetobutylicum*. J. Bacteriol. *185*, 1092–1096.

Saloheimo, M., Paloheimo, M., Hakola, S., Pere, J., Swanson, B., Nyyssönen, E., Bhatia, A., Ward, M., and Penttilä, M. (2002). Swollenin, a *Trichoderma reesei* protein with sequence similarity to the plant expansins, exhibits disruption activity on cellulosic materials. Eur. J. Biochem. *269*, 4202–4211.

Sampedro, J., and Cosgrove, D.J. (2005). The expansin superfamily. Genome Biol. *6*, 242.

Sara, M., and Sleytr, U.B. (2000). S-Layer proteins. J. Bacteriol. *182*, 859–868.

Scheller, H.V., and Ulvskov, P. (2010). Hemicelluloses. Annu. Rev. Plant Biol. *61*, 263–289.

Schubot, F.D., Kataeva, I.A., Chang, J., Shah, A.K., Ljungdahl, L.G., Rose, J.P., and Wang, B.C. (2004). Structural basis for the exocellulase activity of the cellobiohydrolase CbhA from *Clostridium thermocellum*. Biochemistry *43*, 1163–1170.

Scott, J.R., and Barnett, T.C. (2006). Surface proteins of Gram-positive bacteria and how they get there. Annu. Rev. Microbiol. *60*, 397–423.

Selig, M., Adney, W., Himmel, M., and Decker, S. (2009). The impact of cell wall acetylation on corn stover hydrolysis by cellulolytic and xylanolytic enzymes. Cellulose *16*, 711–722.

Sleator, R.D., Shortall, C., and Hill, C. (2008). Metagenomics. Lett. Appl. Microbiol. *47*, 361–366.

Sørensen, H., Jørgensen, C., Hansen, C., Jørgensen, C., Pedersen, S., and Meyer, A. (2006). A novel GH43 α-l-arabinofuranosidase from *Humicola insolens*: mode of action and synergy with GH51 α-l-arabinofuranosidases on wheat arabinoxylan. Appl. Microbiol. Biotechnol. 73, 850–861.

Stals, I., Sandra, K., Geysens, S., Contreras, R., Van Beeumen, J., and Claeyssens, M. (2004). Factors influencing glycosylation of *Trichoderma reesei* cellulases. I: Postsecretorial changes of the O- and N-glycosylation pattern of Cel7A. Glycobiology *14*, 713–724.

Su, X., Mackie, R.I., and Cann, I.K. (2012). Biochemical and mutational analyses of a multidomain cellulase/mannanase from *Caldicellulosiruptor bescii*. Appl. Environ. Microbiol. *78*, 2230–2240.

Sun, Y., and Cheng, J. (2002). Hydrolysis of lignocellulosic materials for ethanol production: a review. Bioresour. Technol. *83*, 1–11.

Sunna, A. (2010). Modular organisation and functional analysis of dissected modular β-mannanase *Cs*Man26 from *Caldicellulosiruptor* Rt8B.4. Appl. Microbiol. Biotechnol. *86*, 189–200.

Tripathi, S.A., Olson, D.G., Argyros, D.A., Miller, B.B., Barrett, T.F., Murphy, D.M., McCool, J.D., Warner, A.K., Rajgarhia, V.B., Lynd, L.R., *et al.* (2010). Development of pyrF-based genetic system for targeted gene deletion in *Clostridium thermocellum* and creation of a *pta* mutant. Appl. Environ. Microbiol. *76*, 6591–6599.

US Department of Energy (2011). US Billion-Ton Update: Biomass Supply for a Bioenergy and Bioproducts Industry, Perlack, R.D., and Stokes, B.J. (Leads), (ORNL/TM-2011/224, Oak Ridge National Laboratory, Oak Ridge, TN).

Upreti, R.K., Kumar, M., and Shankar, V. (2003). Bacterial glycoproteins: functions, biosynthesis and applications. Proteomics 3, 363–379.

VanFossen, A.L., Lewis, D.L., Nichols, J.D., and Kelly, R.M. (2008). Polysaccharide degradation and synthesis by extremely thermophilic anaerobes. Ann. N.Y. Acad. Sci. *1125*, 322–337.

VanFossen, A.L., Verhaart, M.R., Kengen, S.M., and Kelly, R.M. (2009). Carbohydrate utilization patterns for the extremely thermophilic bacterium *Caldicellulosiruptor saccharolyticus* reveal broad growth substrate preferences. Appl. Environ. Microbiol. 75, 7718–7724.

VanFossen, A.L., Ozdemir, I., Zelin, S.L., and Kelly, R.M. (2011). Glycoside hydrolase inventory drives plant polysaccharide deconstruction by the extremely thermophilic bacterium *Caldicellulosiruptor saccharolyticus*. Biotechnol. Bioeng. *108*, 1559–1569.

Vanholme, R., Morreel, K., Ralph, J., and Boerjan, W. (2008). Lignin engineering. Curr. Opin. Plant Biol. *11*, 278–285.

Vocadlo, D.J., and Davies, G.J. (2008). Mechanistic insights into glycosidase chemistry. Curr. Opin. Chem. Biol. *12*, 539–555.

Vodovnik, M., and Marinsek-Logar, R. (2010). Cellulosomes – Promising Supramolecular Machines of Anaerobic Cellulolytic Microorganisms. Acta Chim. Slov. *57*, 767–774.

van de Werken, H.J., Verhaart, M.R., VanFossen, A.L., Willquist, K., Lewis, D.L., Nichols, J.D., Goorissen, H.P., Mongodin, E.F., Nelson, K.E., van Niel, E.W., *et al.* (2008). Hydrogenomics of the extremely thermophilic bacterium *Caldicellulosiruptor saccharolyticus*. Appl. Environ. Microbiol. 74, 6720–6729.

Williamson, G., Kroon, P.A., and Faulds, C.B. (1998). Hairy plant polysaccharides: a close shave with microbial esterases. Microbiology *144 (Pt 8)*, 2011–2023.

Wilson, D.B. (2011). Microbial diversity of cellulose hydrolysis. Curr. Opin. Microbiol. *14*, 259–263.

Wootton, J.C. (1994). Non-globular domains in protein sequences: automated segmentation using complexity measures. Comput. Chem. *18*, 269–285.

Wu, X., McLaren, J., Madl, R., and Wang, D. (2010). Biofuels from lignocellulosic biomass. In Sustainable Biotechnology, Singh, O.V., and Harvey, S.P., eds. (Springer, Netherlands), pp. 19–41.

Xu, F. (2011). Enzymatic degradation of lignocellulosic biomass. In Biocatalysis for Green Chem. and Chemical Process Development (John Wiley & Sons, Inc., Hoboken, NJ, USA), pp. 361–390.

Yadav, S., Yadav, P.K., Yadav, D., and Yadav, K.D.S. (2009). Pectin lyase: a review. Process Biochem. *44*, 1–10.

Yamada, R., Hasunuma, T., and Kondo, A. (2013). Endowing non-cellulolytic microorganisms with cellulolytic activity aiming for consolidated bioprocessing. Biotechnol. Adv. *31*, 754–763.

Yang, S.J., Kataeva, I., Wiegel, J., Yin, Y., Dam, P., Xu, Y., Westpheling, J., and Adams, M.W. (2010). Classification of *'Anaerocellum thermophilum'* strain DSM 6725 as *Caldicellulosiruptor bescii* sp. nov. Int. J. Syst. Evol. Microbiol. *60*, 2011–2015.

Yi, Z., Su, X., Revindran, V., Mackie, R.I., and Cann, I. (2013). Molecular and biochemical analyses of CbCel9A/Cel48A, a highly secreted multi-modular cellulase by *Caldicellulosiruptor bescii* during growth on crystalline cellulose. PLoS One *8*, e84172.

Yip, V.L.Y., and Withers, S.G. (2004). Nature's many mechanisms for the degradation of oligosaccharides. Org. Biomol. Chem. *2*, 2707–2713.

Yip, V.L.Y., and Withers, S.G. (2006). Breakdown of oligosaccharides by the process of elimination. Curr. Opin. Chem. Biol. *10*, 147–155.

You, C., Zhang, X.Z., Sathitsuksanoh, N., Lynd, L.R., and Zhang, Y.H. (2012). Enhanced microbial utilization of recalcitrant cellulose by an *Ex Vivo* cellulosome-microbe complex. Appl. Environ. Microbiol. *78*, 1437–1444.

Zurawski, J.V., Blumer-Schuette, S.E., Conway, J.M., and Kelly, R.M. (2014). The extremely thermophilic genus *Caldicellulosiruptor*: physiological and genomic characteristics for complex carbohydrate conversion to molecular hydrogen. In Microbial BioEnergy: Hydrogen Production, Advances in Photosynthesis and Respiration, De Philippis, R., and Zannoni, D., eds. (Springer Science+Business Media, Dordrecht, Netherlands).

Zverlov, V., Mahr, S., Riedel, K., and Bronnenmeier, K. (1998). Properties and gene structure of a bifunctional cellulolytic enzyme (CelA) from the extreme thermophile *'Anaerocellum thermophilum'* with separate glycosyl hydrolase family 9 and 48 catalytic domains. Microbiology *144 (Pt 2)*, 457–465.

Cellulases from Thermophilic Fungi

5

Duochuan Li

Abstract

Thermophilic fungi are species that grow at high temperature. Because of their thermostable enzymes, thermophilic fungi have received increasing attention. In this recent decade, cellulases from thermophilic fungi have remarkable progress in many aspects, such as purifying and characterizing the cellulase components, understanding the mechanism of cellulose degradation, cloning and expression of cellulase genes, determining the structures of cellulase components, improving the properties of cellulases by mutation and fusion, understanding structure–function relationships in cellulases and demonstrating the industrial potential of cellulases. Especially, recent genomic sequencing, transcriptome data and secreted proteins showed a number of cellulase genes in thermophilic fungi. The present review will summarizes recent advances in these areas, identify the present gaps and give future perspectives in researches on cellulases from thermophilic fungi.

Introduction

Cellulose constitutes bulk of the plant cell wall materials and is the most abundant and renewable non-fossil carbon source on Earth. Its degradation to the constituent monosaccharides has attracted considerable attention for the production of food and biofuels (Sukumaran *et al.*, 2010; Wilson, 2009). The degradation of cellulose to glucose is accomplished by the cooperative action of endocellulases (E.C.3.1.1.4), exocellulases (cellobiohydrolase, CBH, E.C.3.2.1.91) and beta-glucosidases (E.C.3.2.1.21) normally. Endocellulases randomly hydrolyse internal glycosidic linkages, which results in a rapid decrease in polymer length and a gradual increase in the reducing sugar concentration. Exocellulases hydrolyse cellulose chains by removing cellobiose either from the reducing ends or the non-reducing ends, which results in rapid release of reducing sugars but little change in polymer length. Endocellulases and exocellulases act synergistically upon cellulose to produce cellobiose, which is then cleaved by beta-glucosidase to glucose (Zhang *et al.*, 2006; Vlasenko *et al.*, 2010).

Thermophilic fungi are species that grow at a maximum temperature of 50°C or above, and a minimum of 20°C or above (Mouchacca, 1997). Because of their thermostable enzymes for many biotechnological applications including biomass degradation, thermophilic fungi have received increasing attention (Maheshwari *et al.*, 2000). In recent years, cellulases from thermophilic fungi have remarkable progress in many aspects, such as isolation of thermophilic fungi producing cellulases, purifying and characterizing the cellulase components, understanding the mechanism of cellulose degradation, cloning and expression of cellulase genes, determining the structures of cellulase components, improving the properties of cellulases by mutation and fusion, understanding structure–function relationships in cellulases and demonstrating the industrial potential of cellulases. In particular, recent extensive genomic sequencing, transcriptome data analysis and secreted protein identification have revealed a number of cellulase genes in thermophilic fungi. However, the earlier review mainly dealt with production, purification and characterization, application of cellulases from thermophilic fungi (Maheshwari *et al.*, 2000). This review mainly aims at analysing the updated information on molecular, structural, genetic and engineering aspects of cellulases from thermophilic fungi.

Purification and characterization of new cellulases from thermophilic fungi

Since purification and characterization of cellulases from thermophilic fungi was reviewed by Maheshwari *et al.* (2000), some new cellulases has been purified and characterized from thermophilic fungi in recent years, such as two endocellulases from *Melanocarpus albomyces* (Miettinen-Oinonen *et al.*, 2004), two endocellulases from *Melanocarpus* sp. (Kaur et al., 2007), one endocellulase, one exocellulase and two beta-glucosidases from *Chaetomium thermophilum* (Chu *et al.*, 2004; Li *et al.*, 2003; Teng *et al.*, 2006), a beta-glucosidase *Talaromyces thermophilus* (Nakkharat and Haltrich, 2006), a beta-glucosidase from *Scytalidium thermophilum* (Zanoelo *et al.*, 2004), three endo-glucanases from *Talaromyces emersonii* (McCarthy *et al.*, 2003), two beta-glucosidases from *Thermoascus aurantiacus* (De Palma-Fernandez *et al.*, 2002), four endocellulases and two beta-glucosidases from *Myceliophthora* sp. (Badhan *et al.*, 2007) and a beta-glucosidase from *C. thermophilum* var. *coprophilum* (Venturi *et al.*, 2002). The characteristics of purified thermophilic fungal cellulases mainly focus on molecular weight, optimal pH, optimal temperature, thermostability and *glycosylation.* Usually, thermophilic fungal cellulases are single polypeptides, whereas it has been reported that some beta-glucosidases are dimeric. Molecular masses of thermophilic fungal cellulases have a wide range from 30 kDa to 250 kDa with different carbohydrate contents from 2 to 50%. Optimal pH and temperature are similar for the majority of the purified cellulases from thermophilic fungi. They are active in the pH 4.0–7.0 range, and have a high temperature maximum for activity at 50–80°C. Cellulases from thermophilic fungi exhibit remarkable thermal stability, and are stable at 60°C with longer half-life at 70, 80 and 90°C than those from other fungi. It should be emphasised that thermostable cellulases play essential role in adaptation to grow at high temperature for thermophilic fungi. Furthermore, they may find a potential industrial application. In industry, the high temperature active and thermostable cellulase is considered advantageous because at higher temperatures chemical reactions are more rapid, and enzyme has tolerance to higher temperatures. In addition, cellulose swells at higher temperatures, which makes it easier to break down.

Mechanism for thermostability of thermophilic proteins from thermophilic bacteria and hyperthermophilic archaea has been studied more extensively in recent years (Kumar *et al.*, 2000; Pack and Yoo, 2004; Yokota *et al.*, 2006; Trivedi *et al.*, 2006). It is generally believed that amino acid residues at the protein surface and core are involved in thermostability. It was observed that Arg, Glu and Tyr frequently occur on the surface of thermophilic proteins, and Ala, Gly, Val, Leu and Ile in the core of thermophilic proteins. Arg and Glu at the protein surface forms ion pair or ion pair networks, which is related to protein thermostability. Tyr at the protein surface is involved in cation–pi interactions or forms aromatic clusters, which are key factors for protein thermostability. Ala is a hydrophobic residue. It is thought that Ala at the protein core is concerned with compact packing, which would stabilize the core of proteins. Other hydrophobic residue, such as Val, Leu and Ile, may contribute to the hydrophobic interaction, which is the force for maintaining protein stability. Pro residue at the protein beta-turns and random coils plays specific roles in protein thermostability. The rigid pyrrolidine ring can make the conformation of surroundings to be more reasonable by reducing the backbone unfolding entropy of proteins. Cys is a hydrophobic amino acid with a short side chain containing one sulfur atom. It has been proposed that Cys does not favour protein thermostability. The sulfur atom is relatively large and cannot contribute to the compact packing of protein structures. But if two Cys residues are cross-linked and form one disulfide bridge, the disulfide bridges may provide thermostability to proteins.

Compared with thermophilic proteins from thermophilic bacteria and hyperthermophilic archaea, there are few evidences on thermostability of thermophilic proteins from thermophilic fungi. *Humicola grisea* Cel12A is a thermostable endoglucanase. Structure analysis shows that there are three unusual free cysteines in the enzyme: Cys175, Cys206 and Cys216. To determine if these cysteines are important for thermostability, three *H. grisea* Cel12A mutants were constructed by site-directed mutagenesis: C175G, C206P and C216V. It is found that the three free cysteines play a significant role in modulating the stability of the enzyme. Mutating the free cysteine at residue 206 to a proline and the free cysteine at residue 216 to a serine causes a reduction in the T_m of 9.1 and 5.5°C, respectively, compared to the wild-type enzyme. When the free cysteine at residue 175 is mutated to a glycine, the T_m of the enzyme is increased by 1.3°C (Sandgren *et al.*, 2005).

It should be noted that although some structural factors are recognized as important features of thermostable proteins, molecular determinants of thermostability are still debated because of its complexity and inconsistency of findings from different studies.

Cellulase genes from thermophilic fungi

A great number of genes encoding cellulases from thermophilic fungi, containing endocellulases, exocellulases and beta-glucosidases, have been isolated, analysed and expressed with the aim of (i) understanding of gene structure and function, (ii) enzyme overproduction, and (iii) protein structure and molecular improvement. Cellulases are classified as glycosyl hydrolases. According to the glycosyl hydrolase classification system which is based on amino acid sequence similarity of the catalytic domains, cellulases have been placed in families 1, 3, 5, 6, 7, 8, 9, 12, 26, 44, 45, 48 and 61 (www. cazy.org). Cellulases from thermophilic fungi are found in families 1, 3, 5, 6, 7, 12 and 45. A brief summary of the research on cloning of thermophilic fungal cellulase genes is given in Table 5.1. Recent genomic sequencing showed a number of cellulase genes in thermophilic fungi (Table 5.2; Berka *et al.*, 2011). Genome analysis of thermophilic fungi will increase our understanding of the evolutionary relationships of cellulases from thermophilic fungi, and help us find novel cellulase genes and their potential scientific, ecological and industrial importance.

The entire sequence of thermophilic fungal cellulase genes cloned contains an open reading frame (ORF) usually interrupted by introns with consensus 5′ and 3′ intron splice sites, a 3′ untranslated sequence (UTS), a 5′ UTS and an upstream regulatory sequence (URS) or upstream regulatory region (URR). In the URS of thermophilic fungal cellulases, CAAT and TATAA boxes are present. Importantly, some sequences involved in regulation of cellulase expression are found in the URS of thermophilic fungal cellulases, such as consensus motifs (SYGGRG) for binding of the carbon catabolic repressor element (CREI/CREA) (Murray *et al.*, 2003, 2004; Takashima *et al.*, 1999a, 1996; Hong *et al.*, 2003a,b), a putative pH regulator transcript factor PacC consensus sequence (GCCARG) (Murray *et al.*, 2003), and one or more copies of cellulase expression regulator (ACEI, ACEII and XYR1) binding sites (AGGCAAA, GGSTAAA and GGCATT) (Murray *et al.*, 2003, 2004; Collins *et al.*, 2007; Takashima *et al.*, 1999a, 1996; Hong *et al.*, 2003a,b). In addition, there are putative glycosylation sites (Asn-X-Thr/Ser) in most of the deduced amino acid sequences of cellulases from thermophilic fungi (Collins *et al.*, 2007; Li *et al.*, 2009).

Production of cellulases from fungi is commonly induced only in the presence of cellulose as a substrate. Because of its insolubility, cellulose is not able to trigger the induction directly. Therefore, a basal level of constitutive cellulase is required for the induction. The cellulase expressed constitutively might degrade cellulose into soluble cellooligosaccharides, and then the soluble cellooligosaccharides enter the cell and are converted into an inducer that triggers cellulase induction. Cellulase gene expression in fungi has been reported to be controlled by a repressor/inducer system, in which cellulose or other products of degradation act as inducers whereas glucose or easily metabolizable carbon sources act as repressors. The repressor/inducer system has also been described in cellulases of thermophilic fungi (Maheshwari *et al.*, 2000). It has been reported that expression of a cellobiohydrolase gene (*cbh2*) from the thermophilic

Table 5.1 A brief summary of research done on the cloning of thermophilic fungal cellulase genes

Source	Gene	Family	Reference
Acremonium thermophilum	*cel7a*	7	Voutilainen *et al*. (2008)
Chaetomium senegalense	*gh6*	6	Busk and Lange (2013)
Chaetomium senegalense	*gh7*	7	Busk and Lange (2013)
Chaetomium senegalense	*gh45*	45	Busk and Lange (2013)
Chaetomium thermophilum	*cbh1*	7	Liu *et al*. (2006)
Chaetomium thermophilum	*cbh2*	6	Liu *et al*. (2005)
Chaetomium thermophilum	*cbh3*	7	Li *et al*. (2009)
Chaetomium thermophilum	*cel7a*	7	Voutilainen *et al*. (2008)
Chaetomium thermophilum	*bgl*	3	Xu *et al*. (2011)
Chaetomium brasiliense	*cel12a*	12	Goedegebuur *et al*. (2002)
Humicola grisea	*cbh1.2*	7	Pocas-Fonseca *et al*. (1997)
Humicola grisea	*cel12a*	12	Goedegebuur *et al*. (2002)
Humicola grisea	*egl2*	5	Takashima *et al*. (1997)
Humicola grisea	*egl3*	45	Takashima *et al*. (1999a)
Humicola grisea	*egl4*	45	Takashima *et al*. (1999a)
Humicola grisea	*bgl4*	1	Takashima *et al*. (1999b)
Humicola grisea var. *thermoidea*	*cbh1*	7	Takashima *et al*. (1996)
Humicola grisea var. *thermoidea*	*eg1*	7	Takashima *et al*. (1996)
Humicola grisea var. *thermoidea*	*exo1*	7	Takashima *et al*. (1998)
Humicola insolens	*avi2*	6	Moriya *et al*. (2003)
Humicola insolens	*egII*	5	Schou *et al*. (1993)
Humicola insolens	*egIII*	12	Schou *et al*. (1993)
Humicola insolens	*egVI*	6	Schou *et al*. (1993)
Humicola insolens	*cbhI*	7	Schou *et al*. (1993)
Humicola insolens	*cmc3*	5	Dalboge and Hansen (1994)
Malbranchea cinnamomea	*gh6*	6	Busk and Lange (2013)
Malbranchea cinnamomea	*gh7*	7	Busk and Lange (2013)
Malbranchea cinnamomea	*gh45*	45	Busk and Lange (2013)
Melanocarpus albomyces	*cel45a*	45	Haakana *et al*. (2004)
Melanocarpus albomyces	*cel7a*	7	Haakana *et al*. (2004)
Melanocarpus albomyces	*cel7b*	7	Haakana *et al*. (2004)
Remersonia thermophila	*gh45*	45	Busk and Lange (2013)
Scytalidium thermophilum	*bgl4*	3	Wang *et al*. (2013)
Scytalidium indonesiacum	*gh45*	45	Busk and Lange (2013)
Talalaromyces emersonii	*cel3a*	3	Murray *et al*. (2004)
Talaromyces emersonii	*bg1*	1	Collins *et al*. (2007)
Talaromyces emersonii	*aven1*	3	Collins *et al*. (2007)
Talaromyces emersonii	*cbh2*	6	Murray *et al*. (2003)
Talaromyces emersonii	*cel7*	7	Grassick *et al*. (2004)
Thermoascus aurantiacus	*eg1*	5	Hong *et al*. (2003a)
Thermoascus aurantiacus	*cbh1*	7	Hong *et al*. (2003b)
Thermoascus aurantiacus	*bgl1*	3	Hong *et al*. (2007)
Thermoascus aurantiacus	*bgl2*	3	Hong *et al*. (2006)

Table 5.1 continued

Source	Gene	Family	Reference
Thermoascus aurantiacus	*cel7a*	7	Voutilainen *et al.* (2008)
Thermoascus aurantiacus	*bgl2*	3	Jong *et al.* (2006)
Thermoascus aurantiacus	*bgl1*	3	Jong *et al.* (2007)
Thermoascus aurantiacus	*cbh1*	7	McClendon *et al.* (2012)
Thermoascus aurantiacus	*egl*	5	McClendon *et al.* (2012)

Table 5.2 Numbers of cellulase genes of GH families in some sequenced thermophilic fungal genomes (http://fungalgenomics.ca/wiki/Fungal_Genomes; Berka *et al.*, 2011)

Fungus	GH1	GH3	GH5	GH6	GH7	GH12	GH45
Calcarisporium thermophilum	0	3	6	0	0	0	0
Thermomyces lanuginosus	1	7	4	0	0	1	0
Talaromyces thermophilus	1	7	4	0	0	1	0
Myceliophthora ferusii	1	9	5	3	5	1	0
Scytalidium thermophilum	1	8	7	4	5	1	2
Myriococcum thermophilum	1	12	11	3	5	2	1
Myceliophthora thermophila	2	14	9	3	6	1	1
Thielavia terrestris	1	14	16	1	5	3	2
Thermoascus aurantiacus	2	10	8	0	1	0	0

fungus *T. emersonii* was induced by cellulose, and repressed by glucose (Murray *et al.*, 2003). Similar induction patterns are also exhibited by other cellulase genes from thermophilic fungi, such as *T. emersonii cel7* (Grassick *et al.*, 2004), *H. grisea* var. *thermoidea cbh* (Pocas-Fonseca *et al.*, 2000; Paula *et al.*, 2003), *T. emersonii bgl* (Collins *et al.*, 2007) and *T. aurantiacus cbh1/cel7a* (Benko *et al.*, 2007).

Glucose repression is a common phenomenon in the regulation of fungal cellulase genes. At present, the molecular basis of glucose repression to fungal cellulase gene expression has been more understood. It is demonstrated that the URS in the fungal cellulase promoters plays a key role in the regulation of glucose repression. The protein product of the regulatory gene *cre1* in *Trichoderma reesei* is a negatively acting transcription factor that binds to DNA sequence motif with the consensus sequence SYGGRG in the URS. In the presence of glucose, the activated CREI (a Cys_2His_2 zinc finger protein) binds to the consensus motif, and represses transcription of cellulase genes (Ilmen *et al.*, 1997). Recently, three new transcription factors (ACEI, ACEII and XYR1) involved in cellulase regulation have been identified in *T. reesei*. ACEI, binding to AGGCAAA sequences in the promoter regions of cellulase genes, represses expression of all major cellulase genes in the presence of cellulose. ACEII binds to GGSTAAA sequences in the promoter regions of *cbh1* and *cbh2* and induces expression of *cbh1* and *cbh2*. XYR1 is a zinc binuclear cluster protein that binds to a GGCTAA motif in the 5′-upstream region of XYR1-regulated genes and involved in the induction of all major cellulase genes (Furukawa *et al.*, 2009). Unlike the transcription factors of *T. reesei* cellulase genes, the transcription factors of cellulase genes from thermophilic fungi have not been identified so far. However, there are potential binding sites of these genes in 5′ upstream region (Murray *et al.*, 2003, 2004; Collins *et al.*, 2007; Takashima *et al.*, 1999a, 1996; Hong *et al.*, 2003a,b), and CREI genes from two thermophilic fungi (*T. emersonii* and *T. aurantiacus*) were cloned (AY070919, AY604200, AF440004). Therefore, it is likely that cellulase regulation in thermophilic fungi is similar to that in *T. reesei*.

Most cloned cellulase genes of thermophilic

fungi are well expressed in host, such as *E. coli*, yeast, and filamentous fungi. An endocellulase gene and a cellobiohydrolase gene of *H. grisea* var. *thermoidea* were expressed in *Aspergillus oryzae*, and the expressed EGI and CBHI enzymes showed high activity to Avicel and carboxymethyl cellulose, respectively (Takashima *et al.*, 1996). Two endocellulase genes from *H. grisea* were expressed in *A. oryzae*, and the recombinant purified products showed relatively high activity towards carboxymethyl cellulose (Takashima *et al.*, 1999a). Transformation of *T. reesei* with the two endochitinase genes of *M. albomyces* resulted in increase in cellulase activity about several times higher than those of the parental *M. albomyces* (Haakana *et al.*, 2004). A gene of *M. albomyces*, coding for a cellobiohydrolase, was cloned into *Saccharomyces cerevisiae*. The cellobiohydrolase gene was successfully expressed in the transformed *S. cerevisiae* (Voutilainen *et al.*, 2007). Three gene encoding cellobiohydrolase from *C. thermophilum* were overexpressed in *Pichia pastoris*, and their expression level was over 1 mg/ml (Liu *et al.*, 2005, 2006; Li *et al.*, 2009). Introduction of *T. aurantiacus* cellobiohydrolase and endocellulase genes into *S. cerevisiae* resulted in high cellulase levels in the host (Hong *et al.*, 2003a,b), and a gene encoding beta-glucosidase of the fungus was also expressed in *P. pastori* (Hong *et al.*, 2007). When a gene encoding beta-glucosidase of *T. emersonii* was cloned into *T. reesei*, the cells of *T. reesei* secreted the recombinant dimmer subunit enzyme with 17 potential N-glycosylation sites in its functionally active (Murray *et al.*, 2004). A beta-glucosidase gene from *H. grisea* var. *thermoidea* was successfully expressed in *S. cerevisiae* (Benoliel *et al.*, 2010). A cellobiohydrolase gene of *T. emersonii* was transferred to *E. coli*, and highly expressed (Grassick *et al.*, 2004). Because there are putative glycosylation sites in most of the deduced amino acid sequences of cellulases from thermophilic fungi, most of the recombinant cellulases expressed in yeast and filamentous fungi are glycosylated (Takashima *et al.*, 1999a, 1996; Li *et al.*, 2009). The glycosylation of the cellulases helps to improve their thermostability. Expression of thermophilic fungal cellulase genes in the heterologous hosts is summarized in Table 5.3.

Primary and three-dimensional (3-D) structures of cellulases from thermophilic fungi

Primary structure

A characteristic of cellulases is their modular structure. Typically, basic structures of endocellulases and cellobiohydrolases are composed of four domains or regions: a signal peptide with about 20 amino acid residues, a cellulose-binding domain (CBD) containing conserved aromatic amino acid residues, a catalytic domain (CD) containing conserved certain aspartic and glutamic acid residues, and a hinge region (linker) rich in Ser, Thr and Pro residues, such as *T. emersonii* CBHII (Murray *et al.*, 2003), *H. grisea* EGL3 (Takashima *et al.*, 1999a), and *H. grisea* var. *thermoidea* CBHI (Takashima *et al.*, 1996). However, *H. grisea* var. *thermoidea* EGL1 (Takashima *et al.*, 1996), *H. grisea* EGL4 (Takashima *et al.*, 1999a), *T. aurantiacus* CBHI (Hong *et al.*, 2003a), *H. grisea* CBHI (Takashima *et al.*, 1998), *T. emersonii* CBH1 (Grassick *et al.*, 2004), and *M. albomyces* Cel7B (Haakana *et al.*, 2004) have been shown to consist solely of a catalytic domain. In *C. thermophilum* CBHs (CBH1, CBH2 and CBH3), CBH1 and CBH2 show an architecture with a typical cellulose-binding domain, linker and catalytic domain regions, but CBH3 only comprises a catalytic domain and lacks a cellulose-binding domain and a hinge region (Liu *et al.* 2005, 2006; Li *et al.*, 2009)

Cellulase structure domains have different biochemical functions. The signal peptide presumably mediates secretion of the enzyme and it is cleaved off by signal peptidases after the protein has been transported across the membrane. The cellulase catalytic domain, containing catalytic sites and substrate-binding sites, is responsible for the hydrolysis of the substrate. The serine/threonine-rich region, which links the catalytic domain with the cellulose-binding domain, is usually glycosylated with sugar chains post-translationally to yield the mature protein. The glycosylation sites may be necessary for the secretion of the protein and maintenance of its stability.

Cellulases are thought to be anchored to their substrate through cellulase-binding domains (CBDs). The cellulose-binding domains of fungi

Table 5.3 Expression of thermophilic fungal cellulase genes in the heterologous hosts

Source	Gene	Host	Reference
Acremonium thermophilum	*cel7a*	*Trichoderma reesei*	Voutilainen *et al*. (2008)
Chaetomium thermophilum	*cel7a*	*Trichoderma reesei*	Voutilainen *et al*. (2008)
Chaetomium thermophilum	*cbh1*	*Pichia pastoris*	Liu *et al*. (2006)
Chaetomium thermophilum	*cbh2*	*Pichia pastoris*	Liu *et al*. (2005)
Chaetomium thermophilum	*cbh3*	*Pichia pastoris*	Li *et al*. (2009)
Chaetomium thermophilum	*bgl*	*Pichia pastoris*	Xu *et al*. (2011)
Humicola grisea	*egl2*	*Aspergillus oryzae*	Takashima *et al*. (1997)
Humicola grisea	*egl3*	*Aspergillus oryzae*	Takashima *et al*. (1999a)
Humicola grisea	*egl4*	*Aspergillus oryzae*	Takashima *et al*. (1999a)
Humicola grisea	*bgl4*	*Aspergillus oryzae*	Takashima *et al*. (1999b)
Humicola grisea var. *thermoidea*	*bgl4*	*Saccharomyces cerevisiae*	Benoliel *et al*. (2010)
Humicola grisea var. *thermoidea*	*eg1*	*Aspergillus oryzae*	Takashima *et al*. (1996)
Humicola grisea var. *thermoidea*	*cbh1*	*Aspergillus oryzae*	Takashima *et al*. (1996)
Humicola insolens	*avi2*	*Humicola insolens*	Moriya *et al*. (2003)
Humicola insolens	*cbhII*	*Saccharomyces cerevisiae*	Heinzelman *et al*. (2009a,b)
Melanocarpus albomyces	*cel7b*	*Trichoderma reesei*	Haakana *et al*. (2004)
Melanocarpus albomyces	*cel7a*	*Trichoderma reesei*	Haakana *et al*. (2004)
Melanocarpus albomyces	*cel45a*	*Trichoderma reesei*	Haakana *et al*. (2004)
Myceliophthora thermophila	*bgl3a*	*Pichia pastoris*	Karnaouri *et al*. (2013)
Scytalidium thermophilum	*bgl4*	*Candida glabrata*	Wang *et al*. (2013)
Talaromyces emersonii	*cel3a*	*Trichoderma reesei*	Muarry *et al*. (2004)
Talaromyces emersonii	*cel7*	*Escherichia coli*	Grassick *et al*. (2004)
Talaromyces emersonii	*cel7A*	*Saccharomyces cerevisiae*	Voutilainen *et al*. (2010)
Thermoascus aurantiacus	*cbh1*	*Saccharomyces cerevisiae*	Hong *et al*. (2003b)
Thermoascus aurantiacus	*eg1*	*Saccharomyces cerevisiae*	Hong *et al*. (2003a)
Thermoascus aurantiacus	*bgl1*	*Pichia pastoris*	Hong *et al*. (2007)
Thermoascus aurantiacus	*cel7a*	*Trichoderma reesei*	Voutilainen *et al*. (2008)

have high similarity and independently fold. Usually, the cellulase-binding domains are composed of less than 40 amino acid residues and they interact with cellulose though the binding site of the CBDs, a hydrophobic flat surface formed by conserved three aromatic amino acid residues. The planar architecture of the binding site is thought to be complementary to the flat surfaces presented by cellulose crystals. In addition, a conserved cysteine motif, which likely mediates the tertiary structure or protein–protein interaction through the formation of disulfide bridges, is found in the cellulose-binding domain (Hashimoto, 2006; Shoseyov *et al*., 2006). It has been shown that deletion of the CBDs of the *T. reesei* Cel7A and Cel6A and the thermophilic fungus *H. grisea* CBH1 greatly reduces their enzymatic activity towards crystalline cellulose (Takashima *et al*., 1998), indicating that the tight binding to cellulose mediated by the CBD is necessary for the efficient hydrolysis of crystalline cellulose by these enzymes. Further studies show that the three aromatic amino acid residues in the CBD affect its cellulose-binding ability and enzymatic activities. Substitution of the three aromatic amino acid residues of *H. grisea* CBH1 CBD with other amino acid residues by site-directed mutagenesis demonstrates that high activity of *H. grisea* CBH1 on crystalline cellulose correlates with high cellulose-binding ability, which is dependent on the combination and configuration of the three aromatic residues (Takashima *et al.*, 2007). It should be noted that the cellulase-binding domain differs from the substrate binding site in the catalytic domain. It has been shown that CBDs promote the enzyme's action on insoluble substrates but are not needed when the enzyme attacks soluble substrates. It is suggested that CBHs without CBD might play an important role in the hydrolysis of soluble substrates (Li *et al.*, 2009).

Three-dimensional (3-D) structure

Since the first 3-D structure of *T. reesei* Cel6A cellulase was solved in 1990, over a hundred cellulase structures have been determined (www.cazy.org), containing cellulases of families 1, 3, 5, 6, 7, 8, 9, 12, 45 and 48, but so far only several 3-dimensional structures of the cellulases in families 5, 6, 7, 12 and 45, are solved from thermophilic fungi (Table 5.4). These structures of thermophilic fungus cellulases from different families are described below.

Cellulases in family 5 are of the endoglucanase type. The fold of enzymes is found to be a common β/α-barrel. In this family, only a structure solved is that of *T. aurantiacus* Cel5A in thermophilic fungi. *T. aurantiacus* Cel5A consists only of the catalytic domain, and 8-fold β/α barrel in the enzyme is very regular and compact with short loops. A substrate binding cleft (groove, crevice) is visible at the C-terminal end of the barrel. This cleft is the cellulose substrate binding site, suggesting that the cleft has the potential to bind seven glucose residues (−4 to +3). In contrast to a lot of other family 5 cellulase structures, the structure of *T. aurantiacus* Cel5A has only few extra-barrel features, a short two stranded β-sheet in β/α-loop 3 and three one-turn helices. The enzyme contains two cysteine residues (Cys212 and Cys249), implicated in a disulfide bridge (Lo Leggio and Larsen, 2002).

There are endoglucanases and cellobiohydrolases in family 6 cellulases. 3-D structures have been reported for the endoglucanase Cel6B and the cellobiohydrolase Cel6A of this family from the thermophilic fungus *Humicola insolens* (Varrot *et al.*, 2003; Davies *et al.*, 2000). The structure of these two cellulases presents the distorted β/α-barrel fold with a central β-barrel made up of seven, not eight, parallel β-strands. A substrate binding crevice is formed between strands I and strands VII. The crevice of the Cel6A contains at least four substrate-binding sites, −2 to +2, and that of the Cel6B has six substrate-binding sites, −2 to +4. The significant difference between the endoglucanase Cel6B and the cellobiohydrolase

Table 5.4 Cellulases from thermophilic fungi with solved 3-D structures

Source	Name	Family	Fold	Reference
Humicola insolens	Cel6A	6	β/α-barrel	Varrot *et al.* (2003)
Humicola insolens	Cel6B	6	β/α-barrel	Davies *et al.* (2000)
Humicola insolens	EGI	7	β-sandwich	Davies *et al.* (1997)
Humicola insolens	Cel7B	7	β-sandwich	Mackenzie *et al.* (1998)
Humicola insolens	EGV	45	β-barrel	Davies *et al.* (1996)
Humicola grisea	Cel12A	12	β-sandwich	Sandgren *et al.* (2004)
Talaromyces emersonii	CBHIB	7	β-sandwich	Grassick *et al.* (2004)
Thermoascus aurantiacus	Cel5A	5	β/α-barrel	Lo Leggio and Larsen (2002)
Melanocarpus albomyces	maEG	45	β-barrel	Hirvonen and Papageorgou (2003)
Melanocarpus albomyces	Cel7B	7	β-sandwich	Parkkinen *et al.* (2008)
Thermoascus aurantiacus	TaGH61	61[a]	β-sandwich	Quinlan *et al.* (2011)
Thielavia terrestris	GH61E	61[a]	β-sandwich	Harris *et al.* (2010)

[a]GH61 glycoside hydrolases are currently termed Auxiliary Activity Family 9 (AA9), which are characterized as Cu^{2+} ion-dependent polysaccharide monooxygenases (LPMOs) containing highly conserved histidine residues (Levasseur *et al.*, 2013).

Cel6A is that two extended surface loops enclose the active site in the Cel6A, but are absent in the Cel6B, resulting in an enclosed substrate-binding cleft in the Cel6A and an open substrate cleft in the Cel6B. Because of this structural difference, endoglucanase can hydrolyse bonds internally in cellulose chains; cellobiohydrolase act on chain ends.

Like family 6 cellulases, cellulases from family 7 contain endoglucanases and cellobiohydrolases. The 3-D structure of a number of endoglucanases and cellobiohydrolases from this family are reported from different organisms, but only a few 3-D structures of cellulases from thermophilic fungi have been reported, such as *T. emersonii* CBHIB (Grassick *et al.*, 2004), *Humicola insolens* EGI/Cel7B (Davies *et al.*, 1997; Mackenzie *et al.*, 1998) and *M. albomyces* Cel7B (Parkkinen *et al.*, 2008). The 3-D structure of this family is a double β-sandwich. The structure of *M. albomyces* Cel7B (a cellobiohydrolase) is a representative of the family 7 cellobiohydrolases. The structure of *M. albomyces* Cel7B consists of two antiparallel β-sheets packed face-to-face to form a β-sandwich. Both β-sheets contain six β-strands. Because their strong curve, these two sheets form concave and convex surfaces of the sandwich. A characteristic of the Cel7B is that the loops connecting the strands extend from the concave face of the sandwich and form an enclosed substrate-binding tunnel. The extended loops are stabilized by the presence of nine disulfide bonds. The tunnel is about 50 Å long and contains nine substrate-binding sites, −7 to +2 (Parkkinen *et al.*, 2008). The thermophilic fungus *H. insolens* EGI/Cel7B is an endoglucanase in family 7. Structurally, *H. insolens* EGI has an over β-sandwich structure (Davies *et al.*, 1997; Mackenzie *et al.*, 1998), as *Melanocarpus albomyces* Cel7B. The structure is constructed of two large anti-parallel β-sheets consisting of seven and eight strands respectively. A substrate-binding cleft is formed in the centre of a canyon formed by the curvature of the β-strands of the β-sandwich. The cleft contains at least four substrate-binding sites, −2 to +2. However, there is structural difference between *H. insolens* EGI and *M. albomyces* Cel7B. *H. insolens* EGI has an open substrate-binding cleft, and *M. albomyces* Cel7B has an enclosed substrate-binding tunnel (Davies *et al.*, 1997; Parkkinen *et al.*, 2008), which is similar to the endoglucanases and the cellobiohydrolases in family 6.

The structure has been solved for a family 12 fungal cellulase (Cel12A) from a thermophilic fungus *Humicola grisea* (Sandgren *et al.*, 2003, 2004). *H. grisea* Cel12A is an endoglucanase. Its structure is composed of 15 β-strands that fold into two anti-parallel β-sheets, which pack on top of one another to form a compact curved β-sandwich. The convex β-sheet consists of six anti-parallel strands, and the concave β-sheet consists of nine anti-parallel strands. The structure's concave face creates a long and not blocked substrate-binding cleft with seven substrate-binding sites, −4 to +2.

So far, the structures of only two endoglucanases from family 45 were solved, the thermophilic fungus *H. insolens* Cel45A (EGV) (Davies *et al.*, 1996) and the thermophilic fungus *M. albomyces* 20 kDa endoglucanase (Hivonen and Papageorgiou, 2003). Among these two structures, the structure of *M. albomyces* 20 kDa endoglucanase (maEG) is studied in more detail. The molecule of the 20 kDa endoglucanase has the shape of a flattened sphere. The main structural feature is a six-stranded β-barrel. The β-strands are connected with long disulfide-bonded loop structures while the remainder of the structure is completed by three helices. The seven disulfides are located between residues 11 and 134, 16 and 85, 31 and 55, 47 and 12, 155 and 166, 188 and 88, and 198 and 86. A substrate binding groove is formed between the β-barrel and the loop structures. This groove sub-divided into six substrate-binding sites, −4 to +1 (Hivonen and Papageorgiou, 2003).

As stated above, all cellulases have a long substrate-binding groove on the enzyme's surface. This groove is considered as the active centre (containing substrate-binding sites and catalytic sites), which binds sugar units of cellulose that are subsequently cleaved. In the groove, a few aromatic residue side chains are found as hydrophobic platforms (the aromatic residues bind to the hydrophobic faces of the sugar rings) that form substrate-binding sites, and two strictly conserved amino acid residues (Asp and Glu) as catalytic nucleophile and catalytic acid/base are composed

Table 5.5 Substrate-binding residues and catalytic residues of cellulases from thermophilic fungi. *H.*, *Humicola*; *T.*, *Talaromyces*; *M.*, *Melanocarpus*

Source	Name	Catalytic sites	Main substrate-binding sites	Reference
H. insolens	Cel6A	Asp226-Asp405	Trp137,274,277,371	Varrot *et al.* (2003)
H. insolens	Cel6B	Asp139-Asp316	Trp52,282,186,189	Davies *et al.* (2000)
H. insolens	Cel7B	Glu197-Glu202	Tyr147, 171, Trp347,356	Mackenzie *et al.* (1998)
H. insolens	EGV	Asp10-Asp121	Trp18,Tyr8	Davies *et al.* (1996)
H. grisea	Cel12A	Glu120-Glu205	Tyr9,66,132, Trp115	Sandgren *et al.* (2004)
T. emersonii	CBHIB	Glu212-Glu217	Trp38,40,371,380, Tyr47	Grassick *et al.* (2004)
T. aurantiacus	Cel5A	Glu133-Glu240	Trp278,279,273,170,174	Lo Leggio and Larsen (2002)
M. albomyces	maEG	Asp10-Asp120	Trp18, Tyr8	Hirvonen and Papageorgou (2003)
M. albomyces	Cel7B	Glu212-Glu217	Trp40,Tyr247,145,246,144	Parkkinen *et al.* (2008)

of catalytic sites. By the catalytic sites, cellulases cleave glycosidic bonds using acid-base catalysis with inversion or retention mechanism (Zhang *et al.*, 2006). The active sites of the cellulases from thermophilic fungi identified on the basis of their three-dimensional structure are summarized in Table 5.5.

Homology modelling of cellulases of families 1 and 3 from thermophilic fungi by Swiss-Model

Crystal structures of endocellulases and exocellulases (cellobiohydrolase) from thermophilic fungi are available, but crystal structures of beta-glucosidases from thermophilic fungi are not solved so far. Beta-glucosidases are found in families 1 and 3. Structures of these cellulases can be prediction on the basis of the known three-dimensional structure cellulases. Crystal structures are available for cellulases of families 1 and 3 from other fungi, which may provide valuable insights into the structure and the active sites of the thermophilic fungus beta-glucosidases.

H. grisea var. *thermoidea* BGL4 is a thermostable beta-glucosidase of family 1, and its 3-D structure has not been solved so far (Takashima *et al.*, 1999b; Benoliel *et al.* 2010). The beta-glucosidase shares high identity of 74.14% with *Hypocrea jecorina* beta-glucosidase BGL2 of family 1, whose 3-D structure has been reported (Jeng *et al.*, 2011). The high degree of identity of amino acid sequence between BGL4 and BGL2 allows the generation and the prediction of a plausible 3-D model for *H. grisea* var. *thermoidea* BGL4 protein based on the published structure of *H. jecorina* BGL2. The protein sequence of *H. grisea* var. *thermoidea* BGL4 is used for building a 3-D structure by Swiss-Model (an automated comparative protein modelling server) using *H. jecorina* BGL2 as a template. The homology model of BGL4 showed that the over fold of BGL4 has a classical $(\alpha/\beta)(8)$-TIM barrel fold (Fig. 5.1A) and is very similar to the over fold of the template *H. jecorina* BGL2. All the important amino acid residues involved in catalytic sites and substrate-binding sites in *H. jecorina* BGL2, are found in *H. grisea* var. *thermoidea* BGL4.

Recently, crystal structure of glycoside hydrolase family 3 beta-glucosidase 1 (AaBGL1) from *Aspergillus aculeatus* was solved (Suzuki *et al.*, 2013). AaBGL1 is a dimer in solution. The dimer interface is expanded by N-glycans. The AaBGL1 monomer consists of three domains; a catalytic TIM (triosephosphateisomerase) barrel-like domain, an α/β sandwich domain and a FnIII (fibronectin type III) domain. These domains are connected with two linker regions. A family 3 beta-glucosidase (CtBGL1) from *C. thermophilum* (Xu *et al.*, 2011) shares high identity (61.61%) with AaBGL1. The homology model of CtBGL1 obtained using AaBGL1 as a template by Swiss-Model showed that the over fold of CtBGL1 has a catalytic TIM barrel-like domain, an α/β sandwich domain and a FnIII domain (Fig. 5.1B). All the important amino acid residues involved in catalytic sites, substrate-binding sites and formation of disulfide bonds in AaBGL1, are found in CtBGL1. Therefore, it is suggested that

A

B

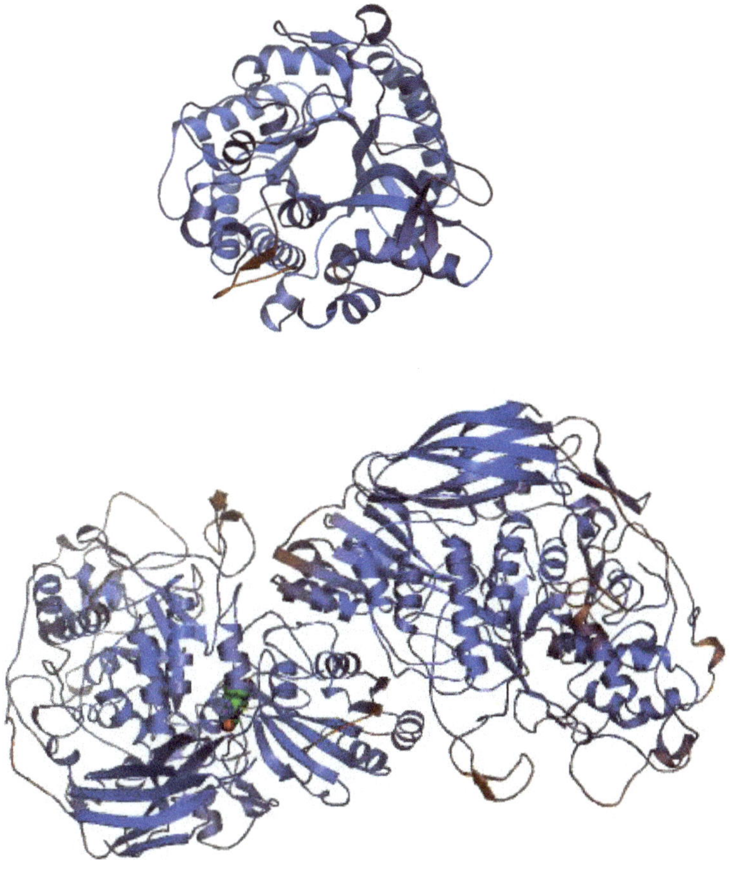

Figure 5.1 Homology model of cellulases of families 1 and 3 by Swiss-Model. (A) The plausible 3-D model of *Humicola grisea* var. *thermoidea* BGL4 (family 1) was produced by using the 3-D structure of Hypocrea jecorina BGL2 as a template (Jeng *et al.*, 2011). (B) The plausible 3-D model of *Chaetomium thermophilum* CtBGL1 was produced by using the 3-D structure of *Aspergillus aculeatus* AaBGL1 as a template (Suzuki *et al.*, 2013).

the structure of CtBGL1 is likely to be similar to that of AaBGL1.

Phylogenetic analysis of cellulases from thermophilic fungi

Amino acid sequence analysis of cloned cellulases from thermophilic fungi shows that they are found in families 1, 3, 5, 6, 7, 12 and 45 (Table 5.1). To understand knowledge of the evolution and relationship between different family cellulases from thermophilic fungi, phylogeny of 43 thermophilic fungal cellulases from different families, which were obtained from the GenBank, was analysed with the CLUSTALW multiple sequence alignment programme to generate a phylogenetic tree (www.ebi.ac.uk/clustalw). In the phylogenetic tree (Fig. 5.2), all the 43 thermophilic fungal cellulases are classified into different clusters which coincide with different families (family 1, 5, 6, 3, 12, 45 and 7). Within each cluster or family, cellulase members are sub-divided into different groups which are in agreement with monophyletic groups, suggesting that cellulases of thermophilic fungi within a family might evolve along with organismal evolution. Furthermore, cellulase members of each cluster or family show significant levels of sequence similarity, and also share important catalytic residues and structure, suggesting that they may have a common evolutionary origin.

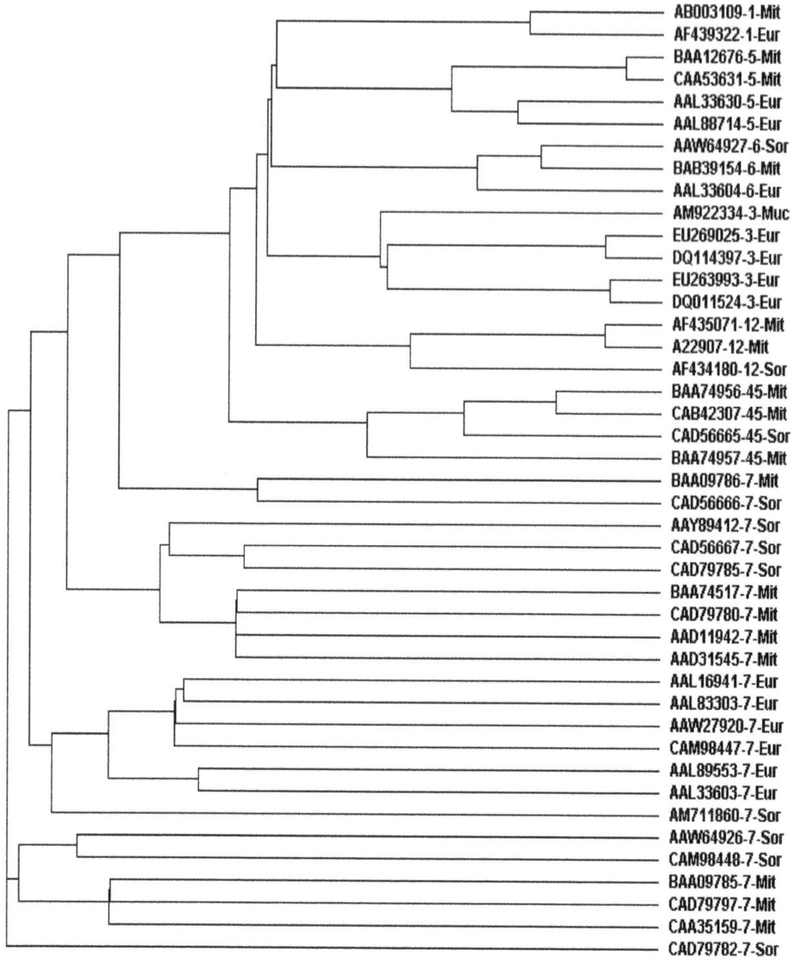

Figure 5.2 Thermophilic fungal cellulases phylogenic tree based on the amino acid sequence of 43 thermophilic fungal cellulases using CLUSTALW. Each entry has a GenBank accession number from AB003109, AF439322 to CAD79782. 1, family 1; 3, family 3; 5, family 5; 6, family 6; 7, family 7; 12, family 12; 45, family 45. Mit, Mitosporic *Ascomycota*; Eur, *Eurotiomycetes*; Sor, *Sordariomycetes*; Muc, *Mucoromycotina*.

Molecular modification of cellulases from thermophilic fungi

In biomass conversion processes of cellulases, the current challenge is how to degrade highly crystalline(?) cellulose both efficiently and cheaply. To increase cellulase efficiencies and to lower their cost, cellulases need to be improved to get characteristics necessary for application, such as higher catalytic efficiency on cellulose, higher stability at elevated temperature and at non-physiological pH, higher tolerance to end-product inhibition and conversion of cellulase to glycosynthase. At present, there are two main research technologies containing structure based rational mutagenesis (site-directed mutagenesis) and random mutation based directed evolution for cellulase improvement or engineering (Zhang *et al.*, 2006).

Cellulases from thermophilic fungi are thermostable. Further increase of their thermostability would be beneficial for industrial applications. *M. albomyces* Cel7B is a cellobiohydrolase. To improve its thermostability, the Cel7B enzyme was randomized by error-prone PCR, and 49 positive mutant clones were screened from 14,600 random clones by a robotic high-throughput thermostability screening method. Two positive thermostable mutants show improved unfolding temperature (T_m) by 1.5°C for the A30T mutant and 3.5°C for the S290T mutant. The optimum temperature on a soluble substrate for the A30T mutant has improved by 5°C. Sequence analysis of amino acid sequences from the A30T mutant and the S290T mutant indicates that there is only one amino acid change in their sequences. The amino acid alterations are located in the β-strands further away from the active site tunnel of the Cel7B enzyme, which could improve protein packing (Voutiainen *et al.*, 2007). It was also reported that a thermostable cellobiohydrolase (Cel7A) from the thermophilic fungus *T. emersonii* was engineered with site-directed mutagenesis in order to further improve its thermostability and activity. Additional disulfide bridges were introduced to the catalytic module of Cel7A. Three mutants' thermostability was clearly improved which was also reflected as improved Avicel hydrolysis efficiency at 75°C (Voutilainen *et al.*, 2010).

Recently, a new approach SCHEMA, which is a computational approach and uses protein structure data to generate new sequences for special purposes, has been used to create thermostable fungal cellulases (Heinzelman *et al.*, 2009a,b). With the high resolution of *H. insolens* CBHII (Varrot *et al.*, 2003) as a template for SCHEMA, new DNA sequences of cellulases were designed and synthesized by DNA2.0. Using the computer-generated sequences, a total of 31 new cellulase genes were expressed in *S. cerevisiae*; every one of these cellulases was found to be more stable than the most stable parent cellulase from the thermophilic fungus *H. insolens*, as measured either by its half-life of inactivation at 63°C or by T_{50} (Heinzelman *et al.* 2009a,b). These findings demonstrate the value of using the structure guided recombination to efficiently generate highly stable cellulases. Because novel and interesting cellulases can be synthesized in the laboratory based on a protein database and our recombination design and not isolated from any organism, this approach is a nice demonstration of the power of synthetic biology.

In addition to improvement of cellulases thermostability, increase of cellulase stability in detergent solutions by Protein Eng. was also reported. *H. insolens* Cel45 (an endoglucanase) is used in detergent industry, but the Cel45 is inactivated by the detergent C12-LAS (an anionic surfactant) by combining with surface positive charge of the enzyme. Therefore, it is thought substitution of surface charge of the Cel45, either by removal of positive amino acids or by introduction of negative changes for its stability in the detergent solution. According to Cel45 crystal structure, different mutations involved in its surface charge were obtained by site-directed mutagenesis. The research data of these mutants shows that the introduction of positive charges or removal of negative charges greatly increases detergent sensitivity. The position Arg158Glu gives the highest increase in stability against C12-LAS (Otzen *et al.*, 1999).

Improving cellulase activity using site-directed mutagenesis and directed evolution has aroused increasing attention among researchers. However, because of lacking general rules for site-directed mutagenesis and limitation of screening methods on solid cellulosic substrates for directed evolution, there are a few successful examples of significantly higher activity cellulase mutants (Zhang *et al.*, 2006). It was reported a 20% improvement in the activity on microcrystalline cellulose of a modified endoglucanase Cel5A from *Acidothermus cellulolyticus* by site-directed mutagenesis. A 5-fold higher specific activity *Bacillus subtilis* endoglucanase mutant was found by directed evolution. An endocellulase gene from the termite *Reticulitermes speratus* was mutated with site-directed mutagenesis, and three mutants, G91A, Y97W and K429A, had higher activities towards carboxymethyl cellulose than the wild type (Ni *et al.*, 2010). Unfortunately, there are few reports on improving cellulase activity of thermophilic fungi using site-directed mutagenesis and directed evolution so far. As stated above, the S290T mutant from *M. albomyces* Cel7B exhibits not only improved thermostability, but also a 2-fold increase in the rate of hydrolysis of Avicel at 70°C (Voutiainen *et al.*, 2007). Similar results were also obtained from *M. albomyces* Cel7A using site-directed mutagenesis (Voutiainen *et al.*, 2010). In our laboratory, a thermostable cellobiohydrolase gene from *C. thermophilum* is mutated with error-prone PCR. The hydrolysis rate of the cellobiohydrolase mutant on cellulose was higher than the parental enzyme (Wang *et al.*, 2012).

Cellulose-binding domains (CBDs) of cellulases play an important role in enzymatic activities on crystalline cellulose. Studies have shown the potential of CBDs for modifying the characteristics of enzymes (Shoseyov *et al.*, 2006). A basic approach in CBD engineering is to add a CBD in order to improve hydrolytic activity. Addition of a CBD from *T. reesei* CBHII to *T. harzianum* chitinase results in increased hydrolytic activity of insoluble substrates (Limon *et al.*, 2001). The thermophilic fungus *H. grisea* contains two endoglucanases, EGL3 with a CBD and EGL4 without a CBD. The fusion protein, EGL4CBD, which consists of the EGL4 catalytic domain and the EGL3 cellulose-binding domain, shows relatively high activity to carboxymethyl cellulose (Takashima *et al.*, 1999a). *M. albomyces* Cel45A, Cel7A and Cel7B lack a consensus CBD and its associated linker (Haakana *et al.*, 2004). To improve their efficiency, these three cellulases naturally harbouring no CBD were genetically modified to carry the CBD of *T. reesei* CBHI/Cel7A. Presence of the CBD improves their hydrolytic potential to crystalline cellulose (Szijarto *et al.*, 2008).

An important development in cellulase engineering is that cellulases can be converted to glycosynthases by site-directed mutagenesis (Zhang *et al.*, 2006). The glycosynthases are retaining glycosidase mutants in which the catalytic nucleophile has been converted to a non-nucleophilic residue. The first glycosynthase reported from thermophilic fungi is *H. insolens* Cel7B. Based on its X-ray structure and active site, *H. insolens* Cel7B was converted by site-directed mutagenesis to glycosynthase which is a Glu197Ala (E197A) catalytic nucleophile mutant. The mutant Cel7B E197A glycosynthase is able to catalyse the regio- and stereoselective glycosylation in high yield (Fort *et al.*, 2000). To widen its substrate specificity, three second generation mutants of the *H. insolens* Cel7B E197A glycosynthase were prepared and characterized by site-directed mutagenesis, Cel7B E209A H209A, Cel7B E197A H209G and Cel7B E197A H209A A211T (Blanchard *et al.*, 2007a,b). The double mutants Cel7B E209A H209A and Cel7B E197A H209G preferentially catalyse the formation of a β-(1,4) linkage between the two disaccharides, while the single mutant Cel7B E197A and triple mutant Cel7B E197A H209A A211T can produce predominantly the β-(1,3)-linked tetrasaccharide. This research shows that active site mutations can modulate the regioselectivity of the glycosylation reaction catalysed by *H. insolens* Cel7B E197A glycosynthase.

Enhancement of cellulase activity by Auxiliary Activity Family 9 (AA9) enzymes from thermophilic fungi

Recent studies have shown that oxidative enzymes are involved in the degradation of plant biomass (Horn *et al.*, 2012). A key study demonstrated that glycoside hydrolases family 61 (GH61) greatly enhances cellulase activity in lignocellulose hydrolysis through a metal ion-dependent mechanism (Harris *et al.*, 2010). The 3D structures of the GH61 enzymes display a predominantly beta-sheet fold with an extended planar face, the centre of which contains the N-terminal histidine where a metal ion is believed to bind, indicating that they are unlikely to be glycoside hydrolases (Harris *et al.*, 2010; Karkehabadi *et al.*, 2008; Quinlan *et al.*, 2011). A recent important functional study further demonstrated that nature of the enhancement of cellulase activity by the GH61 enzymes is an oxidative mechanism (Quinlan *et al.*, 2011), similar to a structurally related chitin-binding protein (CBP21), which belongs to carbohydrate-binding module family 33 (CBM33) and is able to dramatically boost the enzymatic conversion of chitin by chitinases (Vaaje-Kolstad *et al.*, 2010; Aachmann *et al.*, 2012). Currently, GH61 glycoside hydrolases, originally classified as glycoside hydrolases based on very weak endo-1,4-b-D-glucanase activity in one family member (Karlsson *et al.*, 2001; Koseki *et al.*, 2008), are termed Auxiliary Activity Family 9 (AA9), which are characterized as Cu^{2+} ion-dependent polysaccharide monooxygenases (LPMOs) containing highly conserved histidine residues (Levasseur *et al.*, 2013). Up to now, only two AA9 enzymes were reported from thermophilic fungi, an AA9 protein (TaGH61) from the thermophilic fungus *T. aurantiacus* (Quinlan *et al.*, 2011) and an AA9 protein (GH61E) from the thermophilic fungus *Thielavia terrestris* (Harris *et al.*, 2010), which greatly enhance cellulase activity in lignocellulose hydrolysis. The discovery of the AA9 family as a major cellulase-enhancing factor dramatically promotes cellulose degradation and significantly reduces cellulase cost. It is a significant breakthrough in hydrolysis of lignocellulosic biomass to fermentable sugars, and will provide a new avenue towards more efficient enzymatic conversion of biomass.

Fungi play a central role in the degradation of cellulose in terrestrial environments. Fungal cellulolytic systems and oxidative degradation of cellulose are very complex in nature. Studies have revealed that thermophilic fungi produce multiple forms of the cellulase components (Maheshwari *et al.*, 2000, Table 5.2). On the other hand, thermophilic fungi also produce multiple forms of AA9 components for cellulose degradation. It was reported that *T. terrestris* produced at least six representatives of the AA9 family, comprising in total about 10% of the total soluble protein (Harris *et al.*, 2010). Recent genomic sequencing showed a number of AA9 genes in thermophilic fungi. According to current publicly available annotations from thermophilic fungi, there are 24, 20, 4, 4, and 4 AA9 genes in *Thielavia terrestris, Myceliophthora thermophila, Thermomyces lanuginosus, T. aurantiacus* and *C. thermophilum*, respectively. These data indicate the widespread nature of AA9 family of enzymes in cellulose degradation and the phylogenetic diversity of AA9 proteins. The AA9 family of enzymes likely holds major potential for industrial decomposition of cellulosic materials.

Present gaps and future perspectives

Although there has been extensive research on cellulases from thermophilic fungi in recent years, there are still some main research gaps as follows.

1. There are about 40–50 species of known thermophilic fungi, but only a minority of thermophilic fungal cellulases has been characterized in detail, such as the cellulases from *H. grisea, H. grisea* var. *thermoidea, H. insolens, M. albomyces, T. emersonii, T. aurantiacus, T. aurantiacus* var. *levisporus,* and *C. thermophilum*. Characterization of cellulases from other thermophilic fungi is necessary, first because of finding some novel cellulases for specific applications, and secondly because of accumulating more evidences for our understanding of their evolution.
2. Evolution of different family cellulases from thermophilic fungi is interesting. However, our knowledge of the evolution is poor. Up to now, we do not know that where thermophilic fungi originate although they are thought to evolve from mesophilic fungi, and the number of most family cellulase genes from thermophilic fungi is very limited. In addition, further researches on the thermophilic fungal cellulase genes acquired by horizontal gene transfer from other organisms or inherited by continuous vertical descent from an ancient ancestor are necessary.
3. Cellulases from thermophilic fungi are thermostable, but their thermostability mechanism is unclear. There is little information on research of their thermostability mechanism. To investigate thermostability mechanism of cellulases from thermophilic fungi, a preferable approach is to mutate cellulases by site-directed mutagenesis and directed evolution for obtaining thermostable mutations. Using directed molecular evolution, a lot of mutations are created by random mutagenesis. After screening thermostable mutations and analysing their sequences, amino acid residues involved in thermostability will be determined.
4. It is commonly believed that cellulases from thermophilic fungi play an important role in the utilization of cellulose as a carbon and energy source and in the recycling of cellulose in nature. A question is whether they, especially with cell-associated cellulases, have important biological functions, such as cell wall biosynthesis and remodelling during hyphal elongation. It has been found that some beta-glucosidases of thermophilic fungi occur in the cytosol or on the surface of cell (Maheshwari *et al.*, 2000), but their functional role is unknown. Some molecular approaches like targeted gene disruption, gene silencing and insertional mutagenesis

will contribute to the understanding of these gene function.

5 Recently, two non-cellulase proteins have been identified in *Trichoderma* fungi that appear to modify cellulose and enhance its hydrolysis by cellulase (Wilson 2009). The fungal proteins with some homology to plant expansin are called swollenin. Compared with these identified swollenins of *Trichoderma* fungi, there are no reports of swollenin from thermophilic fungi.

6 Although extensive researches on cellulases have been carried out, there are some important scientific questions to be solved in cellulases, containing cellulases from thermophilic fungi. These scientific questions mainly are the mechanism that cellulases hydrolyse crystalline cellulose, the synergism of cellulase system, the role of cellulose-binding domains (CBDs), the 3-D structure of new cellulases and the relationship between structure and function of cellulases. Further research on these questions assist us in developing better and more versatile cellulases for the utilization of biomass.

Conclusions

Significant advances in biochemistry, molecular biology, structure biology and Protein Eng. of cellulases from thermophilic fungi have been made in recent years. The developments have led to speculation and anticipation of their scientific values and commercial potential in biotechnology, and have also added greatly to our understanding of characterization and structure of thermostable cellulases from thermophilic fungi. However, much more work on cellulases from thermophilic fungi is required to further understand their nature and fully explore their industrial potential.

Acknowledgements

This work was supported by Chinese National Programs for High Technology Research and Development, the Chinese Project of Transgenic Organisms, and the Chinese National Nature Science Foundation.

References

Aachmann, F.L., Sorlie, M., Skjak-Brak, G., Eijsink, V.G.H., and Vaaje-Kolstad, G. (2012). NMR structure of a lytic polysaccharide monooxygenase provides insight into copper binding, protein dynamics, and substrate interactions. Proc. Natl. Acad. Sci. U.S.A. *109*, 18779–18784.

Badhan, A.K., Chadha, B.S., Kaur, J., Saini, H.S., and Bhat, M.K. (2007). Production of multiple xylanolytic and cellulolytic enzymes by thermophilic fungus *Myceliophthora* sp. IMI38099. Bioresour. Technol. *98*, 504–510.

Benko, Z., Drahos, E., Szengyel, Z., Puranen, T., Vehmaanpera, J., and Reczey, K. (2007). *Thermoascus aurantiacus* CBHI/Cel7A production in *Trichoderma reesei* on alternative carbon sources. Appl. Biochem. Biotechnol. *137*, 195–204.

Benoliel, B., Pocas-Fonseca, M.J., Torres, F.A., and de Moraes, L.M. (2010). Expression of a glucose-tolerant beta-glucosidase from *Humicola grisea* var. *thermoidea* in *Saccharomyces cerevisiae*. Appl. Biochem. Biotechnol. *160*, 2036–2044.

Berka, R.M., Grigoriev, I.V., Otillar, R., Salamov, A., Grimwood, J., Reid, I., Ishmael, N., John, T., Darmond, C., Moisan, M.C., *et al.* (2011). Comparative genomic analysis of the thermophilic biomass-degrading fungi *Myceliophthora thermophila* and *Thielavia terrestris*. Nat. Biotechnol. *29*, 922–927.

Blanchard, S., Cottaz, S., Coutinho, P.M., Patkar, S., Vind, J., Boer, H., Koivula, A., Driguez, H., and Armand, S. (2007a). Mutation of fungal endoglucanases into glycosynthases and characterization of their acceptor substrate specificity. J. Mol. Catal. B-Enzym. *44*, 106–116.

Blanchard, S., Armand, S., Couthino, P., Patkar, S., Vind, J., Samain, E., Driguez, H., and Cottaz, S. (2007b). Unexpected regioselectivity of *Humicola insolens* Cel7B glycosynthase mutants. Carbohydr. Res. *342*, 710–716.

Busk, P.K., and Lange, L. (2013). Cellulolytic potential of thermophilic species from four fungal orders. AMB Express 3, 47.

Chu, C.X., Li, D.C., and Guo, R.F. (2004). Purification and properties of β-glucosidase from thermophilic *Cheatomium thermophilum*. Mycosystema *23*, 397–402.

Collins, C.M., Murray, P.G., Denman, S., Morrissey, J.P., Byrnes, L., Teeri, T.T., and Tuohy, M.G. (2007). Molecular cloning and expression analysis of two distinct beta-glucosidase genes, *bg1* and *aven1*, with very different biological roles from the thermophilic saprophytic fungus *Talaromyces emersonii*. Mycol. Res. *111*, 840–849.

Dalboge, H., and Heldt-Hansen, H.P. (1994). A novel method for efficient expression cloning of fungal enzyme genes. Mol. Gen. Genet. *243*, 253–260.

Davies, G.J., Dodson, G.G., Moore, M.H., Tolley, S.P., Dauter, Z., Wilson, K.S., Rasmussen, G., and Schuelein, M. (1996). Structure determination and refinement of the *Humicola insolens* endoglucanase V at 1.5-A resolution. Acta Crystallogr. D 52, 7–17.

Davies, G.J., Ducros, V., Lewis, R.J., Borchert, T.V., and Schulein, M. (1997). Oligosaccharide specificity of a family 7 endoglucanase: insertion of potential sugar-binding subdites. J. Biotechnol. *57*, 91–100.

Davies, G.J., Brzozowski, A.M., Dauter, M., Varrot, A., and Schulein, M. (2000). Structure and function of *Humicola insolens* family 6 cellulases: structure of the endoglucanase, Cel6B, at 1.6Å resolution. Biochem. J. *348*, 201–207.

De Palma-Fernandez, E.R., Gomes, E., and da Silva, R. (2002). Purification and characterization of two beta-glucosidases from the thermophilic fungus *Thermoascus aurantiacus*. Folia Microbiol. *47*, 685–690.

Fort, S., Boyer, V., Greffe, L., Davies, G., Moroz, O., Christiansen, L., Schiilein, M., Cottaz, S., and Driguez, H. (2000). Highly efficient synthesis of beta(1–4)-oligo- and -polysaccharides using a mutant cellulase. J. Am. Chem. Soc. *122*, 5429–5437.

Furukawa, T., Shida, Y., Kitagami, N., Mori, K., Kato, M., Kobayashi, T., Okada, H., Ogasawara, W., and Morikawa, Y. (2009). Identification of specific binding sites for XYR1, a transcriptional activator of cellulolytic and xylanolytic genes in *Trichoderma reesei*. Fungal. Genet. Biol. *46*, 564–574.

Goedegebuur, F., Fowler, T., Phillips, J., van der Kley, P., van Solingen, P., Dankmeyer, L., and Power, S.D. (2002). Cloning and relational analysis of 15 novel fungal endoglucanases from family 12 glycosyl hydrolase. Curr. Genet. *41*, 89–98.

Grassick, A., Murray, P.G., Thompson, R., Collins, C.M., Byrnes, L., Birrane, G., Higgins, T.M., and Tuohy, M.G. (2004). Three-dimensional structure of a thermostable native cellobiohydrolase CBH IB and molecular characterization of the cel7 gene from the filamentous fungus, *Talaromyces emersonii*. Eur. J. Biochem. *271*, 4495–4506.

Haakana, H., Miettinen-Oinonen, A., Joutsjoki, V., Mantyla, A., Suominen, P., and Vehmaanpera, J. (2004). Cloning of cellulase genes from *Melanocarpus albomyces* and their efficient expression in *Trichoderma reesei*. Enzyme Microb. Technol. *34*, 159–167.

Harris, P.V., Welner, D., McFarland, K.C., Re, E., Poulsen, J.C.N., Brown, K., Salbo, R., Ding, H., Vlasenko, E., Merino, S., *et al.* (2010). Stimulation of lignocellulosic biomass hydrolysis by proteins of glycoside hydrolase family 61: structure and function of a large, enigmatic family. Biochemistry *49*, 3305–3316.

Hashimoto, H. (2006). Recent structural studies of carbohydrate-binding modules. Cell. Mol. Life Sci. *63*, 2954–2967.

Heinzelman, P., Snow, C.D., Wu, I., Nguyen, C., Villalobos, A., Govindarajan, S., Minshull, J., and Arnold, F.H. (2009a). A family of thermostable fungal cellulases created by structure-guided recombination. Proc. Natl. Acad. Sci. U.S.A. *106*, 5610–5615.

Heinzelman, P., Snow, C.D., Smith, M.A., Yu, X., Kanaan, A., Boulware, K., Villalobos, A., Govindarajan, S., Minshull, J., and Arnold, F.H. (2009b). SCHEMA recombination of a fungal cellulase uncovers a single mutation that contributes markedly to stability. J. Biol. Chem. *284*, 26229–26233.

Hirvonen, M., and Papageorgiou, A.C. (2003). Crystal structure of a family 45 endoglucanase from *Melanocarpus albomyces*: mechanistic implications based on the free and cellobiose-bound forms. J. Mol. Biol. *329*, 403–410.

Hong, J., Tamaki, H., Yamamoto, K., and Kumagai, H. (2003a). Cloning of a gene encoding thermostable cellobiohydrolase from *Thermoascus aurantiacus* and its expression in yeast. Appl. Microbiol. Biotechnol. *63*, 42–50.

Hong, J., Tamaki, H., Yamamoto, K., and Kumagai, H. (2003b). Cloning of a gene encoding a thermo-stable endo-β-1,4-glucanase from *Thermoascus aurantiacus* and its expression in yeast. Biotechnol. Lett. *25*, 657–661.

Hong, J., Tamaki, H., and Kumagai, H. (2006). Unusual hydrophobic linker region of beta-glucosidase (BGLII) from *Thermoascus aurantiacus* is required for hyper-activation by organic solvents. Appl. Microbiol. Biotechnol. *73*, 80–88.

Hong, J., Tamaki, H., and Kumagai, H. (2007). Cloning and functional expression of thermostable beta-glucosidase gene from *Thermoascus aurantiacus*. Appl. Microbiol. Biotechnol. *73*, 1331–1339.

Horn, S.J., Vaaje-Kolstad, G., Westereng, B., and Eijsink, V.G. (2012). Novel enzymes for the degradation of cellulose. Biotechnol. Biofuels *5*,45.

Ilmen, M., Saloheimo, A., Onnela, M.L., and Penttila, M.E. (1997). Regulation of cellulase gene expression in the filamentous fungus *Trichederma reesei*. Appl. Environ. Microbiol. *63*, 1298–1306.

Jeng, W.Y., Wang, N.C., Lin, M.H., Lin, C.T., Liaw, Y.C., Chang, W.J., Liu, C.I., Liang, P.H., and Wang, A.H. (2011). Structural and functional analysis of three β-glucosidases from bacterium *Clostridium cellulovorans*, fungus *Trichoderma reesei* and termite *Neotermes koshunensis*. J. Struct. Biol. *173*, 46–56.

Karkehabadi, S., Hansson, H., Kim, S., Piens, K., Mitchinson, C., and Sandgren, M. (2008). The first structure of a glycoside hydrolase family 61 member, Cel61B from *Hypocrea jecorina*, at 1.6A resolution. J. Mol. Biol. 383, 144–154.

Karlsson, J., Saloheimo, M., Siika-Aho, M., Tenkanen, M., Penttila, M., and Tjerneld, F. (2001). Homologous expression and characterization of Cel61A (EGIV) of *Trichoderma reesei*. Eur. J. Biochem. *268*, 6498–6507.

Karnaouri, A., Topakas, E., Paschos, T., Taouki, I., and Christakopoulos, P. (2013). Cloning, expression and characterization of an ethanol tolerant GH3 β-glucosidase from *Myceliophthora thermophila*. PeerJ. *1*, e46.

Kaur, J., Chadha, B.S., Kumar, B.A., and Saini, H.S. (2007). Purification and characterization of two endoglucanases from *Melanocarpus* sp. MTCC 3922. Bioresour. Technol. *98*, 74–81.

Koseki, T., Mese, Y., Fushinobu, S., Masaki, K., Fujii, T., Ito, K., Shiono, Y., Murayama, T., and Iefuji, H. (2008). Biochemical characterization of a glycoside hydrolase family 61 endoglucanase from Aspergillus kawachii. Appl. Microbiol. Biotechnol. 77, 1279–1285.

Kumar, S., Tsai, C.J., and Nussinov, R. (2000). Factors enhancing protein thermostability. Protein Eng. *13*, 179–191.

Levasseur, A., Drula, E., Lombard, V., Coutinho, P.M., and Henrissat, B. (2013). Expansion of the enzymatic repertoire of the CAZy database to integrate auxiliary redox enzymes. Biotechnol. Biofuels *6*, 41.

Li, D.C., Lu, M., Li, Y.L., and Lu, J. (2003). Purification and characterization of an endocellulase from the thermophilic fungus *Chaetomium thermophilum*. Enzyme Microbial. Technol. *33*, 932–937.

Li, Y.L., Li, H., Li, A.N., and Li, D.C. (2009). Cloning of a gene encoding thermostable cellobiohydrolase from the thermophilic fungi *Chaetomium thermophilum* and its expression in *Pichia pastoris*. J. Appl. Microbiol. *106*, 1867–1875.

Limon, M.C., Margolles-Clark, E., Benitez, T., and Penttila, M. (2001). Addition of substrate-binding domains increases substrate-binding capacity and specific activity of a chitinase from *Trichoderma harzianum*. FEMS Microbiol. Lett. *198*, 57–63.

Liu, S.A., Li, D.C., E, S.J., and Zhang, Y. (2005). Cloning and expressing of cellulase gene (cbh2) from thermophilic fungi *Chaetomium thermophilum* CT2. Chinese J. Biotechnol. *21*, 892–899.

Liu, S.A., Li, D.C., Zhang, Y., Guo, F.X., and Shijin, E. (2006). Effective expression of *Chaetomium thermophilum* CT2 cellobiohydrolase I in *Pichia pastoris*. Mycosystema 25, 256–262.

Lo Leggio, L., and Larsen, S. (2002). The 1.62 A structure of *Thermoascus aurantiacus* endoglucanase: completing the structural picture of subfamilies in glycoside hydrolase family 5. FEBS Lett. *523*, 103–108.

McCarthy, T., Hanniffy, O., Savage, A.V., and Tuohy, M.G. (2003). Catalytic properties and mode of action of three endo-beta-glucanases from *Talaromyces emersonii* on soluble beta-1,4- and beta-1,3;1,4-linked glucans. Int. J. Biol. Macromol. *33*, 141–148.

McClendon, S.D., Batth, T., Petzold, C.J., Adams, P.D., Simmons, B.A., and Singer, S.W. (2012). *Thermoascus aurantiacus* is a promising source of enzymes for biomass deconstruction under thermophilic conditions. Biotechnol. Biofuels *5*, 54.

Mackenzie, L.F., Sulzenbacher, G., Daven, C., Jones, A., Woldike, H.F., Schulein, M., Withers, S.G., and Davies, G.J. (1998). Crystal structure of the family 7 endoglucanase I (Cel7B) from *Humicola insolens* at 2.2 A resolution and identification of the catalytic nucleophile by trapping of the covalent glycosyl-enzyme intermediate. Biochem. J. *335*, 409–416.

Maheshwari, R., Bharadwaj, G., and Bhat, M.K. (2000). Thermophilic fungi: their physiology and enzymes. Microbiol. Mol. Biol. Rev. *64*, 461–488.

Miettinen-Oinonen, A., Londesborough, J., Joutsjoki, V., Lantto, R., Vehmaanpera, J., and Biotecd, P.L. (2004). Three cellulases from *Melanocarpus albomyces* for textile treatment at neutral pH. Enzyme Microb. Technol. *34*, 332–341.

Moriya, T., Watanabe, M., Sumida, N., Okakura, K., and Murakami, T. (2003). Cloning and overexpression of the avi2 gene encoding a major cellulase produced by *Humicola insolens* FERM BP-5977. Biosci. Biotechnol. Boichem. *67*, 1434–1437.

Mouchacca, J. (1997). Thermophilic fungi: biodiversity and taxonomic status. Cryptog. Mycolog. *18*, 19–69.

Murray, P.G., Collins, C.M., Grassick, A., and Tuohy, M.G. (2003). Molecular cloning, transcriptional, and expression analysis of the first cellulase gene (cbh2), encoding cellobiohydrolase II, from the moderately thermophilic fungus *Talaromyces emersonii* and structure prediction of the gene product. Biochem. Biophys. Res. Commun. *301*, 280–286.

Murray, P.G., Aro, N., Collins, C.M., Grassick, A., Penttila, M., Saloheimo, M., and Tuohy, M. (2004). Expression in *Trichoderma reesei* and characterisation of a thermostable family 3 beta-glucosidase from the moderately thermophilic fungus *Talaromyces emersonii*. Protein Expr. Purif. *38*, 248–257.

Nakkharat, P., and Haltrich, D. (2006). Purification and characterisation of an intracellular enzyme with beta-glucosidase and beta-galactosidase activity from the thermophilic fungus *Talaromyces thermophilus* CBS 236.58. J. Biotechnol. *123*, 304–313.

Ni, J.F., Takehara, M., and Watanabe, H. (2010). Identification of activity related amino acid mutations of a GH9 termite cellulase. Bioresour. Technol. *101*, 6438–6443.

Otzen, D.E., Christiansen, L., and Schulein, M. (1999). A comparative study of the unfolding of EGV in denaturant and surfactant. Protein Sci. *8*, 1878–1887.

Pack, S.P., and Yoo, Y.J. (2004). Protein thermostability: structure-based difference of amino acid between thermophilic and mesophilic proteins. J. Biotechnol. *111*, 269–277.

Parkkinen, T., Koivula, A., Vehmaanpera, J., and Rouvinen, J. (2008). Crystal structures of *Melanocarpus albomyces* cellobiohydrolase Cel7B in complex with cello-oligomers show high flexibility in the substrate binding. Protein Sci. *17*, 1383–1394.

Paula, E.H.D., Pocas-Fonseca, M.J., and Azevedo, M.O. (2003). The product of *Humicola grisea* var. *thermoidea* cbh1.2 gene is the major expressed protein under induction by lignocellulosic residues. World J. Microbiol. Biotechnol. *19*, 631–635.

Pocas-Fonseca, M.J., Lima, B.D., Brigido, M.M., Silva-Pereira, I., Felipe, M.S., Radford, A., and Azevedo, M.O. (1997). *Humicola grisea* var. *thermoidea* cbh1.2: A new gene in the family of cellobiohydrolases is expressed and encodes a cellulose-binding domain-less enzyme. J. Gen. Appl. Microbiol. *43*, 115–120.

Pocas-Fonseca, M.J., Silva-Pereira, I., Rocha, B.B., and Azevedo, M.O. (2000). Substrate-dependent differential expression of *Humicola grisea* var. *thermoidea* cellobiohydrolase genes. Can. J. Microbiol. *46*, 749–752.

Quinlan, R.J., Sweeney, M.D., Lo Leggio, L., Otten, H., Poulsen, J.C., Johansen, K.S., Krogh, K.B., Jorgensen, C.I., Tovborg, M., Anthonsen, A., Tryfona, T., *et al.* (2011). Insights into the oxidative degradation of cellulose by a copper metalloenzyme that exploits biomass components. Proc. Natl. Acad. Sci. U.S.A. *108*, 15079–15084.

Sandgren, M., Gualfetti, P.J., Paech, C., Paech, S., Shaw, A., Gross, L.S., Saldajeno, M., Berglund, G.J., Jones, T.A., and Mitchinson, C. (2003). The *Humicola grisea* Cel12A enzyme structure at 1.2 A resolution and the impact of its free cysteine residues on thermal stability. Protein Sci. *12*, 2782–2793.

Sandgren, M., Berglund, G.I., Shaw, A., Stahlberg, J., Kenne, L., Desmet, T., and Mitchinson, C. (2004). Crystal complex structure reveal how substrate is bound in the −4 to the +2 binding sites of *Humicola grisea* Cel12A. J. Mol. Biol. *342*, 1505–1517.

Sandgren, M., Stahlberg, J., and Mitchinson, C. (2005). Structural and biochemical studies of GH family 12 cellulases: improved thermal stability, and ligand complexes. Prog. Biophys. Mol. Bio. *89*, 246–291.

Schou, C., Rasmussen, G., Kaltoft, M.B., Henrissat, B., and Schulein, M. (1993). Stereochemistry, specificity and kinetics of the hydrolysis of reduced cellodextrins by nine cellulases. Eur. J. Biochem. *217*, 947–953.

Shoseyov, O., Shani, Z., and Levy, I. (2006). Carbohydrate binding modules: biochemical properties and novel applications. Microbiol. Mol. Biol. Rev. *70*, 283–295.

Sukumaran, R.K., Surender, V.J., Sindhu, R., Binod, P., Janu, K.U., Sajna, K.V., Rajasree, K.P., and Pandey, A. (2010). Lignocellulosic ethanol in India: prospects, challenges and feedstock availability. Bioresour. Technol. *101*, 4826–4833.

Suzuki, K., Sumitani, J., Nam, Y.W., Nishimaki, T., Tani, S., Wakagi, T., Kawaguchi, T., and Fushinobu, S. (2013). Crystal structures of glycoside hydrolase family 3 β-glucosidase 1 from *Aspergillus aculeatus*. Biochem. J. *452*, 211–221.

Szijarto, N., Siika-Aho, M., Tenkanen, M., Alapuranen, M., Vehmaanpera, J., Reczey, K., and Viikari, L. (2008). Hydrolysis of amorphous and crystalline cellulose by heterologously produced cellulases of *Melanocarpus albomyces*. J. Biotechnol. *136*, 140–147.

Takashima, S., Nakamura, A., Hidaka, M., Masaki, H., and Uozumi, T. (1996). Cloning, sequencing, and expression of the cellulase genes of *Humicola grisea* var. *thermoidea*. J. Biotechnol. *50*, 137–147.

Takashima, S., Nakamura, A., Masaki, H., and Uozumi, T. (1997). Cloning, sequencing, and expression of a thermostable cellulase gene of *Humicola grisea*. Biosci. Biotechnol. Biochem. *61*, 245–250.

Takashima, S., Iikura, H., Nakamura, A., Hidaka, M., Masaki, H., and Uozumi, T. (1998). Isolation of the gene and characterization of the enzymatic properties of a major exoglucanase of *Humicola grisea* without a cellulose-binding domain. J. Biochem. *124*, 715–725.

Takashima, S., Iikura, H., Nakamura, A., Hidaka, M., Masaki, H., and Uozumi, T. (1999a). Comparison of gene structures and enzymatic properties between two endoglucanases from *Humicola grisea*. J. Biotechnol. *67*, 85–97.

Takashima, S., Nakamura, A., Hidaka, M., Masaki, H., and Uozumi, T. (1999b). Molecular cloning and expression of the novel fungal beta-glucosidase genes from *Humicola grisea* and *Trichoderma reesei*. J. Biochem. *125*, 728–736.

Takashima, S., Ohono, M., Hidaka, M., Nakamura, A., Masaki, H., and Uozumi, T. (2007). Correlation between cellulose binding and activity of cellulose-binding domain mutants of *Humicola grisea* cellobiohydrolase. FEBS Lett. *581*, 5891–5896.

Teng, F.C., Li, D.C., Li, Y.L., and Li, Z.J. (2006). Purification and properties of β-glucosidase from *Chaetomium thermophilum*. Mycosystema 25, 481–487.

Trivedi, S., Gehlot, H.S., and Rao, S.R. (2006). Protein thermostability in *Archaea* and *Eubacteria*. Genet. Mol. Res. *5*, 816–827.

Vaaje-Kolstad, G., Westereng, B., Horn, S.J., Liu, Z., Zhai, H., Sorlie, M., and Eijsink, V.G.H. (2010). An oxidative enzyme boosting the enzymatic conversion of recalcitrant polysaccharides. Science *330*, 219–222.

Varrot, A., Frandsen, T.P., von Ossowski, I., Boyer, V., Cottaz, S., Driguez, H., Schulein, M., and Davies, G.J. (2003). Structure basis for ligand binding and processivity in cellobiohydrolase Cel6A from *Humicola insolens*. Structure *11*, 855–864.

Venturi, L.L., Polizeli, Mde, L., Terenzi, H.F., Furriel, R.P., and Jorge, J.A. (2002). Extracellular beta-D-glucosidase from *Chaetomium thermophilum* var. *coprophilum*: production, purification and some biochemical properties. J. Basic. Microbiol. *42*, 55–66.

Vlasenko, E., Schülein, M., Cherry, J., and Xu, F. (2010). Substrate specificity of family 5, 6, 7, 9, 12, and 45 endoglucanases. Bioresour. Technol. *101*, 2405–2411.

Voutilainen, S.P., Boer, H., Linder, M.B., Puranen, T., Rouvinen, J., Vehmaanpera, J., and Koivula, A. (2007). Heterologous expression of *Melanocarpus albomyces* cellobiohydrolase Cel7B, and random mutagenesis to improve its thermostability. Enzyme Microb. Technol. *41*, 234–243.

Voutilainen, S.P., Puranen, T., Siika-Aho, M., Lappalainen, A., Alapuranen, M., Kallio, J., Hooman, S., Viikari, L., Vehmaanperä, J., and Koivula, A. (2008). Cloning, expression, and characterization of novel thermostable family 7 cellobiohydrolases. Biotechnol. Bioeng. *101*, 515–528.

Voutilainen, S.P., Murray, P.G., Tuohy, M.G., and Koivula, A. (2010). Expression of *Talaromyces emersonii* cellobiohydrolase Cel7A in *Saccharomyces cerevisiae* and rational mutagenesis to improve its thermostability and activity. Protein Eng. Des. Sel. *23*, 69–79.

Wang, X.J., Peng, Y.J., Zhang, L.Q., Li, A.N., and Li, D.C. (2012). Directed evolution and structural prediction of cellobiohydrolase II from the thermophilic fungus *Chaetomium thermophilum*. Appl. Microbiol. Biotechnol. *95*, 1469–1478.

Wang, X., Ike, M., Shiroma, R., Tokuyasu, K., and Sakakibara, Y. (2013). Expression of neutral β-glucosidase from *Scytalidium thermophilum* in *Candida glabrata* for ethanol production from alkaline-pretreated rice straw. J. Biosci. Bioeng. *116*, 362–365.

Wilson, D.B., 2009. Cellulases and biofuels. Curr. Opin. Biotechnol. *20*, 295–299.

Xu, R.Y., Teng, F.C., Zhang, C., and Li, D.C. (2011). Cloning of a gene encoding β-glucosidase from *Chaetomium thermophilum* CT2 and its expression in *Pichia pastoris*. J. Mol. Microbiol. Biotechnol. *20*, 16–23.

Yokota, K., Satou, K., and Ohki, S.Y. (2006). Comparative analysis of protein thermostability: differences in amino acid control and substitution at the surfaces and in the core regions of thermophilic and mesophilic proteins. Sci. Tech. Adv. Mater. *7*, 255–262.

Zanoelo, F.F., Polizeli, Mde, L., Terenzi, H.F., and Jorge, J.A. (2004). Beta-glucosidase activity from the thermophilic fungus *Scytalidium thermophilum* is stimulated by glucose and xylose. FEMS Microbiol. Lett. *240*, 137–143.

Zhang, Y.H.P., Himmel, M.E., and Mielenz, J.R. (2006). Outlook for cellulase improvement: screening and selection strategies. Biotechnol. Adv. *24*, 452–481.

Alcohol Dehydrogenases and Their Physiological Functions in Hyperthermophiles

6

Kesen Ma and Ching Tse

Abstract

Alcohol dehydrogenases (ADHs) are enzymes that catalyse the inter-conversion of alcohols to corresponding aldehydes/ketones with the concomitant reduction of NAD^+ or $NADP^+$. NAD(P)-dependent ADHs from mesophilic and thermophilic bacteria/archaea are orderly clustered as three distinct groups which are zinc-dependent, metal-free short-chain, and iron-containing/activated. ADHs from extreme thermophiles and hyperthermophiles are highly thermostable with optimal temperature of 65°C and above, which is advantageous for industrial applications. To date, ADHs from 10 extreme thermophiles and 15 hyperthermophiles are well characterized with respects to their thermostabilities and catalytic activities. This review aims mainly to provide a review of the most common use of ADHs, physiological functions of ADHs in hyperthermophiles and their potential applications in biotechnology. Relevant aspects including the principal procedures for purification, biophysical, biochemical and catalytic properties of ADHs from extreme thermophiles and hyperthermophiles are also discussed. These information will provide further insight into the importance of the ubiquitously expressed ADHs in microorganisms, particularly hyperthermophiles and their prospective contributions to advancing industrial processes.

Introduction

Alcohol dehydrogenases (ADHs) are a group of enzymes belonging to the versatile family of oxidoreductases, which catalyse the reversible inter-conversion of primary and secondary alcohols into corresponding aldehydes and ketones, respectively with the concomitant reduction of $NAD(P)^+$ to NAD(P)H (Fang *et al.*, 1995; Leonida, 2001; Kroutil *et al.*, 2004). ADHs have been extensively studied since the first purification of the enzyme from *Saccharomyces cerevisiae* in 1937 (Negelein and Wulff, 1937). They can be classified into three distinct groups/types according to their molecular size, metal content and cofactor requirement (Reid and Fewson, 1994; Guy *et al.*, 2003). Type I also known as the medium-chain ADHs contains approximately 370 amino acid residues per chain and may be dimeric (in higher eukaryotes) or tetrameric (commonly in prokaryotes) and is zinc-dependent (Eklund *et al.*, 1976; Guy *et al.*, 2003). Type II also known as short-chain ADHs rarely contains metals and has approximately 250 amino acid residues per chain (Persson *et al.*, 1991). Type III also known as long-chain ADHs contains over 380 to as many as 900 amino acid residues and is either iron-containing or activated by iron (Scopes, 1983). Type III ADHs are usually oxygen-sensitive and show little sequence homology with type I and type II ADHs (Guy *et al.*, 2003). Phylogenetic analysis using amino acid sequences of ADHs from mesophiles and thermophiles available from database (Swiss-Prot, NCBI, TrEMBL, OWL, PIR, PDB and PRF) reveals that NAD(P)-dependent ADHs are clustered in consistent with the three types of enzymes (Fig. 6.1). It also shows that ADHs are classified into the distinct types mentioned above based on metal content, which is irrespective of growth temperature, location of isolation, and origin of archaea or bacteria.

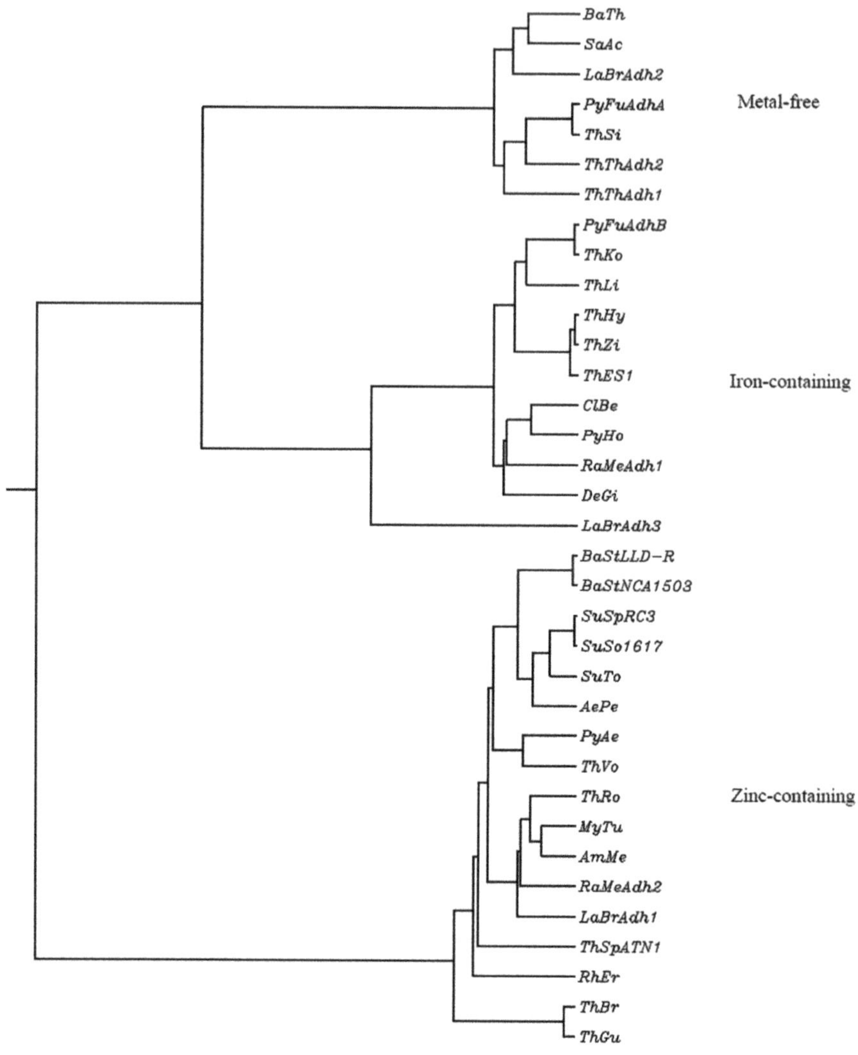

Figure 6.1 Three types of ADHs are identified using phylogenetic analysis of amino acid sequences from mesophiles and thermophiles. All sequences are obtained from the database, Swiss-Prot, NCBI, TrEMBL, OWL, PIR, PDB and PRF. The sequences of ADHs from mesophiles are randomly selected whereas other sequences include all purified ADHs of thermophiles with optimal growth temperatures above 65°C. F_{420}-dependent ADH from *Methanoculleus thermophilicus* and the long-chain multifunctional ADHE from *Clostridium acetobutylicum* and *Escherichia coli* are not included. The sequences were aligned using ClustalW (http://www.genome.jp/tools/clustalw/) using average distance method with 100 bootstrap with adjustment of gap insertion before clustered. Abbreviations represent each of the source organisms from which the corresponding ADH is identified: AePe, *Aeropyrum pernix;* AmMe, *Amycolatopsis methanolica;* BaTh, *Bacillus thermoleovorans;* BaStLLD-R, *Bacillus stearothermophilus* strain LLD-R; BaStNCA1503, *Bacillus stearothermophilus* strain NCA 1503; ClBe, *Clostridium brijerinckii;* DeGi, *Desulforibrio gigas;* LaBrAdh1, *Lactobacillus brevis* Adh1; LaBrAdh2, *Lactobacillus brevis* Adh2; LaBrAdh3, *Lactobacillus brevis* Adh3; MyTu, *Mycobacterium tuberculosis;* PyFuAdhA, *Pyrococcus furiosus* AdhA; PyFuAdhB, *Pyrococcus furiosus* AdhB; PyHo, *Pyrococcus horikoshii;* RaMeAdh1, *Ralstonia metallidurans* Adh1; RaMeAdh2, *Ralstonia metallidurans* Adh2; RhEr, *Rhodococcus erythropolis;* SaAc, *Sulfolobus acidocaldarius;* SuSo1617, *Sulfolobus solfataricus* 1617; SuTo, *Sulfolobus tokodaii;* SuSpRC3, *Sulfolobus* sp. RC3; ThBr, *Thermoanaerobacter brockii;* ThGu, *Thermococcus guaymasensis;* ThHy, *Thermococcus hydrothermalis;* ThKo, *Thermococcus kodakarensis;* ThLi, *Thermococcus litoralis;* ThRo, *Thermomicrobium roseum;* ThSi, *Thermococcus sibiricus;* ThThAdh1, *Thermus thermophilus* HB27 Adh1; ThThAdh2, *Thermus thermophilus* HB27 Adh2; ThZi, *Thermococcus zilligii;* ThES1, *Thermococcus* strain ES1*;* ThVo, *Thermoplasma volcanium;* ThSpATN1, *Thermus* sp. ATN1.

The most extensively studied of type I ADH is horse liver ADH (Eklund *et al.*, 1976). It is a dimer of identical subunits and each subunit has two zinc atoms and a coenzyme NAD-binding site (Jörnvall, 1970; Eklund *et al.*, 1976). The zinc atoms have both catalytic and structural roles, and the one that is involved in the catalytic process has three ligands with two cysteine and one histidine residues whereas the other one is coordinated with four cysteine residues that contribute to the structural stability of the ADH (Branden *et al.*, 1975). Horse liver ADH is a primary/secondary ADH, and it catalyses the oxidation of various primary and secondary alcohols into corresponding aldehydes/ketones (Jörnvall, 1970; Branden *et al.*, 1975). The representative examples of type II ADH are those from *Drosophila melanogaster* (Thatcher, 1980). Genetically polymorphic isozymes of ADHs in *D. melanogaster* have been investigated with great interest in understanding the evolution of enzymes in eukaryotes (Day *et al.*, 1974). ADH2 from *Zymomonas mobilis* is a type III ADH, which resembles yeast enzyme closely in protein sequence and is a tetramer of identical subunit and is activated by Fe^{2+} but inactivated by Zn^{2+} (Neale *et al.*, 1986). It is responsible for 90% of total ethanol oxidizing activity in *Z. mobilis* and has relatively low acetaldehyde reducing activity (Neale *et al.*, 1986).

Microbial ADHs have played considerably important roles in industrial applications (Lamed *et al.*, 1981; Atomi, 2005). One of the earliest use is in wine and beer production, which exploits the natural process known as ethanol fermentation in *S. cerevisiae* (Evans, 1990; Dequin, 2001). Although physiological functions of ethanol fermentation in *S. cerevisiae* are for the regeneration of NAD^+ to maintain redox balance and the production of ATP when there is a lack of oxygen, the ADH is the key enzyme to convert acetaldehyde into ethanol in this process (Lutstorf and Megnet, 1968; Heick *et al.*, 1969; Wills and Phelps, 1975; Ciriacy, 1979; Beier *et al.*, 1985; Young and Pilgrim, 1985; Lieber, 1997; Smidt *et al.*, 2012). The second largest scale fermentation is the acetone–butanol–ethanol (ABE) fermentation (Lütke-Eversloh and Bahl, 2011), in which *Clostridium acetobutylicum* is the commonly used mesophilic bacterium (Bahl *et al.*, 1982a,b). Similarly, ADHs in *C. acetobutylicum* are responsible for the conversion of acetaldehyde into ethanol and also butyraldehyde into butanol (Ross, 1960; Jones and Woods, 1986). The production of ethanol and butanol has a ratio of 1:6 and both are used as biofuel/solvent (Ross, 1960). Moreover, ADH1 from *Z. mobilis* is also applicable for biofuel production and it converts acetaldehyde into ethanol at a rate of 1800 μmol per minute per gram of dry weight of cells at 30°C (Neale *et al.*, 1986). ADHs in thermophilic *Clostridium thermocellum*, *Thermoanaerobacter saccharolyticum*, *Thermoanaerobacter ethanolicus*, *Thermoanaerobacter mathranii*, *Thermoanaerobium brockii*, *Bacillus stearothermophilus* and *Geobacillus thermoglucosidasius* also convert acetaldehyde into ethanol with potentials for biofuel production (Lamed and Zeikus, 1980b; Zeikus, 1980; Ng *et al.*, 1981; Kannan and Mutharasan, 1985; Shaw *et al.*, 2008; Cripps *et al.*, 2009; Taylor *et al.*, 2009). Purified ADHs are also applicable in biosensors for monitoring alcohol concentrations in the processes of producing solvent alcohol and acetic acid in vinegar through the oxidation of ethanol (Lamed *et al.*, 1981; Drysdale and Fleet, 1988; Chen, 1995). Isolated enantioselective ADHs also have promising applications in synthesis of chiral organic compounds for the chemical or pharmaceutical industries (Schmid *et al.*, 2001).

ADHs from mesophiles generally have a narrow substrate specificity and low stability at high temperatures. These properties hamper their industrial applications (Hirakawa *et al.*, 2004). An interesting group of microorganisms known as thermophiles contain ADHs with higher thermostability and broad range of substrate specificities, which has generated great interest in searching for their potential biotechnological applications (Radianingtyas and Wright, 2003). These microorganisms have the advantages of tolerance to pH, temperature and environmental change and utilization of a wider range of substrates including complex carbohydrates hemicelluloses and lignocelluloses (Zaldivar *et al.*, 2001; Sommer *et al.*, 2004; Gerday and Glansdorff, 2007). In comparison with optimal growth temperature range, mesophiles are organisms that grow best between 20 to 45°C (Willey, 2008), while thermophiles are organisms that can grow above

50°C (Kristjánsson, 1992). Thermophiles can be further divided into different groups, which are moderate thermophiles with optimal growth temperatures between 50–64°C (Madigan *et al.*, 1997), extreme thermophiles with optimal growth temperatures between 65–79°C (Wagner and Wiegel, 2008), and hyperthermophiles that grow optimally at 80°C and above or capable of growing at 90°C (Kelly and Adams, 1994; Stetter, 1996).

This review aims to compare and contrast the biochemistry, biophysical and catalytic properties of ADHs from extreme thermophiles and hyperthermophiles, to examine the physiological roles of ADHs in hyperthermophiles, and to comment on the current use and future potential applications of ADHs from hyperthermophiles in pharmaceutical and industrial sectors.

Principal procedures for the purification of ADHs

ADHs purified and characterized from extreme thermophiles and hyperthermophiles are summarized and general methods used for the purification of both native and recombinant ADHs are described below.

ADHs purified from thermophiles

Up to today, many ADHs have been purified and characterized including 10 from extreme thermophiles and 15 from hyperthermophiles. Those extreme thermophiles are *Bacillus stearothermophilus* strain LLD-R (Guagliardi *et al.*, 1996; Ceccarelli *et al.*, 2004), *Bacillus thermoleovorans* (Kato *et al.*, 2001; Yao *et al.*, 2013), *Carboxydothermus hydrogenoformans* (Zhou *et al.*, 2013), *Sulfolobus solfataricus* Gθ (Ammendola *et al.*, 1992; Cannio *et al.*, 1999), *Sulfolobus* sp. RC3 (Cannio *et al.*, 1996; Raia *et al.*, 2001), *Thermoanaerobacter brockii* (Ismaiel *et al.*, 1993; Bogin *et al.*, 1997, 1998; Peretz *et al.*, 1997; Korkhin *et al.*, 1998; Kleifeld *et al.*, 2000), *Thermoanaerobacter ethanolicus* strain JW 200 and 39E (Bryant and Ljungdahl, 1981; Bryant *et al.*, 1988, 1992; Burdette and Zeikus, 1994; Burdette *et al.*, 1996, 2002; Holt *et al.*, 2000), *Thermomicrobium roseum* (Yoon *et al.*, 2002), *Thermus thermophilus* strain HB27 and MTCC 1494 (Pennacchio *et al.*, 2008; Raghava and Gupta, 2010; Rocha-Martín *et al.*, 2012) and *Thermus* sp. ATN1 (Höllrigl *et al.*, 2008). Those hyperthermophiles are *Aeropyrum pernix* (Guy *et al.*, 2002, 2003; Hirakawa *et al.*, 2004), *Pyrobaculum aerophilum* (Vitale *et al.*, 2010, 2013), *Pyrococcus furiosus* (Ma and Adams, 1999; van der Oost *et al.*, 2001; Machielsen *et al.*, 2006), *Sulfolobus acidocaldarius* (Pennacchio *et al.*, 2010, 2012), *Sulfolobus solfataricus* strain MT4 and strain 1617 (Rella *et al.*, 1987; Ammendola *et al.*, 1992; Cannio *et al.*, 1996; Raia *et al.*, 1996, 2001; Giordano *et al.*, 1999; Esposito *et al.*, 2002; Contursi *et al.*, 2003), *Sulfolobus tokodaii* (Kawarabayasi *et al.*, 2001; Yanai *et al.*, 2009), *Thermococcus* strain ES-1 (Ma *et al.*, 1994, 1995; Ying *et al.*, 2009), *Thermococcus guaymasensis* (Ying and Ma, 2011), *Thermococcus hydrothermalis* (Quintela *et al.*, 1995; Antoine *et al.*, 1999), *Thermococcus kodakarensis* (Wu *et al.*, 2013), *Thermococcus litoralis* (Ma *et al.*, 1994; Ma and Adams, 2001; Pennacchio *et al.*, 2008), *Thermococcus sibiricus* (Lyashenko *et al.*, 2010; Stekhanova *et al.*, 2010; Bezsudnova *et al.*, 2012), *Thermococcus zilligii* (Li and Stevenson, 1997, 2001), *Thermotoga hypogea* (Ying *et al.*, 2007), and *Thermotoga maritima* (Brinen *et al.*, 2003; Beauchamp *et al.*, 2014).

General methods for the purification of ADHs

Native and/or recombinant ADHs have been successfully purified from all the extreme thermophiles and hyperthermophiles mentioned above. As an initial step of the general procedures for the purification of native ADHs, microorganisms are grown under conditions of optimal pHs and temperatures. The cells are then harvested by centrifugation followed by cell lysis, which is performed by using sonication or French Pressure Cells Press or lysozyme for obtaining a crude cell extract. The cell debris in the crude cell extract is removed using centrifugation and the supernatant is the cell-free extract that is then loaded onto different columns equilibrated with corresponding buffers at a specific flow rate.

DEAE-Sepharose (anion exchange), affinity/hydrophobic interaction, and size exclusion are among the commonly used columns, from which ADH is eluted out and purified (Table 6.1).

Table 6.1 Commonly used chromatography columns for the purification of ADHs from hyper/thermophiles

Chromatography	Columns	Selected sources[a]
Anion exchange	DEAE-cellulose/sepharose fast flow	Pharmacia Biotech
	Resource Q	Pharmacia
	Q-Sepharose	Fisher Scientific
	Mono Q	Amersham-Pharmacia
	CIM QA disk	BIA Separation
Affinity	Hydroxyapatite	Amersham GE Health
	Matrex gel	Amicon
	Blue-dextran-sepharose	Pharmacia
	HisTrap HP	GE Healthcare
Gel filtration	Superdex	Amersham Biosciences
	Sephadex	GE Healthcare
Hydrophobic interaction	Phenyl sepharose	Pharmacia
	Phenyl superose	GE Healthcare
	Butyl sepharose	Sigma-Aldrich
Metallochelating	Nickel affinity chromatography	GE Healthcare
	HisTrap chelating, HP affinity	Thomas Scientific

[a]Selected manufactures/companies that supply those chromatography column materials listed.

The purity of the ADH is normally judged using SDS-PAGE. Therefore, the separation of ADHs from other proteins can be accomplished using ion-exchange chromatography by surface charge (Regnier, 1984; Yamamoto *et al.*, 1988; Biotechnology, 1991; Sofer and Hagel, 1997), affinity chromatography by surface ligand or antigen (Cuatrecasas, 1970; Porath *et al.*, 1975), gel filtration chromatography by size (Hagel, 1998; Stanton, 2004; Ó'Fágáin *et al.*, 2011), or hydrophobic interaction chromatography by hydrophobic interaction between the surface and the stationary phase (Roettger and Ladisch, 1989; Eriksson and Belew, 1998). The elution of ADHs from columns can be achieved by applying an increasing or a decreasing gradient of NaCl, KCl, NH_4SO_4 or KH_2PO_4. Fractions eluted from each column are tested for ADH activity and only those with ADH activity and less contamination with other proteins determined by SDS-PAGE are loaded onto another column and corresponding chromatography procedures are then followed. Finally, the peak fractions with ADH activities and without contamination of any other protein are collected and concentrated by ultrafiltration using an Amicon Millipore membrane.

To obtain recombinant ADHs, cloning of the gene encoding the ADH is required using PCR as an initial step. The interested genes are then ligated to a pET expression vector that is then transformed into an *E. coli* host. The selection is performed using antibiotics-containing plates. Protein expression is then induced using IPTG (isopropyl β-D-1-thiogalactopyranoside) when the growth of the host cells reaches the required density measured by turbidity. The cells are harvested using centrifugation and are then subjected to cell-lysis and purification, for which similar steps described above for the purification of native ADHs are followed. However, recombinant ADHs are commonly his-tagged at the N- or C-terminus for a one-step purification using HisTrap chelating HP affinity column.

Biophysical properties of thermostable ADHs

Biophysical properties including high temperature dependent activity, thermostability, major factors contributing to the thermostability, subunit composition, determination of coenzyme specificity and metal-binding motifs of thermostable ADHs are described below.

High-temperature-dependent activity and thermostability

ADHs from extreme thermophiles and hyperthermophiles have high-temperature dependent activities and extremely high thermostabilities (Table 6.2). Activities of *P. furiosus* AdhD (Machielsen *et al.*, 2006) and *T. sibiricus* ADH (Stekhanova *et al.*, 2010) increase along with the rise of temperatures up to 100°C while the activities of *S. solfataricus* strain MT4 (Rella *et al.*, 1987) and strain 1617 ADHs (Ammendola *et al.*, 1992), *Thermococcus* strain ES-1 ADH (Ying *et al.*, 2009), *T. guaymasensis* ADH (Ying and Ma, 2011) and *T. hypogea* ADH (Ying *et al.*, 2007) increase along with the rise of temperatures up to 95°C. Optimal temperatures for these hyperthermophilic ADHs are not determined as their enzymatic activities are not tested above the mentioned temperatures due to instrument limitation. *P. furiosus* AdhC (Ma and Adams, 1999) has optimal temperature above 90°C. *A. pernix* ADH (Hirakawa *et al.*, 2004), *P. aerophilum* ADH (Vitale *et al.*, 2010, 2013), *P. furiosus* AdhA (van der Oost *et al.*, 2001), *T. kodakarensis* ADH (Wu *et al.*, 2013) have optimal temperature at 90°C. *Thermococcus* sp. AN1 ADH (Li and Stevenson, 1997), *S. tokodaii* ADH (Yanai *et al.*, 2009) and *T. thermophilus* strain HB27 ADH2 have optimal temperatures at 85°C, whereas *T. litoralis* ADH (Ma *et al.*, 1994; Ma and Adams, 2001), *T. hydrothermalis* ADH (Antoine *et al.*, 1999), *T. thermophilus* strain MTCC 1494 ADH (Raghava and Gupta, 2010), *S. acidocaldarius* ADH2 (Pennacchio *et al.*, 2012) and glycerol dehydrogenase from *T. maritima* have optimal temperatures at 80°C (Beauchamp *et al.*, 2014). Short chain ADHs including *T. thermophilus* strain HB27 ADH1 (Pennacchio *et al.*, 2008; Rocha-Martín *et al.*, 2012) and *S. acidocaldarius* ADH1 (Pennacchio *et al.*, 2010, 2012) have relatively lower optimum temperatures, which are 73°C and 75°C respectively. It should be noted that different ADHs isolated from the same stain and same species of hyperthermophiles have different optimal temperatures for their activities (Table 6.2). *P. furiosus* AdhD, *S. acidocaldarius* ADH2 and *T. thermophilus* strain HB27 ADH2 have the same optimal temperatures for enzymatic activities as the optimal growth temperatures of the corresponding organisms (Machielsen *et al.*, 2006; Pennacchio *et al.*, 2012; Rocha-Martín *et al.*, 2012). This suggests one form of ADH is a dominant one over the other and they may have different physiological roles for the organism. Interestingly, *P. furiosus* AdhB, which is encoded in the same operon Lam A as AdhA lost 80% of its activity after 24 hours of storage at 4°C and the activity can not be restored by incubation at elevated temperatures, or the addition of divalent cations such as iron, zinc or magnesium (van der Oost *et al.*, 2001). One possible explanation for the lost activity is the loss of iron ion, which is known for other archaeal iron-containing ADHs (Ma *et al.*, 1994, 1995). In comparison, ADHs from extreme thermophiles *C. hydrogenoformans* (Zhou *et al.*, 2013) and *T. roseum* (Yoon *et al.*, 2002) have optimal temperatures at 70°C whereas *B. stearothermophilus* strain LLD-R has optimal temperatures at 65°C (Sheehan *et al.*, 1988; Sakoda and Imanaka, 1992; Guagliardi *et al.*, 1996). ADHs from *T. ethanolicus* strain JW 200 and 39E and *Sulfolobus* sp. RC3 have optimal temperatures close to that of ADHs from hyperthermophiles, which are all above 75°C, and particularly for *Sulfolobus* sp. RC3 ADH the activity increases up to 95°C (Bryant and Ljungdahl, 1981; Bryant *et al.*, 1988; Burdette and Zeikus, 1994; Cannio *et al.*, 1996).

Major factors contributing to the thermostability of ADHs

Enzymes including ADHs from thermophiles and mesophiles can share high levels of similarity in amino acid sequences, for example, ADHs from *T. brockii* and *C. beijerinckii* have 75% identity (Bogin *et al.*, 1997, 1998; Korkhin *et al.*, 1998) with similar three-dimensional structures, and they also share the same catalytic mechanisms (Vieille and Zeikus, 2001). One of the factors correlated with thermostability is a more hydrophobic core, and this property is shown clearly in the ADHs of *T. brockii* and *T. sibiricus*, and the corresponding ADHs have a half life of 1 hours and 2 hours at 90°C, respectively (Bogin *et al.*, 1997, 1998; Korkhin *et al.*, 1998; Lyashenko *et al.*, 2010;

Stekhanova *et al.*, 2010). Another factor contributing to their thermostability is the composition of amino acids, which includes the substitution of Gly or Lys with Arg for forming more α-helix structures or having more charged residues such as Glu compared to their mesophilic counterparts (Vieille and Zeikus, 2001). These adaptations are observed in the ADH of *S. solfataricus* strain 1617 with the substitution of Cys with Glu and a two-times higher ratio of Arg to Lys than mesophilic ADHs (Ammendola *et al.*, 1992). These modification leads to more ionic interactions and a more compact folding structure. Moreover, the role of prolines by either insertion of proline residues or proline substitution into loops has been hypothesized to contribute to protein stabilization (Matthews *et al.*, 1987; Herning *et al.*, 1991, 1992; Watanabe *et al.*, 1994). ADH of *T. brockii* contains eight more prolines than that of *C. beijerinckii*, and the proline residues located at the N-termini of two helices and on b-turn contribute to the increase in thermostability (Bogin *et al.*, 1998; Korkhin *et al.*, 1998). The substitution of proline residues is also observed in *T. ethanolicus* strain 39E secondary ADH (2° ADH), which shares 85% sequence homology with ADH from *C. beijerinckii*, and 9 of the non-conservative regions between the two sequences are substitution of proline residues (Burdette *et al.*, 1996). However, ADH in *Thermococcus* sp. AN1 increases its thermostability by replacing proline residues with smaller residues (Ala, Ser, Thr) to change the protein conformation (Höllrigl *et al.*, 2008). Other modifications include placement of structural determinants to strengthen the interface between subunits by the formation of disulfide bonds, hydrogen bonds or by hydrophobic interactions to increase in thermostability (Bogin *et al.*, 1998; Korkhin *et al.*, 1998; Vieille and Zeikus, 2001). Since large variations are in the amino acids compositions of ADHs within hyper/thermophiles, there is no single factor contributing to the overall increase in thermostabilities and each of the factors may have specific contributions to and adaptations in the protein structures to cope with environmental changes.

Common biophysical properties within each type of ADHs

Most ADHs characterized belong to one of the three groups (types) of ADHs and each group is further divided as NAD or NADP dependent. It is also believed that ADHs within one group share similar structures due to the same metal ion binding sites and similar molecular sizes and amino acid compositions.

Type I Zn-containing ADHs that are NAD dependent include ADHs from *A. pernix* (Hirakawa *et al.*, 2004), *S. solfataricus* strain MT4 (Rella *et al.*, 1987) and strain 1617 (Ammendola *et al.*, 1992), *S. tokodaii* (Yanai *et al.*, 2009), *Sulfolobus* sp. RC3 (Cannio *et al.*, 1996), *T. roseum* (Yoon *et al.*, 2002), *T. thermophilus* strain MTCC1494 (Raghava and Gupta, 2010), *B. stearothermophilus* strain LLD-R (Guagliardi *et al.*, 1996) and glycerol dehydrogenase from *T. maritima* (Beauchamp *et al.*, 2014) whereas ADHs from *P. aerophilum* (Vitale *et al.*, 2010, 2013), *T. guaymasensis* (Ying and Ma, 2011), and *T. brockii* (Korkhin *et al.*, 1998), *Thermus* sp. ATN1 (Höllrigl *et al.*, 2008), *T. ethanolicus* strain JW200 and 2° ADH from *T. ethanolicus* strain 39E are NADP dependent (Bryant and Ljungdahl, 1981; Bryant *et al.*, 1988). Primary ADH from *T. ethanolicus* 39E has higher catalytic activities for $NADP^+$-dependent ethanol oxidation and NADH-dependent acetaldehyde reduction, and it is able to utilize both coenzymes for catalytic activities (Burdette and Zeikus, 1994). All ADHs except *S. solfataricus* strain MT4 are homotetrameric with subunit size between 34 kDa and 44.5 kDa, and *S. solfataricus* strain MT4 is dimeric with a 35.5 kDa subunit size (Bryant and Ljungdahl, 1981; Rella *et al.*, 1987; Bryant *et al.*, 1988; Ammendola *et al.*, 1992; Sakoda and Imanaka, 1992; Burdette and Zeikus, 1994; Cannio *et al.*, 1996; Guagliardi *et al.*, 1996; Bogin *et al.*, 1997; Korkhin *et al.*, 1998; Suzuki *et al.*, 2002; Yoon *et al.*, 2002; Brinen *et al.*, 2003; Guy *et al.*, 2003; Hirakawa *et al.*, 2004; Höllrigl *et al.*, 2008; Yanai *et al.*, 2009; Raghava and Gupta, 2010; Vitale *et al.*, 2010, 2013; Ying and Ma, 2011).

Type II ADHs are metal-free. Those from *T. thermophilus* strain HB27 (Pennacchio *et al.*, 2008; Rocha-Martín *et al.*, 2012), *S. acidocaldarius* (Pennacchio *et al.*, 2010, 2012) are NAD

Table 6.2 Physical properties of alcohol dehydrogenases from hyper/thermophiles

Organisms	Growth[a] T_{opt}, °C (minimum–maximum)	Subunit (kDa)	Metal (g-atom/ subunit)	Function[b]	Thermostability $t_{1/2}$, hours (at°C)	O_2 sensitivity (+/–)	Optimal temperature[c] (°C)	References
P. aerophilum	100 (75–104)	38.5 (α_4)	Zn[d]	S	nd	–	90	Völkl *et al*. (1993), Vitale *et al*. (2010, 2013)
A. pernix	90–95 (70–97)	39.5 (α_4)	Zn (2)	C+S	0.5 (90)	–	90	Sako *et al*. (1996), Guy *et al*. (2003), Hirakawa *et al*. (2004)
T. guaymasensis	88 (56–90)	40 (α_4)	Zn (0.9)	C (putative)	24 (95), 70 (80)	+	>95	Canganella *et al*. (1998), Ying and Ma (2011)
S. solfataricus 1617	78–87	37.5 (α_4)	Zn (2)	C+S	5 (70), 3 (85)	–	>95	Cannio *et al*. (1996), Esposito *et al*. (2002)
S. solfataricus MT4	80	35.5 (α_2)	Zn (2)	C+S	20 (60), 5 (70)	–	>95	Rella *et al*. (1987)
T. maritima	80 (55–90)	39.7 (α_4)	Zn	nd	7 (50)	–	80	Huber *et al*. (1986), Brinen *et al*. (2003), Beauchamp *et al*. (2014)
S. tokodaii	80 (75–80)	38 (α_4)	Zn (2)	C (putative)	2 (70)	–	85	Suzuki *et al*. (2002), Yanai *et al*. (2009)
Sulfolobus sp. RC3	75	37 (α_4)	Zn[d]	nd	3 (85), 1 (90)	–	>95	Cannio *et al*. (1996)
T. roseum	70–75 (70–85)	43 (α_2)	Zn[d]	nd	1 (77)	–	70	Jackson (1973), Yoon *et al*. (2002)
B. stearothermophilus LLD-R	70	37 (α_4)	Zn (2)	C+S	5 (65), 3 (70)	–	65	Guagliardi *et al*. (1996), Ceccarelli *et al*. (2004)
T. brockii	70 (65–70)	38 (α_4)	Zn (1.0)	C	1 (93)	–	nd	Zhang *et al*. (1993), Bogin *et al*. (1997, 1998), Kleifeld *et al*. (2000)
Thermus sp. ATN1	70	40 (α_4)	Zn[d]	nd	nd	–	nd	Höllrigl *et al*. (2008)
T. thermophilus MTCC 1494	65–72	43 (α_4)	Zn (1)	S	nd	–	80	Henne *et al*. (2004), Raghava and Gupta (2010)
T. ethanolicus JW 200 (1°ADH)	69 (38–78)	44.5 (α_4)	Zn (3.2–4.6)	C+S	nd	–	85	Wiegel and Ljungdahl (1981), Bryant *et al*. (1988)
T. ethanolicus JW 200 (2°ADH)	69 (38–78)	42 (α_4)	Zn (3.2–5.54)	C+S	>2 (80)[e]	–	75	Bryant and Ljungdahl (1981), Wiegel and Ljungdahl (1981), Bryant *et al*. (1988)
T. ethanolicus 39E (1° ADH)	69 (38–78)	42 (α_4)	Zn[d]	nd	nd	+	nd	Bryant and Ljungdahl (1981), Burdette and Zeikus (1994)
T. ethanolicus 39E (2° ADH)	69 (38–78)	40 (α_4)	Zn[d]	C	1.7 (90)	–	90	Bryant and Ljungdahl (1981), Burdette and Zeikus (1994)
P. furiosus (AdhC)	100 (70–103)	48 (α_6)	Zn (0.9), Fe (0.8)	nd	160 (85), 7 (95)	+	>90	Fiala and Stetter (1986), Ma and Adams (1999)

Thermococcus ES1	91 (81–91)	46 (α_4)	Fe (1.0)	C (putative)	35 (85), 4 (95)	+	>95	Pledger and Baross (1989), Ma *et al*. (1995), Ying *et al*. (2009)
T. litoralis	88 (55–98)	48 (α_4)	Fe (0.45)	C+S	5 (85), 0.25 (96)	+	85	Neuner *et al*. (1990), Ma *et al*. (1994), Ma and Adams (2001)
T. kodakarensis	85 (60–100)	31 (α_1)	Fe[d]	S (putative)	16.2 (85), 4.5 (95)	+	90	Atomi *et al*. (2004), Wu *et al*. (2013)
T. hydrothermalis	85 (55–100)	45 (α_2/α_4)	Fe (0.5)	C	0.5 (80), 2.5 (70)	+	80	Godfroy *et al*. (1997), Antoine *et al*. (1999)
T. zilligii	75–80 (55–85)	46 (α_4)	Fe[d]	nd	0.27 (80)	+	85	Klages and Morgan (1994), Li and Stevenson (1997, 2001)
T. hypogea	70 (56–90)	40 (α_2)	Fe (1.0)	C+S	10 (70), 2 (90)	+	>95	Fardeau *et al*. (1997), Ying *et al*. (2007)
T. ethanolicus JW200	69 (38–78)	44 (α_4)	Fe[d]	nd	>3 (70)[f]	+	nd	Wiegel and Ljungdahl (1981), Holt *et al*. (2000)
P. furiosus (AdhA)	100 (70–103)	26 (α_2/α_4/α_6)	No		150 (80), 22.5 (90), 0.42 (100)	–	90	Fiala and Stetter (1986), van der Oost *et al*. (2001)
P. furiosus (AdhD)	100 (70–103)	32 (α_1)	No		2.16 (100)	–	>100	Fiala and Stetter (1986), Machielsen *et al*. (2006)
S. acidocaldarius (Adh1)	80 (75–80)	29 (α_4)	No		0.5 (90)	–	75	Chen *et al*. (2005), Pennacchio *et al*. (2010)
S. acidocaldarius (Adh2)	80 (75–80)	27 (α_4)	No		0.5 (88)	–	80	Chen *et al*. (2005), Pennacchio *et al*. (2012)
T. sibiricus	78 (40–88)	26.2 (α_2)	No		2 (90), 1 (100)	–	>100	Miroshnichenko *et al*. (2001), Stekhanova *et al*. (2010)
C. hydrogenoformans	78	30 (α_1)	No		12 (50)	–	70	Wu *et al*. (2005), Zhou *et al*. (2013)
T. thermophilus HB27 (Adh1)	65–72	27 (α_4)	No		0.5 (90)	–	73	Henne *et al*. (2004), Pennacchio *et al*. (2008)
T. thermophilus HB27 (Adh2)	65–72	27 (α_4)	No		nd	–	85	Henne *et al*. (2004), Rocha-Martín *et al*. (2012)

[a]Optimal growth temperature (minimum–maximum = range of growth temperatures).
[b]Function of metal: C, catalytic role; S, structural role.
[c]Optimal temperature for enzyme activity.
[d]Metal contained was deduced by sequence analysis with metal-binding motif among closely related ADHs.
[e]77% activity remained after incubated at 80°C for 2 hours.
[f]No loss in activity for 3 hours at 70°C.
nd, not determined; 1° ADH, primary alcohol dehydrogenase; 2° ADH, secondary alcohol dehydrogenase; No, metal free.

dependent while *P. furiosus* AdhA (Machielsen *et al.*, 2006), *T. sibiricus* ADH (Lyashenko *et al.*, 2010; Stekhanova *et al.*, 2010), *B. thermoleovorans* ADH and *C. hydrogenoformans* ADH are NADP dependent (Kato *et al.*, 2001). *P. furiosus* AdhA is dimeric but may be adopted to tetrameric or octameric structure with a subunit size of 26 kDa (Machielsen *et al.*, 2006). Similarly, ADH of *T. sibiricus* is dimeric with a subunit size of 26.2 kDa (Lyashenko *et al.*, 2010; Stekhanova *et al.*, 2010). The structure of ADHs in *T. thermophilus* strain HB27 and *S. acidocaldarius* are homotetrameric but with a smaller subunit size of 27–29 kDa (Pennacchio *et al.*, 2008, 2010, 2012; Rocha-Martín *et al.*, 2012). The subunit size of *B. thermoleovorans* ADH is calculated from its amino acid sequence to be 27 kDa (Kato *et al.*, 2001). *C. hydrogenoformans* ADH is monomeric with a subunit size of 30 kDa (Zhou *et al.*, 2013).

A characteristic of type III Fe-containing ADHs is that all of them from hyperthermophiles including *T. hydrothermalis*, *T. litoralis*, *Thermococcus* strain ES-1 and *T. zilligii* and *T. hypogea* are NADP dependent (Ma *et al.*, 1994; Li and Stevenson, 1997; Antoine *et al.*, 1999; Ma and Adams, 2001; Ying *et al.*, 2007). All ADHs except *T. hypogea* ADH is homotetrameric with sizes ranging from 45 to 48 kDa (Ma *et al.*, 1994; Li and Stevenson, 1997; Antoine *et al.*, 1999; Ma and Adams, 2001; Ying *et al.*, 2007). *T. hypogea* ADH is a homodimer with a subunit size of 40 kDa (Ying *et al.*, 2007). The structure of ADH from *T. hydrothermalis* is pH-dependent, being a tetramer (197 kDa) at pH 10.5 and a dimer (80.5 kDa) at pH 7.5, and the two pH values are corresponding to the optimal pH values for alcohol oxidation and aldehyde reduction respectively (Antoine *et al.*, 1999). However, *P. furiosus* AdhD and *T. kodakarensis* ADH does not show similarity to the ADHs above, but rather to the aldo-keto reductase superfamily (Machielsen *et al.*, 2006; Wu *et al.*, 2013). *P. furiosus* AdhD is NAD dependent and *T. kodakarensis* ADH is NADP dependent (Machielsen *et al.*, 2006; Wu *et al.*, 2013). Both *P. furiosus* AdhD and *T. kodakarensis* ADH are monomeric with a subunit size of 32 kDa and 31 kDa, respectively (Machielsen *et al.*, 2006; Wu *et al.*, 2013).

It is shown that ADHs belong to type I and type III regardless of extreme thermophiles or hyperthermophiles share more similar structure than type II ADHs. This might be because type II ADHs are metal free and does not need to have specific structures of conserved motifs for metal binding.

Determination of coenzyme specificity for NAD(H) and NADP(H)

Studies have also been done to investigate amino acid sequences that determine NAD or NADP dependency of ADHs. Amino acid residues that form stable hydrogen bonds with the cofactor are predicted to be determinants of cofactor specificity (Guy *et al.*, 2003). In type I ADHs, it is shown that NAD(H)-dependent enzymes have aspartate in different positions of an individual amino acid chain but in equivalent to position Asp223 of horse liver ADH (Reid and Fewson, 1994). The Asp residue forms hydrogen bonds to two hydroxyl groups of the adenosine ribose of NAD^+, and examples include Asp218 in ADH of *A. pernix*, Asp203 in ADH of *S. solfataricus* strain 1617, Asp195 in ADH of *B. stearothermophilus* (Esposito *et al.*, 2002; Guy *et al.*, 2003; Ceccarelli *et al.*, 2004). In NADP-dependent ADHs, it is suggested that four amino acid residues including Gly198, Ser199, Arg200 and Tyr218 (*T. brockii* ADH numbering) are conserved and the residues are capable of making contacts with 2′-phosphate oxygen atom of $NADP^+$ (Peretz and Burstein, 1989). It is also found that an aspartate residue at position 42 (*S. acidocaldarius* ADH1 numbering) is conserved in the amino acid sequences of ADHs from short-chain dehydrogenases/reductases (SDRs) family including *Leifsonia* ADHs, *T. thermophilus* ADHs, *S. acidocaldarius* ADHs, *C. hydrogenoformans* ADH, suggesting that this aspartate residue plays a critical role in determining the preference for NAD(H) in the SDR family (Schlieben *et al.*, 2005; Höffken *et al.*, 2006; Inoue *et al.*, 2006; Pennacchio *et al.*, 2008; Zhou *et al.*, 2013). Type III ADHs from hyperthermophiles are $NADP^+$-dependent and a typical coenzyme-binding site is glycine-rich (Jornvall *et al.*, 1987). A conserved region GGGSXXD is found in ADHs from *Thermococcus* strain ES1, *T. hydrothermalis* and *T. zilligii* and this region is responsible for coenzyme binding (Antoine *et al.*, 1999; Ying *et al.*, 2009).

Metal binding motifs of zinc- and iron-dependent ADHs

Signature amino acid residues of binding motifs are identified in Zn- or Fe-containing ADHs. Zinc metal in ADHs can have structural, catalytic or both functions. The structural zinc is usually coordinated by four cysteine residues, Cys111, Cys103, Cys97, and Cys100 (numbered according to horse liver ADH) in a tetrahedral manner (Eklund *et al.*, 1976; Ceccarelli *et al.*, 2004). However, this is not conserved for all hyperthermophilic ADHs, for example the fourth cysteine is replaced by aspartic acid residue in *A. pernix* ADH (Guy *et al.*, 2002) and glutamate residues in *S. solfataricus* strain 1617 (Esposito *et al.*, 2002). The catalytic zinc in primary ADHs is most commonly coordinated with four ligands, three of which are Cys48, His68 and Cys174 (horse liver ADH numbering) (Eklund *et al.*, 1976). The C-terminal Cys ligand is replaced by Asp residue (Asp150) in 2° ADHs, for instance, the ones from *T. ethanolicus* strain 39E and *T. brockii* (Burdette *et al.*, 1996; Peretz *et al.*, 1997). Similarly, iron in type III ADHs has catalytic, structural or both functions. Histidine residues are proposed to be involved in iron-binding motifs and a conserved sequence Asp209His213His278His293 (*Thermococcus* strain ES-1 ADH amino acid sequence numbering) is found in type III ADHs of hyperthermophiles (Ying *et al.*, 2009). It is known that there are eight histidine residues in *Thermococcus* strain ES1, 11 in *T. zilligii* and 10 in *T. hydrothermalis*, and these histidine residues are predicted to be involved in iron-binding (Li and Stevenson, 1997; Antoine *et al.*, 1999; Ying *et al.*, 2009). Owing to the lack of crystal structures of type III ADHs from hyperthermophiles, the amino acid residues of binding motifs for structural or catalytic iron are not confirmed.

Biochemical and catalytic properties of thermostable ADH

The most important biochemical and catalytic properties of thermostable ADHs including optimal pH, substrate specificity and enzyme kinetic parameters are reviewed below.

Optimal pH values for both activities of oxidation and reduction

In general, optimal pH values of ADHs for catalysing the oxidation of alcohols are higher than those of aldehydes/ketones reduction. This trend is the same for ADHs studied from hyper/thermophiles (Table 6.3). The optimal pH range for oxidation of alcohols is between 8.2 and 11 and mostly at 10.5, whereas for the reduction of aldehydes/ketones is between 6 and 8 and mostly at 7.5. However, *T. zilligii* ADH has an optimal pH for oxidation of alcohols around 6.8–7 (Li and Stevenson, 1997) and glycerol dehydrogenase from *T. maritima* has optimal pH for oxidation of glycerol at 7.9 (Beauchamp *et al.*, 2014). ADHs of *T. kodakarensis* and *S. acidocaldarius* have optimal pHs for reduction of aldehydes/ketones at 3 and 5, respectively (Pennacchio *et al.*, 2010, 2012; Wu *et al.*, 2013). *T. thermophilus* strain HB27 ADH2 and *S. solfataricus* strain 1617 ADH show no difference in optimal pH for reduction of aldehydes/ketones and oxidation of alcohols, although the former is neutral at pH 7 and the latter is around pH 8.8–9.6 (Raia *et al.*, 2001; Rocha-Martín *et al.*, 2012). Generally, ADHs from extreme thermophiles have lower optimal pHs for oxidation of alcohols than hyperthermophiles, but same optimal pHs for reduction of aldehydes/ketones as hyperthermophiles. The optimal pH range for the oxidation of alcohols is between 8 and 9 with the exception of *T. roseum* ADH, which is 10 (Yoon *et al.*, 2002).

Substrate specificity and enzyme kinetic parameters

The substrate specificity and enzyme kinetics are central biochemical and catalytic properties of ADHs, and they may suggest their physiological roles in extreme thermophiles and hyperthermophiles and provide information of their suitability for pharmaceutical or industrial applications. The substrate specificities of ADHs in reduction and oxidation have been examined using a range of primary and secondary alcohols, aldehydes and ketones including aliphatic, cyclic, branched and aromatic (Plapp *et al.*, 1993).

Table 6.3 Biochemical and kinetics properties of alcohol dehydrogenases from hyper/thermophiles

			Alcohol oxidation				Aldehyde/ketone reduction					
Organisms	ADH (1°/2°)	Co-enzyme	pH[a] (optimal)	Apparent K_m (mM)[b]	Apparent k_{cat} (s^{-1})	Apparent k_{cat}/apparent K_m (s^{-1} mM^{-1})	pH[c] (optimal)	Apparent K_m (mM)[d]	Apparent k_{cat} (s^{-1})	Apparent k_{cat}/apparent K_m (s^{-1} mM^{-1})	Maximum temperature[e] (°C)	Reference
P. aerophilum	2°	NADP	na[f]				7.5	2.4 (α-tetralone)	1.7×10^2	716.4	90	Vitale *et al.* (2010)
A. pernix	1° and 2°	NAD	10.5	13.7 (ethanol)	0.23	0.017	8.0	nd (acetaldehyde)	nd	nd	>95	Guy *et al.* (2003), Hirakawa *et al.* (2004)
				5.43 (benzyl alcohol)	1.02	0.189		0.333 (benzaldehyde)	1.22	3.66		
T. guaymasensis	2°	NADP	10.5	0.38 (2-butanol)	833	2.2×10^2	7.5	0.21 (diacetyl)	202	962	>95	Ying and Ma (2011)
S. solfataricus 1617	1°	NAD	(8.8–9.6)	0.7 (ethanol)	0.5	0.714	(8.8–9.6)	nd (acetaldehyde)	nd	nd	>95	Raia *et al.* (2001)
				0.8 (benzyl alcohol)	1.4	1.7		0.3 (benzaldehyde)	3.5	11.6		
S. solfataricus MT4	1°	NAD	8.5	0.260 (ethanol)	1.833	7.05	7.5	0.0025 (anisaldehyde)	0.933	373.3	>95	Rella *et al.* (1987)
T. maritima	1°	NAD	7.9	39 (glycerol)	15.84	0.406	6.0	30 (DHA)	8.866	0.296	90	Beauchamp *et al.* (2014)
S. tokodaii	1°	NAD	10.5	9.08 (ethanol)	0.849	0.093	7.0	29.7 (acetaldehyde)	10.20	0.343	95	Yanai *et al.* (2009)
Sulfolobus sp. RC3	1°	NAD	nd	17.2 (ethanol)	nd	nd	nd	nd	nd	nd	>95	Cannio *et al.* (1996)
T. roseum	1°	NAD	10	24.2 (ethanol)	0.917	0.0379	nd	nd	nd	nd	90	Yoon *et al.* (2002)
B. stearothermophilus LLD-R	2°	NAD	8	nd (ethanol)	0.262	nd	6	nd (acetaldehyde)	0.972	nd	70	Guagliardi *et al.* (1996)
T. brockii	2°	NADP	nd	nd	nd	nd	(7.5–8.0)	nd (butan-2-one)	5.32	nd	85	Keinan *et al.* (1987), Korkhin *et al.* (1998)

Thermus sp. ATN1	1°	NAD	9	2.09 (cyclohexanol)	51.62	24.7	6	3.68 (cyclohexanone)	37.54	10.2	nd	Höllrigl *et al.* (2008)
T. thermophilus MTCC 1494	2°	NAD	8.8	35 (ethanol)	18.06	0.516	nd	0.07 (DL-glyceraldehyde)	8.88	123.0	89	Raghava and Gupta (2010)
T. ethanolicus JW 200 (1°ADH)	1°	NADP	8.9	16.7 (ethanol)	33.82	2.025	7.6	210 (acetaldehyde)	278	1.324	88	Bryant *et al.* (1988)
T. ethanolicus JW 200 (2°ADH)	2°	NADP	8.9	49 (ethanol)	9.642	0.197	7.6	44.8 (acetaldehyde)	19.28	0.43	90	Bryant and Ljungdahl (1981), Bryant *et al.* (1988)
T. ethanolicus 39E (1°ADH)	1°	NAD(P)[g]	9	25 (ethanol)	7.7	0.308	6.5	0.25 (acetaldehyde)	6.61	0.035	nd	Burdette and Zeikus (1994)
T. ethanolicus 39E (2°ADH)	2°	NADP	9	66 (ethanol)	5.367	0.080	6.5	2.1 (acetaldehyde)	47.33	22.540	>90	Burdette and Zeikus (1994), Burdette *et al.* (1996)
P. furiosus (Adh C)	1°	NADP	9.4–10.2	29.4 (ethanol)	19.17	0.652	nd	0.17 (acetaldehyde)	5.48	32.24	>95	Ma and Adams (1999)
Thermococcus. ES1	1°	NADP	10.4	10.4 (ethanol)	84	8.077	7.0	1.0 (acetaldehyde)	23.6	23.6	>95	Ma *et al.* (1995), Ying *et al.* (2009)
T. litoralis	1°	NADP	8.8	11 (ethanol)	26	2.36	nd	0.4 (acetaldehyde)	nd	nd	85	Ma *et al.* (1994), Ma and Adams (2001)
T. kodakarensis	2°	NADP	9	61.3 (meso-2,3-butanediol)	43.8	0.71	3	30.2 (acetoin)	13.7	0.45	95	Wu *et al.* (2013)
T. hydrothermalis	1°	NADP	10.5	2.0 (benzyl-alcohol)	23	11.5	7.5	0.01 (benzaldehyde)	1.7	170	90	Antoine *et al.* (1999)
T. zilligii	1°	NADP	6.8–7.0	10 (ethanol)	nd	nd	nd	nd (acetylaldehyde)	nd	nd	100	Li and Stevenson (1997, 2001)
T. hypogea	1°	NADP	11	9.7 (ethanol)	14	1.5	8	3.1 (acetylaldehyde)	15	4.84	>95	Ying *et al.* (2007)
P. furiosus (Adh A)	2°	NADP	10–11	60.8 (2-pentanol)	18.24	0.3	7.5	3.3 (pyruvaldehyde)	13.86	4.2	102.7	van der Oost *et al.* (2001)
P. furiosus (AdhD)	2°	NAD	8.8	86.8 (2,3-butane-diol)	60.76	0.7	6.1	6.5 (acetoin)	11.7	1.8	>100	Machielsen *et al.* (2006)

Table 6.3 continued

Organisms	ADH (1°/2°)	Co-enzyme	Alcohol oxidation: pH[a] (optimal)	Apparent K_m (mM)[b]	Apparent k_{cat} (s^{-1})	Apparent k_{cat}/apparent K_m ($s^{-1}mM^{-1}$)	Aldehyde/ketone reduction: pH[c] (optimal)	Apparent K_m (mM)[d]	Apparent k_{cat} (s^{-1})	Apparent k_{cat}/apparent K_m ($s^{-1}mM^{-1}$)	Maximum temperature[e] (°C)	Reference
S. acidocaldarius (Adh1)	1°	NAD	8.2	6.4 ((S)-indanol)	13.7	2.14	5.1	0.71 (isatin)	22	31	85	Pennacchio *et al.* (2010)
S. acidocaldarius (Adh2)	2°	NAD	10	0.81 (isoborneol)	16.6	20.5	5	0.11 (2,2-dichloro-acetophenone)	0.65	5.91	80	Pennacchio *et al.* (2012)
T. sibiricus	2°	NADP	10.5	54.4 (D-xylose)	0.70	0.0129	7.5	17.75 (pyruvaldehyde)	2.04	0.115	>100	Stekhanova *et al.* (2010)
C. hydrogeno-formans	nd	NADP	nd	nd	nd	nd	6	5.3 (ethyl pyruvate)	3.97	0.749	80	Zhou *et al.* (2013)
T. thermophilus HB27 (Adh1)	2°	NAD	10	4.2 ((S)-(α)–1-indanol)	61.4	14.6	6	1.0 (ethyl benzoyl-formate)	50.1	50.1	80	Pennacchio *et al.* (2008)
T. thermophilus HB27 (Adh2)	2°	NAD	7	22.6 (1-phenyl-ethanol)	1.7	0.075	7	0.58 (2,2',2'trifluoro-acetophenon)	39	67.2	>90	Rocha-Martín *et al.* (2012)

[a]Optimal pH for alcohol oxidation activity.
[b]Apparent K_m values with the corresponding substrate (alcohol) indicated in the bracket.
[c]Optimal pH for aldehydes/ketones reduction.
[d]Apparent K_m values with the corresponding substrate (aldehydes/ketones) indicated in the bracket.
[e]Maximal temperature for enzyme activity.
[f]No catalytic reaction towards all alcohol tested.
[g]Both NAD and NADP serve as co-enzyme with higher catalytic efficiencies for $NADP^+$-dependent ethanol oxidation and NADH-dependent acetaldehyde reduction than those for NAD^+-dependent ethanol oxidation and NADPH-dependent acetaldehyde reduction respectively.
1°, primary alcohol dehydrogenase; 2°, secondary alcohol dehydrogenase; nd, not determined.

ADH of *A. pernix*

A. pernix ADH oxidizes aliphatic primary, secondary and aromatic alcohols with no significant differences in the reactivity between primary and secondary alcohols (Hirakawa *et al.*, 2004). It prefers aliphatic ketones as substrates for reduction as the reduction rate for acetophenone is difficult to quantify due to the very small substrate conversion (Hirakawa *et al.*, 2004). However, benzaldehyde having the acetophenone structure without the methyl group is a preferable substrate (Hirakawa *et al.*, 2004). The highest k_{cat}/K_m value is obtained for 1-hexanol and 2-nonanone for oxidation and reduction respectively (Hirakawa *et al.*, 2004).

ADHs from *Sulfolobus* species

S. solfataricus strains MT4 and 1617 ADHs have a broad substrate specificity that includes linear and branched alcohols, linear and cyclic secondary alcohols, linear and cyclic ketones (Rella *et al.*, 1987; Raia *et al.*, 2001). *S. solfataricus* strain MT4 ADH shows the highest k_{cat}/K_m value with 3-methyl-cyclohexanol and anisaldehyde for oxidation and reduction respectively, whereas they are cyclohexanol and methoxybenzaldehyde for *S. solfataricus* strain 1617 (Wu *et al.*, 2013). Mutation for ADH of *S. solfataricus* strain 1617 with a single substitution of Asn249 with Tyr in an attempt for improving the catalytic efficiency results in a 7-fold higher activity for benzyl alcohol, and higher transformation rates for aliphatic alcohols and aromatic aldehydes (Giordano *et al.*, 1999). However, K_m values for all other alcohols and aldehydes are increased, suggesting no overall improvement in the catalytic efficiency (Giordano *et al.*, 1999). Moreover, the ADH from *Sulfolobus* sp. RC3, which shares 88.4% amino acid sequence identity with that of *S. solfataricus* DSM 1617 is compared for catalytic efficiency and substrates specificity (Cannio *et al.*, 1996). In the comparative kinetics analysis, representative substrates ethanol and benzyl-alcohol are used (Cannio *et al.*, 1996). It is observed that ADH of *S. solfataricus* DSM1617 shows 3.5 times higher V_{max} for aromatic alcohol and lower K_m to the same extent but these differences are not observed in *Sulfolobus* sp. RC3 ADH (Cannio *et al.*, 1996). Although only 17 amino acid residues are replaced in *Sulfolobus* sp. RC3 ADH (total ~348 amino acids), 6 of them are clustered into a narrow stretched region (a.a. 17–26) and 7 of them are located in the amino-terminal regions (a.a. 14–77) (Cannio *et al.*, 1996). In medium chain ADHs, the catalytic efficiency and substrates specificity are highly influenced by the residues in the amino-terminal regions especially when charge introduction and shift occur as in this case (Jörnvall *et al.*, 1978, 1987). The prediction of segmental flexibility with normalized B values for comparing proteins with different spreads and averages (Karplus and Schulz, 1985) and study of possible secondary structure clearly indicate that the replacement of all seven amino acid residues in *Sulfolobus* sp. RC3 ADH contributes to decreased flexibility and the formation of coil-flexible turns, which is a reasonable argument for the differences of K_m values between the two *Sulfolobus* enzymes (Cannio *et al.*, 1996).

ADH from hyperthermophilic *S. tokodaii* of the same genus, shows different substrate specificity for reduction but similar substrate specificity for oxidation (Yanai *et al.*, 2009). The enzyme has reactions towards broad substrates including primary, secondary alcohol and benzyl alcohol and higher catalytic efficiencies towards 1-pentanol, 1-hexanol, and benzyl alcohol (Yanai *et al.*, 2009). Also, the catalytic efficiencies are lower for secondary alcohols. In contrast, unlike *S. solfataricus* the enzyme does not catalyse the reduction of ketones such as acetone, 2-butanone, 2-pentanone, 2-hexanone, and cyclohexanone (Yanai *et al.*, 2009). It catalyses the reduction of various aldehydes except formaldehyde and the highest catalytic efficiency is achieved when benzaldehyde is used as substrate (Yanai *et al.*, 2009). The V_{max} values for the reduction of aldehydes are much higher than the oxidation of substrates but the K_m values for better substrates such as 1-pentanol, benzyl alcohol and 1-hexanol are lower than those for the corresponding aldehydes, and hence it is difficult to conclude whether oxidation or reduction is the more efficient reaction (Yanai *et al.*, 2009).

Short-chain ADH from *S. acidocaldarius* shows different substrate specificity. The enzyme has no or poor activity on aliphatic linear and branched alcohols, but shows a discrete activity on aliphatic cyclic and bicyclic alcohols (Pennacchio *et al.*, 2010). For reduction, it has no activity toward aliphatic and aromatic aldehydes, and on aliphatic linear, branched and cyclic ketones except for 3-methylcyclohexanone alcohol, but is very effective in reducing aliphatic and aryl α-keto esters, halogenated aryl a-keto esters, and halogenated β-keto esters (Pennacchio *et al.*, 2010). Similarly, ADH2 purified from the same species, shows poor activities on a discrete number of aliphatic linear and branched alcohols as well as aliphatic cyclic, bicyclic and aromatic alcohols (Pennacchio *et al.*, 2012). Moreover, it displays a preference for the reduction of alicyclic, bicyclic and aromatic ketones and α-ketoesters, but no activity on aldehydes (Pennacchio *et al.*, 2012). Both enzymes have high activity toward 1-indanol and α-tetralol (Pennacchio *et al.*, 2010, 2012).

ADHs of *P. furiosus*

Three of the four ADHs (AdhA, AdhC and AdhD) from *P. furiosus* are characterized (Ma and Adams, 1999; van der Oost *et al.*, 2001; Machielsen *et al.*, 2006). AdhA and AdhD have a higher reaction rate of oxidation than reduction, whereas AdhC shows the opposite observation (Ma and Adams, 1999; van der Oost *et al.*, 2001; Machielsen *et al.*, 2006). AdhA oxidizes a range of aliphatic and aromatic primary (C1-C10) and secondary (C3-C8) alcohols, and reactions on secondary alcohols are faster than primary alcohols (van der Oost *et al.*, 2001). Moreover, reaction on alcohols with secondary or third hydroxyl group is less effective (van der Oost *et al.*, 2001). The highest k_{cat}/K_m value in oxidation is obtained when 2-pentanol is used as substrate. In reduction, the highest rate is observed when pyruvaldehyde is used as substrate, where other substrates tested only show 10% or less activities (van der Oost *et al.*, 2001). In comparison, AdhD also has a preference for secondary alcohol and oxidizes a wide range of alcohols similar to AdhA (Machielsen *et al.*, 2006). In contrary, it reduces various ketones and aldehydes (Machielsen *et al.*, 2006). The highest k_{cat}/K_m value is observed with 2,3-butanediol and diacetyl-acetoin for oxidation and reduction respectively (Machielsen *et al.*, 2006). AdhC oxidized a range of aliphatic (C_2–C_8) and aromatic primary alcohols with the highest activity for 1-butanol but shows no activity with methanol and little with polyols (Ma and Adams, 1999). It is able to reduce acetaldehyde, phenylacetaldehyde, and methylglyoxal (Ma and Adams, 1999). It is more efficient in catalysing aldehyde reduction as shown by the K_m values (Ma and Adams, 1999).

ADHs from *Thermococcus* species

ADHs from seven species of *Thermococcus* have been purified and characterized including *T. hydrothermalis, T. litoralis, Thermococcus* strain ES-1, *T. kodakarensis, T. sibiricus, T. guaymasensis* and *Thermococcus* sp. AN1 (Ma *et al.*, 1994; Li and Stevenson, 1997, 2001; Antoine *et al.*, 1999; Ma and Adams, 2001; Ying *et al.*, 2009; Stekhanova *et al.*, 2010; Ying and Ma, 2011; Wu *et al.*, 2013). ADHs from *T. hydrothermalis, T. litoralis, Thermococcus* strain ES-1 and *Thermococcus* sp. AN1 are primary ADHs as they oxidize a broad range of primary alcohols and show no catalysing abilities in oxidizing secondary alcohols (Ma *et al.*, 1994; Li and Stevenson, 1997, 2001; Antoine *et al.*, 1999; Ma and Adams, 2001; Ying *et al.*, 2009). Moreover, they have lower K_m values for substrates in reduction reactions than that in oxidation reaction, which is also shown by the higher k_{cat}/K_m value for NAD(P)H than NAD (Ma *et al.*, 1994; Li and Stevenson, 1997, 2001; Antoine *et al.*, 1999; Ma and Adams, 2001; Ying *et al.*, 2009). *T. hydrothermalis* ADH oxidizes a series of primary aliphatic and aromatic alcohols preferentially from C_2 to C_8, methanol, cyclohexanol and glycerol (Antoine *et al.*, 1999). *T. litoralis* ADH exhibits a very broad substrate specificity for primary alcohols (C_2 to C_8) but glycerol and sorbitol are not oxidized (Ma *et al.*, 1994; Ma and Adams, 2001). ADH of *Thermococcus* strain ES-1 catalyses a range of aliphatic (C_2 to C_8) and aromatic primary alcohols but it does not oxidize methanol, glycerol and has little if any activity with polyols or hydroxycarboxylic acids (Ma *et al.*, 1994; Ying *et al.*, 2009). ADH of *Thermococcus* sp. AN1 oxidizes a series of primary linear alcohols but not methanol, isopropanol, or glycerol (Li and Stevenson, 1997, 2001). ADHs from *T. kodakarensis* and *T. guaymasensis* are 2° ADHs as they have preference for the oxidation of secondary alcohol (Ying and Ma, 2011; Wu *et al.*, 2013). For *T. kodakarensis* ADH, 1-phenyl-2-propanol is preferred to be oxidized over 2-phenyl-1-propanol, and 2-alkanols is preferred over 1-alkanols. The highest activity in oxidation is towards meso-2,3-butanediol and towards pyruvic aldehyde in reduction. It is also enatio-selective and stereo-selective for oxidation and has high preference towards the (*S*)-1-phenylethanol or the *meta*- and *para*-substituted derivatives (Wu *et al.*, 2013). *T. guaymasensis* ADH shows broad substrate specificity for oxidation including polyol with low activity, primary alcohols but not methanol, and it has higher activities with secondary alcohols and diol (2,3-butanediol and 1,4-butanediol) with *R*-specificity. The highest activity is shown towards 2-butanol. It is also able to reduce various aldehyde and ketones, but diacetyl to acetoin is irreversible. It is highly efficient in the reduction of ketones and this process can be coupled with excess isopropanol to regenerate NADPH. The apparent K_m values and catalytic efficiency for NADPH are 40 times lower and 5 times higher than those for $NADP^+$, respectively (Ying and Ma, 2011). *T. sibiricus* ADH shows broad substrate specificity for oxidation of sugar with the highest oxidation rates towards pentoses, D-arabinose and D-xylose. However, it does not reduce sugar and the highest reduction rates are found with dimethylglyoxal and pyruvaldehyde (Stekhanova *et al.*, 2010).

ADHs from *Thermotoga* species

ADH from *T. hypogea* is a primary alcohol dehydrogenase, which is able to oxidize a broad range of primary alcohols and diols with the highest activity using 1-butanol, but not secondary alcohols and polyols such as glycerol. It can reduce acetaldehyde, methylglyoxal and butyraldehyde with the highest activity (Ying *et al.*, 2007). Glycerol dehydrogenase from *T. maritima* has the highest activity in the conversion of glycerol to dihydroxyacetone (DHA) (Beauchamp *et al.*, 2014). At their optimal pH for reactions, the apparent k_{cat} for oxidation of glycerol oxidation was 1.8 times higher than that of DHA reduction (Beauchamp *et al.*, 2014). It also has low activities towards ethylene glycol, isopropanol, and 1,3-propanediol (Beauchamp *et al.*, 2014). However, no activity is shown towards 1-propanol. It is enantioselective with enantiomeric excess above 99% on racemic 1,2-propanediol towards the production of (*R*)-1,2-propanediol from hydroxyacetone (Beauchamp *et al.*, 2014).

ADH of *P. aerophilum*

P. aerophilum ADH does not show activity towards all ketones, alcohols or aldehydes and is active only when aromatic ketone α-tetralone is used as a substrate (Vitale *et al.*, 2010). The lack of activities towards most substrates is explained by the fact that its catalytic domain differs from well-studied and well-characterized ADHs (Vitale *et al.*, 2010). Secondly, key amino acid residues (e.g. Ser 40 and His 43 in *S. solfataricus* ADH) that are involved in the binding of alcohol substrates are lacking (Esposito *et al.*, 2002; Vitale *et al.*, 2010). Moreover, key residues (e.g. Cys 38, His 68 and Cys 154 in *S. solfataricus* ADH) involved in the coordination of the catalytic Zn ion is also lacking in the enzyme active site (Esposito *et al.*, 2002; Vitale *et al.*, 2013).

ADHs from *B. stearothermophilus* and *Thermus* species

ADHs from two extreme thermophiles *B. stearothermophilus* and *Thermus* sp. ATN1 are studied for the oxidation and reduction substrate specificity. ADH-hT from *B. stearothermophilus* strain LLD-R oxidizes primary and secondary alcohols, with a higher efficiency for secondary alcohols and the highest reaction rate using butan-2-ol (Guagliardi *et al.*, 1996). The size of the two substituent groups on the carbon with hydroxyl group affects the activity, but in general, the activity is the highest when at least one of the group is methyl, and it decreases dramatically when the size of the substitutes is larger than methyl (Guagliardi *et al.*, 1996). This is an indication that the existence of a 'small site' and a 'large site' in the catalytic centre (Guagliardi *et al.*, 1996). The substrate specificity for reduction are also broad, and it catalyses the reduction of linear aldehydes and ketones except pentan-3-one and the highest activity with acetaldehyde (Guagliardi *et al.*, 1996). The overall reaction is the highest for aldehydes and decreases progressively for alcohols then ketones (Guagliardi *et al.*, 1996).

The substrate specificity of *Thermus* sp. ATN1 ADH is studied in detail. It prefers oxidation of primary alcohols over secondary alcohols and corresponding aliphatic aldehydes (Höllrigl *et al.*, 2008). Its activity steeply increases with chain length until 1-propanol and then decreases slightly with continue increase in chain length and the similar trend is observed for secondary alcohols (Höllrigl *et al.*, 2008). In reduction reaction, higher conversion rate is shown towards aldehydes followed by cyclic ketones then aliphatic ketones, and butyraldehyde is the best substrate (Höllrigl *et al.*, 2008). The activity decreases towards aldehyde with increase in chain length and bulkier substituents in both aliphatic ketones and aldehyde hinder reactions (Höllrigl *et al.*, 2008).

Narrow substrate specificity is observed for *T. thermophilus* HB27 ADH1, and it has no activity on aliphatic linear, branched, and cyclic alcohols except for a discrete activity on cyclohexanol and three methoxy-substituted benzyl alcohols are found to be highly active substrates. The enzyme is also not active on aliphatic aldehydes, and aliphatic linear, branched and cyclic ketones. However, it shows discrete activity with aryl aldehydes, such as benzaldehyde or methoxybenzaldehydes, and high reduction rates with halogenated aryl ketones, such as 2,2,2-trifluoroacetophenone, 2-chloroacetophenone, and 4-chlorobutyrophenone, and with aryl diketones, such as 1-phenyl-1,2-propanedione (Pennacchio *et al.*, 2008). *T. thermophilus* HB27 ADH2 shows similar substrate specificity profile in oxidation but the profile for reduction is different. It can reduce acyl aldehyde and efficiently reduce racemic mixture of 2-phenylpropionaldehyde into *R*-isomers, which is very rare for ADHs. Moreover, it reduces non-halogenated cyclic and halogenated aryl ketones (Rocha-Martín *et al.*, 2012).

ADHs from *C. hydrogenoformans* and *T. roseum*

ADHs from *C. hydrogenoformans* and *T. roseum* only have their substrate specificity determined for reduction and oxidization respectively. *C. hydrogenoformans* ADH exhibits a broad substrate specificity towards aliphatic linear, branched, and cyclic and aromatic ketones with cycloheptannone showing the highest activity (Zhou *et al.*, 2013). *T. roseum* ADH oxidizes mainly primary aliphatic alcohols and exhibits high substrate specificity towards ethanol, *n*-propanol and crotyl alcohol. The highest reaction rate is observed when ethanol is used as substrate (Yoon *et al.*, 2002). A gradual decrease in rate is observed by increasing the length and branching of the carbon chain of the alcohol and it does not show activity towards long-chain primary alcohols, short chain secondary alcohols and tertiary alcohol (Yoon *et al.*, 2002).

ADHs of *T. ethanolicus*

Five ADHs are characterized from *T. ethanolicus*, three from *T. ethanolicus* strain JW200 and two from *T. ethanolicus* strain 39E (Bryant and Ljungdahl, 1981; Bryant *et al.*, 1988; Burdette and Zeikus, 1994; Burdette *et al.*, 1996; Holt *et al.*, 2000). Primary ADH (1° ADH) from *T. ethanolicus* strain JW200 strictly oxidizes primary alcohol with 1-pentanol at the highest rate. It reduces aldehydes with propionaldehyde at a higher rate than acetaldehyde, but ketones such as acetone or 2,3-butanedione are not reduced (Bryant *et al.*, 1988). The 2° ADH is less specific, besides secondary alcohols, it also oxidizes ethanol, and primary alcohols at a lower rate. The maximal activity is observed with 2-propanol that is 15 times higher than ethanol. It reduces both aldehydes and ketones with higher activities for propionaldehyde, acetone and 2,3-butanedione than acetaldehyde (Bryant *et al.*, 1988). The 1° ADH from *T. ethanolicus* strain 39E, oxidizes solely primary alcohol and reduced aldehydes but not ketones (Burdette and Zeikus, 1994). Similar to strain JW200, the 2° ADH has wider substrate specificities and oxidizes both primary and secondary alcohol but with a much higher catalytic rate with secondary alcohols (Burdette and Zeikus, 1994). It reduces both aldehydes and ketones and generally has higher catalytic activity with ketones and the highest towards propanone (Burdette and Zeikus, 1994).

Physiological functions of ADHs

ADHs are present in all three domains of life (Woese *et al.*, 1990). The widely expressed profile of ADHs suggests that they play important roles in many physiological processes. In mesophiles and thermophiles, the primary function of ADHs is proposed to be involved in ethanol metabolism (Reid and Fewson, 1994), whereas in hyperthermophiles due to the fact that only few of them can produce ethanol, many enzymes are proposed to have functions in reduction of aldehydes (Eram and Ma, 2013).

Regeneration of coenzyme $NAD(P)^+$

The physiological functions of ADHs are studied in detail in representative of mesophiles and extreme thermophiles, which are *S. cerevisiae* and *T. ethabolicus* respectively. The physiological functions of ADHs from *S. cerevisiae*, which play important roles in fermentation, have been successfully exploited for commercial applications including wine-making, baking and brewing (Evans, 1990; Dequin, 2001; Leskovac *et al.*, 2002). *S. cerevisiae* ADHs play an important role in linking the reactions of fermentation of sugar and oxidative respiratory (De Smidt *et al.*, 2008). In carbohydrate metabolism, ADHs regenerate NAD^+ from NADH produced in glycolysis and restore the redox balance through the reduction of acetaldehyde to ethanol, and upon depletion of carbon sugar, fermentation of ethanol that is accumulated, and then is oxidized to acetaldehyde, which is also catalysed by an ADH. Acetaldehyde can be further oxidized to acetic acid which can be metabolized via the tricarboxylic acid cycle or acts as intermediate metabolite in gluconeogensis (Lutstorf and Megnet, 1968; Heick *et al.*, 1969; Wills and Phelps, 1975; Ciriacy, 1979; Beier *et al.*, 1985; Young and Pilgrim, 1985; Lieber, 1997; Smidt *et al.*, 2012). ADHs from hyperthermophiles can also catalyse the regeneration of coenzyme $NAD(P)^+$; however, it is very limited as the concentration of alcohol produced is very low (~mM) (Kengen *et al.*, 1994; Ma *et al.*, 1995; Fardeau *et al.*, 1997; Ying and Ma, 2011).

Oxidation of alcohols to aldehydes

ADHs from *T. ethanolicus* strain JW200 and 39E have different physiological roles. It is suggested that in *T. ethanolicus* strain JW200, the 2° ADH is responsible for ethanol formation and primary ADH is involved in converting alcohols to aldehydes that can be used as alternative carbon sources for *T. ethanolicus* (Bryant *et al.*, 1988). This is supported by several facts, and firstly 2° ADH is produced earlier during the growth and 1° ADH is expressed late in the fermentation (Bryant *et al.*, 1988). Secondly, 2° ADH shows a relatively 4.5 times lower K_m towards acetaldehyde than the 1° ADH, and its activity is enhanced by pyruvate (Bryant and Ljungdahl, 1981; Bryant *et al.*, 1988). However, the K_m values of $NADP^+$ for both ADHs are low comparing to K_m values of NADPH when ethanol and acetaldehyde are used as substrates respectively, indicating reduction is the favourite direction of the catalytic reactions (Bryant *et al.*, 1988). It is not known whether alcohols can be effectively used for producing carbon substrates, but it might be used for maintaining a high NADPH/NADP ratio and a low redox state (Bryant *et al.*, 1988). In *T. ethanolicus* strain 39E, the 1° ADH is responsible for ethanol consumption and nicotinamide-cofactor recycling based on several kinetics data (Burdette and Zeikus, 1994). It displays eight times higher catalytic efficiency for NAD^+-dependent oxidation than $NADP^+$-dependent oxidation, and in contrary, 30 times higher catalytic efficiency for NADPH-dependent reduction than NADH-dependent reduction (Burdette and Zeikus, 1994). Moreover, it displays similar activities for both acetaldehyde reduction and ethanol oxidation (Burdette and Zeikus, 1994). Hence, the 1° ADH may play a role in maintaining the NAD(H) to NADP(H) ratio and function in parallel with the ferredoxin-linked nicotinamide-dinucleotide oxidoreductase activities that interconvert NADH and NADPH in *T. ethanolicus* strain 39E (Lovitt *et al.*, 1988). It is also suggested that the 1° ADH is solely responsible for the NAD(H)-dependent interconversion of acetaldehyde and ethanol, since the 2° ADH does not show any activity towards NAD(H)-dependent reactions (Burdette and Zeikus, 1994). On the other hand, the 2° ADH possesses a dual activity of both acetyl-CoA reductive thioesterase and acetaldehyde dehydrogenase (Burdette and Zeikus, 1994). It catalyses the formation of ethanol from acetyl-CoA, and the activity is NADPH dependent (Burdette and Zeikus, 1994). In cell free extract, the NADPH-linked activity was 10-fold higher than the NADH-linked activity and most catalysis is carried out by 2° ADH as 1° ADH has 0.5 times lower rate for NADH-linked activity (Burdette and Zeikus, 1994). This shows that in both strains JW200 and 39E, the 2° ADHs are principally responsible for ethanol formation (Smith and Kaplan, 1980; Bryant *et al.*, 1988; Burdette and Zeikus, 1994). Enzyme kinetics data suggest that general functions of ADHs from hyperthermophiles are probably involved in aldehyde reduction, but not in alcohol oxidation (Table 6.3).

Alcohol production and detoxification of aldehyde

Among the hyperthermophiles from which ADHs are purified and characterized, only a few of them are reported to be able to produce ethanol, and the physiological roles of their ADHs are proposed mainly based on enzyme kinetics data. *P. furiosus* is found to be able to produce H_2, CO_2, acetate, alanine and small amounts of ethanol and the physiological role of its AdhC is suggested to catalyse the production of alcohol (Kengen *et al.*, 1994). *Thermococcus* strain ES-1 produces some ethanol and butanol when cultures are grown at low concentrations of elemental sulfur and its ADH shows 10 times lower K_m values for aldehydes than alcohols (Ma *et al.*, 1994, 1995; Ying *et al.*, 2009). Hence, the enzyme may play a role in aldehyde reduction rather than alcohol oxidation. It is also suggested that the ADH is responsible for disposal of excess reductant under S°-limited conditions (Ma *et al.*, 1995). *T. guaymasensis* also produces ethanol as an end product of fermentation and the apparent K_m value and catalytic efficiency of its ADH towards NADPH are 40 times lower and 5 times higher than those for $NADP^+$, respectively (Ying and Ma, 2011). The higher catalytic efficiency on NADPH than on $NADP^+$ as a coenzyme suggests its physiological function is more likely to be associated with NADPH-oxidizing metabolism. Based on these data, the physiological roles of the enzyme are proposed to be in catalysing the formation of alcohols for instance ethanol or acetoin concomitant to the NADPH oxidation (Ying and Ma, 2011). Hyperthermophilic bacterium *T. hypogea* produces trace amount of ethanol (Fardeau *et al.*, 1997). The physiological function of this enzyme is suggested to be catalysing the reduction of aldehydes based on its enzyme kinetics results. The activity of aldehyde reduction is about 17 times higher than alcohol oxidation, and the apparent K_m values are three times lower for NADPH and reduction substrates (Ying *et al.*, 2007). ADH of *T. litoralis* has much lower K_m value (~20× lower) for acetaldehyde than ethanol. The generation of aldehyde from pyruvate *in vivo* is possibly catalysed by a bifunctional 2-keto acid ferredoxin oxidoreductase (Ma *et al.*, 1994, 1997; Eram and Ma, 2013). It is proposed that the ADH may also play a role in the detoxification of aldehydes produced at high temperatures (Ma *et al.*, 1994, 1997).

The change in holoenzyme structure of *T. hydrothermalis* ADH is accompanied by the change in enzymatic activity. The dominant form of this ADH is dimeric, which is the form existed at optimal pH for the reduction of aldehyde and it is deduced that this enzyme is involved in aldehyde detoxification than alcohol turnover (Antoine *et al.*, 1999). ADH of *T. thermophilus* shows higher k_{cat} and k_{cat}/K_m values for NADH than NAD^+ and this suggests the physiological direction of the catalytic reaction is the reduction of aldehyde rather than oxidation (Pennacchio *et al.*, 2008; Rocha-Martín *et al.*, 2012). A distinctive feature of *T. thermophilus* peptidoglycan is the presence of mureopeptides with phenylacetic acid residues (Quintela *et al.*, 1995). It is speculated that the ADH may play a role in the hypothetical phenylacetate synthesis pathway through the conversion of benzoylformate into mandelate (Pennacchio *et al.*, 2008). The gene encoding the short-chain ADH from *T. sibiricus* is located in the 'saccharolytic gene island' containing the complete set of genes responsible for the hydrolysis of cellulose and beta-linked polysaccharides and it is proposed that this ADH plays a role in the carbohydrate metabolism (Bezsudnova *et al.*, 2012). *P. furiosus* AdhD shows high catalytic efficiency for the reduction reaction and lower K_m values for the substrate and cofactor involved in this reaction, which suggests that the physiological function of AdhD is catalysing the reduction of aldehydes or ketones (Machielsen *et al.*, 2006).

Applications of ADHs in biotechnology

ADHs can be purified and used in biosensors and analytical tools (Mello and Kubota, 2002), which are used for the detection of ethanol in wine, alcoholic beverages and synthetic samples (Leca and Marty, 1997; Katrlik *et al.*, 1998; Tobalina *et al.*, 1999). They can also be used for cofactor NAD(P) recycling (Kroutil *et al.*, 2004). This is achieved through either a coupled-enzyme systems with hydrogenases and NADP-oxidase or a single-enzyme system with a chemo-stable ADHs (Kroutil *et al.*, 2004). In fact, there is a wide range of applications of ADHs in biotechnology, which include but not limited to the alcohol fermentation and synthesis of chiral chemicals for pharmaceutical and industrial processes.

Alcohol fermentation for wineries and biofuel production

Ethanol fermentation and acetone–butanol–ethanol (ABE) fermentation are the two largest scales and most important fermentations in the world today (Nimcevic and Gapes, 2000). Ethanol fermentation is a microbial process to regenerate NAD^+ to maintain redox potential from fermentable sugars with end products of carbon dioxide and ethanol (Lin and Tanaka, 2006). ABE fermentation produces acetone, butanol and ethanol in a ratio of 3:6:1 (Lütke-Eversloh and Bahl, 2011). In both fermentation, ADH plays an important role in catalysing the conversion of acetaldehyde into ethanol and butyraldehyde into butanol in ABE fermentation (Lütke-Eversloh and Bahl, 2011).

S. cerevisiae

The earliest industrial applications of ethanol fermentation is using yeast *S. cerevisiae* in the production of alcoholic beverages and leavening bread (Lin and Tanaka, 2006). Pure cultures of *S. cerevisiae* var. *ellipsoideus* and var. *vini* are commonly used in wine production (Mocke, 2005). *S. cerevisiae* is ideal for making alcoholic beverages due to several reasons including high end-product ethanol concentration of 18% of fermentation broth, high tolerance to ethanol and a safe food additive for human consumption (Berry and Slaughter, 2003; Lin and Tanaka, 2006).

Z. mobilis

Z. mobilis is another microorganism capable of high ethanol production of approximately 1.7–1.9 mol ethanol produced per mol glucose metabolized (Miyamoto, 1997; Lin and Tanaka, 2006). It is also used for alcoholic beverages and bioethanol production (Osunkoya and Okwudinka, 2011). In comparison, it has the advantages of higher ethanol yield (5–10% more ethanol per fermented glucose), high tolerance to ethanol of up to 12%, and 2.5 times higher specific ethanol production rate. However, it ferments only glucose, fructose and sucrose, whereas *S. cerevisiae* utilizes a wider range of substrates including but not limited to maltose, starch, galactose, fructose (Zimmermann and Entian, 1997; Osunkoya and Okwudinka, 2011). The first recombinant strain of *Z. mobilis* was engineered to ferment xylose by introducing and expressing xylose utilizing genes from *E. coli* about two decades ago (Zhang *et al.*, 1995). This overcomes the major disadvantage of *Z. mobilis* in its ability to grow on limited substrates.

Genetically engineered *E. coli*

Owing to the increasing demands for renewable energy sources, ethanol fermentation is explored for the possibilities of biofuel production. An ideal microorganism for biofuel production should have the following characteristics: Firstly, high ethanol production of approximately 10% of fermentation broth and with few side products (Dien *et al.*, 2003). Secondly, high ethanol tolerance of over 40 g/l (Dien *et al.*, 2003). Thirdly, metabolize major sugars including glucose, xylose, arabinose, galactose and mannose (Wyman, 1996). Lastly, resistance to acidic pH or high temperatures (Dien *et al.*, 2003). Currently, the enthanologenic microorganisms that are most suitable for industrial exploitation are *Z. mobilis*, *S. cerevisiae*, *E. coli*, and *Klebsiella oxytoca* (Dien *et al.*, 2003; Clomburg and Gonzalez, 2010; Jang *et al.*, 2012; Buschke *et al.*, 2013; Eram and Ma, 2013). Genetic engineering has been used for maximizing the suitability of the above four microorganisms in large scale biofuel production (Dien *et al.*, 2003). Native ADHs from *E. coli* are not sufficient for achieving high ethanol yields, and the first successfully genetically engineered *E. coli* has the plasmid expressing pyruvate decarboxylase and ADH2 from *Z. mobilis* to produce ethanol exclusively through the conversion of pyruvate into acetaldehyde and then ethanol (Ingram *et al.*, 1987). Moreover, *E. coli* strain ATCC 11303 is determined to be the most promising strain of its capability to utilize xylose and high ethanol tolerance of 53–56 g/l (Alterthum and Ingram, 1989).

K. oxytoca

K. oxytoca is able to grow on a varies sugar including hexoses, pentoses, cellobiose and cellotriose (Dien *et al.*, 2003). Similar to *E. coli*, the PET operon (production of ethanol operon) that is under the control of *lac* promoter containing *adh2* and *pdc* genes from *Z. mobilis* was introduced to *K. oxytoca* strain M5A1 and P2 to increase ethanol production (Wood and Ingram, 1992).

C. acetobutylicum

In ABE fermentation, *C. acetobutylicum* is the most intensively used microorganism in industrial setting for the production of acetone and recently it has been re-studied for potential use in biofuel production (Bahl *et al.*, 1982a,b). AdhE from *C. acetobutylicum* has dual functions of converting acetyl-CoA into acetaldehyde then to ethanol and converting butyryl-CoA into butyraldehyde then to butanol (Lütke-Eversloh and Bahl, 2011).

Thermophilic alcohol producers

Members from thermophiles of *Clostridium*, *Thermoanaerobacter* and *Geobacillus* genera are also known to have high ethanol production with ADHs as the enzyme to catalyst the conversion of acetaldehyde into ethanol, and metabolic engineering is performed to explore their potential use in bioethanol production (Lamed and Zeikus, 1980a; Hartley and Shama, 1987; Demain *et al.*, 2005; Shaw *et al.*, 2008; Barnard *et al.*, 2010; Yao and Mikkelsen, 2010; Svetlitchnyi *et al.*, 2013; Tomás *et al.*, 2013). Three species including *Thermoanaerobacter mathranii* (patent number WO2007134607A1), *Thermoanaerobacter saccharolyticum* (patent number WO/2007/130984) and *Geobacillus thermoglucosidasius* (patent number WO/2008/038019A3) have been investigated, and their commercialization has been initiated for bioethanol production (Mikkelsen *et al.*, 2007; Shaw *et al.*, 2007; Atkinson *et al.*, 2008). The optimal temperatures for the growth of *T. mathranii, T. saccharolyticum and G. thermoglucosidasius* are 70–75°C, 40–50°C and 61–63°C respectively (Lee *et al.*, 1993; Larsen *et al.*, 1997; Nazina *et al.*, 2001). Mutant strains are produced for maximizing ethanol production, broadening substrates utilization and increasing ethanol tolerance (Taylor *et al.*, 2009). The following mutant strains are examples of the ones that process the characteristics mentioned, and hence beneficial to industrial bioprocesses.

T. mathranii BG1L1, a mutant strain with deletion of the lactate dehydrogenase gene, is being investigated for biofuel production by the Danish company Biogasol (Mikkelsen *et al.*, 2007). In laboratory-scale, fluidized bed reactor containing un-detoxified wheat straw hydrolysate or maize stover hydrolysate, which contains a mixture of glucose and xylose from 12 to 41 g/l, produces 0.39–0.42 g of ethanol per gram of sugar consumed (Georgieva and Ahring, 2007; Georgieva *et al.*, 2008). It is also reported to be resistant to acetic acid up to 10 g/l and other metabolic inhibitors in wheat straw hydrolysate (Georgieva and Ahring, 2007).

T. saccharolyticum is announced by the US-based company Mascoma for commercialized ethanol production (Shaw *et al.*, 2007). The commercialized strain is the mutant strain *T. saccharolyticum* ALK2 with deletion of the acetate kinase and lactate dehydrogenase genes, and it is reported to produce 0.38 g of ethanol per gram of sugar consumed when 50 g/l of mixed sugars including glucose, xylose, galactose and mannose are used as substrates (Shaw *et al.*, 2008). This mutant strain has several advantages over the other mutant strains. Firstly, it has a stable homoethanologenic phenotype of over 150 generations in continuous culture; secondly, it co-utilizes D-xylose, D-glucose, mannose and L-arabinose and lastly it has a high ethanol tolerance of 37 g/l (Shaw *et al.*, 2008).

G. thermoglucosidasius has been genetically engineered to enhance the production of ethanol and significant progress has been made in commercial development of *Geobacillus* sp. from the progeny of genetically altered *G. thermoglucosidasius* with high ethanol yields (Atkinson *et al.*, 2006, 2007). A promising candidate is the mutant strain *G. thermoglucosidasius* TM242, with deletion of lactate dehydrogenase and pyruvate formate lyase and up-regulation of pyruvate dehydrogenase, which is reported to produce

0.42 g of ethanol per gram of glucose consumed (Cripps *et al.*, 2009).

In comparison with mesophiles, thermophiles have the benefits of working at high temperatures, which will result in less contamination and higher rates of production (Hartley and Shama, 1987). Also, they utilize wider ranges of substrates including pentose sugars and lignocelluloses (Hartley and Shama, 1987). On top of these advantages, the mutants strains have greater catabolic flexibility with enhanced ethanol production than wild types (Taylor *et al.*, 2009). Hyperthermophiles are known to have the metabolic capacity of producing alcohols using a variety of substrates including lignocellulosic materials; however, further investigation for enhancing their productivity is required for their industrial applications.

Stereoselective catalysis of ADHs for synthesis of chiral chemicals

Although extreme thermophiles and hyperthermophiles possessing ADHs and producing trace amount or no ethanol are not applicable for bioethanol production, their ADHs have attracted increasing attention in the synthesis of enantiomerically pure stereoisomers of chiral alcohols for industrial or pharmaceutical applications. In addition to the advantages of optimal growth at elevated temperatures for increasing production yield and high tolerance to organic solvent, enzymes including ADHs different from chemocatalysts also have the advantages of higher enantioselectivity, capable of accepting a wide array of complex molecules as substrates and catalysing reactions with unparalleled chiral (enantio) selectivity (Schmid *et al.*, 2001). These provide more efficient reactions with less by-product and reduce the need to block and de-block steps commonly used in organic synthesis (Schmid *et al.*, 2001).

ADHs from several extreme thermophiles and hyperthermophiles show broad substrate specificity spectrum and produce exclusively (*S*)/(*R*)-enantiomer in high enantiomeric excess of ≥90%. ADHs that catalyse the reduction of aromatic ketones into aromatic chiral alcohols are specially interested in synthesis of useful building blocks in pharmaceutical applications. From the economic perspective, ADHs that are NAD dependent, which is a less expensive co-factor than $NADP^+$ are highly desirable (Hirakawa *et al.*, 2004).

Enantioselective catalysis of ADHs from *Pyrococcus* and *Thermococcus* species

Two NAD-dependent ADHs from the aldo-keto-reductase superfamily including short-chain AdhD from *P. furiosus* and *T. kodakarensis* ADH are also highly enantioselective (Machielsen *et al.*, 2006; Wu *et al.*, 2013). *P. furiosus* AdhD reduces 2-pentanol to (*S*)-2-pentanol with e.e. value of 89.4%, and *T. kodakarensis* ADH reduces 2,2,2-trifluoroacetopheone to (*R*)-alcohol with an optical purity of over 99% e.e. (Machielsen *et al.*, 2006; Wu *et al.*, 2013). ADH from *T. guaymasensis* is stereoselective and obeys the anti-Prelog rule. In the oxidation reaction, (2*R*,3*R*)-2,3-butanediol is converted to (*R*)-2-acetoin with e.e value of 94%. More significantly, in the reduction of racemic acetoin, (2*R*,3*R*)-2,3-butanediol (preferably from (*R*)-acetoin) is produced with the e.e. value of >99%, and meso-2,3-butanediol is also produced, which is probably from (*S*)-acetoin. Similar compounds including (*R*)-1,3-butanediol and acetoin are reported to be important building blocks in the synthesis of antibiotics (Matsuyama *et al.*, 2001; Nestl *et al.*, 2007).

Enantioselective catalysis of ADHs from *T. brockii* and *Thermus* species

Extreme thermophiles including *T. brockii* and *Thermus* sp. ATN1 also have ADHs that reduce ketones into alcohols with high enantioselectivity (Keinan *et al.*, 1987; Höllrigl *et al.*, 2008). ADH in *T. brockii* reduces a wide range of aliphatic, halogenated aliphatic, non-conjugated unsaturated, and bifunctional ketones into corresponding (*S*)-alcohols with optical purity of over 90% (Keinan *et al.*, 1987). ADH in *Thermus* sp. ATN1 reduces ketones including 2-pentanone, 2-heptanone, benzylacetone, and phenylacetone to their corresponding (*S*)-alcohol in optical purities of greater than 99% enantiomeric excess (e.e.). In racemic 3-methylcyclohexanone, only the (*S*)-form is converted and yields (1*S*,3*S*)-3-methylcyclohexanol with diastereomeric excess of 98.7% (Höllrigl *et al.*, 2008). Besides the highly valued reaction of stereoselective reduction of ketones to form chiral alcohols for the production of important precursors for pharmaceutical applications, it is noteworthy that an NAD^+-dependent ADH2 from *S. acidocaldarius* catalyses the asymmetric reduction of benzil to (*R*)-benzoin with both excellent conversion percentage of 98% and optical purity of 98% (Pennacchio *et al.*, 2012). Optically pure (*R*)-benzoin is a useful urease inhibitor (Tanaka *et al.*, 2004), and widely used for preparing amino alcohols in important chiral synthons or synthesis of different heterocycles (Aoyagi *et al.*, 2000).

The NAD-dependent ADHs from *T. thermophilus* show a broad substrate specificity spectrum for reduction and are highly enantioselective (Pennacchio *et al.*, 2008; Rocha-Martín *et al.*, 2012). *T. thermophilus* ADH1 reduces acetophenone, 2,2,2-trifluoroacetophenone, α-tetralone, and α-methyl and α-ethyl benzoylformates to (*S*)-1-phenylethanol (>99% e.e.), (*R*)-α-(trifluoromethyl)benzyl alcohol (93% e.e.), (*S*)-α-tetralol (>99% e.e.), methyl (*R*)-mandelate (92% e.e.), and ethyl (*R*)-mandelate (95% e.e.) respectively, and its ADH2 reduces dynamic reductive kinetic resolution of *rac*-2-phenyl propionaldehyde to its corresponding primary alcohol with high e.e. value towards *R* isomers and this is not found in other ADHs (Pennacchio *et al.*, 2008; Rocha-Martín *et al.*, 2012). *T. thermophilus* ADH3 is stereospecific in oxidizing glycerol into D-glyceraldehyde and opens up several synthetic opportunities (Raghava and Gupta, 2010).

Enantioselective catalysis of ADHs from *T. maritima*

Glycerol dehydrogenase from *T. maritima* is stereoselective and active on racemic 1,2-propanediol. It converts hydroxyacetone to (*R*)-1,2-propanediol with e.e. of > 99%, and it is suggested that the enzyme can be used for chiral synthesis (Beauchamp *et al.*, 2014).

It will be increasingly important to exploit more ADHs that have the various properties including dependence on inexpensive cofactor NAD(H), high thermal stability, reversibility, broad substrate specificity, high stability in organic solvent and high enantioselecivity for potential usages as biocatalyst for organic synthesis.

Conclusions and future perspectives

ADHs are ubiquitously expressed in all three domains of life, and they play important roles in physiological processes that are not limited to alcohol oxidation, aldehyde reduction and detoxification. ADHs exhibit a variety of biophysical, biochemical and catalytic properties, but there are only limited studies of ADHs from extreme thermophiles and hyperthermophiles. A large number of genes encoding ADHs are identified from the genome sequences of hyperthermophiles, which are available from NCBI (http://www.ncbi.nlm.nih.gov/), but only a small portion of them is characterized (Table 6.4), which may be novel ones providing great opportunities for advancing our understanding and exploiting their pharmaceutical and industrial applications.

It appears that more than one ADHs present in hyper/thermophiles (Table 6.4). It is plausible to assume that they have different physiological functions, which warrants further investigation. Different metal ions such as zinc and iron have either catalytic and/or structural functions, but the predominant presence of iron-containing ADHs in hyperthermophiles is a unique feature. Moreover, their wider substrate specificity, high catalytic efficiency and excellent thermostability are valuable properties for further exploitation for potential applications in biotechnology.

Table 6.4 Abundance of genes encoding ADHs in completed genome sequences of hyperthermophilic archaea and bacteria

Organisms	Growth temperature[a] optimum (range, °C)	Phenotype[b]	Genome size (Mbp)	ADH			SDR[f]	References
				Zn[c]	Fe[d]	Metal-free[e]		
Archaea								
Aeropyrum pernix	90–95 (70–100)	Ae	1.67	4 (1)	1			Kawarabayasi *et al.* (1999)
Aeropyrum camini	85 (70–97)	Ae	1.60	1	1		1	Nakagawa *et al.* (2004), Daifuku *et al.* (2013)
Archaeoglobus fulgidus	83 (60–95)	An	2.18	1	3			Klenk *et al.* (1997)
Archaeoglobus profundus	82 (65–90)	An	1.56		1			von Jan *et al.* (2010)
Caldicellulosiruptor bescii	78–80 (42–90)	An	2.93	4	2		4	Kataeva *et al.* (2009), Yang *et al.* (2010)
Desulfurococcus fermentans	80–82 (63–89)	An	1.38	2	1		1	Perevalova *et al.* (2005), Susanti *et al.* (2012)
Desulfurococcus kamchatkensis	85 (65–87)	An	1.37	2	1		2	Ravin *et al.* (2009)
Desulfurococcus mucosus	85 (76–93)	An	1.31	2	1		1	Wirth *et al.* (2011)
Ferroglobus placidus	85 (65–95)	An	2.2		1		1	Hafenbradl *et al.* (1996), Anderson *et al.* (2011a)
Ignicoccus hospitalis	90 (75–98)	An	1.30	2	1		1	Jahn *et al.* (2008), Podar *et al.* (2008)
Pyrobaculum arsenaticum	90 (68–100)	An	2.12	5	1		3	Huber *et al.* (2000), Ehrlich and Newman (2008)
Pyrobaculum aerophilum	100 (75–104)	Fac Ae	2.22	5 (1)			2	Fitz-Gibbon *et al.* (2002)
Pyrobaculum calidifontis	90–95 (75–100)	Fac Ae	2.01	3			5	Amo *et al.* (2002)
Pyrobaculum islandicum	100 (74–102)	An	1.83	5	1		3	Huber *et al.* (1987)
Pyrobaculum neutrophilum	98	Fac Ae	1.77	3			2	Fitz-Gibbon *et al.* (1997)
Pyrobaculum oguniense	90–94 (70–97)	Fac Ae	2.44	5				Sako *et al.* (2001), Bernick *et al.* (2012)
Pyrococcus abyssi	103 (67–103)	An	1.77		1		1	Cohen *et al.* (2003)
Pyrococcus furiosus	100 (70–103)	An	1.91	2 (1)	2 (1)	2 (1)	5	Robb *et al.* (2001), Bridger *et al.* (2012)
Pyrococcus horikoshii	98 (85–105)	An	1.74	1	1		1	Kawarabayasi *et al.* (1998)
Pyrococcus yayanosii	98 (80–108)	An	1.72	1	1		1	Birrien *et al.* (2011), Jun *et al.* (2011)
Pyrolobus fumarii	106 (90–113)	Fac Ae	1.84	2	1		2	Blöchl *et al.* (1997), Göker *et al.* (2011)
Staphylothermus hellenicus	85 (70–90)	An	1.58	2	1		1	Anderson *et al.* (2011b)
Staphylothermus marinus	85/92	An	1.57	2	1	1	1	Anderson *et al.* (2009)

Table 6.4 continued

Organisms	Growth temperature[a] optimum (range,°C)	Phenotype[b]	Genome size (Mbp)	ADH Zn[c]	ADH Fe[d]	ADH Metal-free[e]	SDR[f]	References
Sulfolobus acidocaldarius DSM 639	80 (75–80)	Ae	2.23	11			6 (2)	Chen *et al*. (2005)
Sulfolobus islandicus HVE 10/4	80	Ae	2.66	17	1		10	Guo *et al*. (2011)
Sulfolobus islandicu L.D.8.5	80	Ae	2.75	17			10	Reno *et al*. (2009)
Sulfolobus islandicu LAL14/1	80	Ae	2.47	8	1		7	Jaubert *et al*. (2013)
Sulfolobus islandicu L.S.2.15	80	Ae	2.74	19			7	Reno *et al*. (2009)
Sulfolobus islandicus M.14.25	80	Ae	2.61	17			6	Reno *et al*. (2009)
Sulfolobus islandicus M.16.4	80	Ae	2.54	19			6	Reno *et al*. (2009)
Sulfolobus islandicus M.16.27	80	Ae	2.69	17	2		7	Reno *et al*. (2009)
Sulfolobus islandicus REY15A	80	Ae	2.52	17	1		10	Guo *et al*. (2011)
Sulfolobus islandicus Y.N.15.51	80	Ae	2.85	20	2		6	Reno *et al*. (2009)
Sulfolobus islandicus Y.G.57.14	80	Ae	2.7	17	1		6	Reno *et al*. (2009)
Sulfolobus solfataricus P2	80	Ae	2.99	13			2	She *et al*. (2001)
Sulfolobus solfataricus 98/2	85	Ae	2.67	16			9	Lucas *et al*. (unpublished)
Sulfolobus tokodaii	80 (75–80)	Ae	2.69	9 (1)	1			Kawarabayasi *et al*. (2001)
Thermococcus barophilus	85 (48–100)	An	2.06	1	1	1		Vannier *et al*. (2011)
Thermococcus gammatolerans	88 (55–95)	An	2.05	1	1		1	Jolivet *et al*. (2003), Zivanovic *et al*. (2009)
Thermococcus litoralis	88 (55–98)	An	2.22		3 (1)		1	Gardner *et al*. (2012)
Thermococcus kodakaraensis	85 (60–100)	An	2.09		2 (1)	2	1	Fukui *et al*. (2005)
Thermococcus onnurineus	80	An	1.85		2	3	2	Lee *et al*. (2008)
Thermococcus sibiricus	78 (40–88)	An	1.85		1	2 (1)	1	Mardanov *et al*. (2009)
Thermofilum pendens	85–90	An	1.81	2	1		2	Anderson *et al*. (2008)
Thermoproteus tenax	86	An	1.84	2			2	Siebers *et al*. (2011)
Thermoproteus uzoniensis	85	An	1.94	8			3	Mardanov *et al*. (2011)

Table 6.4 continued

Organisms	Growth temperature[a] optimum (range, °C)	Phenotype[b]	Genome size (Mbp)	ADH Zn[c]	ADH Fe[d]	ADH Metal-free[e]	SDR[f]	References
Thermosphaera aggregans	85 (65–90)	An	1.32	3	1	2	2	Spring *et al.* (2010)
Bacteria								
Aquifex aeolicus	90	Fac Ae	1.55	1		1		Deckert *et al.* (1998)
Thermocrinis albus	85 (55–98)	Ae	1.5	1	1		3	Wirth *et al.* (2010)
Thermotoga naphthophila	80 (47–88)	An	1.81	2	4			Tanaka *et al.* (2004)
Thermotoga maritima	80 (55–90)	An	1.86	5 (1)	6 (1)		5	Nelson *et al.* (1999)
Thermotoga petrophila	80 (47–88)	An	1.82	3	4		6	Takahata *et al.* (2001), Zhaxybayeva *et al.* (2009)
Thermotoga neapolitana	80 (55–90)	An	1.88	3	3		5	Jannasch *et al.* (1988), Lee *et al.* (2009)

[a]Optimal and range of growth temperatures (°C).
[b]Ae, strictly aerobic; An, strictly anaerobic; Fac Ae, facultative aerobic; Fac An, facultative anaerobic.
[c]Zn, predicted zinc-containing alcohol dehydrogenase.
[d]Fe, predicted iron-containing alcohol dehydrogeanse.
[e]Metal-free, predicted short chain alcohol dehydrogenase without metal.
[f]SDR, predicted short-chain dehydrogenase.

Acknowledgements

This work was supported by research grants from Natural Sciences and Engineering Research Council (NSERC, Canada), Genome Canada/Genome Prairie to KM.

References

Alterthum, F., and Ingram, L.O. (1989). Efficient ethanol production from glucose, lactose, and xylose by recombinant *Escherichia coli*. Appl. Environ. Microbiol. *55*, 1943–1948.

Ammendola, S., Raia, C.A., Caruso, C., Camardella, L., D'Auria, S., De Rosa, M., and Rossi, M. (1992). Thermostable NAD+-dependent alcohol dehydrogenase from *Sulfolobus solfataricus*: gene and protein sequence determination and relationship to other alcohol dehydrogenases. Biochemistry *31*, 12514–12523.

Amo, T., Paje, M.L.F., Inagaki, A., Ezaki, S., Atomi, H., and Imanaka, T. (2002). *Pyrobaculum calidifontis* sp. nov., a novel hyperthermophilic archaeon that grows in atmospheric air. Archaea *1*, 113–121.

Anderson, I., Rodriguez, J., Susanti, D., Porat, I., Reich, C., Ulrich, L.E., Elkins, J.G., Mavromatis, K., Lykidis, A., and Kim, E. (2008). Genome sequence of *Thermofilum pendens* reveals an exceptional loss of biosynthetic pathways without genome reduction. J. Bacteriol. *190*, 2957–2965.

Anderson, I.J., Dharmarajan, L., Rodriguez, J., Hooper, S., Porat, I., Ulrich, L.E., Elkins, J.G., Mavromatis, K., Sun, H., and Land, M. (2009). The complete genome sequence of *Staphylothermus marinus* reveals differences in sulfur metabolism among heterotrophic Crenarchaeota. BMC Genomics *10*, 145.

Anderson, I., Risso, C., Holmes, D., Lucas, S., Copeland, A., Lapidus, A., Cheng, J.-F., Bruce, D., Goodwin, L., and Pitluck, S. (2011a). Complete genome sequence of *Ferroglobus placidus* AEDII12DO. Stand. Genomic Sci. *5*, 50.

Anderson, I., Wirth, R., Lucas, S., Copeland, A., Lapidus, A., Cheng, J.F., Goodwin, L., Pitluck, S., Davenport, K., and Detter, J.C. (2011b). Complete genome sequence of *Staphylothermus hellenicus* P8. Stand. Genomic Sci. *5*, 12–20.

Antoine, E., Rolland, J.L., Raffin, J.P., and Dietrich, J. (1999). Cloning and over-expression in *Escherichia coli* of the gene encoding NADPH group III alcohol dehydrogenase from *Thermococcus hydrothermalis*. Eur. J. Biol. Chem. *264*, 880–889.

Aoyagi, Y., Agata, N., Shibata, N., and Horiguchi, M. (2000). Lipase TL-mediated kinetic resolution of benzoin: facile synthesis of (1*R*,2*S*)-*erythro*-2-amino-1, 2-diphenylethanol. Tetrahedron Lett. *41*, 10159–10162.

Atkinson, A., Cripps, R., Eley, K., Martin, S., Mercier, C., Milner, P., and Rudd, R. (2008). TMO Renewables Ltd. Thermophilic microorganisms for ethanol production, WO/2008/038019A3.

Atomi, H. (2005). Recent progress towards the application of hyperthermophiles and their enzymes. Curr. Opin. Chem. Biol. *9*, 166–173.

Atomi, H., Fukui, T., Kanai, T., Morikawa, M., and Imanaka, T. (2004). Description of *Thermococcus kodakaraensis* sp. nov., a well studied hyperthermophilic archaeon previously reported as *Pyrococcus* sp. KOD1. Archaea *1*, 263–267.

Bahl, H., Andersch, W., Braun, K., and Gottschalk, G. (1982a). Effect of pH and butyrate concentration on the production of acetone and butanol by *Clostridium acetobutylicum* grown in continuous culture. Eur. J. Appl. Microbiol. Biotechnol. *14*, 17–20.

Bahl, H., Andersch, W., and Gottschalk, G. (1982b). Continuous production of acetone and butanol by *Clostridium acetobutylicum* in a two-stage phosphate limited chemostat. Eur. J. Appl. Microbiol. Biotechnol. *15*, 201–205.

Barnard, D., Casanueva, A., Tuffin, M., and Cowan, D. (2010). Extremophiles in biofuel synthesis. Environ. Technol. *31*, 871–888.

Beauchamp, J., Gross, P.G., and Vieille, C. (2014). Characterization of *Thermotoga maritima* glycerol dehydrogenase for the enzymatic production of dihydroxyacetone. Appl. Environ. Microbiol. *98*, 1–12.

Beier, D.R., Sledziewski, A., and Young, E.T. (1985). Deletion analysis identifies a region, upstream of the ADH2 gene of *Saccharomyces cerevisiae,* which is required for ADR1-mediated derepression. Mol. Cell. Biol. *5*, 1743–1749.

Bernick, D.L., Karplus, K., Lui, L.M., Coker, J., Murphy, J.N., Chan, P.P., Cozen, A.E., and Lowe, T.M. (2012). Complete genome sequence of *Pyrobaculum oguniense.* Stand. Genomic Sci. *6*, 336–345.

Berry, D., and Slaughter, J. (2003). Alcoholic beverage fermentations. In Fermented Beverage Production, Lea, A.G.H, and Piggot, J.R., eds. (Springer, New York, USA), pp. 25–39.

Bezsudnova, E.Y., Boyko, K.M., Polyakov, K.M., Dorovatovskiy, P.V., Stekhanova, T.N., Gumerov, V.M., Ravin, N.V., Skryabin, K.G., Kovalchuk, M.V., and Popov, V.O. (2012). Structural insight into the molecular basis of polyextremophilicity of short-chain alcohol dehydrogenase from the hyperthermophilic archaeon *Thermococcus sibiricus*. Biochimie *12*, 2628–2638.

Biotechnology, P.L. (1991). Ion Exchange Chromatography: Principles and Methods, 3rd edition (Pharmacia, Uppsala, Sweden).

Birrien, J.-L., Zeng, X., Jebbar, M., Cambon-Bonavita, M.-A., Quérellou, J., Oger, P., Bienvenu, N., Xiao, X., and Prieur, D. (2011). *Pyrococcus yayanosii* sp. nov., an obligate piezophilic hyperthermophilic archaeon isolated from a deep-sea hydrothermal vent. Int. J. Syst. Evol. Microbiol. *61*, 2827–2881.

Blöchl, E., Rachel, R., Burggraf, S., Hafenbradl, D., Jannasch, H.W., and Stetter, K.O. (1997). *Pyrolobus fumarii*, gen. and sp. nov., represents a novel group of archaea, extending the upper temperature limit for life to 113°C. Extremophiles *1*, 14–21.

Bogin, O., Peretz, M., and Burstein, Y. (1997). *Thermoanaerobacter brockii* alcohol dehydrogenase: characterization of the active site metal and its ligand amino acids. Protein Sci. *6*, 450–458.

Bogin, O., Peretz, M., Hacham, Y., Burstein, Y., Korkhin, Y., and Frolow, F. (1998). Enhanced thermal stability of *Clostridium beijerinckii* alcohol dehydrogenase after strategic substitution of amino acid residues with prolines from the homologous thermophilic *Thermoanaerobacter brockii* alcohol dehydrogenase. Protein Sci. 7, 1156–1163.

Branden, C., Jornvall, H., Eklund, H., Furugren, B., Jornvall, H., Negelein, E., Wulff, H., Sund, H., and Theorell, H. (1975). Alcohol Dehydrogenase (The enzymes XI Academic Press, New York, USA).

Bridger, S.L., Lancaster, W.A., Poole, F.L., Schut, G.J., and Adams, M.W. (2012). Genome sequencing of a genetically tractable *Pyrococcus furiosus* strain reveals a highly dynamic genome. J. Bacteriol. *194*, 4097–4106.

Brinen, L.S., Canaves, J.M., Dai, X., Deacon, A.M., Elsliger, M.A., Eshaghi, S., Floyd, R., Godzik, A., Grittini, C., and Grzechnik, S.K. (2003). Crystal structure of a zinc-containing glycerol dehydrogenase (TM0423) from *Thermotoga maritima* at 1.5 Å resolution. Proteins Struct. Funct. Bioinf. *50*, 371–374.

Bryant, F., and Ljungdahl, L.G. (1981). Characterization of an alcohol dehydrogenase from *Thermoanaerobacter ethanolicus* active with ethanol and secondary alcohols. Biochem. Biophys. Res. Commun. *100*, 793–799.

Bryant, F.O., Wiegel, J., and Ljungdahl, L.G. (1988). Purification and properties of primary and secondary alcohol dehydrogenases from *Thermoanaerobacter ethanolicus*. Appl. Environ. Microbiol. *54*, 460–465.

Bryant, F.O., Wiegel, J., and Ljungdahl, L.G. (1992). Comparison of alcohol dehydrogenases from wild-type and mutant strain, JW200 Fe 4, of *Thermoanaerobacter ethanolicus*. Appl. Microbiol. Biotechnol. *37*, 490–495.

Burdette, D., and Zeikus, J. (1994). Purification of acetaldehyde dehydrogenase and alcohol dehydrogenases from *Thermoanaerobacter ethanolicus* 39E and characterization of the secondary-alcohol dehydrogenase (2 degrees Adh) as a bifunctional alcohol dehydrogenase—acetyl-CoA reductive thioesterase. Biochem. J. *302*, 163–170.

Burdette, D., Vieille, C., and Zeikus, J. (1996). Cloning and expression of the gene encoding the *Thermoanaerobacter ethanolicus* 39E secondary-alcohol dehydrogenase and biochemical characterization of the enzyme. Biochem. J. *316*, 115–122.

Burdette, D., Jung, S.H., Shen, G.J., Hollingsworth, R., and Zeikus, J. (2002). Physiological function of alcohol dehydrogenases and long-chain (C30) fatty acids in alcohol tolerance of *Thermoanaerobacter ethanolicus*. Appl. Environ. Microbiol. *68*, 1914–1918.

Buschke, N., Schäfer, R., Becker, J., and Wittmann, C. (2013). Metabolic engineering of industrial platform microorganisms for biorefinery applications – optimization of substrate spectrum and process robustness by rational and evolutive strategies. Bioresour. Technol. *135*, 544–554.

Canganella, F., Jones, W.J., Gambacorta, A., and Antranikian, G. (1998). *Thermococcus guaymasensis* sp. nov. and *Thermococcus aggregans* sp. nov., two novel thermophilic archaea isolated from the Guaymas Basin hydrothermal vent site. Int. J. Syst. Evol. Microbiol. *48*, 1181–1185.

Cannio, R., Fiorentino, G., Carpinelli, P., Rossi, M., and Bartolucci, S. (1996). Cloning and overexpression in *Escherichia coli* of the genes encoding NAD-dependent alcohol dehydrogenase from two *Sulfolobus* species. J. Bacteriol. *178*, 301–305.

Cannio, R., Fiorentino, G., Rossi, M., and Bartolucci, S. (1999). The alcohol dehydrogenase gene: distribution among *Sulfolobales* and regulation in *Sulfolobus solfataricus*. FEMS Microbiol. Lett. *170*, 31–39.

Ceccarelli, C., Liang, Z.X., Strickler, M., Prehna, G., Goldstein, B.M., Klinman, J.P., and Bahnson, B.J. (2004). Crystal structure and amide H/D exchange of binary complexes of alcohol dehydrogenase from *Bacillus stearothermophilus*: insight into thermostability and cofactor binding. Biochemistry *43*, 5266–5277.

Chen, J.S. (1995). Alcohol dehydrogenase: multiplicity and relatedness in the solvent-producing clostridia. FEMS Microbiol. Rev. *17*, 263–273.

Chen, L., Brügger, K., Skovgaard, M., Redder, P., She, Q., Torarinsson, E., Greve, B., Awayez, M., Zibat, A., and Klenk, H.-P. (2005). The genome of *Sulfolobus acidocaldarius*, a model organism of the Crenarchaeota. J. Bacteriol. *187*, 4992–4999.

Ciriacy, M. (1979). Isolation and characterization of further cis-and trans-acting regulatory elements involved in the synthesis of glucose-repressible alcohol dehydrogenase (ADHII) in *Saccharomyces cerevisiae*. Mol. Gen. Genet. *176*, 427–431.

Clomburg, J.M., and Gonzalez, R. (2010). Biofuel production in *Escherichia coli*: the role of metabolic engineering and synthetic biology. Appl. Microbiol. Biotechnol. *86*, 419–434.

Cohen, G.N., Barbe, V., Flament, D., Galperin, M., Heilig, R., Lecompte, O., Poch, O., Prieur, D., Quérellou, J., and Ripp, R. (2003). An integrated analysis of the genome of the hyperthermophilic archaeon *Pyrococcus abyssi*. Mol. Microbiol. *47*, 1495–1512.

Contursi, P., Cannio, R., Prato, S., Fiorentino, G., Rossi, M., and Bartolucci, S. (2003). Development of a genetic system for hyperthermophilic Archaea: expression of a moderate thermophilic bacterial alcohol dehydrogenase gene in *Sulfolobus solfataricus*. FEMS Microbiol. Lett. *218*, 115–120.

Cripps, R., Eley, K., Leak, D.J., Rudd, B., Taylor, M., Todd, M., Boakes, S., Martin, S., and Atkinson, T. (2009). Metabolic engineering of *Geobacillus thermoglucosidasius* for high yield ethanol production. Metab. Eng. *11*, 398–408.

Cuatrecasas, P. (1970). Protein purification by affinity chromatography derivatizations of agarose and polyacrylamide beads. J. Biol. Chem. *245*, 3059–3065.

Daifuku, T., Yoshida, T., Kitamura, T., Kawaichi, S., Inoue, T., Nomura, K., Yoshida, Y., Kuno, S., and Sako, Y. (2013). Variation of the virus-related elements within syntenic genomes of the hyperthermophilic Archaeon *Aeropyrum*. Appl. Environ. Microbiol. *79*, 5891–5898.

Day, T.H., Hillier, P., and Clarke, B. (1974). Properties of genetically polymorphic isozymes of alcohol dehydrogenase in *Drosophila melanogaster*. Biochem. Genet. *11*, 141–153.

De Smidt, O., Du Preez, J.C., and Albertyn, J. (2008). The alcohol dehydrogenases of *Saccharomyces cerevisiae*: a comprehensive review. FEMS Yeast Res. *8*, 967–978.

Deckert, G., Warren, P.V., Gaasterland, T., Young, W.G., Lenox, A.L., Graham, D.E., Overbeek, R., Snead, M.A., Keller, M., and Aujay, M. (1998). The complete genome of the hyperthermophilic bacterium *Aquifex aeolicus*. Nature *392*, 353–358.

Demain, A.L., Newcomb, M., and Wu, J.D. (2005). Cellulase, *clostridia*, and ethanol. Microbiol. Mol. Biol. Rev. *69*, 124–154.

Dequin, S. (2001). The potential of genetic engineering for improving brewing, wine-making and baking yeasts. Appl. Microbiol. Biotechnol. *56*, 577–588.

Dien, B., Cotta, M., and Jeffries, T. (2003). Bacteria engineered for fuel ethanol production: current status. Appl. Microbiol. Biotechnol. *63*, 258–266.

Drysdale, G., and Fleet, G. (1988). Acetic acid bacteria in winemaking: a review. Am. J. Enol. Viticult 39, 143–154.

Ehrlich, H.L., and Newman, D.K. (2008). Geomicrobiology (CRC Press, Boca Raton, Florida, USA).

Eklund, H., Nordström, B., Zeppezauer, E., Söderlund, G., Ohlsson, I., Boiwe, T., Söderberg, B.-O., Tapia, O., Brändén, C.-I., and Åkeson, Å. (1976). Three-dimensional structure of horse liver alcohol dehydrogenase at 2.4 Å resolution. J. Mol. Biol. *102*, 27–59.

Eram, M.S., and Ma, K. (2013). Decarboxylation of pyruvate to acetaldehyde for ethanol production by hyperthermophiles. Biomolecules 3, 578–596.

Eriksson, K., and Belew, M. (1998). Hydrophobic Interaction Chromatography. Protein Purification: Principles High-Resolution Methods and Applications (John Wiley and Sons, Inc., New York, USA), pp. 283–309.

Esposito, L., Sica, F., Raia, C.A., Giordano, A., Rossi, M., Mazzarella, L., and Zagari, A. (2002). Crystal Structure of the Alcohol Dehydrogenase from the Hyperthermophilic Archaeon *Sulfolobus solfataricus* at 1.85 Å Resolution. J. Mol. Biol. *318*, 463–477.

Evans, I. (1990). Yeast strains for baking: recent developments. In Yeast Technology, Spencer, J.F.T., and Spender, D.M., eds. (Springer-Verlag, Berlin, Germany), pp. 13–53.

Fang, J.M., Lin, C.H., Bradshaw, C.W., and Wong, C.H. (1995). Enzymes in organic synthesis: oxidoreductions. J. Chem. Soc. Perkin Trans. *8*, 967–978.

Fardeau, M.L., Ollivier, B., Patel, B., Magot, M., Thomas, P., Rimbault, A., Rocchiccioli, F., and Garcia, J.-L. (1997). *Thermotoga hypogea* sp. nov., a xylanolytic, thermophilic bacterium from an oil-producing well. Int. J. Syst. Evol. Microbiol. *47*, 1013–1019.

Fiala, G., and Stetter, K.O. (1986). *Pyrococcus furiosus* sp. nov. represents a novel genus of marine heterotrophic archaebacteria growing optimally at 100°C. Arch. Microbiol. *145*, 56–61.

Fitz-Gibbon, S., Choi, A.J., Miller, J.H., Stetter, K.O., Simon, M.I., Swanson, R., and Kim, U.J. (1997). A fosmid-based genomic map and identification of 474 genes of the hyperthermophilic archaeon *Pyrobaculum aerophilum*. Extremophiles *1*, 36–52.

Fitz-Gibbon, S.T., Ladner, H., Kim, U.J., Stetter, K.O., Simon, M.I., and Miller, J.H. (2002). Genome sequence of the hyperthermophilic crenarchaeon *Pyrobaculum aerophilum*. Proc. Natl. Acad. Sci. U.S.A. *99*, 984–989.

Fukui, T., Atomi, H., Kanai, T., Matsumi, R., Fujiwara, S., and Imanaka, T. (2005). Complete genome sequence of the hyperthermophilic archaeon *Thermococcus kodakaraensis* KOD1 and comparison with *Pyrococcus* genomes. Genome Res. *15*, 352–363.

Gardner, A.F., Kumar, S., and Perler, F.B. (2012). Genome sequence of the model hyperthermophilic archaeon *Thermococcus litoralis* NS-C. J. bacteriol. *194*, 2375–2376.

Georgieva, T.I., and Ahring, B.K. (2007). Evaluation of continuous ethanol fermentation of dilute-acid corn stover hydrolysate using thermophilic anaerobic bacterium *Thermoanaerobacter* BG1L1. Appl. Microbiol. Biotechnol. *77*, 61–68.

Georgieva, T.I., Mikkelsen, M.J., and Ahring, B.K. (2008). Ethanol production from wet-exploded wheat straw hydrolysate by thermophilic anaerobic bacterium *Thermoanaerobacter* BG1L1 in a continuous immobilized reactor. In Biotechnology for Fuels and Chemicals, Adney, W.S., McMillan, J.D., Mielenz, J.R., and Klasson, K.T., eds. (Humana Press, New York, USA), pp. 99–110.

Gerday, C., and Glansdorff, N. (2007). Physiology and biochemistry of extremophiles (ASM Press, Washington, USA).

Giordano, A., Cannio, R., La Cara, F., Bartolucci, S., Rossi, M., and Raia, C. (1999). Asn249Tyr substitution at the coenzyme binding domain activates *Sulfolobus solfataricus* alcohol dehydrogenase and increases its thermal stability. Biochemistry *38*, 3043–3054.

Godfroy, A., Lesongeur, F., Raguénès, G., Quérellou, J., Antoine, E., Meunier, J.-R., Guezennec, J., and Barbier, G. (1997). *Thermococcus hydrothermalis* sp. nov., a new hyperthermophilic archaeon isolated from a deep-sea hydrothermal vent. Int. J. Syst. Evol. Microbiol. *47*, 622–626.

Göker, M., Nolan, M., Lucas, S., Hammon, N., Deshpande, S., Cheng, J.F., Tapia, R., Han, C., Goodwin, L., and Pitluck, S. (2011). Complete genome sequence of *Pyrolobus fumarii* type strain (1AT). Stand. Genomic Sci. *4*, 381–392.

Guagliardi, A., Martino, M., Iaccarino, I., Rosa, M.D., Rossi, M., and Bartolucci, S. (1996). Purification and characterization of the alcohol dehydrogenase from a novel strain of *Bacillus stearothermophilus* growing at 70°C. Int. J. Biochem. Cell Biol. *28*, 239–246.

Guo, L., Brügger, K., Liu, C., Shah, S.A., Zheng, H., Zhu, Y., Wang, S., Lillestøl, R.K., Chen, L., and Frank, J. (2011). Genome analyses of Icelandic strains of *Sulfolobus islandicus,* model organisms for genetic and virus–host interaction studies. J. Bacteriol. *193*, 1672–1680.

Guy, J.E., Isupov, M.N., and Littlechild, J.A. (2002). Crystallization and preliminary X-ray diffraction studies of a novel alcohol dehydrogenase from the hyperthermophilic archaeon *Aeropyrum pernix*. Acta Crystallogr. Sect. D Biol. Crystallogr. *59*, 174–176.

Guy, J.E., Isupov, M.N., and Littlechild, J.A. (2003). The structure of an alcohol dehydrogenase from the hyperthermophilic archaeon *Aeropyrum pernix*. J. Mol. Biol. *331*, 1041–1051.

Hafenbradl, D., Keller, M., Dirmeier, R., Rachel, R., Roßnagel, P., Burggraf, S., Huber, H., and Stetter, K.O. (1996). *Ferroglobus placidus* gen. nov., sp. nov., a novel hyperthermophilic archaeum that oxidizes Fe^{2+} at neutral pH under anoxic conditions. Arch. Microbiol. *166*, 308–314.

Hagel, L. (1998). Gel-filtration chromatography. In Current Protocols in Molecular Biology, Ausubel, F.M., Brent, R., Robert, E.K., Kingston, R.E., Moore, D.D., Seidman, J.G., Smith, J.A., and Struhl, K., eds. (John Wiley and Sons Inc., New York, USA) pp. 10.9.1–10.9.32.

Hartley, B., and Shama, G. (1987). Novel ethanol fermentations from sugar cane and straw. Philos. Trans. Roy. Soc. London Ser. A *321*, 555–568.

Heick, H., Willemot, J., and Begin-Heick, N. (1969). The subcellular localization of alcohol dehydrogenase activity in barker's yeast. Biochim. Biophys. Acta Enzymol. *191*, 493–501.

Henne, A., Brüggemann, H., Raasch, C., Wiezer, A., Hartsch, T., Liesegang, H., Johann, A., Lienard, T., Gohl, O., and Martinez-Arias, R. (2004). The genome sequence of the extreme thermophile *Thermus thermophilus*. Nat. Biotechnol. *22*, 547–553.

Herning, T., Yutani, K., Taniyama, Y., and Kikuchi, M. (1991). Effects of proline mutations on the unfolding and refolding of human lysozyme: the slow refolding kinetic phase does not result from proline cis–trans isomerization. Biochemistry *30*, 9882–9891.

Herning, T., Yutani, K., Inaka, K., Kuroki, R., Matsushima, M., and Kikuchi, M. (1992). Role of proline residues in human lysozyme stability: a scanning calorimetric study combined with X-ray structure analysis of proline mutants. Biochemistry *31*, 7077–7085.

Hirakawa, H., Kamiya, N., Kawarabayashi, Y., and Nagamune, T. (2004). Properties of an alcohol dehydrogenase from the hyperthermophilic archaeon *Aeropyrum pernix* K1. J. Biosci. Bioeng. *97*, 202–206.

Höffken, H.W., Duong, M., Friedrich, T., Breuer, M., Hauer, B., Reinhardt, R., Rabus, R., and Heider, J. (2006). Crystal structure and enzyme kinetics of the (S)-specific 1-phenylethanol dehydrogenase of the denitrifying bacterium strain EbN1. Biochemistry *45*, 82–93.

Höllrigl, V., Hollmann, F., Kleeb, A.C., Buehler, K., and Schmid, A. (2008). TADH, the thermostable alcohol dehydrogenase from *Thermus* sp. ATN1: a versatile new biocatalyst for organic synthesis. Appl. Microbiol. Biotechnol. *81*, 263–273.

Holt, P.J., Williams, R.E., Jordan, K.N., Lowe, C.R., and Bruce, N.C. (2000). Cloning, sequencing and expression in Escherichia coli of the primary alcohol dehydrogenase gene from *Thermoanaerobacter ethanolicus* JW200. FEMS Microbiol. Lett. *190*, 57–62.

Huber, R., Langworthy, T.A., König, H., Thomm, M., Woese, C.R., Sleytr, U.B., and Stetter, K.O. (1986).

Thermotoga maritima sp. nov. represents a new genus of unique extremely thermophilic eubacteria growing up to 90°C. Arch. Microbiol. *144*, 324–333.

Huber, R., Kristjansson, J., and Stetter, K. (1987). *Pyrobaculum* gen. nov., a new genus of neutrophilic, rod-shaped archaebacteria from continental solfataras growing optimally at 100°C. Arch. Microbiol. *149*, 95–101.

Huber, R., Sacher, M., Vollmann, A., Huber, H., and Rose, D. (2000). Respiration of arsenate and selenate by hyperthermophilic archaea. Syst. Appl. Microbiol. *23*, 305–314.

Ingram, L., Conway, T., Clark, D., Sewell, G., and Preston, J. (1987). Genetic engineering of ethanol production in *Escherichia coli*. Appl. Environ. Microbiol. *53*, 2420–2425.

Inoue, K., Makino, Y., Dairi, T., and Itoh, N. (2006). Gene cloning and expression of *Leifsonia* alcohol dehydrogenase (LSADH) involved in asymmetric hydrogen-transfer bioreduction to produce (R)-form chiral alcohols. Biosci. Biotechnol. Biochem. *70*, 418–426.

Ismaiel, A.A., Zhu, C., Colby, G., and Chen, J.-S. (1993). Purification and characterization of a primary-secondary alcohol dehydrogenase from two strains of *Clostridium beijerinckii*. J. Bacteriol. *175*, 5097–5105.

Jackson, T.J. (1973). *Thermomicrobium*, a new genus of extremely thermophilic bacteria. Int. J. Syst. Evol. Microbiol. *23*, 28–36.

Jahn, U., Gallenberger, M., Junglas, B., Eisenreich, W., Stetter, K.O., Rachel, R., and Huber, H. (2008). *Nanoarchaeum equitans* and *Ignicoccus hospitalis*: new insights into a unique, intimate association of two archaea. J. Bacteriol. *190*, 1743–1750.

von Jan, M., Lapidus, A., Glavina Del Rio, T., Copeland, A., Tice, H., Cheng, J.-F., Lucas, S., Chen, F., Nolan, M., and Goodwin, L. (2010). Complete genome sequence of *Archaeoglobus profundus* type strain (AV18T). Stand. Genomic Sci. *2*, 327–346.

Jang, Y.S., Park, J.M., Choi, S., Choi, Y.J., Seung, D.Y., Cho, J.H., and Lee, S.Y. (2012). Engineering of microorganisms for the production of biofuels and perspectives based on systems metabolic engineering approaches. Biotechnol. Adv. *30*, 989–1000.

Jannasch, H.W., Huber, R., Belkin, S., and Stetter, K.O. (1988). *Thermotoga neapolitana* sp. nov. of the extremely thermophilic, eubacterial genus *Thermotoga*. Arch. Microbiol. *150*, 103–104.

Jaubert, C., Danioux, C., Oberto, J., Cortez, D., Bize, A., Krupovic, M., She, Q., Forterre, P., Prangishvili, D., and Sezonov, G. (2013). Genomics and genetics of *Sulfolobus islandicus* LAL14/1, a model hyperthermophilic archaeon. Open Biol. 3, 130010.

Joe Shaw, A., Jenney Jr, F.E., Adams, M.W., and Lynd, L.R. (2008). End-product pathways in the xylose fermenting bacterium, *Thermoanaerobacterium saccharolyticum*. Enzyme Microb. Technol. 42, 453–458.

Jolivet, E., L'Haridon, S., Corre, E., Forterre, P., and Prieur, D. (2003). *Thermococcus gammatolerans* sp. nov., a hyperthermophilic archaeon from a deep-sea hydrothermal vent that resists ionizing radiation. Int. J. Syst. Evol. Microbiol. 53, 847–851.

Jörnvall, H. (1970). Horse liver alcohol dehydrogenase. Eur. J. Biochem. *16*, 25–40.

Jörnvall, H., Eklund, H., and Brändén, C. (1978). Subunit conformation of yeast alcohol dehydrogenase. J. Biol. Chem. *253*, 8414–8419.

Jörnvall, H., Persson, B., and Jeffery, J. (1987). Characteristics of alcohol/polyol dehydrogenases. Eur. J. Biochem. *167*, 195–201.

Jun, X., Lupeng, L., Minjuan, X., Oger, P., Fengping, W., Jebbar, M., and Xiang, X. (2011). Complete genome sequence of the obligate piezophilic hyperthermophilic archaeon *Pyrococcus yayanosii* CH1. J. Bacteriol. *193*, 4297–4298.

Kannan, V., and Mutharasan, R. (1985). Ethanol fermentation characteristics of *Thermoanaerobacter ethanolicus*. Enzyme Microb. Technol. *7*, 87–89.

Karplus, P., and Schulz, G. (1985). Prediction of chain flexibility in proteins. Naturwiss. Tech. *72*, 212–213.

Kataeva, I.A., Yang, S.-J., Dam, P., Poole, F.L., Yin, Y., Zhou, F., Chou, W., Xu, Y., Goodwin, L., and Sims, D.R. (2009). Genome sequence of the anaerobic, thermophilic, and cellulolytic bacterium *Anaerocellum thermophilum* DSM 6725. J. Bacteriol. *191*, 3760–3761.

Kato, T., Miyanaga, A., Haruki, M., Imanaka, T., Morikawa, M., and Kanaya, S. (2001). Gene Cloning of an alcohol dehydrogenase from thermophilic alkane-degrading B23. J. Biosci. Bioeng. *91*, 100–102.

Katrlik, J., Švorc, J., Stred'anský, M., and Miertuš, S. (1998). Composite alcohol biosensors based on solid binding matrix. Biosens. Bioelectron. *13*, 181–191.

Kawarabayasi, Y., Sawada, M., Horikawa, H., Haikawa, Y., Hino, Y., Yamamoto, S., Sekine, M., Baba, S.-i., Kosugi, H., and Hosoyama, A. (1998). Complete sequence and gene organization of the genome of a hyper-thermophilic archaebacterium, *Pyrococcus horikoshii* OT3. DNA Res. *5*, 55–76.

Kawarabayasi, Y., Hino, Y., Horikawa, H., Yamazaki, S., Haikawa, Y., Jin-no, K., Takahashi, M., Sekine, M., Baba, S.I., and Ankai, A. (1999). Complete genome sequence of an aerobic hyper-thermophilic crenarchaeon, *Aeropyrum pernix* K1. DNA Res. *6*, 83–101.

Kawarabayasi, Y., Hino, Y., Horikawa, H., Jin-no, K., Takahashi, M., Sekine, M., Baba, S.-i., Ankai, A., Kosugi, H., and Hosoyama, A. (2001). Complete genome sequence of an aerobic thermoacidophilic crenarchaeon, *Sulfolobus tokodaii* strain7. DNA Res. *8*, 123–140.

Keinan, E., Seth, K.K., and Lamed, R. (1987). Synthetic applications of alcohol-dehydrogenase from *Thermoanaerobium brockiia*. Ann. N.Y. Acad. Sci. *501*, 130–149.

Kelly, R.M., and Adams, M.W. (1994). Metabolism in hyperthermophilic microorganisms. A. Van. Leeuw. J. Microb. *66*, 247–270.

Kengen, S., De Bok, F., Van Loo, N., Dijkema, C., Stams, A., and De Vos, W. (1994). Evidence for the operation of a novel Embden-Meyerhof pathway that involves ADP-dependent kinases during sugar fermentation by *Pyrococcus furiosus*. J. Biol. Chem. *269*, 17537–17541.

Klages, K.U., and Morgan, H.W. (1994). Characterization of an extremely thermophilic sulphur-metabolizing archaebacterium belonging to the *Thermococcales*. Arch. Microbiol. *162*, 261–266.

Kleifeld, O., Frenkel, A., Bogin, O., Eisenstein, M., Brumfeld, V., Burstein, Y., and Sagi, I. (2000). Spectroscopic studies of inhibited alcohol dehydrogenase from *Thermoanaerobacter brockii*: proposed structure for the catalytic intermediate state. Biochemistry *39*, 7702–7711.

Klenk, H.P., Clayton, R.A., Tomb, J.F., White, O., Nelson, K.E., Ketchum, K.A., Dodson, R.J., Gwinn, M., Hickey, E.K., and Peterson, J.D. (1997). The complete genome sequence of the hyperthermophilic, sulphate-reducing archaeon *Archaeoglobus fulgidus*. Nature *390*, 364–370.

Korkhin, Y., Peretz, M., Bogin, O., Burstein, Y., and Frolow, F. (1998). NADP-dependent bacterial alcohol dehydrogenases: crystal structure, cofactor-binding and cofactor specificity of the ADHs of *Clostridium beijerinckii* and *Thermoanaerobacter brockii*. J. Mol. Biol. *278*, 967–981.

Kristjánsson, J.K., and Stetter, K.O. (1992). Thermophilic bacteria. In Thermophilic Bacteria, Kristjánsson, J.K., ed (CRC Press, Boca Raton, Florida, USA). pp. 1–20.

Kroutil, W., Mang, H., Edegger, K., and Faber, K. (2004). Recent advances in the biocatalytic reduction of ketones and oxidation of *sec*-alcohols. Curr. Opin. Chem. Biol. *8*, 120–126.

Lamed, R., and Zeikus, J. (1980a). Ethanol production by thermophilic bacteria: relationship between fermentation product yields of and catabolic enzyme activities in *Clostridium thermocellum* and *Thermoanaerobium brockii*. J. Bacteriol. *144*, 569–578.

Lamed, R., and Zeikus, J. (1980b). Glucose fermentation pathway of *Thermoanaerobium brockii*. J. Bacteriol. *141*, 1251–1257.

Lamed, R., Keinan, E., and Zeikus, J. (1981). Potential applications of an alcohol-aldehyde/ketone oxidoreductase from thermophilic bacteria. Enzyme Microb. Technol. 3, 144–148.

Larsen, L., Nielsen, P., and Ahring, B.K. (1997). *Thermoanaerobacter mathranii* sp. nov., an ethanol-producing, extremely thermophilic anaerobic bacterium from a hot spring in Iceland. Arch. Microbiol. *168*, 114–119.

Leca, B., and Marty, J.-L. (1997). Reusable ethanol sensor based on a NAD^+-dependent dehydrogenase without coenzyme addition. Anal. Chim. Acta. *340*, 143–148.

Lee, D., Seo, H., Park, C., and Park, K. (2009). WeGAS: a web-based microbial genome annotation system. Biosci. Biotechnol. Biochem. *73*, 213.

Lee, H.S., Kang, S.G., Bae, S.S., Lim, J.K., Cho, Y., Kim, Y.J., Jeon, J.H., Cha, S.-S., Kwon, K.K., and Kim, H.-T. (2008). The complete genome sequence of *Thermococcus onnurineus* NA1 reveals a mixed heterotrophic and carboxydotrophic metabolism. J. Bacteriol. *190*, 7491–7499.

Lee, Y.-E., Jain, M.K., Lee, C., and Zeikus, J.G. (1993). Taxonomic distinction of saccharolytic thermophilic anaerobes: description of *Thermoanaerobacterium xylanolyticum* gen. nov., sp. nov., and *Thermoanaerobacterium saccharolyticum* gen. nov., sp. nov.; reclassification of *Thermoanaerobium brockii*, *Clostridium thermosulfurogenes*, and *Clostridium thermohydrosulfuricum* E100–69 as *Thermoanaerobacter brockii* comb. nov., *Thermoanaerobacterium thermosulfurigenes* comb. nov., and *Thermoanaerobacter thermohydrosulfuricus* comb. nov., respectively; and transfer of *Clostridium thermohydrosulfuricum* 39E to *Thermoanaerobacter ethanolicus*. Int. J. Syst. Evol. Microbiol. *43*, 41–51.

Leonida, M.D. (2001). Redox enzymes used in chiral syntheses coupled to coenzyme regeneration. Curr. Med. Chem. *8*, 345–369.

Leskovac, V., Trivić, S., and Peričin, D. (2002). The three zinc-containing alcohol dehydrogenases from baker's yeast, *Saccharomyces cerevisiae*. FEMS Yeast Res. *2*, 481–494.

Li, D., and Stevenson, K.J. (1997). Purification and sequence analysis of a novel NADP (H)-dependent type III alcohol dehydrogenase from *Thermococcus* strain AN1. J. Bacteriol. *179*, 4433–4437.

Li, D., and Stevenson, K.J. (2001). [17] Alcohol dehydrogenase from *Thermococcus* strain AN1. Methods Enzymol. *331*, 201–207.

Lieber, C.S. (1997). Ethanol metabolism, cirrhosis and alcoholism. Clin. Chim. Acta. *257*, 59–84.

Lin, Y., and Tanaka, S. (2006). Ethanol fermentation from biomass resources: current state and prospects. Appl. Microbiol. Biotechnol. *69*, 627–642.

Lovitt, R., Shen, G., and Zeikus, J. (1988). Ethanol production by thermophilic bacteria: biochemical basis for ethanol and hydrogen tolerance in *Clostridium thermohydrosulfuricum*. J. Bacteriol. *170*, 2809–2815.

Lütke-Eversloh, T., and Bahl, H. (2011). Metabolic engineering of *Clostridium acetobutylicum*: recent advances to improve butanol production. Curr. Opin. Chem. Biol. *22*, 634–647.

Lutstorf, U., and Megnet, R. (1968). Multiple forms of alcohol dehydrogenase in *Saccharomyces cerevisiae*: I. Physiological control of ADH-2 and properties of ADH-2 and ADH-4. Arch. Biochem. Biophys. *126*, 933–944.

Lyashenko, A., Bezsudnova, E., Gumerov, V., Lashkov, A., Mardanov, A., Mikhailov, A., Polyakov, K., Popov, V., Ravin, N., and Skryabin, K. (2010). Expression, purification and crystallization of a thermostable short-chain alcohol dehydrogenase from the archaeon *Thermococcus sibiricus*. Acta Crystallogr., Sect. F: Struct. Biol. Cryst. Commun. *66*, 655–657.

Ma, K., and Adams, M.W. (1999). An unusual oxygen-sensitive, iron-and zinc-containing alcohol dehydrogenase from the hyperthermophilic archaeon *Pyrococcus furiosus*. J. Bacteriol. *181*, 1163–1170.

Ma, K., and Adams, M.W. (2001). [16] Alcohol dehydrogenases from *Thermococcus litoralis* and *Thermococcus* strain ES-1. Methods Enzymol. *331*, 195–201.

Ma, K., Robb, F.T., and Adams, M. (1994). Purification and characterization of NADP-specific alcohol dehydrogenase and glutamate dehydrogenase from the hyperthermophilic archaeon *Thermococcus litoralis*. Appl. Environ. Microbiol. *60*, 562–568.

Ma, K., Loessner, H., Heider, J., Johnson, M.K., and Adams, M. (1995). Effects of elemental sulfur on the metabolism of the deep-sea hyperthermophilic archaeon *Thermococcus* strain ES-1: characterization of a sulfur-regulated, non-heme iron alcohol dehydrogenase. J. Bacteriol. *177*, 4748–4756.

Ma, K., Hutchins, A., Sung, S.J.S., and Adams, M.W. (1997). Pyruvate ferredoxin oxidoreductase from the hyperthermophilic archaeon, Pyrococcus furiosus, functions as a CoA-dependent pyruvate decarboxylase. Proc. Natl. Acad. Sci. U.S.A. *94*, 9608–9613.

Machielsen, R., Uria, A.R., Kengen, S.W., and van der Oost, J. (2006). Production and characterization of a thermostable alcohol dehydrogenase that belongs to the aldo-keto reductase superfamily. Appl. Environ. Microbiol. 72, 233–238.

Madigan, M.T., Martinko, J.M., Parker, J., and Brock, T.D. (1997). Biology of Microorganisms (Prentice hall, New Jersey, USA).

Mardanov, A.V., Ravin, N.V., Svetlitchnyi, V.A., Beletsky, A.V., Miroshnichenko, M.L., Bonch-Osmolovskaya, E.A., and Skryabin, K.G. (2009). Metabolic versatility and indigenous origin of the archaeon *Thermococcus sibiricus*, isolated from a Siberian oil reservoir, as revealed by genome analysis. Appl. Environ. Microbiol. *75*, 4580–4588.

Mardanov, A.V., Gumerov, V.M., Beletsky, A.V., Prokofeva, M.I., Bonch-Osmolovskaya, E.A., Ravin, N.V., and Skryabin, K.G. (2011). Complete genome sequence of the thermoacidophilic crenarchaeon *Thermoproteus uzoniensis* 768–20. J. Bacteriol. *193*, 3156–3157.

Matsuyama, A., Yamamoto, H., Kawada, N., and Kobayashi, Y. (2001). Industrial production of (*R*)-1, 3-butanediol by new biocatalysts. J. Mol. Catal. B: Enzym. *11*, 513–521.

Matthews, B., Nicholson, H., and Becktel, W. (1987). Enhanced protein thermostability from site-directed mutations that decrease the entropy of unfolding. Proc. Natl. Acad. Sci. U.S.A. *84*, 6663–6667.

Mello, L.D., and Kubota, L.T. (2002). Review of the use of biosensors as analytical tools in the food and drink industries. Food Chem. 77, 237–256.

Mikkelsen, M.J., and Ahring, B.K. (2007). Biogasol. *Thermoanaerobacter mathranii* strain BGl, WO/2007/134607A1.

Miroshnichenko, M.L., Hippe, H., Stackebrandt, E., Kostrikina, N.A., Chernyh, N.A., Jeanthon, C., Nazina, T.N., Belyaev, S.S., and Bonch-Osmolovskaya, E.A. (2001). Isolation and characterization of *Thermococcus sibiricus* sp. nov. from a Western Siberia high-temperature oil reservoir. Extremophiles 5, 85–91.

Miyamoto, K. (1997). Renewable Biological Systems for Alternative Sustainable Energy Production (Food and Agricultue Organization of the United Nations, Rome, Italy).

Mocke, B.A. (2005). The breeding of yeast strains for novel oenological outcomes. Available online: https://scholar.sun.ac.za/handle/10019.1/2544. Accessed 19th Feburary 2014.

Nakagawa, S., Takai, K., Horikoshi, K., and Sako, Y. (2004). *Aeropyrum camini* sp. nov., a strictly aerobic, hyperthermophilic archaeon from a deep-sea hydrothermal vent chimney. Int. J. Syst. Evol. Microbiol. 54, 329–335.

Nazina, T., Tourova, T., Poltaraus, A., Novikova, E., Grigoryan, A., Ivanova, A., Lysenko, A., Petrunyaka, V., Osipov, G., and Belyaev, S. (2001). Taxonomic study of aerobic thermophilic *bacilli*: descriptions of *Geobacillus subterraneus* gen. nov., sp. nov. and *Geobacillus uzenensis* sp. nov. from petroleum reservoirs and transfer of *Bacillus stearothermophilus, Bacillus thermocatenulatus, Bacillus thermoleovorans, Bacillus kaustophilus, Bacillus thermodenitrificans* to *Geobacillus* as the new combinations *G. stearothermophilus, G. th.* Int. J. Syst. Evol. Microbiol. *51*, 433–446.

Neale, A.D., Scopes, R.K., Kelly, J.M., and Wettenhall, R.E. (1986). The two alcohol dehydrogenases of *Zymomonas mobilis*. Eur. J. Biochem. *154*, 119–124.

Negelein, E., and Wulff, H. (1937). Diphosphopyridinproteid ackohol, acetaldehyd. Biochem. Z. *293*, 351–389.

Nelson, K.E., Clayton, R.A., Gill, S.R., Gwinn, M.L., Dodson, R.J., Haft, D.H., Hickey, E.K., Peterson, J.D., Nelson, W.C., and Ketchum, K.A. (1999). Evidence for lateral gene transfer between Archaea and bacteria from genome sequence of *Thermotoga maritima*. Nature *399*, 323–329.

Nestl, B.M., Bodlenner, A., Stuermer, R., Hauer, B., Kroutil, W., and Faber, K. (2007). Biocatalytic racemization of synthetically important functionalized α-hydroxyketones using microbial cells. Tetrahedron: Asymmetry *18*, 1465–1474.

Neuner, A., Jannasch, H.W., Belkin, S., and Stetter, K.O. (1990). *Thermococcus litoralis* sp. nov.: a new species of extremely thermophilic marine archaebacteria. Arch. Microbiol. *153*, 205–207.

Ng, T.K., Ben-Bassat, A., and Zeikus, J. (1981). Ethanol production by thermophilic bacteria: fermentation of cellulosic substrates by cocultures of *Clostridium thermocellum* and *Clostridium thermohydrosulfuricum*. Appl. Environ. Microbiol. *41*, 1337–1343.

Nimcevic, D., and Gapes, J.R. (2000). The acetone-butanol fermentation in pilot plant and pre-industrial sclae. J. Mol. Microbiol. Biotechnol. *2*, 15–20.

Ó'Fágáin, C., Cummins, P.M., and O'Connor, B.F. (2011). Gel-filtration chromatography. In Protein Chromatography, Walls, D., and Loughram, S.T., eds. (Humana Press, New York, USA), pp. 25–33.

van der Oost, J., Voorhorst, W.G., Kengen, S.W., Geerling, A., Wittenhorst, V., Gueguen, Y., and de Vos, W.M. (2001). Genetic and biochemical characterization of a short-chain alcohol dehydrogenase from the hyperthermophilic archaeon *Pyrococcus furiosus*. Eur. J. Biochem. *268*, 3062–3068.

Osunkoya, O., and Okwudinka, N. (2011). Utilization of sugar refinery waste (molasses) for ethanol production-using *Saccharomyces Cervicae*. Am. J. Sci. Ind. Res *2*, 694–706.

Pennacchio, A., Pucci, B., Secundo, F., La Cara, F., Rossi, M., and Raia, C.A. (2008). Purification and characterization of a novel recombinant highly enantioselective short-chain NAD (H)-dependent alcohol dehydrogenase from *Thermus thermophilus*. Appl. Environ. Microbiol. *74*, 3949–3958.

Pennacchio, A., Giordano, A., Pucci, B., Rossi, M., and Raia, C.A. (2010). Biochemical characterization of a recombinant short-chain NAD (H)-dependent dehydrogenase/reductase from *Sulfolobus acidocaldarius*. Extremophiles *14*, 193–204.

Pennacchio, A., Sannino, V., Sorrentino, G., Rossi, M., Raia, C.A., and Esposito, L. (2012). Biochemical and structural characterization of recombinant short-chain NAD (H)-dependent dehydrogenase/reductase from *Sulfolobus acidocaldarius* highly enantioselective on diaryl diketone benzil. Appl. Microbiol. Biotechnol., 1–16.

Peretz, M., and Burstein, Y. (1989). Amino acid sequence of alcohol dehydrogenase from the thermophilic bacterium *Thermoanaerobium brockii*. Biochemistry *28*, 6549–6555.

Peretz, M., Bogin, O., Tel-Or, S., Cohen, A., Li, G., Chen, J.-S., and Burstein, Y. (1997). Molecular cloning, nucleotide sequencing, and expression of genes encoding alcohol dehydrogenases from the thermophile *Thermoanaerobacter brockii* and the mesophile *Clostridium beijerinckii*. Anaerobe *3*, 259–270.

Perevalova, A., Svetlichny, V., Kublanov, I., Chernyh, N., Kostrikina, N., Tourova, T., Kuznetsov, B., and Bonch-Osmolovskaya, E. (2005). *Desulfurococcus fermentans* sp. nov., a novel hyperthermophilic archaeon from a Kamchatka hot spring, and emended description of the genus *Desulfurococcus*. Int. J. Syst. Evol. Microbiol. *55*, 995–999.

Persson, B., Krook, M., and Jornvall, H. (1991). Characteristics of short-chain alcohol dehydrogenases and related enzymes. Eur. J. Biochem. *200*, 537–543.

Plapp, B.V., Green, D.W., Sun, H.W., Park, D.H., and Kim, K. (1993). Substrate specificity of alcohol dehydrogenases. In Enzymology and Molecular Biology of Carbonyl Metabolism 4, Weiner, H., ed. (Plenum Press, New York, USA), pp. 391–400.

Pledger, R.J., and Baross, J.A. (1989). Characterization of an Extremely thermophilic archaebacterium isolated from a black smoker polychaete (*Paralvinella* sp.) at the Juan de Fuca Ridge. Syst. Appl. Microbiol *12*, 249–256.

Podar, M., Anderson, I., Makarova, K.S., Elkins, J.G., Ivanova, N., Wall, M.A., Lykidis, A., Mavromatis, K., Sun, H., and Hudson, M.E. (2008). A genomic analysis of the archaeal system *Ignicoccus hospitalis-Nanoarchaeum equitans*. Genome Biol. *9*, R158.

Porath, J., Carlsson, J., Olsson, I., and Belfrage, G. (1975). Metal chelate affinity chromatography, a new approach to protein fractionation. Nature *258*, 598–599.

Quintela, J.C., Pittenauer, E., Allmaier, G., Aran, V., and De Pedro, M. (1995). Structure of peptidoglycan from *Thermus thermophilus* HB8. J. Bacteriol. *177*, 4947–4962.

Radianingtyas, H., and Wright, P.C. (2003). Alcohol dehydrogenases from thermophilic and hyperthermophilic archaea and bacteria. FEMS Microbiol. Rev. *27*, 593–616.

Raghava, S., and Gupta, M.N. (2010). Purification and characterization of an alcohol dehydrogenase with an unusual specificity towards glycerol from *Thermus thermophilus*. Bioresour. Technol. *101*, 2554–2557.

Raia, C.A., Caruso, C., Marino, M., Vespa, N., and Rossi, M. (1996). Activation of *Sulfolobus solfataricus* alcohol dehydrogenase by modification of cysteine residue 38 with iodoacetic acid. Biochemistry *35*, 638–647.

Raia, C.A., Giordano, A., and Rossi, M. (2001). Alcohol dehydrogenase from *Sulfolobus solfataricus*. Methods Enzymol. *331*, 176–195.

Ravin, N.V., Mardanov, A.V., Beletsky, A.V., Kublanov, I.V., Kolganova, T.V., Lebedinsky, A.V., Chernyh, N.A., Bonch-Osmolovskaya, E.A., and Skryabin, K.G. (2009). Complete genome sequence of the anaerobic, protein-degrading hyperthermophilic crenarchaeon *Desulfurococcus kamchatkensis*. J. Bacteriol. *191*, 2371–2379.

Regnier, F.E. (1984). High-performance ion-exchange chromatography. Methods Enzymol. *104*, 170.

Reid, M.F., and Fewson, C.A. (1994). Molecular characterization of microbial alcohol dehydrogenases. Crit. Rev. Microbiol. *20*, 13–56.

Rella, R., Raia, C.A., Pensa, M., Pisani, F.M., Gambacorta, A., Rosa, M., and Rossi, M. (1987). A novel archaebacterial NAD^+-dependent alcohol dehydrogenase. Eur. J. Biochem. *167*, 475–479.

Reno, M.L., Held, N.L., Fields, C.J., Burke, P.V., and Whitaker, R.J. (2009). Biogeography of the *Sulfolobus islandicus* pan-genome. Proc. Natl. Acad. Sci. U.S.A. *106*, 8605–8610.

Robb, F.T., Maeder, D.L., Brown, J.R., DiRuggiero, J., Stump, M.D., Yeh, R.K., Weiss, R.B., and Dunn, D.M. (2001). Genomic sequence of hyperthermophile, *Pyrococcus furiosus*: implications for physiology and enzymology. Methods Enzymol. *330*, 134.

Rocha-Martín, J., Vega, D., Bolivar, J.M., Hidalgo, A., Berenguer, J., Guisán, J.M., and López-Gallego, F. (2012). Characterization and further stabilization of a new anti-prelog specific alcohol dehydrogenase from *Thermus thermophilus* HB27 for asymmetric reduction of carbonyl compounds. Bioresour. Technol. *103*, 343–350.

Roettger, B.F., and Ladisch, M.R. (1989). Hydrophobic interaction chromatography. Biotechnol. Adv. *7*, 15–29.

Ross, D. (1960). The acetone-butanol fermentation. Prog. Ind. Microbiol. *3*, 71–90.

Sako, Y., Nomura, N., Uchida, A., Ishida, Y., Morii, H., Koga, Y., Hoaki, T., and Maruyama, T. (1996). *Aeropyrum pernix* gen. nov., sp. nov., a novel aerobic hyperthermophilic archaeon growing at temperatures up to 100°C. Int. J. Syst. Evol. Microbiol. *46*, 1070–1077.

Sako, Y., Nunoura, T., and Uchida, A. (2001). *Pyrobaculum oguniense* sp. nov., a novel facultatively aerobic and hyperthermophilic archaeon growing at up to 97 degrees°C. Int. J. Syst. Evol. Microbiol. *51*, 303–309.

Sakoda, H., and Imanaka, T. (1992). Cloning and sequencing of the gene coding for alcohol dehydrogenase of *Bacillus stearothermophilus* and rational shift of the optimum pH. J. Bacteriol. *174*, 1397–1402.

Schlieben, N.H., Niefind, K., Müller, J., Riebel, B., Hummel, W., and Schomburg, D. (2005). Atomic Resolution Structures of *R*-specific Alcohol Dehydrogenase from

Lactobacillus brevis Provide the Structural Bases of its Substrate and Cosubstrate Specificity. J. Mol. Biol. *349*, 801–813.

Schmid, A., Dordick, J., Hauer, B., Kiener, A., Wubbolts, M., and Witholt, B. (2001). Industrial biocatalysis today and tomorrow. Nature *409*, 258–268.

Scopes, R. (1983). An iron-activated alcohol dehydrogenase. FEBS Lett. *156*, 303–306.

Shaw, A.J., Desai, A.G., Lynd, L.R., Tyurin, M.V., Podkaminer, K., Bardsley, J., and Hogsett, D.A. (2007). Mascoma and Dartmouth College. Thermophilic organisms for conversion of lignocellulosic biomass to ethanol, WO/2007/130984.

Shaw, A.J., Podkaminer, K.K., Desai, S.G., Bardsley, J.S., Rogers, S.R., Thorne, P.G., Hogsett, D.A., and Lynd, L.R. (2008). Metabolic engineering of a thermophilic bacterium to produce ethanol at high yield. Proc. Natl. Acad. Sci. U.S.A. *105*, 13769–13774.

She, Q., Singh, R.K., Confalonieri, F., Zivanovic, Y., Allard, G., Awayez, M.J., Christina, C.-Y., Clausen, I.G., Curtis, B.A., and De Moors, A. (2001). The complete genome of the crenarchaeon *Sulfolobus solfataricus* P2. Proc. Natl. Acad. Sci. U.S.A. *98*, 7835–7840.

Sheehan, M., Bailey, C., Dowds, B., and McConnell, D. (1988). A new alcohol dehydrogenase, reactive towards methanol, from *Bacillus stearothermophilus*. Biochem. J. *252*, 661–666.

Siebers, B., Zaparty, M., Raddatz, G., Tjaden, B., Albers, S.-V., Bell, S.D., Blombach, F., Kletzin, A., Kyrpides, N., and Lanz, C. (2011). The complete genome sequence of *Thermoproteus tenax*: a physiologically versatile member of the Crenarchaeota. PloS One *6*, e24222.

Smidt, O., du Preez, J.C., and Albertyn, J. (2012). Molecular and physiological aspects of alcohol dehydrogenases in the ethanol metabolism of *Saccharomyces cerevisiae*. FEMS Yeast Res. *12*, 33–47.

Smith, L.T., and Kaplan, N.O. (1980). Purification, properties, and kinetic mechanism of coenzyme A-linked aldehyde dehydrogenase from *Clostridium kluyveri*. Arch. Biochem. Biophys. *203*, 663–675.

Sofer, G., and Hagel, L. (1997). Handbook of Process Chromatography: A Guide to Optimization, Scale Up, and Validation (Academic Press, San Diego, USA).

Sommer, P., Georgieva, T., and Ahring, B.K. (2004). Potential for using thermophilic anaerobic bacteria for bioethanol production from hemicellulose. Biochem. Soc. Trans. *32*, 283–289.

Spring, S., Rachel, R., Lapidus, A., Davenport, K., Tice, H., Copeland, A., Cheng, J.F., Lucas, S., Chen, F., and Nolan, M. (2010). Complete genome sequence of *Thermosphaera aggregans* type strain (M11TL T). Stand. Genomic Sci. *2*, 245–259.

Stanton, P. (2004). Gel filtration chromatography. In HPLC of Peptides and Proteins: Methods and Protocol, M.I. Aguilar (Humana Press Inc., New Jersey, USA), pp. 55–73.

Stekhanova, T.N., Mardanov, A.V., Bezsudnova, E.Y., Gumerov, V.M., Ravin, N.V., Skryabin, K.G., and Popov, V.O. (2010). Characterization of a thermostable short-chain alcohol dehydrogenase from the hyperthermophilic archaeon *Thermococcus sibiricus*. Appl. Environ. Microbiol. *76*, 4096–4098.

Stetter, K.O. (1996). Hyperthermophilic procaryotes. FEMS Microbiol. Rev. *18*, 149–158.

Stetter, K.O. (2013). A brief history of the discovery of hyperthermophilic life. Biochem. Soc. Trans. *41*, 416–420.

Susanti, D., Johnson, E.F., Rodriguez, J.R., Anderson, I., Perevalova, A.A., Kyrpides, N., Lucas, S., Han, J., Lapidus, A., and Cheng, J.-F. (2012). Complete genome sequence of *Desulfurococcus fermentans*, a hyperthermophilic cellulolytic crenarchaeon isolated from a freshwater hot spring in Kamchatka, Russia. J. Bacteriol. *194*, 5703–5704.

Suzuki, T., Iwasaki, T., Uzawa, T., Hara, K., Nemoto, N., Kon, T., Ueki, T., Yamagishi, A., and Oshima, T. (2002). *Sulfolobus tokodaii* sp. nov.(f. *Sulfolobus* sp. strain 7), a new member of the genus *Sulfolobus* isolated from Beppu Hot Springs, Japan. Extremophiles *6*, 39–44.

Svetlitchnyi, V.A., Kensch, O., Falkenhan, D.A., Korseska, S.G., Lippert, N., Prinz, M., Sassi, J., Schickor, A., and Curvers, S. (2013). Single-step ethanol production from lignocellulose using novel extremely thermophilic bacteria. Biotechnol. Biofuels *6*, 31–46.

Takahata, Y., Nishijima, M., Hoaki, T., and Maruyama, T. (2001). *Thermotoga petrophila* sp. nov. and *Thermotoga naphthophila* sp. nov., two hyperthermophilic bacteria from the Kubiki oil reservoir in Niigata, Japan. Int. J. Syst. Evol. Microbiol. *51*, 1901–1909.

Tanaka, T., Kawase, M., and Tani, S. (2004). α-Hydroxyketones as inhibitors of urease. Bioorg. Med. Chem. *12*, 501–505.

Taylor, M.P., Eley, K.L., Martin, S., Tuffin, M.I., Burton, S.G., and Cowan, D.A. (2009). *Thermophilic ethanologenesis*: future prospects for second-generation bioethanol production. Trends Biotechnol. *27*, 398–405.

Thatcher, D.R. (1980). The complete amino acid sequence of three alcohol dehydrogenase alleloenzymes (AdhN-11, AdhS and AdhUF) from the fruitfly *Drosophila melanogaster*. Biochem. J. *187*, 875–883.

Tobalina, F., Pariente, F., Hernandez, L., Abruna, H., and Lorenzo, E. (1999). Integrated ethanol biosensors based on carbon paste electrodes modified with [Re (phen-dione)$(CO)_3$Cl] and [Fe (phen-dione)$_3$] $(PF_6)_2$. Anal. Chim. Acta *395*, 17–26.

Tomás, A.F., Karagöz, P., Karakashev, D., and Angelidaki, I. (2013). Extreme thermophilic ethanol production from rapeseed straw: Using the newly isolated *Thermoanaerobacter pentosaceus* and combining it with *Saccharomyces cerevisiae* in a two-step process. Biotechnol. Bioeng. *110*, 1574–1582.

Völkl, P., Huber, R., Drobner, E., Rachel, R., Burggraf, S., Trincone, A., and Stetter, K.O. (1993). *Pyrobaculum aerophilum* sp. nov., a novel nitrate-reducing hyperthermophilic archaeum. Appl. Environ. Microbiol. *59*, 2918–2926.

Vannier, P., Marteinsson, V.T., Fridjonsson, O.H., Oger, P., and Jebbar, M. (2011). Complete genome sequence of the hyperthermophilic, piezophilic, heterotrophic, and carboxydotrophic archaeon *Thermococcus barophilus* MP. J. Bacteriol. *193*, 1481–1482.

Vieille, C., and Zeikus, G.J. (2001). Hyperthermophilic enzymes: sources, uses, and molecular mechanisms for thermostability. Microbiol. Mol. Bio. Rev. *65*, 1–43.

Vitale, A., Rosso, F., Barbarisi, A., Labella, T., and D'Auria, S. (2010). Properties and evolution of an alcohol dehydrogenase from the Crenarchaeota *Pyrobaculum aerophilum*. Gene *461*, 26–31.

Vitale, A., Thorne, N., Lovell, S., Battaile, K.P., Hu, X., Shen, M., D'Auria, S., and Auld, D.S. (2013). Physicochemical Characterization of a Thermostable Alcohol Dehydrogenase from *Pyrobaculum aerophilum*. PloS One *8*, e63828.

Wagner, I.D., and Wiegel, J. (2008). Diversity of thermophilic anaerobes. Ann. N.Y. Acad. Sci. *1125*, 1–43.

Watanabe, K., Masuda, T., Ohashi, H., Mihara, H., and Suzuki, Y. (1994). Multiple proline substitutions cumulatively thermostabilize *Bacillus Cereus* ATCC7064 oligo-1, 6-glucosidase. Eur. J. Biochem. *226*, 277–283.

Wiegel, J., and Ljungdahl, L.G. (1981). *Thermoanaerobacter ethanolicus* gen. nov., spec. nov., a new, extreme thermophilic, anaerobic bacterium. Arch. Microbiol. *128*, 343–348.

Willey, J., Sherwood, L.K., and Prescott, L.M. (2008). Prescott, Harley, and Klein's Microbiology (McGraw-Hill Higher Education, New York, USA).

Wills, C., and Phelps, J. (1975). A technique for the isolation of yeast alcohol dehydrogenase mutants with altered substrate specificity. Arch. Biochem. Biophys. *167*, 627–637.

Wirth, R., Sikorski, J., Brambilla, E., Misra, M., Lapidus, A., Copeland, A., Nolan, M., Lucas, S., Chen, F., and Tice, H. (2010). Complete genome sequence of *Thermocrinis albus* type strain (HI 11/12T). Stand. Genomic Sci. 2, 194.

Wirth, R., Chertkov, O., Held, B., Lapidus, A., Nolan, M., Lucas, S., Hammon, N., Deshpande, S., Cheng, J.F., and Tapia, R. (2011). Complete genome sequence of *Desulfurococcus mucosus* type strain (O7/1T). Stand. Genomic Sci. *4*, 173.

Woese, C.R., Kandler, O., and Wheelis, M.L. (1990). Towards a natural system of organisms: proposal for the domains Archaea, Bacteria, and Eucarya. Proc. Natl. Acad. Sci. U.S.A. *87*, 4576–4579.

Wood, B.E., and Ingram, L. (1992). Ethanol production from cellobiose, amorphous cellulose, and crystalline cellulose by recombinant *Klebsiella oxytoca* containing chromosomally integrated *Zymomonas mobilis* genes for ethanol production and plasmids expressing thermostable cellulase genes from *Clostridium thermocellum*. Appl. Environ. Microbiol. *58*, 2103–2110.

Wu, M., Ren, Q., Durkin, A.S., Daugherty, S.C., Brinkac, L.M., Dodson, R.J., Madupu, R., Sullivan, S.A., Kolonay, J.F., and Nelson, W.C. (2005). Life in hot carbon monoxide: the complete genome sequence of *Carboxydothermus hydrogenoformans* Z-2901. PLoS Genet. *1*, e65.

Wu, X., Zhang, C., Orita, I., Imanaka, T., Fukui, T., and Xing, X.H. (2013). Thermostable alcohol dehydrogenase from *Thermococcus kodakarensis* KOD1 for enantioselective bioconversion of aromatic secondary alcohols. Appl. Environ. Microbiol. *79*, 2209–2217.

Wyman, C. (1996). Handbook on Bioethanol: Production and Utilization (CRC Press, Boca Raton, Florida, USA).

Yamamoto, S., Nakanishi, K., and Matsuno, R. (1988). Ion-exchange Chromatography of Proteins (CRC Press, Boca Raton, Florida, USA).

Yanai, H., Doi, K., and Ohshima, T. (2009). *Sulfolobus tokodaii* ST0053 produces a novel thermostable, NAD-dependent medium-chain alcohol dehydrogenase. Appl. Environ. Microbiol. *75*, 1758–1763.

Yang, S.J., Kataeva, I., Wiegel, J., Yin, Y., Dam, P., Xu, Y., Westpheling, J., and Adams, M.W. (2010). Classification of *Anaerocellum thermophilum* strain DSM 6725 as *Caldicellulosiruptor bescii* sp. nov. Int. J. Syst. Evol. Microbiol. *60*, 2011–2015.

Yao, N., Ren, Y., and Wang, W. (2013). Genome sequence of a thermophilic *bacillus, Geobacillus thermodenitrificans* DSM465. Genome announcements *1*, e01046-01013.

Yao, S., and Mikkelsen, M.J. (2010). Identification and overexpression of a bifunctional aldehyde/alcohol dehydrogenase responsible for ethanol production in *Thermoanaerobacter mathranii*. J. Mol. Microbiol. Biotechnol. *19*, 123–133.

Ying, X., Wang, Y., Badiei, H.R., Karanassios, V., and Ma, K. (2007). Purification and characterization of an iron-containing alcohol dehydrogenase in extremely thermophilic bacterium *Thermotoga hypogea*. Arch. Microbiol. *187*, 499–510.

Ying, X., Grunden, A.M., Nie, L., Adams, M.W., and Ma, K. (2009). Molecular characterization of the recombinant iron-containing alcohol dehydrogenase from the hyperthermophilic Archaeon, *Thermococcus strain ES1*. Extremophiles *13*, 299–311.

Ying, X., and Ma, K. (2011). Characterization of a zinc-containing alcohol dehydrogenase with stereoselectivity from the hyperthermophilic archaeon *Thermococcus guaymasensis*. J. Bacteriol. *193*, 3009–3019.

Yoon, S.Y., Noh, H.S., Kim, E.H., and Kong, K.H. (2002). The highly stable alcohol dehydrogenase of *Thermomicrobium roseum*: purification and molecular characterization. Comp. Biochem. Physiol. B: Biochem. Mol. Biol. *132*, 415–422.

Young, E.T., and Pilgrim, D. (1985). Isolation and DNA sequence of ADH3, a nuclear gene encoding the mitochondrial isozyme of alcohol dehydrogenase in *Saccharomyces cerevisiae*. Mol. Cell. Biol. *5*, 3024–3034.

Zaldivar, J., Nielsen, J., and Olsson, L. (2001). Fuel ethanol production from lignocellulose: a challenge for metabolic engineering and process integration. Appl. Microbiol. Biotechnol. *56*, 17–34.

Zeikus, J. (1980). Chemical and fuel production by anaerobic bacteria. Annu. Rev. Microbiol. *34*, 423–464.

Zhang, M., Eddy, C., Deanda, K., Finkelstein, M., and Picataggio, S. (1995). Metabolic engineering of a pentose metabolism pathway in ethanologenic *Zymomonas mobilis*. Science *267*, 240–243.

Zhang, Z., Djebli, A., Shoham, M., Frolow, F., Peretz, M., and Burstein, Y. (1993). Crystal Parameters of an Alcohol Dehydrogenase from the Extreme Thermophile *Thermoanaerobium brockii*. J. Mol. Biol. *230*, 353–355.

Zhaxybayeva, O., Swithers, K.S., Lapierre, P., Fournier, G.P., Bickhart, D.M., DeBoy, R.T., Nelson, K.E., Nesbø, C.L., Doolittle, W.F., and Gogarten, J.P. (2009). On the chimeric nature, thermophilic origin, and phylogenetic placement of the *Thermotogales*. Proc. Natl. Acad. Sci. U.S.A. *106*, 5865–5870.

Zhou, S., Zhang, S.C., Lai, D.Y., Zhang, S.L., and Chen, Z.M. (2013). Biocatalytic characterization of a short-chain alcohol dehydrogenase with broad substrate specificity from thermophilic *Carboxydothermus hydrogenoformans*. Biotechnol. Lett. 35, 359–365.

Zimmermann, F.K., and Entian, K.D. (1997). Yeast Sugar Metabolism (CRC Press, Boca Raton, Florida, USA).

Zivanovic, Y., Armengaud, J., Lagorce, A., Leplat, C., Guérin, P., Dutertre, M., Anthouard, V., Forterre, P., Wincker, P., and Confalonieri, F. (2009). Genome analysis and genome-wide proteomics of Thermococcus gammatolerans, the most radioresistant organism known amongst the Archaea. Genome Biol. *10*, R70.

Roles of Polyamines in Thermophiles

7

Tairo Oshima

Abstract

Extreme thermophiles often produce unusual polyamines such as longer polyamines (for instance caldopentamine) and branched polyamines (for instance, tetrakis(3-aminopropyl)ammonium). These polyamines are essential for life at higher temperature and the highest growth temperatures of mutants which lack a gene involved in unusual polyamine synthesis, are lower than those of wild type cells. Unusual polyamines protect DNA and RNA from thermal damages more efficiently than the standard polyamines. Enzymes and metabolic pathways of unusual polyamine biosynthesis are unique and new pathways and new enzymes have been clarified.

Polyamines

Polyamines are essential components of living cells (Cohen, 1998; Tabor and Tabor, 1984, 1985). Polyamines are non-cyclic (that is, nucleic acid bases are excluded), aliphatic compounds containing two or more amino nitrogen atoms. Though linguistically 'poly' means three or more, in the academic field of polyamine research, diamines such as putrescine, cadaverine and 1,3-diaminopropane are included in polyamines. Also agmatine is often included in polyamines.

Most of organisms, higher animals including mammals, plants, yeast, fungi and most of bacteria, produce three 'standard' polyamines, putrescine, spermidine and spermine. *Escherichia coli* and many other bacteria are able to produces only putrescine and spermidine (Tabor and Tabor, 1985). However, often rich mediums such as LB broth contain a considerable amount of spermine and the standard polyamines are generally incorporated into living cells, thus *E. coli* grown in a rich medium contains spermine. Some of bacteria belonging into alpha-proteobacteria such as *Caulobacter* species produce sym-homospermidine as the major polyamine component of the cell (Hamana *et al.*, 2001). Often sym-homospermidine is a starting material for the secondary metabolism of plants and many plants produce this polyamine (Ober and Hartmann, 1999; Nurhayati *et al.*, 2009; Niemuller *et al.*, 2012). Thermospermine (Oshima, 1979), a structural isomer of spermine, which was discovered for the first time in cells of an extremely thermophilic bacterium, *Thermus thermophilus,* is an important growth factor for many plants and plants can produce both spermine and thermospermine (Knott at al., 2007; Kakehi *et al.*, 2008; Muniz *et al.*, 2008; Imai *et al.*, 2004; Yamaguchi *et al.*, 2007). In plants, roles of these structurally related compounds are different from each other; spermine protects plants from environmental stresses, but cannot stimulate the growth.

Since polyamines are strongly basic compounds, they interact with acidic compounds and acidic residues of biopolymers such as DNA, RNA and proteins. Likewise polyamines bind with acidic parts of membrane lipids, especially phospholipids. Thus polyamines induce structural changes of biopolymers, stabilize them, and activate biopolymer syntheses such as DNA replication, RNA synthesis and protein synthesis. Some enzymatic reactions such as Staphylococal nuclease, require the presence of polyamines for the activity. In short, polyamines are important for many biological phenomena, especially those related to cell multiplication, differentiation, and protection against environmental stresses (Chattopadhyay and Tabor, 2013).

Thermophiles and unusual polyamines

Thermophiles which are capable of growing at 75°C or higher temperatures, are classified into 'extreme thermophiles'. Among extreme thermophiles, those which are capable of growing at 80°C or higher, are called as hyperthermophiles. Extreme thermophiles and hyperthermophiles, including both bacterial and archaeal species, generally produce unusual polyamines in addition to the standard polyamines (Oshima, 1989, 2007, 1983; Kneifel *et al.*, 1986; Hamana *et al.*, 1996; Hosoya *et al.*, 2004). The unusual polyamines produced by extreme thermophiles and hyperthermophiles can be categorized in two groups; one is long polyamines such as caldopentamine and caldohexamine, and the other is branched polyamines such as N^4-aminopropylspermine and tetrakis(3-aminopropyl)ammonium.

Typical example is cellular polyamine composition of *T. thermophilus*. The extreme thermophile produces more than 18 different polyamines including long polyamines such as caldopentamine and branched polyamines such as tetrakis(3-aminopropyl)ammonium. Chemical structures and systematic names of polyamines found in the extreme thermophile are listed in Table 7.1. Some polyamines listed in the table are detected in cells of mutant strains which lack one of genes coding for an enzyme involved in polyamine metabolism. *Thermus* and taxonomically related thermophilic species such as *Meiothermus* produce both long and branched polyamines. Some extreme thermophiles do not produce long polyamines and branched polyamines are major components of the cellular polyamines. Examples are *Aquifex, Hydrogenothermus* (bacteria), *Pyrococcus, Archaeoglobus, Thermococcus* and *Thermodesulfobacterium* (archaea) species. In contrast, some extreme thermophiles cannot produce

Table 7.1 Systematic names, trivial names and chemical structures of polyamines found in cells of *T. thermophilus*

Systematic name	Trivial name	Chemical structure
1,3-Diaminopropane		$NH_2(CH_2)_3NH_2$
1,4-Diaminobutane	Putrescine	$NH_2(CH_2)_4NH_2$
1,5-Diaminopentane	Cadaverine	$NH_2(CH_2)_5NH_2$
N^1-Amino-4-guanidobutane	Agmatine	$(NH_2)(NH)CNH(CH_2)_4NH_2$
1,7-Diamino-4-azaheptane	Norspermidine, caldine	$NH_2(CH_2)_3NH(CH_2)_3NH_2$
1,8-Diamino-4-azaoctane	Spermidine	$NH_2(CH_2)_3NH(CH_2)_4NH_2$
1,9-Diamino-5-azanonane	*sym*-Homospermidine	$NH_2(CH_2)_4NH(CH_2)_4NH_2$
N^8-Guanido-1,8-diamino-4-azaoctane	Aminopropylagmatine	$(NH_2)(NH)CNH(CH_2)_4NH(CH_2)_3NH_2$
N^9-Guanido-1.9-diamino-5-azanonae	Aminobutylagmatine	$(NH_2)(NH)CNH(CH_2)_4NH(CH_2)_4NH_2$
1,11-Diamino-4,8-diazaundecane	Thermine	$NH_2(CH_2)_3NH(CH_2)_3NH(CH_2)_3NH_2$
1,12-Diamino-4,9-diazadodecane	Spermine	$NH_2(CH_2)_3NH(CH_2)_4NH(CH_2)_3NH_2$
1,12-Diamino-4,8-diazadodecane	Thermospermine	$NH_2(CH_2)_3NH(CH_2)_3NH(CH_2)_4NH_2$
1,13-Diamino-4,9-diazatridecane	Homospermine	$NH_2(CH_2)_3NH(CH_2)_4NH(CH_2)_4NH_2$
1,15-Diamino-4,8,12-triazapentadecane	Caldopentamine	$NH_2(CH_2)_3NH(CH_2)_3NH(CH_2)_3NH(CH_2)_3NH_2$
1,16-Diamino-4,8,12-triazahexadecane	Homocaldopentamine	$NH_2(CH_2)_3NH(CH_2)_3NH(CH_2)_3NH(CH_2)_4NH_2$
1,19-Diamino-4,8,12,16-tetraazanonadecane	Caldohexamine	$NH_2(CH_2)_3NH(CH_2)_3NH(CH_2)_3NH(CH_2)_3NH(CH_2)_3NH_2$
Tris(3-aminopropyl)amine	Mitsubishine	$[NH_2(CH_2)_3]_3N$
Tetrakis(3-aminopropyl)ammonium		$[NH_2(CH_2)_3]_4N^+$

branched polyamines. Examples are *Thermotoga* (bacteria), *Pyrodictium* and *Aeropyrum* species.

Cellular concentrations of unusual polyamines are positively correlated with the growth temperature. In *T. thermophilus*, long and branched polyamines are abundant when the cells are gown at 75°C or higher (Oshima, 1989). Thermine which is one of tetraamines is also abundant in the cells grown at higher temperatures. At lower temperatures, spermine and spermidine are major polyamines. The major polyamine of *Calditerricola satsumensis*, an extreme thermophile isolated from compost soil, is a branched polyamine, N^4-aminopropylspermine, if grown at 75°C, but in cells grown at 60°C, only small amount of the branched polyamine was found, in which spermidine and spermine were the major components of cellular polyamines (Moriya *et al.*, 2011). Likewise, major component of cellular polyamines in *Thermococcus kodakarensis*, a hyperthermophilic archaeon, is a branched polyamine, N^4-bis(aminopropyl)spermidine, but in cells grown at lower temperatures, cellular concentration of the branched polyamine is very low (Morimoto *et al.*, 2011). Long and/or branched polyamines have been never found in moderate thermophiles which cannot grow at 75°C or higher. These findings strongly imply that the unusual polyamines are essential for life at higher temperature extremes.

Unusual polyamines found in extreme thermophiles, are rarely detected in mesophilic animals and plants. For instance, thermine (= norspermine) is produced by shrimps (Oshima, 1983; Zappia *et al.*, 1978; Stillway and walle, 1977) and insects (Oshima, 1983). Branched polyamines are found in plants (Hamana *et al.*, 2000).

Polyamine analysis using HPLC

Polyamines can be analysed by using TLC, GC and HPLC. Though GC and GC-MS were used in some experiments (Hamana *et al.*, 2011), two types of HPLC are generally employed in routine assays: one is HPLC using a cation exchange resin column (Oshima, 1983; Oshima *et al.*, 2011) and the other uses reverse-phase column chromatography equipped with a C-18 column, following benzoylation of polyamines (Redmond and Teseng, 1979; Flores and Galston, 1982; Slocum *et al.*, 1989). HPLC analysis using a cation exchange resin column is easier, simpler and more sensitive. No derivatization of polyamines is required and TCA extract can be applied on the chromatographic column directly after appropriate dilution. The analytical equipment for the cation exchange HPLC can be constructed using commercially available replacement parts sold for automated amino acid analysers, such as high performance pump, automated sample injector, fluorescence detector, column oven and pre-packed column. In the author's laboratory, a polyamine analyser was constructed by combining a short cation exchange resin column, two HPLC pumps, fluorescence detector, and recorder (and/or a data logger). The schematic diagram and details of analytical procedures are described in the literature (Oshima *et al.*, 2011). For ion exchange resin, any ion exchange resin prepared for an amino acid analyser can be used, one example is CK-10S made by Mitsubish Chemical Co., of which particle size is 11 μm diameter.

The column is washed with 2 N KOH prior to use. Samples (0.1 ml) containing more than 5–100 pmol of each polyamine are applied to the column every 20–60 minutes continuously. The column is developed at a rate of 2 ml/min with a citrate-HCl buffer containing 1.68 – 2.0 M KCl, 5% isopropanol, and 0.3% Brij. The column is heated to 55°C in a column oven. The polyamines eluted from the column are detected using a fluorescence detector after mixed with a borate buffer containing *o*-phthalaldehyde. Under the standard conditions (30-minute run per sample), 30–40 samples can be analysed automatically overnight. One example of cellular polyamine composition of *T. thermophilus* grown at 78°C is illustrated in Fig. 7.1.

Only serious problem of the polyamine analysis using the cation exchange resin column chromatography is separation between spermine and thermospermine. These polyamines are structural isomers to each other, and never separated on the chromatography. If the separation is important, that is, if it is necessary to measure these two polyamines separately, only way is to use C_{18} column chromatography. On a C_{18} column, benzoylated spermine and benzoylated thermospermine are separated clearly. Generally

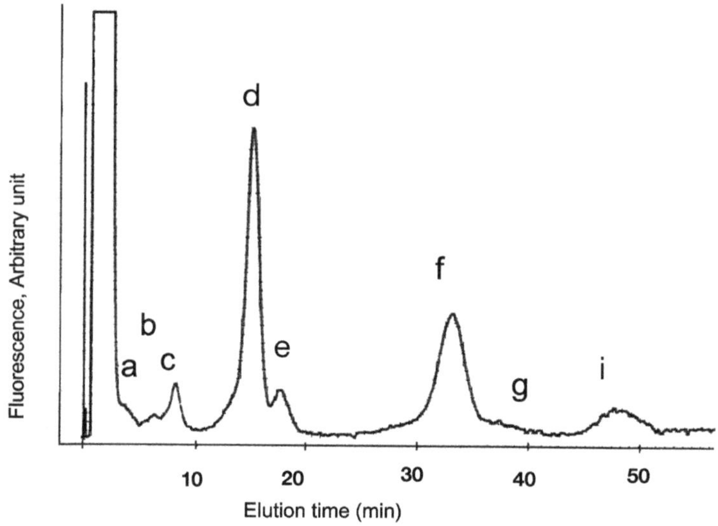

Figure 7.1 Polyamine composition of *T. thermophilus* cells grown at 78°C in a synthetic medium. (a) putrescine, (b) caldine (=nor-spermidine), (c) spermidine, (d) thermine, (e) spermine+thermospermine, (f) caldopentamine, (g) homocaldopentamine, (h) tetrakis (3-aminopropyl) ammonium.

speaking, separation is much better in the case of the reverse-phase HPLC than the ion-exchange HPLC.

The elution time of a polyamine is correlated to number of nitrogen atoms in the molecule when straight chain polyamines are compared (see Fig. 7.1). Thus tetraamines such as thermine (=nor-spermine), spermine and thermospermine are eluted after triamines such as spermidine, caldine (=norspermidine), and sym-homospermidine were eluted, and before caldopentamine and homocaldopentamine are eluted. If number of nitrogen atoms in a molecule is same, the elution time is correlated with number of carbon atoms; thus spermidine is eluted after caldine was eluted, and before sym-homospermidine is eluted. Likewise spermine is eluted after thermine (=norspermine) is eluted. If a polyamine contains tertiary amino nitrogen atoms, the elution time for the polyamine is longer. For example, retention time of mitsubishine (contains a tertiary amino nitrogen atom) is later than that of thermine though the chemical formula of these two polyamines are same, $C_9N_4H_{24}$. Likewise the retention time of tetrakis(3-aminopropyl)ammonium is larger than that of caldopentamine though numbers of carbon and nitrogen atoms of these two polyamines are same, C_{12} and N_5.

Metabolic pathways of polyamines in *T. thermophilus*

In higher animals, polyamines are synthesized from ornithine, a non-protein amino acid (Fig. 7.2). The first step is decarboxylation of ornithine to putrescine and this step is catalysed by ornithine decarboxylase. The enzyme is usually abbreviated as ODC, and this enzyme is precisely regulated by ODC-antizyme and ODC-antizyme inhibitor (Hayashi *et al.*, 1996; Coffino, 2001; Fujita *et al.*, 1982). Then putrescine is converted to spermidine and spermine by the actions of spermidine synthase and spermine synthase, respectively. These enzymes are aminopropyltransferases and catalyse aminopropylation using decarboxylated S-adenosylmethionine (SAM) as an aminpropyl donor. Decarboxylated S-adenosylmethionine is produced by an enzyme 'SAM decarboxylase' and again this enzyme is highly regulated to control metabolism of polyamines. In a sense, polyamine metabolism in higher animals is a one-way reaction from ornithine to putrescine to spermidine to spermine, and this one-way metabolism is finely tuned in order to maintain proper concentrations of cellular polyamines.

In contrast, polyamines in plants and prokaryotes are derived from ornithine and/or arginine. Putrescine can be synthesized from ornithine

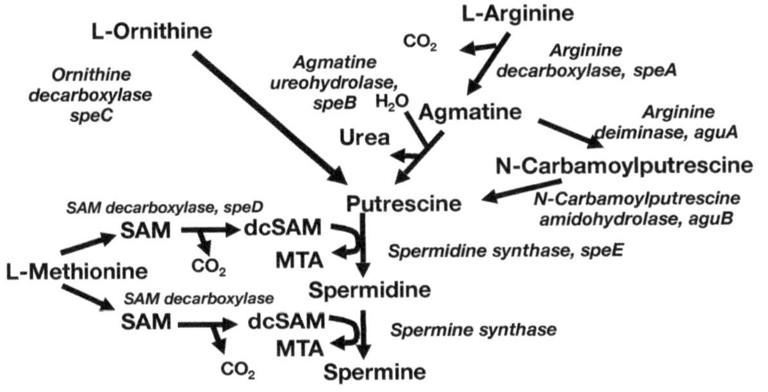

Figure 7.2 Generalized metabolic map of polyamine metabolism.

like in animal cells, and also from arginine via agmatine (see Fig 7.2). For putrescine production from arginine, two metabolic pathways are known. The first step is decarboxylation of argnine to agmatine in the both pathways, then agmatine is hydrolysed to putrescine in one pathway, or converted to *N*-carbamoylputrescine in the other pathway as shown in Fig. 7.2. In the later pathway, the formed intermediate compound is further converted to putrescine catalysed by another enzyme (Cohen, 1998; Tabor and Tabor, 1984, 1985; Lucas *et al.*, 2007; Landete *et al.*, 2010). Spermidine and spermine are synthesized from putrescine like in animal cells. Fig. 7.2 illustrates generalized polyamine metabolism in prokaryotes and plants. In this figure, gene names in *E. coli* genome which encode the enzymes catalysing the corresponding reaction, are listed. For instance, *E. coli* gene which codes ODC, is called to be *speC* as shown in Fig. 7.2.

Metabolic pathways for polyamine biosynthesis in an extreme thermophile, *T. thermophilus*, has been extensively studied in the author's laboratories (Ohnuma *et al.*, 2005, 2011). This microorganism is the first extreme thermophile which was used as host cell for DNA manipulation at high temperature. Thus it was not so difficult to create deletion mutants of the desired gene and reverse genetics is a powerful tool to clarify metabolic pathways. Fig. 7.3 summarizes conclusions so far obtained.

In the genome of the extreme thermophile, a gene coding for ornithine decarboxylase is missing, that is, putrescine is not produced from ornithine. Instead, all polyamines are derived from arginine. In the extreme thermophile, arginine is converted to agmatine by the action of arginine decarboxylase coded by *speA* gene. Then formed agmatine is not converted to putrescine as shown in Fig. 7.2, but directly aminopropylated to aminopropylagmatine by the action of an enzyme coded by *speE* gene as illustrated in Fig. 7.3; the compound is a new natural substance and has never been found in the biosphere. The *speE* is homologous to the gene for spermidine synthase of *E. coli*. The enzyme acts on spermidine and caldine (= norspemidine) in addition to agmatine and produces spermine and thermine (= norspermine), respectively. However, the enzyme could not accept putrescine as a substrate. In *T. thermophilus* spermidine is never produced from putrescine.

Aminopropylagmatine is then converted to spermidine by the action of an enzyme coded by a gene homologous to *E. coli speB* (or a gene annotated to be *speB*). Usually *speB* encodes agmatine ureohydrolase, an enzyme which converts agmatine to putrescine. Trivial name of agmatine ureohydrolase is agmatinase and the enzyme is a sister enzyme to arginase. However, a protein coded by *T. thermophilus speB* gene has weak 'agmatinase' activity which explains the presence of trace amount of putrescine in the cells grown under the standard conditions. Recently the author found that *T. thermophilus* SpeB (= expressed product of *speB*) has very, very weak arginase activity (unpublished). Instead the

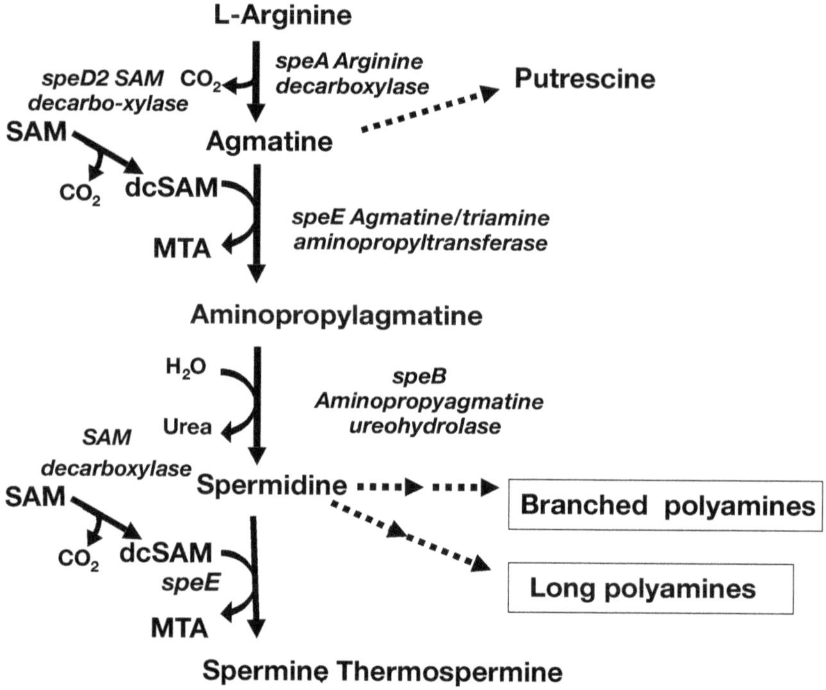

Figure 7.3 Polyamine metabolic pathway of an extreme thermophile, *T. thermophilus*.

recombinant protein coded by *speB* showed powerful 'aminopropylagmatinase' activity.

The pathway shown in Fig. 7.2 is new and unique, but may be widely used by thermophilic bacteria and archaea, since at least two hyperthermophilic archaea (*Pyrococuss furiosus* and *T. kodakarensis*) utilize the same pathway for production of polyamines (Morimoto *et al.*, 2011; Cacciapuoti *et al.*, 2007)

Unfortunately metabolic pathways for unusual polyamines in *T. thermophilus*, especially long and branched polyamines, have not yet been clarified. However, very recently a gene coding for 'branched polyamine synthase' has been identified in the genome of *T. kodakarensis* (Okada *et al.*, 2014). The homologous genes have been detected in a variety of thermophiles which are able to produce branched polyamines as a major polyamine component though chemical structures of the produced branched polyamines differ from species to species (Table 7.2). In *T. thermophilus*, *TTHA0539* gene is highly homologous to the gene in *T. kodakarensis*. The knockout mutant of *T. thermophilus* could not produce tetrakis(3-aminopropyl)ammonium suggesting that the gene codes a branched polyamine synthase of *T. thermophilus*. Unfortunately, the gene product showed no enzymatic activity, suggesting that the *T. thermophilus* enzyme requires the presence of other protein(s) and/or unique coenzyme or cofactor for the catalytic activity.

Table 7.2 Genes homologous to *T. kodakarensis TK1691* found in genomes of extreme thermophiles

Bacterial or archaeal names	Annoation	Homology
Thermococcus kodakaraensis	Hypothetical	100%
Thermus thermophilus	Hypothetical	42
Aquifex aeolicus	Hypothetical	47
Thermoanaerobacter ethanolicus	Methyltransferase	41
Thermotoga thermarum	Hypothetical	68
Methanococcus aeolicus	Hypothetical	33

Reverse genetics

When *speB* gene was knocked out, the mutant produced a large amount of aminopropylagmatine and a small amount of agmatine as cellular polyamines as speculated from polyamine metabolic pathways shown in Fig. 7.3. Long and branched polyamines such as caldopentamine and tetrakis(3-aminopropyl)ammonium were not produced. The mutant could not grow at 75°C or higher temperatures. When spermidine was added, long and branched polyamines were detected in the cells and the growth at 75°C or higher temperatures was recovered. These results strongly suggest that the unusual polyamines are essential for life at high temperature extremes. The growth at 75°C was also recovered when caldopentamine or tetrakis(3-aminopropyl) ammonium was added into the growth mediums. Another important conclusion is that spermidine is key intermediate compound for biosynthesis of long and branched polyamines as shown in Fig. 7.3.

When *speE* gene was deleted, the resulted mutant produced sym-homospermidine as the major polyamine along with moderate amounts of putrescine and agmatine. This result is hard to predict from the metabolic pathways shown in Fig. 7.3, and suggests that path to sym-homospermidine is hidden under normal growth conditions. Interestingly, when spermidine is added to the culture medium, *speE* deficient mutant cannot produce sym-homospermidine, suggesting that the biosynthetic pathway for sym-homospermidine is strongly suppressed by spermidine. Cellular polyamine composition of *speD2* deficient mutant is very similar to that of *speE*- mutant; that is, the major polyamine is sym-homospermidine and small amounts of putrescine and agmatine were detected. Both *speE*- and *speD2*-knockout mutants could not grow at higher temperatures like *speB* deficient mutant, and the growth was recovered by adding either caldopentamine or tetrakis(3-aminopropyl)ammonium into the growth mediums.

Polyamine composition of *speD1* knockout mutant is interesting. The gene is annotated to be 'S-adenosylmethionine decarboxylase', and its amino acid sequence is highly homologous to those of SpeD2 which is true S-adenosylmethionine decarboxylases of *T. thermophilus*. All S-adenosylmethionine decarboxylases of both mesophilic and thermophilic organisms so far studied, are 'pyruvoyl enzymes' which contain a pyruvate residue at the active site which is produced by a unique self-splicing reaction of the polypeptide coded by the gene (Hackert and Pegg, 1984; Recsei and Snell, 1984). However, the protein coded by *speD1* gene lacks unique amino acid sequence which is necessary for the self-splicing reaction and thus cannot produce the essential pyruvate residue for the catalytic function. The author and his colleagues confirmed that the unique self-splicing reaction does not take place in the recombinant protein coded by *T. thermophilus* *speD1* gene. The gene product of *speD2* contains the amino acid sequence for the self-splicing and upon heating the splicing reaction takes place and a pyruvoyl enzyme is formed. At moment, the function of *speD1* gene is unknown. However, the knockout mutant could not produce both long and branched polyamines even in the presence of spermidine indicating that the gene product is required for biosynthesis of unusual polyamines. Since unusual polyamines are missing, the *speD1*-deficient mutant could not grow at 75°C or higher temperatures and the growth was recovered by the addition of caldopentamine and/or tetrakis(3-aminopropyl)ammonium like other knockout mutants.

In summary, reverse genetic studies imply that unusual polyamines, both long and branched polyamines are essential for life at high temperature extremes. Spermidine is key intermediate compound for biosynthesis of unusual polyamines. Spe D1 protein seems to be required for the syntheses of both long and branched polyamines in *T. thermophilus*.

Stability of nucleic acid at high temperatures

Polyamines interact with nucleic acids, induce conformation changes such as B to Z transformation of DNA, and protect DNA and RNA from thermal degradations such as hydrolysis and especially depurination of DNA (Base and Marton, 1987; Vijayanathan *et al.*, 2001; Behe and Felsenfeld, 1981).

In the presence of a polyamine, Tm (thermal denaturation temperature) of DNA or RNA increases. For DNA, straight chain polyamines are more effective and ΔT_m is proportional to number of nitrogen atoms of the polyamine used if the same concentrations of straight chain polyamines are compared. The increment of Tm is also proportional to number of butyl groups in the molecule if number of nitrogen atoms is same (Terui *et al.*, 2005).

DNA depurination (Andreev and Kaboev, 1982; Greer and Zamenhof, 1962; Lindahl and Nyberg, 1972; Baik *et al.*, 2002) is most serious genetic damage even for humans. Since the reaction is faster under acidic conditions or high temperature, the protection of DNA from depurination would be extremely critical for extreme thermophiles. Again longer polyamines protect DNAs more efficiently (Base and Marton, 1987). Polyamines also induce the condensation of DNA in aqueous solution. It has been known that polyamines, especially longer polyamines, induce the condensation if polyamine concentration is more than 300 μM (D'Agostino and Luccia, 2002).

Apparent Tm of RNA is also raised by the addition of polyamines, especially by the addition of branched polyamine, tetrakis(3-aminopropyl) ammonium (Terui *et al.*, 2005). However, the above mentioned empirical rules cannot be applicable for RNAs.

In addition to the binding with polyamines, RNA molecules are thermally stabilized by post-transcriptional modifications. One example is thiolation of nucleoside base at certain positions. Typical example is thiolated U at position 55 of the initiation tRNA of *T. thermophilus* (Watanabe *et al.*, 1974). Methylation at ribose moiety also stabilizes tRNA molecules (Kumagai *et al.*, 1982). In short, tRNA can be stabilized by combination of binding with polyamines and other polycationic substances, and various modifications at bases and 2′-hydroxyl group of ribose (Kumagai *et al.*, 1982; Grosjean and Oshima, 2007; Hayrapetyan *et al.*, 2009).

Protein synthesis *in vitro*

Cell-free protein synthesis at high temperature requires the presence of polyamine (Ohno-Iwashita *et al.*, 1975; Uzawa *et al.*, 1993a). When *T. thermophilus* cell-free extract is used, almost no activity was found at 65°C though amino acid incorporation was recorded at lower temperatures. The highest activity for polypeptide synthesis was observed when a branched polyamine, tetrakis(3-aminopro;yl)ammonium is added to the reaction mixture (Uzawa *et al.*, 1993b). Interestingly, this branched polyamine inhibits phenyalanyl-tRNA formation reaction (Uzawa *et al.*, 1994). The inhibition was suppressed by the addition of straight chain polyamines such as spermine, thermine, and caldopentamine. Thus if overall reaction from phenylalanine to polypeptide formation is recorded, the highest activity was observed in the presence of both the branched polyamine and spermine.

Polyamines are also essential for *in vitro* protein synthesis using a cell-free extract of *Sulfolobus* species, an acidophilic, extremely thermophilic archaeon (Friedman and Oshima, 1989).

References

Andreev, O.A., and Kaboev, O.K. (1982). The influence of spermidine on DNA depurination *in vitro*. Biochim. Biophys. Acta *698*, 100–101.

Baik, M.H., Friesner, R.A., and Lippard, S.J. (2002). Theoretical study on the stability of N-glycosyl bonds: why does N7-platination not promote depurination? J. Am. Chem. Soc. *124*, 4495–4503.

Base, H.S., and Marton, L.J. (1987). The interaction of spermine and pentamines with DNA. Biochem. J. *244*, 243–246.

Behe, M., and Felsenfeld, G. (1981). Effects of methylation on a synthetic polynucleotide: the B-Z transition in poly(dG-m5dC)poly(dG-m5dC). Proc. Natl. Acad. Sci. U.S.A. *78*, 1619–1623.

Cacciapuoti, G., Porcelli, M., Moretti, M.A., Sorrentino, F., Concilio, L., Zappia, V., Liu, Z.J., Tempeel, W., Schubot, F., Rose, J.P., *et al.* (2007). The first agmatine/cadaverin aminopropyl transferase: Biochemical and structural characterization of an enzyme involved in polyamine biosynthesis in the hyperthermophilic archaeon *Pyrococcus furiosus*. J. Bacteriol. *189*, 6057–6067.

Chattopadhyay, M.K., and Tabor, H. (2013). Polyamines are critical for the induction of the glutamate decarboxylase-dependent acid resistance system in *Eschrichia coli*. J. Biol. Chem. *288*, 33559–33570.

Coffino, P. (2001). Regulation of cellular polyamines by antizyme. Nat. Rev. Mol. Cell. Biol. *2*, 188–194.

Cohen, S.S. (1998). A Guide to the Polyamines (Oxford University Press, New York, USA).

D'Agostino, L., and Di Luccia, A. (2002). Polyamiones interact with DNA as molecular aggregates. Eur. J. Biochem. *269*, 4317–4325.

Flores, H.E., and Galston, A.W. (1982). Analysis of polyamines in higher plants by high performance liquid chromatography. Plant Physiol. *69*, 701–706.

Friedman, S.M., and Oshima, T. (1989). Polyamines of sulfur-dependent archaebacteria and their role in protein synthesis. J. Biochem. *105*, 1030–1033.

Fujita, K., Murakami, Y., and Hayashi, S. (1982). A macromolecular inhibitor of the antizyme to ornithine decarboxylase. Biochem. J. *204*, 647–652.

Greer, S., and Zamenhof, S. (1962). Studies on depurination of DNA by heat. J. Mol. Biol. *4*, 123–141.

Grosjean, H., and Oshima, T. (2007). How nucleic acids cope with high temperature. In Physiology and Biochemistry of Extremophiles, Gerday, C., and Glansdorff, N., eds. (ASM Press, Washington, DC), pp. 39–56.

Hackert, M.I., and Pegg, A.I. (1998). Pyruvoyl-dependent enzymes. In Comprehensive Biological Catalysis, Vol. 2, Sinnott, M., ed. (Academic Press, San Diego, USA) pp. 201–216.

Hamana, K., Hamana, H., Niitsu, M., Samejima, K., and Itoh, T. (1996). Polyamines of hyperthermophilic archaebacteria, *Archaeoglobus*, *Thermococcus*, *Pyrobaculum* and *Sulfolobus*., Microbes *87*, 69–76.

Hamana, K., Niitsu, M., and Samejima, K. (2000). Occurrence of tertiary branched tetraamines in two aquatic plants. Can. J. Bot. *78*, 266–269.

Hamana, K., Saito, T., and Okada, M. (2001). Distribution profiles of spermidine and homospermidine within the alpha subclass of the class Proteobacteria. Microbiol. Cult. Coll. *17*, 3–12.

Hamana, K., Hayashi, H., Niitsu, M., Sugata, D., Higuchi, K., and Itoh, T. (2011). Cellular distribution of unusual long and branched polyamines within the newly validated thermophiles belonging to the bacterial orders Thermoanaerobacterales and Clostridiales. J. Jpn. Soc. Extremoophiles *10*, 83–89.

Hayashi, S., Murakami, Y., and Matsufuji, S. (1996). Ornithine decarboxylase antizyme: a novel type of regulatory protein. Trends Biochem. Sci. *21*, 27–30.

Hayrapetyan, A., Grosjean, H., and Helm, M. (2009). Effect of a quaternary pentaamine on RNA stabilization and enzymatic methylation, Biol. Chem., 390, 851–861.

Hosoya, R., Hamana, K., Niitsu, M., and Itoh, T. (2004). Polyamine analysis for chemotaxonomy of thermophilic eubacteria: polyamine distribution profiles within the orders Aquificales, Thermotogales, Thermodesulfobacteriales, Thermales, Thermoanaerobacteriales, Clostridiales and Bacillales. J. Gen. Appl. Microbiol. *50*, 271–287.

Imai, A., Akiyama, T., Kato, T., Sato, S., Tabata, S., Yamamoto., K.T., and Takahashi, T. (2004). Spermine is not essential for survival of Arabidopsis. FEBS Lett. *556*, 148–152.

Kakehi, J., Kuwashiro, Y., Niitsu, M., and Takahashi, T. (2008). Thermospermine is required for stem elongation in *Arabidopsis thaliana*. Plant Cell Physiol. *49*, 1342–1349.

Kneifel, H., Stetter, K.O., Andreesen, J.R., Wiegel, J., Konig, H., and Schobeth, S.M. (1986). Distribution of polyamines in representative species of archaebacteria. System Appl. Microbiol. *7*, 241–245.

Knott, J.M., Romer, P., and Sumper, M. (2007). Putative spermine synthases from *Thalassiosira pseudonana* and *Arabidopsis thaliana*synthesize thermospermine rather than spermine FEBS Lett. *581*, 3081–3086.

Kumagai, I., Watanabe, K., and Oshima, T. (1982). A Thermostable tRNA (guanosine-2′-)-methyltransferase from *Thermus thermophilus* HB27 and the effect of ribose methylation on the conformational stability of tRNA. J. Biol. Chem. *257*, 7388–7395.

Landete, J.M., Arena, M.E., Pardo, I., Manea de Nadra, M.C., and Ferrer, S. (2010). The role of two families of bacterial enzymes in putrescine synthesis from agmaine via agmatine deiminase. Int. Microbiol. *13*, 169–177.

Lindahl, T., and Nyberg, B. (1972). Rate of depurination of native deoxyribonucleic acid. Biochemistry *19*, 3610–3618.

Lucas, P.M., Blancato, V.S., Claisse, O., Magni, C., Lolkema, J.S., and Lonvaud-Funel, A. (2007). Agmatine deiminase pathway genes in *Lactobacillus brevis* are linked to the tyrosine decarboxylation operon in a putative acid resistastance locus. Microbil. *153*, 2221–2230.

Morimoto, N., Fukuda, W., Nakajima, N., Masuda, T., Terui, Y., Kanai, T., Oshima, T., Imanaka, T., and fujiwara, S. (2011). Dual biosynthesis pathway for long-chain polyamines in the hyperthermophilic archaeon *Thermococcus kodakarensis*. J. Bactereiol. *192*, 4991–5001.

Moriya, T., Hikota, T., Yumoto, I., Ito, T., Terui, Y., Yamagishi, A., and Oshima, T. (2011). *Calditerricola satsumensis* gen. nov., sp. nov. and *Calditerricola yamamurae* sp. nov., extreme thermophiles isolated from a high temperature compost. Int. J. Syst. Evol. Microbiol. *61*, 631–636.

Muniz, L., Minguet, E.G., Singh, S.K., Pesquet, E., Vera-Sirera, F., Moreau-Courtois, C.L., Carbonell, J., Blazquez, M.A., and Touminen, H. (2008). ACAULIS5 controls *Arabidopsis xylem* specification through the prevention of premature cell death. Development *135*, 2573–2582.

Niemuller, D., Reimann, A., and Ober, D. (2012). Distinct cell-specific expression of homospermidine synthase involved in pyrrolizidine alkaloid biosynthesis in three species of the boraginales. Plant Physiol. *159*, 920–929.

Nurhayati, N., Gonde, D., and Ober, D. (2009). Evolution of pyrrolizidine alkaloids in *Phalaenopsis orchids* and other monocotyledons: identification of deoxyhypusine synthase, homospermidine synthase and related pseudogenes. Phytochemistry *70*, 508–516.

Ober, D., and Hartmann, T. (1999). Homospermidine synthase, the first pathway-specific enzyme of pyrrolizidine alkaloid biosynthesis, evolved from deoxyhypusine synthase. Proc. Natl. Acad. Sci. U.S.A. *96*, 14777–14782.

Ohno-Iwashita, Y., Oshima, T., and Imahori, K. (1975). *In vitro* protein synthesis at elevated temperature by an extract of an extreme thermophile: Effects of polyamines on the polyuridylic acid-directed reaction. Arch. Biochem. Biophys. *171*, 490–499.

Ohnuma, M., Terui, Y., Tamakoshi, M., Mitome, H., Niitsu, M., Samejima, K., Kawashima, E., and Oshima, T. (2005). N^1-Aminopropylagmatine, a new polyamine produced as a key intermediate in polyamine biosynthesis of an extreme thermophile, *Thermus thermophilus*. J. Biol. Chem. *280*, 30073–30082.

Ohnuma, M., Gambe, T., Terui, Y., Niitsu, M., Sato, T., Tanaka, N., Tamakoshi, M., Samejima, K., Kumasaka, T., and Oshima, T. (2011). Crystal Structures and Enzymatic Properties of a Triamine/Agmatine Aminopropyltransferase from *Thermus thermophilus*. J. Mol. Biol. *408*, 971–986.

Okada, K., Hidese, R., Fukuda, W., Niitsu, M., Takao, K., Horai, Y., Umezawa, N., Higuchi, T., Oshima, T., Yoshikawa, Y., Imanaka, T., and Fujiwara, S. (2014). Identification of a novel aminopropyltransferase involved in the synthesis ofbranched-chain polyamines in hyperthermophiles. J. Bacteriol. *196*, 1866–1876.

Oshima, T. (1979). A new polyamine, thermospermine, 1,12-diamino-4,8-diazadodecane, from an extreme thermophile. J. Biol. Chem. *254*, 8720–8722.

Oshima, T. (1983). Novel polyamines in *Thermus thermophilus*: Isolation, identification and chemical synthesis. Methods in Enzymology *94*, 401–411.

Oshima, T. (1989). Polyamines in thermophiles. In The Physiology of Polyamines, Bachrach, U., and Heimer, Y.M., eds. (CRC Press, Boca Raton, Florida, USA), pp. 36–46.

Oshima, T. (2007). Unique polyamines produced by an extreme thermophile, *Thermus thermophilus*, Amino Acids 33, 367–372.

Oshima, T., Moriya, T., and Terui, Y. (2011). Identification, Chemical Synthesis, and Biological Functions of Unusual Polyamines Produced by Extreme Thermophiles. In Methods in Molecular Biology 720, A.E. Pegg and R.A. Casero, Jr. eds. (New York, USA: Humana Press), pp 81–111.

Reccsei, P.A., and Snell, E.E. (1984). Pyruvoyl enzymes. Annu. Rev. Biochem. *53*, 357–387.

Redmond, J.W., and Teseng, A. (1979). High-pressure liquid chromatographic determination of putrescine, cadaverine, spermidine and spermine. J. Chromatogr. *170*, 479–481.

Stillway, L.W., and Walle, T. (1977). Identification of the unusual polyamines 3,3'diaminodipropylamine and *N,N'*-bis(3-aminopropyl)–1,3-propanediamine in the white shrimp, *Penaeus estiferus*. Biochem. Biophys. Res. Commun. 77, 1103–1107.

Slocum, R.D., Flores, H.E., Galston, A.W., and Weinstein, L.H. (1989). Improved method for HPLC analysis of polyamines, agmatine and aromatic monoamines in plant tissue. Plant Physiol. *89*, 512–517.

Tabor, C.W., and Tabor, H. (1984). Polyamines. Ann. Rev. Biochem. *53*, 749–790.

Tabor, C.W., and Tabor, H. (1985). Polyamines in microorganisms. Microbiol. Rev. *49*, 81–99.

Terui, Y., Ohnuma, M., Hiraga, K., Kawashima, E., and Oshima, T. (2005). Stabilization of nucleic acids by unusual polyamines produced by an extreme thermophile, *Thermus thermophilus*. Biochem. J. *388*, 427–433.

Uzawa, T., Hamasaki, N., and Oshima, T. (1993a). Effects of novel polyamines on cell-free polypeptide synthesis catalyzed by *Thermus thermophilus* HB8 extracts. J. Biochem. *114*, 478–486.

Uzawa, T., Yamagishi, A., Ueda, T., Chikazumi, N., Watanabe, K., and Oshima, T. (1993b). Effects of polyamines on a continuous cell-free protein synthesis system of an extreme thermophile, *Thermus thermophilus*. J. Biochem. *114*, 732–734.

Uzawa, T., Yamagishi, A., Nishikawa, K., and Oshima, T. (1994). Effect of unusual polyamines on pheylalanyl-tRNA formation. J. Biochem. *115*, 830–832.

Vijayanathan, V., Thomas, T., Shirahata, A., and Thomas, T.J. (2001). DNA condensation by polyamines: a laser light scattering study of structural effects. Biochemistry *40*, 13644–13651.

Watanabe, K., Oshima, T., Saneyoshi, M., and Nishimura, S. (1974). Replacement of ribothymidine by 5-methyl-2-thiouridine in sequence GT C in tRNA of an extreme thermophile. FEBS Lett. *43*, 59–63.

Yamaguchi, K., Takahashi, Y., Berberich, T., Imai, A., Takahashi, T., Michael, A.J., and Kusano, T. (2007). A protective role for the polyamine spermine against drought stress in arabidopsis. Biochem. Biophys. Res. Commun. *352*, 486–490.

Zappia, V., Porta, R., Carteni-Farina, M., DeRosa, M., and Gambacorta, A. (1978). Polyamine distribution in Eukaryotes: occurrence of sym-norspermidine and sym-norspermine in Anthropods. FEBS Lett. *94*, 161–165.

DNA Replication in Thermophilic Microorganisms

8

Sonoko Ishino and Yoshizumi Ishino

Abstract

DNA replication is essential for maintaining genetic information and transferring it from ancestor to descendant. To protect the genetic information from mutations, biological organisms have acquired several types of DNA repair systems. The extreme thermophiles living on the earth are microorganisms from the Bacteria and Archaea domains. The structures of the bacterial and archaeal genomes are circular, and the mechanism of replication initiation, by the binding of the initiator protein to the replication origin (*oriC*), is conserved in the two domains. The elongation process is also conserved, because similar primase, helicase, polymerase, and ligase functions are observed in the two domains. However, the proteins involved in the DNA replication process are quite different between Bacteria and Archaea, and their replication machineries seem to have evolved independently. The DNA repair system in extreme thermophiles should work efficiently to maintain their genome integrity at high temperatures. However, the DNA repair systems are diverse and still not comprehensively understood yet, although research in this field has been actively pursued in the systems from *Escherichia coli* to human. In this chapter, we focus on DNA replication in the thermophilic bacteria and archaea, and summarize the current understanding of the molecular mechanisms of replication in thermophilic microorganisms.

Introduction

Accurate duplication and transmission of genetic information are essential and crucially important for living organisms. The molecular mechanism of DNA replication is one of the central themes of molecular biology, and continuous efforts to elucidate the precise molecular mechanism of DNA replication have been made since the discovery of the double helix DNA structure in 1953 (Watson and Crick, 1953). The protein factors that function in the DNA replication process, starting from replication origin recognition to replication fork progression in the semiconservative replication manner, using continuous and discontinuous strand syntheses, have been identified in the three domains of life, Bacteria, Eukarya, and Archaea (Fig. 8.1).

Thermophiles are classified into extreme thermophiles (growing at 75°C) and moderate thermophiles (growing at 55–75°C). The organisms that grow at temperatures above 80°C are especially called hyperthermophiles. Most of the identified hyperthermophiles are archaea, and a very few are bacteria (Stteter, 2013). Analyses of the molecular biology of bacterial DNA replication started in about 1960, and were followed by eukaryotic studies since 1980. Archaeal DNA replication research became active after 1990, after Archaea was recognized as the third domain of life (Ishino *et al.*, 2013). Archaeal cells have a unicellular ultrastructure without a nucleus, and thus resemble bacterial cells, but the proteins involved in the genetic information processing pathways, including DNA replication, transcription, and translation, share strong similarities with those of Eukarya. Most of the archaeal replication proteins obviously differed from the bacterial proteins, and were identified as homologues of the eukaryotic replication proteins, and thus were designated

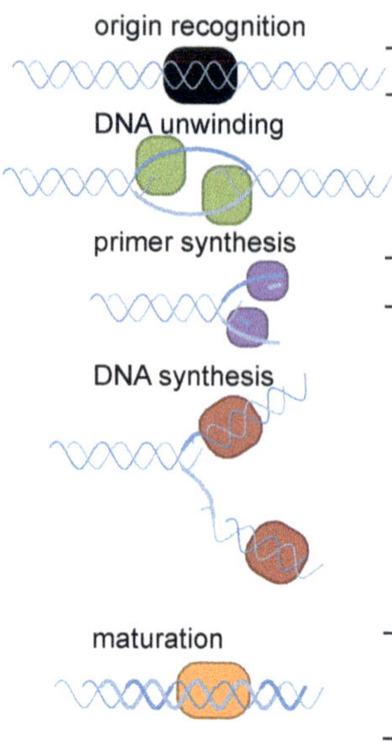

Bacteria	Archaea	Eukarya
DnaA	Cdc6/Orc1	ORC
DnaC DnaB	Cdc6/Orc1 MCM GINS	Cdc6 Cdt1 MCM GINS Cdc45
DnaG	DNA primase	Pol α/ primase
family C DNA polymerase (Pol III)	family B DNA polymerase (Pol B) family D DNA polymerase (Pol D)	family B DNA polymerase (Pol δ) (Pol ε)
clamp loader (γ-complex) clamp (β-clamp)	clamp loader (RFC) clamp (PCNA)	clamp loader (RFC) clamp (PCNA)
Pol I RNaseH DNA ligase	Fen1 Dna2 DNA ligase	Fen1 Dna2 DNA ligase

Figure 8.1 Comparison of the proteins working on each event in the DNA replication process of the three biological domains. Most of the proteins are same names between Archaea and Eukarya, and they are different from those in Bacteria based on the amino acid sequence similarity.

as ORC (origin recognition complex), Cdc (cell division cycle) 6, GINS, MCM (minichromosome maintenance), RPA (replication protein A), PCNA (proliferating cell nuclear antigen), RFC (replication factor C), and FEN (flap endonuclease) 1 (Ishino *et al.*, 2013). These proteins were biochemically characterized (Kelman and White, 2005; Barry and Bell, 2006; Ishino and Ishino, 2012). These facts indicated that the DNA replication machineries of Archaea and Eukarya evolved from a common ancestor, which was different from that of Bacteria (Leipe *et al.*, 1999). Therefore, Archaea is now recognized as a good model to elucidate the functions of each component of the eukaryotic replication machinery complex. Due to the smaller genome size and the lower number of genes, the archaeal replication machinery is probably a simplified form of that in Eukarya. On the other hand, Bacteria and Archaea share the circular genome structure, which is different from the linear form of the eukaryotic genome. Based on these features, studies on archaeal DNA replication will provide fundamental insights into its molecular mechanisms and machinery from an evolutional perspective. With the increasing availability of total genome sequences, the progress in archaeal DNA replication research has been rapid.

Among archaea, the genome sizes of the hyperthermophiles are especially small. In addition, the stable proteins from the hyperthermophiles are suitable for biochemical research. For these reasons, DNA replication studies with hyperthermophiles have progressed rapidly. Another fundamental interest in life science is how life is maintained and the genetic information is protected at extremely high temperatures, which can seriously damage DNA, as compared with our normal environment. Thus, hyperthermophilic organisms are expected to have unusually effective DNA repair systems (Grogan, 2000). However, the differences in the DNA repair mechanisms employed by the hyperthermophiles have not been clarified. In this chapter, the DNA replication systems in thermophilic bacteria and archaea, especially in hyperthermophiles, are summarized.

Replication origin

The genomes are circular in both Bacteria and Archaea. The basic mechanism of DNA replication initiation was predicted as the 'replicon theory' by Jacob and Brenner (Jacob and Brenner, 1963). They proposed that a specific factor, an initiator, recognizes the replicator, which actually means the replication origin, to start the replication of the chromosomal DNA. The replication origin of *E. coli* DNA was identified as *oriC* (origin of chromosome), and the principles of initiation are now well understood (Costa *et al.*, 2013). The DnaA protein is the initiator of bacterial replication (as described below), and it is also conserved in thermophilic bacteria, including *Thermus* (Schaper *et al.*, 2000) and *Thermotoga* (Ozaki *et al.*, 2006). The *oriC* region is generally located upstream of the DnaA-encoding gene, *dnaA*, on the genomes. A unique feature of the *oriC* regions is the repeated sequence motifs, called the DnaA-box, for DnaA binding. The *oriC*-binding abilities of the DnaA proteins from *Thermotoga* and *Aquifex* have been characterized, as described below.

The first archaeal replication origin was identified in the hyperthermophile, *Pyrococcus abyssi*, in 2001. The origin was located just upstream of the gene encoding the Cdc6 and Orc1-like sequences, which are probably the initiator proteins, in the *Pyrococcus* genome (Matsunaga *et al.*, 2001). The eukaryotic-like initiator was originally identified in the genome of *Pyrococcus furiosus*, with the discovery of DNA polymerase II (now called PolD, as described below) (Uemori *et al.*, 1997). It was named Cdc6/Orc1, due to its roughly equal homology with regions of eukaryotic Orc1 and Cdc6, after confirmation of its ability to bind to the *oriC* region on the chromosomal DNA, as described below (Fig. 8.2). The location of *oriC* upstream of the initiator gene is generally conserved in Bacteria and Archaea (Wu *et al.*, 2014). A characteristic of the archaeal *oriC* is the conserved 13 bp repeats, as predicted previously by bioinformatics (Lopez *et al.* 1999), and two of the repeats are longer and surround a predicted DNA unwinding element (DUE) with an AT-rich sequence in *Pyrococcus* genomes (Fig. 8.2) (Matsunaga *et al.*, 2003). The longer repeated sequence was designated as an Origin Recognition Box (ORB), and it was actually recognized by Cdc6/Orc1 in a *Sulfolobus solfataricus* study (Robinson *et al.*, 2004). The 13-base repeat is called a miniORB, as a minimal version of the ORB. A whole genome microarray analysis of *P. abyssi* revealed that Cdc6/Orc1 binds to the *oriC* region with extreme specificity, and the specific binding of the highly purified *P. furiosus* Cdc6/Orc1 to ORB and miniORB was confirmed *in vitro* (Matsunaga *et al.*, 2007).

One remarkable feature of Archaea is that multiple replication origins were identified, first in the *Sulfolobus* genomes (Robinson *et al.*, 2004) and then in others. It is now well recognized that *Sulfolobus* has three origins that work at the same time in the cell cycle (Lundgren *et al.*, 2004; Robinson *et al.*, 2007). Four origins were identified in the *Pyrobaculum* and *Haloferax* genomes (Norais *et al.*, 2007; Perve *et al.*, 2012). Analyses of the mechanism by which the multiple origins are utilized for genome replication are an interesting

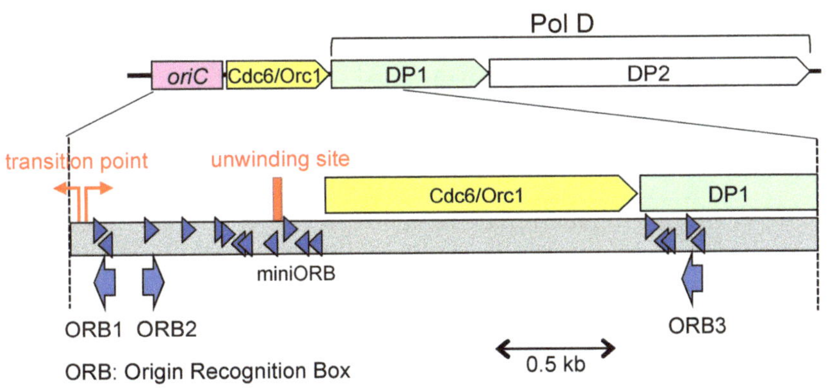

Figure 8.2 The *oriC* region on the *Pyrococcus* genome. The region surrounding *oriC* is schematically presented. The *oriC* is located in the upstream region of the *cdc6/orc1* gene, which codes for the initiator protein. The first unwinding site, identified by P1 nuclease assays, and the initiation site for primer synthesis, identified by a RIP assay, are indicated in the *oriC* region, including the ORB1, ORB2, ORB3 (large arrows), and many miniORB (small arrowheads) are also indicated.

subject in the archaeal DNA replication research field. The main questions are how the initiations of replication from multiple origins are regulated, and how the replication forks progress after the collision of two forks from opposite directions.

A recent popular topic in archaeal replication is that the identified four *oriC*s with differing activities can be deleted individually from the genome, although the cell growth was retarded. Furthermore, it was surprising that a strain lacking all *oriC*s had no apparent defects, and actually grew significantly faster than the wild type. In the origin-less cells, DNA replication started from dispersed sites, and it is now understood to be a homologous recombination-dependent mechanism because of its absolute requirement for the RadA recombinase (Hawkins *et al.*, 2013). The necessity of the replication mechanism at discrete origins is now an interesting discussion topic.

How do DnaA and Cdc6/Orc1 recognize *oriC*?

The origin DNA must be unwound before the replication machinery is assembled, because DNA synthesis occurs on single-stranded DNA. The binding of the initiator proteins to *oriC* seems to be critical for unwinding the origin. An important question is how the DnaA and Cdc6/Orc1 proteins recognize the *oriC* region in the bacterial and archaeal genomes, respectively. These proteins belong to the AAA$^+$ family, based on their amino acid sequences (Kawakami and Katayama, 2010; Duderstadt and Berger, 2013). The current knowledge of the structure–function relationships for DnaA was obtained from *E. coli* studies (Ozaki and Katayama, 2009). The DnaA protein consists of 4 functional domains (Fig. 8.3) (Messer 2002; Kaguni, 2006). Domain I is conserved in some ssDNA/RNA-binding proteins, although its ssDNA-binding activity is weak (Abe *et al.*, 2007), and it bears specific binding activity for several proteins, including the DnaB helicase and the initiation stimulator DiaA (Weigel *et al.*, 1999; Felczak *et al.*, 2005; Abe *et al.* 2007; Keyamura *et al.*, 2007). Domain II is likely to function as a flexible linker (Abe *et al.*, 2007; Nozaki and Ogawa, 2008). Domain III, containing the AAA$^+$ motifs (Neuwald *et al.*, 1999; Iyer *et al.*, 2004b), plays a major role in ATP and ADP binding. DnaA has high affinity (K_d of 10–100 nM) for ATP and ADP (Sekimizu *et al.*, 1987). ATP–DnaA is the active form for replication initiation, whereas ADP–DnaA is the inactive form. The intrinsic ATPase activity of DnaA is DNA-dependent and very weak. Domain IV has a typical helix–turn–helix fold for binding to the DnaA box (Roth and Messer, 1995; Erzberger *et al.*, 2002; Fujikawa *et al.*, 2003), composed of the sequence TTATNCACA (Messer, 2002). Multiple copies of DnaA protomers associate with each other to form a helical nucleoprotein filament that opens the DUE. Studies on the thermophilic DnaA proteins have enhanced our understanding of the nucleofilament structure of DnaA. A recent structural analysis of *Aquifex aeolicus* DnaA revealed the mechanism of the replication origin opening by the direct ATP-dependent stretching of DnaA (Duderstadt *et al.*, 2011). The ATP-dependent assembly of *A. aeolicus* DnaA

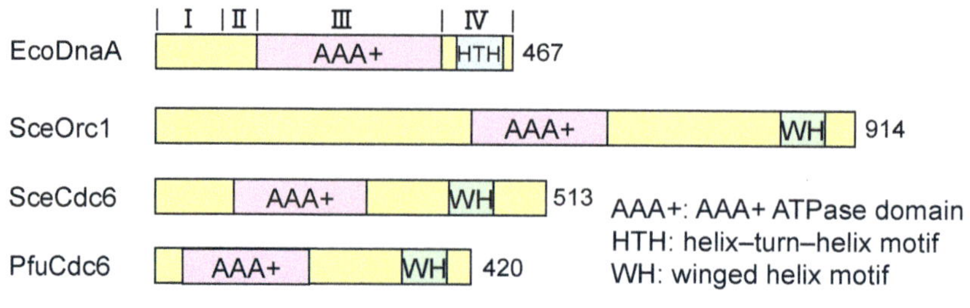

Figure 8.3 Schematic drawing of the initiator proteins. The domain structures of *E. coli* DnaA, *S. cerevisiae* Orc1 and Cdc6, and *P. furiosus* Orc1/Cdc6 are shown with their total amino acid numbers. Domain structure and important functional motifs are indicated in each protein. For DnaA, domains I–IV are indicated above the protein. AAA$^+$, AAA$^+$ domain; HTH, helix–turn–helix motif; WH, winged helix motif.

into a spiral oligomer extends single-stranded DNA. This binding mechanism is similar to that of RecA, but the important difference is that the DnaA-bound nucleofilament prevents the pairing of a complementary strand. Biochemical studies of *Thermotoga maritima* DnaA showed that the protein forms open complexes with the predicted origin region *in vitro*, and suggested that the ATP-dependent molecular switch of the initiator and the replication initiation complex mechanisms are conserved from thermophiles to mesophiles in eubacterial evolution (Ozaki *et al.*, 2006). Following origin opening, DnaA collaborates with the auxiliary DnaC helicase loader to recruit and deposit two hexameric DnaB helicases onto the melted origin.

A protein with a sequence similar to those of eukaryotic Orc1 and Cdc6 was found in *P. furiosus* as the first such protein identified in Archaea, as described above (Uemori *et al.*, 1997). The *oriC* is located in the intergenic region upstream of the *cdc6/orc1* gene in the *Pyrococcus* genome, and it was experimentally confirmed that Cdc6/Orc1 binds to the *oriC* region in *Pyrococcus* cells (Myllykallio, 2000; Matsunaga *et al.*, 2001). In several more archaeal organisms, the homologues of eukaryotic Orc1 and Cdc6 were shown to recognize the origin of DNA replication in *Methanothermobacter thermoautotrophicus* (Capaldi and Berger, 2004), *Aeropyrum pernix* (Grainge *et al.*, 2006), and *S. solfataricus* (Robinson *et al.*, 2004). The repeated sequence, ORB, in the *oriC* region is recognized by Cdc6/Orc1, and the specific binding of Cdc6/Orc1 to the ORB was confirmed *in vitro*. Further studies revealed that one of the two homologues of Cdc6/Orc1 in *A. pernix*, called ORC1 (the two homologues of the initiation protein in this organism are referred to as ORC1 and ORC2 in this study), binds to each ORB as a dimer, and a transition to a higher-order assembly occurs when all four ORBs in the *oriC* region are bound with the ORC1 protein (Grainge *et al.*, 2006). It was assumed that the wrapping of the DNA around ORC1 created distortions in the outer surface of the DNA duplex. The crystal structures of *S. solfataricus* Cdc6-1 and Cdc6-3 forming a heterodimer bound to *ori2* DNA (one of the three origins in this organism) (Dueber *et al.*, 2007), and that of *A. pernix* ORC1 bound to an origin sequence (Gaudier *et al.*, 2007), revealed that both the N-terminal AAA$^+$ ATPase domain and C-terminal Winged-helix (WH) domain contribute to origin DNA binding. This structural information defined the polarity of initiator assembly on the origin and also indicated the induction of substantial distortion of the DNA strands, which probably triggers the unwinding of the duplex DNA to start replication.

Detailed analyses of the *oriC* region in *Pyrococcus* sp. enhanced our understanding of the initiation process of archaeal DNA replication. *Pyrococcus oriC* consists of two ORB repeats (ORB1 and ORB2), several miniORB repeats, and an A/T-rich region, predicted to be a DUE, as described above (Matsunaga *et al.*, 2003). Another cluster of ORB (ORB3) and miniORB repeats exists approximately 1.5 kb away from *oriC* and Cdc6/Orc1 (Matsunaga *et al.*, 2007). An *in vitro* experiment using the *oriC*-plasmid, which contains the *oriC* region of *P. furiosus* in a pUC-derived plasmid, showed that Cdc6/Orc1 binding to the *oriC*-plasmid introduced localized melting of the DNA duplex in the *oriC* region, without any other replication proteins. These experiments also identified the exact site of unwinding in the *oriC* region (Matsunaga *et al.*, 2010). The local unwound site is about 670 bp away from the transition site between leading and lagging syntheses, which was determined previously by an *in vivo* RIP (replication initiation point) assay (Matsunaga *et al.*, 2003).

The replication machinery must be established at the unwound site. However, it is still unclear how the origin is unwound with a width sufficient to load the replicative helicase, DNA primase, and DNA polymerase. The ATPase activity of the Cdc6/Orc1 protein was completely suppressed by binding to the ORB. In addition, limited proteolysis and DNase I-footprint experiments suggested that conformational changes occur in Cdc6/Orc1 bound to the ORB sequence in the presence of ATP. The conformational change should have an important function to start the initiation process (Akita *et al.*, 2010), as in the case of the bacterial DnaA protein. An *in vitro* recruiting assay indicated that Mcm, the replicative DNA helicase, is introduced onto the *oriC* region in a Cdc6/Orc1-dependent, but not ATP-dependent, manner

(Akita *et al.*, 2010). However, this recruitment is not sufficient for the unwinding function of Mcm, and some other function remains to be identified for the functional loading of this helicase to start the progression of the DNA replication fork.

Replicative helicase complex

To start the replication process, the replicative helicase unwinds the double-stranded DNA to form the replication forks (Fig. 8.4). The replicative helicases universally form either homohexameric or heterohexameric ring structures, and encircle the DNA strand using energy from ATP hydrolysis to drive the processive separation of paired duplex substrates (Singleton *et al.*, 2007). The replicative hexameric helicases are classified into two families, the DnaB-family proteins (Fig. 8.4A) and the MCM proteins (Fig. 8.4B), which share a conserved ATPase fold but differ in their quaternary structures (Iyer *et al.*, 2004a; Leipe *et al.*, 2003; Lyubimov *et al.*, 2011). It is interesting that the direction of translocation is opposite between DnaB and MCM. The DnaB helicases, which are present in all bacteria (Leipe *et al.*, 2000), assemble into two-tiered homohexameric rings. In each DnaB protomer, the N-terminal domain is connected to the C-terminal domain by a linker helix, which anchors the adjoining subunits together. A C-terminal RecA-like ATPase domain from each of the six subunits comprises one tier, and works as the helicase motor by hydrolysing ATP to unwind DNA strands (Biswas *et al.*, 1994, 1999b; Leipe *et al.*, 2000). The N-terminal structural domain can bind single-stranded DNA (ssDNA) (Bailey *et al.*, 2007; Lo *et al.*, 2009), and also serves as a scaffold that stabilizes the DnaB hexamers (Biswas *et al.*, 1994) and aids in ring closure (Wang *et al.*, 2008). Biochemical studies also showed that the N-terminal domain is important for binding specific partner proteins (Bailey *et al.*, 2007b; Bird *et al.*, 2000; Biswas and Biswas, 1999b; Biswas *et al.*, 1994). The crystal structure of the DnaB helicase from *A. aeolicus* in complex with ADP revealed a hexamer ring structure distinct from other previously observed closed-ring states, with a large conformational change from that seen in the substrate-free structures (Strycharska *et al.*, 2013).

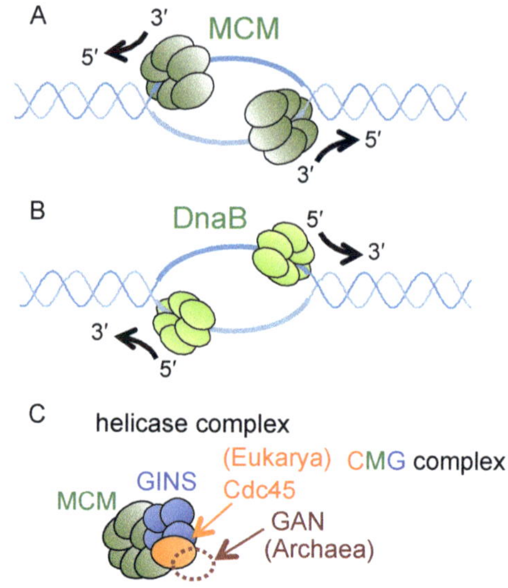

Figure 8.4 Loading of the Replicative helicase onto unwound *oriC*. The replicative helicase is loaded onto a single-strand region made by unwinding at a specific site. The common helicase core is the hexameric ring (A and B). However, more higher complex are formed at the replication fork. The CMG (Cdc45-MCM-GINS) complex is formed for the template DNA unwinding reaction in eukaryotes. On the other hand, a RecJ-like exonuclease, with weak sequence homology to that of Cdc45, may work as a helicase complex with MCM and GINS, although the exact role of its 5′–3′ exonuclease remains unknown.

In the cases of Archaea and Eukarya, MCM, a heterohexameric complex (Mcm2-7), is the catalytic component of the replicative helicase, based on evidence that the Mcm4-6-7 subcomplex from HeLa cells exhibited oligonucleotide-displacement activity *in vitro* (Ishimi, 1997). However, *in vitro* studies of the Mcm proteins did not confirm its role as a replicative helicase for a long time. The Cdc45 protein and the GINS complex, consisting of Sld5, Psf1, Psf2, and Psf3, were found as members of the replicative helicase complex. Cdc45 was originally identified as being involved in the pre-replication complex (pre-RC), but with an unknown function, in *Saccharomyces cerevisiae* (Aparicio *et al.*, 1997), and the GINS complex was also identified in *S. cerevisiae* as essential protein factors for the initiation of DNA replication (Takayama *et al.*, 2003). Immunoprecipitation experiments using *Xenopus* oocyte extracts detected a physical interaction between MCM, Cdc45, and the GINS complex proteins (Kubota *et al.*, 2003). The complex of Cdc45, MCM, and GINS, isolated by immunoaffinity, showed the activity to unwind a 40 nt-oligonucleotide annealed to single-stranded DNA in an ATP-dependent manner (Moyer *et al.*, 2006). This helicase complex is now called the 'CMG complex', in which Cdc45 and GINS are thought to play structural roles in the replicative helicase core (Fig. 8.4C).

In contrast to the heterohexameric MCM complex in Eukarya, most archaeal organisms contain only a single Mcm homologue, and its homohexameric complex exhibits distinct DNA helicase activity by itself *in vitro* (Sakakibara *et al.*, 2009). The amino acid sequence conservation among the four subunits, Sld5, Psf1, Psf2, and Psf3, of the eukaryotic GINS complex suggests that they are ancestral paralogues. The ORFs encoding the homologues of these Gins proteins were predicted in the archaeal genomes by detailed bioinformatic analyses (Makarova *et al.*, 2005). Most of the archaeal genomes contain only one homologue, but some archaea possess two homologues, one more similar to Psf2 and Psf3 and the other more similar to Sld5 and Psf1 (Makarova *et al.*, 2005). A yeast two-hybrid screen detected one Gins as an interaction partner of MCM in *S. solfataricus*, and it was designated as Gins23. Another subunit was identified by the mass spectrometry analysis of an immunoaffinity-purified native GINS complex isolated from a *S. solfataricus* cell extract using an anti-Gins23 antibody, and was designated as Gins15. The archaeal Gins formed a tetrameric structure with a 2:2 molar ratio of Gins23 and Gins15 (Marinsek *et al.*, 2006). *S. solfataricus* GINS interacts with the primase as well as MCM, but no stimulation or inhibition of either the helicase or primase activity was observed by their interactions *in vitro* (Marinsek *et al.*, 2006). *P. furiosus* also includes Gins23 and Gins51 as the complex with a 2:2 ratio (Yoshimochi *et al.*, 2008). *P. furiosus* GINS (PfuGINS) is located preferentially at the *oriC* region of *P. furiosus* chromosomal DNA during the exponential growth phase, as also reported for *P. furiosus* MCM (PfuMCM) (Matsunaga *et al.*, 2007; Yoshimochi *et al.*, 2008). Furthermore, the stimulation of the DNA helicase activity of PfuMCM by the addition of the PfuGINS complex was clearly shown *in vitro* (Yoshimochi *et al.*, 2008).

The archaeal equivalent of Cdc45 had been missing after the identification of the eukaryotic CMG complex, until recently. The GAN (GINS-associated nuclease) exonuclease, which is stimulated by interactions with GINS *in vitro* and shares a similar amino acid sequence and biochemical properties with those of the bacterial RecJ nuclease, was identified (Li *et al.*, 2011). A detailed sequence analysis also revealed that GAN, RecJ, and the eukaryotic Cdc45 protein contain a DHH phosphodiesterase domain, supporting the notion that they are homologues (Sanchez-Pulido and Ponting, 2011; Makarova *et al.*, 2012). All of the data related to the replicative helicases in Archaea have been obtained from thermophiles. It is now a hot topic to debunk the replicative helicase complex in Archaea (Fig. 8.4C).

Another interesting feature is that several archaea have multiple genes encoding Mcm homologues on the genome. Recent comprehensive genomic analyses revealed that thirteen archaeal species have more than one *mcm* gene. However, many of the *mcm* genes in the archaeal genomes seem to reside within mobile elements, originating from viruses (Krupovic *et al.*, 2010). For example, two of the three genes in the *T. kodakarensis* genome are located in regions where genetic elements have presumably been integrated (Fukui *et al.*, 2005). The genetic manipulation system of *T. kodakarensis* (Sato *et al.*, 2003, 2005) is advantageous to investigate the functions of these Mcm proteins. Gene disruption experiments isolated the deletion mutants of *mcm1* and *mcm2* (the names of the *mcm* genes are determined according to the order of the TK number), but the knockout strain for *mcm3* could not be isolated. Mcm3 is relatively abundant in *T. kodakarensis* cells, and among the purified Mcm proteins, only Mcm3 forms a stable hexameric structure in solution. These results support the proposal that Mcm3 is the main helicase core protein in the normal DNA replication process in *T. kodakarensis* (Ishino *et al.*, 2011). The genes encoding Mcm1 and Mcm2 are stably inherited, and their gene products may perform some important functions in DNA metabolism in *T. kodakarensis*. The Mcm1 protein exhibits strong helicase activity *in vitro,* and is stimulated by the *T. kodakarensis* GINS complex. Furthermore, Mcm1 co-precipitated with Mcm3 and GINS, although they did not form a heterohexameric complex, suggesting that Mcm1 is involved in the replisome or repairsome and shares some function in *T. kodakarensis* cells. Although a western blot analysis could not detect TkoMcm2 in the extract from exponentially growing *T. kodakarensis* cells, an RT-PCR experiment detected the *mcm2* gene transcript in the cells (Ishino *et al.,* unpublished). The recombinant Mcm2 protein also has ATPase and helicase activities *in vitro* (Pan *et al.,* 2011b). Therefore, the *mcm2* gene is expressed under normal growth conditions, and may work in some process with a rapid turnover. The exact functions of the Mcm1 and Mcm2 proteins remain to be elucidated. In the case of *Methanococcus maripaludis,* four *mcm* genes are present in its genome, and three of them seem to be derived from phages. The four gene products co-expressed in *E. coli* cells were co-purified in the same fraction, supporting the proposal that multiple Mcm proteins are functional in *M. maripaludis* cells (Walters *et al.,* 2010).

Recruitment of Mcm to the *oriC* region

In the helicase loading process in *Bacteria,* two DnaB hexamers should be positioned in opposite directions on the two strands of the unwound DUE, to create a bidirectional replication fork. The unwinding of a specific region in *oriC* is processed by DnaA, and subsequently, two DnaB hexamers are loaded onto the single-stranded DNA bubble by direct interactions between DnaB and DnaA, and the loading is facilitated by ATP-bound DnaC (Kobori and Kornberg, 1982; Marszalek and Kaguni, 1994; Marszalek *et al.,* 1996; Seitz *et al.,* 2000; Wickner and Hurwitz, 1975). DnaC is also an AAA^+ superfamily protein, and the crystal structure of the AAA^+ domain from *A. aeolicus* DnaC clearly showed that DnaC and DnaA are structural paralogues. DnaC can assemble into a right-handed helical structure, similar to that of DnaA. DnaC and DnaA interact directly with each other in a nucleotide-dependent manner. DnaC may bind the assembled DnaA in an ATP-dependent manner, to assist with the orientation-specific recruitment of DnaB onto an unwound DUE. It is important to understand how the DnaA complex subsequently remodels DNA and regulates the DnaC-dependent loading of DnaB onto *oriC.* The crystal structure analysis revealed that DnaC can form right-handed spiral oligomers, in spite of the more common assemblies of the AAA^+ family proteins into closed-ring hexamers or heptamers (Erzberger and Berger, 2006; Neuwald *et al.,* 1999; Ogura and Wilkinson, 2001), or notched-ring pentameric particles (Bowmanet *et al.,* 2004; Jeruzalmi *et al.,* 2001). The isolated ATPase domain of DnaC can bind to a DnaA mutant lacking its N-terminal DnaB-binding domain, but retaining its ATPase and duplex DNA binding regions. Moreover, this interaction is specific for an ATP-like state, and relies on the integrity of the DnaC AAA^+ oligomerization interface. No stable $DnaC_{AAA+}$ association is observed with ADP or under nucleotide-free conditions. DnaC works as a molecular adaptor that specifically recognizes an ATP-activated DnaA initiator, most likely through direct interactions between their AAA^+/AAA^+ domains. The association of DnaC with DnaB should preclude the filament formation.

During the *oriC*-unwinding process, DnaA binds to the upper strand of the melted region (Speck and Messer, 2001), and different faces of DnaB are recognized by DnaA and DnaC. The N-terminal domain of DnaA binds to the N-terminal side of the DnaB helicase, while the N-terminal domain of DnaC binds to the C-terminal face (Barcena *et al.,* 2001; Galletto *et al.,* 2000; Ludlam *et al.,* 2001; Saluja and Godson, 1995; Seitz *et al.,* 2000; Sutton *et al.,* 1998). The direct interaction between DnaA and DnaB is probably responsible for positioning one helicase on the unwound site. However, the identical interaction between the two proteins on the other side of the DnaA filament results in the opposite orientation of the helicase relative to the unwound position. The direct contact between DnaC and DnaA may be important to allow the second helicase to be placed with the proper geometry on the DNA. The mechanism of the DnaB ring opening for the entry of ssDNA remains to be elucidated. The spiral structure of the DnaC oligomer on the surface of DnaB may provide some force to open the ring. DnaC is actually not conserved in all bacterial genomes, and therefore, it is speculated that DnaA may fulfil the role of both initiator and helicase loader in the bacterial cells lacking DnaC. The dual functions of another protein are also speculated in Archaea and Eukarya, as described below.

How is Mcm recruited onto the unwound region of *oriC*? The detailed loading mechanism of the Mcm helicase has not been elucidated in either Archaea or Eukarya. An *in vitro* recruiting assay using the purified proteins and the *oriC* region from *P. furiosus* revealed that Mcm is recruited to the *oriC* DNA in a Cdc6/Orc1-dependent manner (Akita *et al.*, 2010). The Cdc6/Orc1 protein specifically binds to the ORB sequences in the *oriC* region as the initiator, as described above, thus suggesting the dual functions of Cdc6/Orc1 (Fig. 8.5). A critical question is how the two tasks, origin recognition and Mcm recruitment, are performed by the Cdc6/Orc1 protein. The preloading of Cdc6/Orc1 onto the ORB DNA resulted in a clear reduction in the efficiency of Mcm recruitment to *oriC*. Therefore, free Cdc6/Orc1 prefers to associate with Mcm as a helicase recruiter that brings it to *oriC*. Another important question is whether the ATP binding and hydrolysis activities of Cdc6/Orc1 regulate the Mcm recruitment onto *oriC* in the cells. The DNA binding activities of the Cdc6/Orc1 proteins from *P. furiosus*, *A. fulgidus* (Grainge *et al.*, 2003), *S. solfataricus* (Robinson *et al.*, 2004), and *A. pernix* (Grainge *et al.*, 2006) were not drastically different in the presence and absence of ATP. However, *P. furiosus* Cdc6/Orc1 changes its conformation on the DNA in the presence of ATP. This conformational change may be related to the functions of this initiator protein on the *oriC* region in Archaea. The development of a more precise *in vitro* replication system is needed to analyse the regulatory functions of Cdc6/Orc1 during replication initiation. In *M. thermoautotrophicus*, the Cdc6-2 protein, one of the two Cdc6s, can dissociate the Mcm multimers (Shin *et al.*, 2008). Cdc6-2 might be required to load the Mcm helicase in this organism. On the other hand, one of the two Cdc6s in *Thermoplasma acidophilum* stimulates the helicase activity of MCM *in vitro* with GINS, as described below (Ogino *et al.*, 2014a). Our current understanding is that

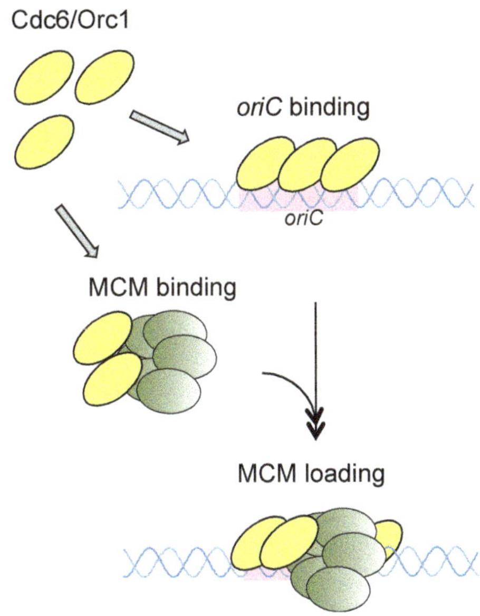

Figure 8.5 Two functions of Cdc6/Orc1, recognition of ORB and recruitment of MCM helicase onto unwound *oriC*. Based on the *in vitro* biochemical analyses described in the text, the Cdc6/Orc1 protein from *P. furiosus* may act as the origin recognition protein and Mcm helicase recruiter.

the interaction between Cdc6/Orc1 and Mcm is probably a general phenomenon in Archaea. However, the effect of Cdc6/Oric1 on the Mcm helicase activity differs among various organisms. Some distinct protein factors may function in various Archaea.

Along the same lines, the eukaryotic Cdc6 protein acts in concert with the ORC to help load the MCM2-7 helicase onto replication origins (Bell and Dutta, 2002; Donovan *et al.*, 1997; Mendez and Stillman, 2003). Experiments in yeast showed that the ATP-bound Orc1 subunit of ORC specifically recruits Cdc6 (Klemm and Bell, 2001; Randell *et al.*, 2006; Speck *et al.*, 2005), and both proteins are also AAA$^+$ ATPases (Neuwald *et al.*, 1999; Speck *et al.*, 2005).

GINS and Cdc6 as helicase enhancers

It was interesting to investigate the GINS complexes from the archaeal species with only one gene encoding a sequence homologous to Gins51 in their genomes. The recombinant Gins51 protein from *T. acidophilum* forms a homotetramer, as confirmed by gel filtration and electron microscopy analyses (Ogino *et al.*, 2011). Furthermore, its physical interaction with *T. acidophilum* Mcm was detected by a surface plasmon resonance analysis (SPR), and the helicase activity of TaMCM was clearly stimulated by TaGins51 *in vitro* (Ogino *et al.*, 2014a). This showed that a homotetrameric GINS, composed of a single protein, stimulates its cognate MCM helicases, as well as the GINS with the heterotetramers of Gins51 and Gins23 dimers from *P. furiosus* (Yoshimochi *et al.*, 2008) and *T. kodakarensis* (Ishino *et al.*, 2011). All Gins proteins from Archaea and Eukarya contain α-helical A- and β-stranded B-domains. A long intrinsically disordered linker region is found in the two domains of the Gins51 proteins from archaea that have only one Gins homologue, and this region probably allow the conserved A- and B-domains to adopt a flexible spatial arrangement (Oyama *et al.*, 2011). Experimental results showed that TaGins51 forms the tetramer and stimulated TaMCM activity without the B-domain (Ogino *et al.*, 2014b). However, the A-domain without the linker region could not form a stable tetramer. These results suggest the disordered linker region between the two domains is critical for the functional complex formation for a single Gins51 protein.

It is interesting that Cdc6-2, one of the two Cdc6/Orc1-like proteins in *T. acidophilum*, stimulated the MCM helicase *in vitro*, as described above (Haugland *et al.*, 2006). The DNA-bound TaMCM can form a complex with TaCdc6–2 and TaGINS, and the three proteins form a more stable complex on DNA (Ogino *et al.*, 2014a).

A stable eukaryotic CMG complex is formed on single-stranded DNA, as compared with Mcm2–7 alone (Ilves *et al.*, 2010). The structures of Mcm2-7 and the CMG complex from *Drosophila*, determined by electron microscopy and single particle image analysis, suggested that Cdc45 and GINS bridge the gap between Mcm2 and Mcm5 in the gapped-ring structure of the Mcm2-7 hexamer (Costa *et al.*, 2011). It is unclear whether the archaeal MCMs have gapped ring-structures when they are loaded onto DNA, although a spiral state is observed (Chen *et al.*, 2005; Costa *et al.*, 2008; Fu *et al.*, 2013; Slaymaker *et al.*, 2013). TaCdc6-2 and TaGINS are probably loaded on the TaMCM-bound DNA and form, at least, a temporary TaMCM–TaGINS–TaCdc6-2 complex. Possible models for the TaMCM activation in *T. acidophilum* are presented in Fig. 8.6. Based on the immunoprecipitation experiment from the *T. acidophilum* cells, TaCdc6-2 may leave the complex after activating the MCM conformation by assembling with TaGINS at the *oriC* region. Regarding the functions of the two Cdc6 proteins in *Thermoplasma*, the other protein, Cdc6-1, may be a recruiter for MCM to the *oriC* region, as in the case of *P. furiosus* (Akita *et*

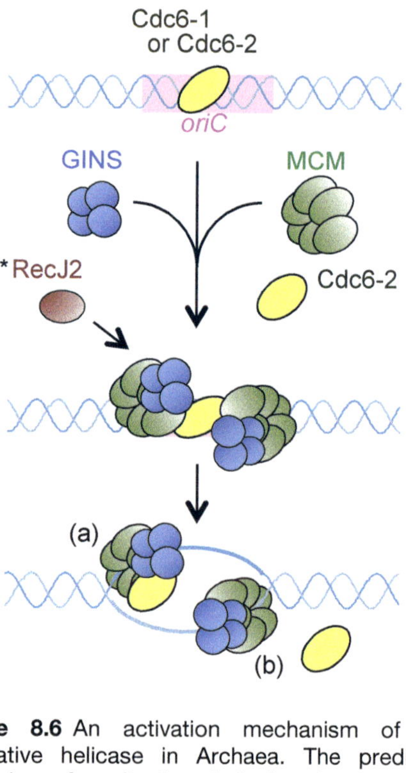

Figure 8.6 An activation mechanism of the replicative helicase in Archaea. The predicted dynamics of replication fork for the activation and progression of the replicative helicase in *T. acidophilum* is shown. Cdc6-1 or Cdc6-2 recognizes the replication origin, and Cdc6-2, MCM, and GINS form the 'CMG like' complex at the origin to activate the MCM helicase. The Cdc6-2–MCM–GINS (a) or MCM–GINS complex (b) unwinds the duplex DNA, as a replication fork progression complex. It is also possible that GAN (Cdc45) participates as part of the active helicase.

al., 2010). Cdc6-1 is autophosphorylated more efficiently than Cdc6-2 *in vitro*, suggesting that a functional difference exists between the two Cdc6 proteins (Haugland *et al.*, 2008). If phosphorylation is involved in the regulation of initiation in the archaeal cells, as in the CDK and DDK-dependent activation in Eukarya, then Cdc6-1 and Cdc6-2 may share their functions in initiation and fork progression, respectively (Fig. 8.6). On the other hand, the Cdc6-1 and Cdc6-3 proteins form a heterodimer binding to *ori2*, one of the three origins in the *S. solfataricus* genome (Dueber *et al.*, 2007). The functional purposes of the multiple Cdc6 proteins may be different in the crenarchaeota with multiple origins.

Primer synthesis

A primase is required for the synthesis of a short oligonucleotide primer to initiate DNA strand synthesis. The primers are usually ribooligonucleotides in living cells. DnaG and the p41–p46 complex are the primases in Bacteria and Archaea, respectively (Fig. 8.7). The catalytic subunit p41 of the archaeal primase shares low, but distinct sequence homology with the eukaryotic primase subunit p48 in the DNA polymerase α-primase complex, as well as with the family X DNA polymerases (Kirk *et al.*, 1999). An *in vitro* study revealed that the p41 protein from *P. furiosus* could not synthesize short RNA by itself, but unexpectedly preferentially synthesized DNA strands up to several kilobases in length (Bocquier *et al.*, 2001). A gene encoding a protein with extensive, but weak similarity to the p58 subunit of the eukaryotic primase was found adjacent to that encoding p41 in the *P. furiosus* genome. The gene product, designated as p46, actually forms a stable complex with p41, and this complex can synthesize a short RNA primer *in vitro* (Liu et *al.*, 2001). The short RNA primer, but not DNA, was identified in *Pyrococcus* cells, and therefore, DNA replication in archaea uses an RNA primer to start nascent DNA synthesis (Matsunaga *et al.*, 2003). Further research on the homologues from *S. solfataricus* (Lao-Sirieix *et al.*, 2004, 2005; De Falco *et al.*, 2004), *P. horikoshii* (Ito *et al.*, 2003; Matsui *et al.*, 2003; Ito *et al.*, 2007), and *P. abyssi* (Le Breton *et al.*, 2007) revealed similar properties *in vitro*. These results showed that the archaeal primases conserve a small p41 protein, the catalytic subunit, and the large subunit modulating the activity in the heterodimeric complex (Fig. 8.7). These subunits are also called PriS and PriL, respectively. The large subunit may adjust the primer length to a 7–14 mer, by interacting with the synthesized primer. Crystallographic studies of *S. solfataricus* primase revealed that PriL does not directly contact the active site of PriS.

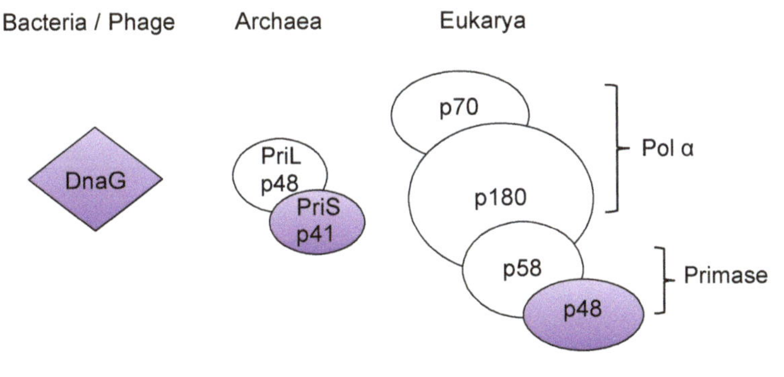

Figure 8.7 Two families of primase to synthesize RNA primer. The primase proteins markedly differ in Bacteria and Archaea/Eukarya. Bacterial DnaG works as a monomeric protein, but the primases in Archaea and Eukarya are heterocomplexes. The archaeal PriS and PriL are very weak sequence homologies to p48 and p58, respectively, in the eukaryotic primase, which forms higher complex with two other subunits of DNA polymerase α. No direct interaction with DNA polymerase has been observed in the archaeal primase. The catalytic proteins are shown in purple.

It is interesting that some archaeal genomes also encode a sequence similar to the bacterial type DnaG primase. However, the DnaG homologue from *P. furiosus* does not show any primer synthesis activity *in vitro*, and thus the DnaG-like protein does not seem to be a primase in *Pyrococcus* cells (Fujikane *et al.*, unpublished). This protein may participate in RNA degradation, as an exosome component (Evguenieva-Hackenberg *et al.*, 2003; Walter *et al.*, 2006). Genetic deletion studies confirmed that only the eukaryotic-like primase is essential for cell viability (Le Breton *et al.*, 2007). However, in the case of *S. solfataricus*, the DnaG homologue actually synthesizes 13 nucleotide primers (Zuo *et al.*, 2010). Both PriSL and DnaG probably evolved from the last universal common ancestor (LUCA) (Hu *et al.*, 2011). It would be expected that the eukaryotic-like dimeric enzyme is the DNA primase in the archaeal cells, and some crenarchaeal cells have a different system. An interesting hypothesis is that two different primases share the primer synthesis for leading and lagging strand replication, respectively, in *Sulfolobus* cells (Zuo *et al.*, 2010).

Further analyses of the PriSL primase from *Sulfolobus* provided the important information that the primase interacts with MCM through Gins23, probably to ensure the coupling of DNA unwinding and priming during the replication fork progression (Marinsek *et al.* 2006). Another direct interaction of PriSL with the clamp loader RFC may regulate the primer synthesis and its transfer to DNA polymerase in archaeal cells (Wu *et al.*, 2007).

Nascent strand synthesis

DNA polymerase catalyses phosphodiester bond formation between the terminal 3′-OH of the primer and the α-phosphate of the incoming triphosphate, to extend the DNA strand. DNA polymerases have been classified into seven families, A, B, C, D, E, X, and Y, based on the amino acid sequence similarity (Table 8.1) (Braithwaite and Ito, 1993; Cann and Ishino, 1999; Ishino and Cann, 1998; Lipps *et al.*, 2003; Ohmori *et al.*, 2001). For example, *E. coli* has five DNA polymerases. Pol I, II, and III belong to families A, B, and C, respectively, and Pol III is essential for DNA replication. Pol IV and Pol V, classified in family Y, function in translesion synthesis (TLS). In bacterial DNA replication, Pol III participates in a complex called the 'holoenzyme', with the clamp and clamp loader (McHenry, 2011). Thermophilic bacteria also have Pol III, and it is probably the replicative DNA polymerase. The α-subunit is responsible for the catalysis of phosphodiester bond formation, as suggested by the crystal structure analysis of the protein from *Thermus aquaticus* (Bailey *et al.*, 2006). The eukaryotic replicative DNA polymerases, Pol α, δ, and ε, belong to family B, and no family C enzyme exists in eukaryotes. The enzymes for TLS, including Pol η, ι, and κ, belong to family Y, as do those in the bacterial domain (Guo *et al.*, 2009). The fundamental ability to incorporate deoxymononucleotides into a DNA strand is widely conserved, but the specific properties, including processivity, synthesis accuracy, and substrate nucleotide selectivity, differ depending on the family.

Table 8.1 Distribution of DNA polymerases in the three domains of life

	A	B	C	D	E	X	Y
Bacteria	I	II	III				IV, V
Archaea							
Crenarchaeota		BI, BII			E[a]		Y
Euryarchaeota		BI		D			Y
Thaumarchaeota		BII		D			Y
Korarchaeota		BI, BII		D			
Aigarchaeota		BII		D			Y
Eukarya	θ, γ[b]	α, δ, ε, ζ				β, λ, μ, σ	η, ι, κ

[a]Plasmid coded.
[b]Mitochondrial.

One of the most fascinating features in the early age of archaeal molecular biology was that the archaea have the eukaryotic Polα-like (Family B) DNA polymerases, because no family B enzyme had been found in Bacteria (until the discovery of *E. coli* Pol II, the bacterial Polα-like enzyme) (Pisani *et al.*, 1992; Perler *et al.*, 1992; Uemori *et al.*, 1993). Molecular studies of the archaeal DNA polymerases became more interesting upon the discovery of at least two family B DNA polymerases in one archaeal cell (Uemori *et al.*, 1995; Cann *et al.*, 1999b). However, this is not true in all of the cases. Some archaea have only one family B DNA polymerase, and instead, their genomes encode a family D DNA polymerase, which seems to be specific for these archaeal organisms and has never been found in bacteria and eukaryotes (Cann and Ishino, 1999; Cann *et al.*, 1998).

The first family D DNA polymerase was identified from *P. furiosus*, by screening for DNA polymerase activity in the cell extract (Imamura *et al.*, 1995), and the corresponding gene was cloned (Uemori *et al.*, 1997). This new DNA polymerase consists of two proteins, named DP1 and DP2, and their deduced amino acid sequences are not conserved in the DNA polymerase families (Uemori *et al.*, 1997). Therefore, this DNA polymerase was proposed as PolD, from family D (Cann *et al.*, 1998; Cann and Ishino, 1999). *P. furiosus* PolD exhibits efficient strand extension activity and strong proof-reading activity (Uemori *et al.*, 1997; Ishino and Ishino 2001). Other family D DNA polymerases were also characterized (Gueguen *et al.*, 2001; Shen *et al.*, 2001, 2003; Tang *et al.*, 2004; Jokela *et al.*, 2004; Henneke *et al.*, 2005; Castrec *et al.*, 2010). PolD genes had previously only been found in Euryarchaeota. However, recent environmental genomics and cultivation efforts revealed novel phyla in Archaea: Thaumarchaeota, Korarchaeota, and Aigarchaeota, and their genome sequences harbour genes encoding PolD. A genetic study on *Halobacterium* sp. NRC-1 revealed that both PolB and PolD are essential for its viability (Berquist *et al.*, 2007). It is possible that PolB and PolD work together at the replication fork for the synthesis of the leading and lagging strands, respectively. PolD can extend the RNA primer and possesses strand displacement activity *in vitro*, and therefore it is more suitable for lagging strand synthesis (Henneke *et al.*, 2005; Ishino and Ishino, 2006). This would make archaeal organisms similar to eukaryotic organisms, in which Polε replicates the leading strand while Polδ replicates the lagging strand. However, gene knockout experiments revealed that PolB is not required for cell viability in some species (*Methanococcus*, Cubonova *et al.*, 2013; *Thermococcus*, Sarmiento *et al.*, 2013). These results suggested that PolD is the only replicative enzyme in those species.

Thaumarchaeota and Aigarchaeota harbour the genes encoding PolD and crenarchaeal PolBII (Brochier-Armanet *et al.*, 2008; Nunoura *et al.*, 2011), while Korarchaeota encodes PolBI, PolBII and PolD (Elkins *et al.*, 2008). Biochemical characterizations of these gene products will contribute to research on the evolution of DNA polymerases in living organisms. A hypothesis was proposed, in which the archaeal ancestor of eukaryotes encoded three DNA polymerases, two distinct family B polymerases and a family D polymerase, which all contributed to the evolution of the eukaryotic replication machinery, consisting of Pol α, δ, and ε (Tahirov *et al.*, 2009). The molecular evolution of PolD is mysterious so far, but it might have evolved from the common ancestor with PolB (Makarova *et al.*, 2014).

The RepA protein (ORF904) has primase and DNA polymerase activities in the N-terminal domain and helicase activity in the C-terminal domain, and is likely to be essential for the replication of pRN1 (Zillig *et al.*, 1993; Lipps *et al.*, 2003; Lipps, 2009). The amino acid sequence of the N-terminal domain lacks homology to any known DNA polymerases or primases, and therefore, it is now classified into family E. Similar proteins are encoded by various archaeal and bacterial plasmids, as well as by some bacterial viruses (Lipps, 2004). Recently, the protein tn2–12p, encoded in the plasmid pTN2 isolated from *Thermococcus nautilus*, was experimentally identified as a DNA polymerase in this family (Soler *et al.*, 2010). This enzyme should be responsible for plasmid replication in Archaea.

Genes encoding family Y-like DNA polymerases have been found in several, but not all, archaeal genomes. Translesion activity has actually been confirmed *in vitro* for some of them. A important question is whether the archaeal cells that lack the PolY homologue have any unknown translesion enzyme or whether PolB or PolD has translesion activity with other accessory factor *in vivo*. Further analyses are required to answer this question.

In summary, our current knowledge of DNA polymerases in Archaea comes from hyperthermophiles. As described above, the distribution of DNA polymerases is becoming clearer (Table 8.1) (Ishino and Ishino, 2006), but the role of each DNA polymerase in the archaeal cells is still somewhat vague. Further analyses including genetic studies will clarify this issue.

Processive DNA synthesis

A molecule with the doughnut-shaped ring structure, called the sliding clamp, is conserved among living organisms to provide processivity to the replicative process, by retaining the DNA polymerases on the template DNA strand for continuous DNA strand synthesis (Fig. 8.8). The bacterial replicase, PolIII, is referred to as a holoenzyme, containing the clamp (β-subunit) and the clamp loader (γ-complex) with the catalytic α-subunit. These proteins form an initiation complex in an ATP-dependent reaction by the clamp loader, in which a β-clamp ring is loaded around the DNA. The ATP binding subunit that drives assembly is encoded by the DnaX protein, which directs the synthesis of two proteins, τ and γ, by programmed ribosomal frameshifting. As compared with *E. coli*, the molecule corresponding to the clamp loader in thermophilic bacteria has not been detected yet, and the clamp loading mechanism has been more actively studied in Archaea and Eukarya. The eukaryotic and archaeal PCNAs form a homotrimeric ring structure (Moldovan *et al.*, 2007; Pan *et al.*, 2011a), while the ring of the bacterial β-clamps is formed by a homodimer (Fig. 8.8). A clamp loader is required to interact with the clamp and open its ring to introduce the DNA strand into the central hole of the clamp ring. In Archaea, the archaeal clamp loader is called RFC (Fig. 8.8). The best characterized archaeal PCNA and RFC molecules to date are the *P. furiosus* PCNA (Cann *et al.*, 1999a; Ishino *et al.*, 2001; Matsumiya *et al.*, 2002, 2003; Tori *et al.*, 2007) and RFC (Cann *et al.*, 2001; Mayanagi *et al.*, 2001; Ishino *et al.*, 2003; Oyama *et al.*, 2001). The molecular mechanism of the clamp loading process has been actively investigated (Indiani *et al.*, 2006). An intermediate PCNA–RFC–DNA complex, in which the PCNA ring is opened with an out-of-plane mode, was detected by a single particle analysis of electron microscopic images of the *P. furiosus* proteins (Fig. 8.9) (Miyata *et al.*, 2005). The crystal structure of the complex, including the ATP-bound clamp loader, the ring-opened clamp, and the template-primer DNA, using proteins from bacteriophage T4, has been reported (Kelch *et al.*, 2011), and our knowledge about the clamp loading mechanism is continuously improving.

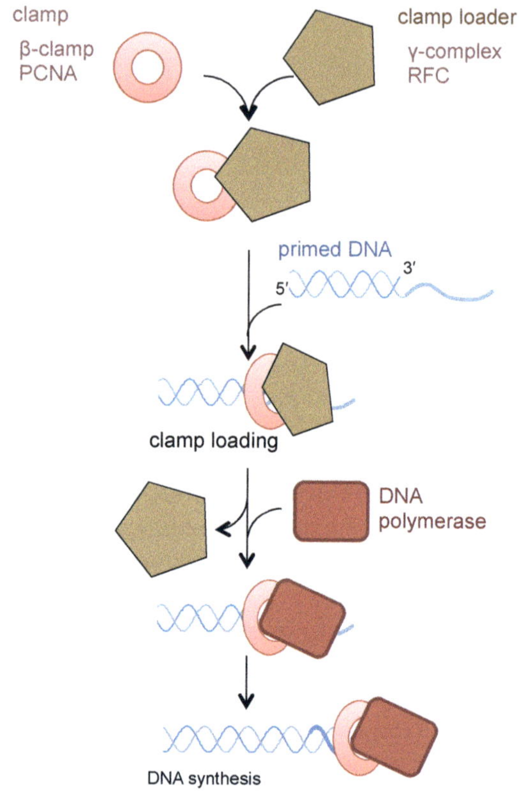

Figure 8.8 Processive DNA synthesis by the help of sliding clamp. The clamp loader (γ-complex or RFC) opens and recruits the clamp (β-clamp or PCNA) ring onto the primer terminus of the DNA strand in an ATP-dependent manner. The clamp loader is then replaced by DNA polymerase, which can synthesize the DNA strand processively without falling off.

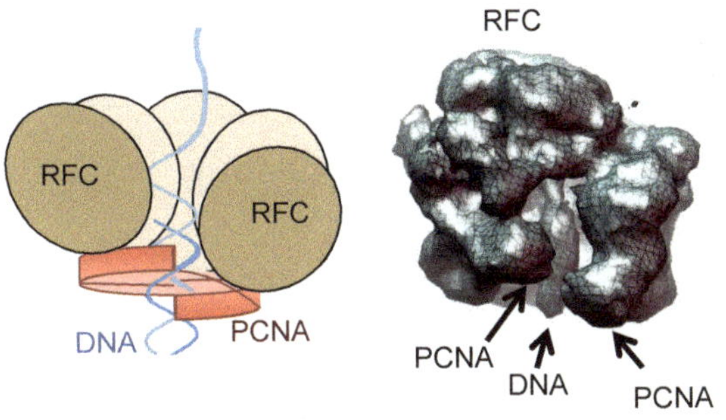

Figure 8.9 Opening mechanism of the sliding clamp ring. To load the clamp molecule onto the DNA strand, the clamp loader opens the clamp ring in an ATP-dependent manner. The structure of the clamp loading complex (PCNA-RFC-DNA ternary complex of the *P. furiosus* proteins), determined by EM single particle analysis, showed that PCNA ring retains an open conformation by extensive interactions with RFC. This spring washer-like conformation of PCNA ring appears to represent the intermediate.

To understand the overall mechanisms of replication fork progression, the DNA polymerase–PCNA complex is the next target for structural and functional analyses, because DNA polymerase accesses the clamp to form the polymerase-clamp complex after clamp loading. The PCNA interacting proteins contain a small conserved sequence motif, called the PIP box, which binds to a common site on PCNA (Warbrick, 2000). The PIP box is 'Qxxhxxaa', where x, h, and a represent any amino acid, a hydrophobic residue (e.g. L, I or M), and an aromatic residue (e.g. F, Y or W), respectively. The DNA polymerases and the clamp loaders both have PIP box-like motifs in their sequences (Vivona and Kelman, 2003). The crystal structure of *P. furiosus* PolB complexed with a monomeric PCNA mutant was determined, and a convincing model of the polymerase-PCNA ring interaction was constructed (Nishida *et al.*, 2009). This study revealed that a novel interaction is formed between a stretched loop of PCNA and the thumb domain of PolB, in addition to the authentic PIP box. A comparison of the model structure with the crystal structures of a family B DNA polymerase from RB69 phage, complexed with DNA (Shamoo *et al.*, 1999; Franklin *et al.*, 2001), suggested that the second interaction may induce a PCNA–PolB complex configuration that favours synthesis over editing. This putative mechanism for the fidelity control of replicative DNA polymerases is supported by experiments, in which mutations at the second interaction site enhanced the exonuclease activity in the presence of PCNA (Franklin *et al.*, 2001). Furthermore, the three-dimensional structure of the DNA polymerase-PCNA-DNA ternary complex was analysed by electron microscopic (EM) single particle analysis. This structural view revealed the entire domain configuration of the trimeric ring of PCNA and DNA polymerase, including the protein-protein or protein-DNA contacts (Mayanagi *et al.*, 2011a).

In contrast to most euryarchaeal organisms, which have a single PCNA homologue forming a homotrimeric ring structure, the majority of crenarchaea have multiple PCNA homologues, and they are capable of forming heterotrimeric rings for their functions (Daimon *et al.*, 2002; Dionne *et al.*, 2003). The three PCNAs have binding specificities for each of the DNA polymerases, DNA ligases, and the FEN-1 endonuclease, and therefore, these three proteins may be loaded onto a single PCNA ring to process Okazaki fragment maturation (Dionne *et al.*, 2003; Imamura *et al.*, 2007). Detailed structural studies of the heterologous PCNA from *S. solfataricus* revealed that the interaction modes between the subunits are conserved with those of the homotrimeric PCNAs (Williams *et al.*, 2006; Doré *et al.*, 2006).

In addition to the crenarchaeal organisms, the euryarchaea, *T. kodakarensis*, has two genes encoding PCNA homologues on its genome (Fukui *et al.*, 2005). Characterizations of the highly purified gene products, PCNA1 and PCNA2, revealed that each PCNA forms a homotrimeric, but not heterotrimeric, ring. PCNA1 stimulated the DNA synthesis by both PolB and PolD *in vitro*, but PCNA2 only had an effect on PolB. Furthermore, the *T. kodakarensis* strain with the *pcna2* disruption was isolated, whereas the strain with the *pcna1* gene disruption was not viable. These results suggested that PCNA1 is essential for DNA replication, and PCNA2 may play a different role in *T. kodakarensis* cells. The sensitivities of the Δ*pcna2* mutant strain to ultraviolet irradiation (UV), methyl methanesulfonate (MMS) and mitomycin C (MMC) were indistinguishable from those of the wild type strain (Kuba *et al.*, 2012). The crystal structures of the two PCNAs revealed the different interactions at the subunit–subunit interfaces (Ladner *et al.*, 2011).

The RFC molecule is conserved and forms a pentameric complex in Eukarya and Archaea. However, the eukaryotic RFC is a heteropentameric complex, consisting of five different proteins, RFC1 to 5, in which RFC1 is larger than the other four RFCs. On the other hand, the archaeal RFC consists of two proteins, RFCS (small) and RFCL (large), in a 4:1 ratio. A different form of RFC, consisting of three subunits, RFCS1, RFCS2, and RFCL, in a 3:1:1 ratio, was also identified in *M. acetivorans* (Chen *et al.*, 2005). The three subunits of RFC may represent an intermediate stage in the evolution of the more complex RFC in eukaryotes from the less complex RFC in Archaea (Chen *et al.*, 2005, 2009). The subunit organization and the spatial distribution of the subunits in the *M. acetivorans* RFC complex were analysed and compared with those of the *E. coli* γ-complex, which is also a pentamer consisting of three different proteins. These two clamp loaders adopt similar subunit organizations and spatial distributions, but the functions of the individual subunits are likely to be distinct (Chen *et al.*, 2009).

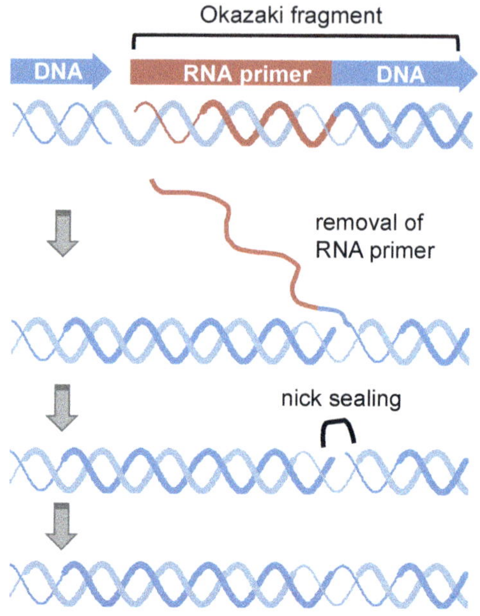

Figure 8.10 Process of Okazaki fragment maturation. The lagging strand needs to be processed to form a continuous strand. The lagging strand synthesis is commonly initiated by RNA primers, which are removed by strand displacement of the primer into a flap, flap removal, and ligation of the two discontinuous strands. Multiple proteins and pathways responsible for maturation of the lagging strand have been identified in both prokaryotes and eukaryotes.

Okazaki fragment maturation

Efficient processing of Okazaki fragments to make a continuous strand is essential for the lagging strand synthesis in the asymmetric DNA replication of all living organisms (Fig. 8.10). DNA ligase universally exists in all living organisms, to connect the Okazaki fragments of the discontinuous strand synthesis during DNA replication. The mechanism of the phosphodiester bond formation via three nucleotidyl transfer steps is well known (Wilkinson *et al.*, 2001; Tomkinson *et al.*, 2006). In the first step, DNA ligase forms a covalent enzyme-AMP intermediate, by using ATP or NAD^+ as a cofactor. In the second step, DNA ligase recognizes the substrate DNA, and the AMP is subsequently transferred from the enzyme to the 5′-phosphate terminus of the DNA, to form a DNA-adenylate intermediate (AppDNA). In the final step, the 5′-AppDNA is attacked by the adjacent 3′-hydroxy group of the DNA for the formation of a phosphodiester bond between the two DNA strands. DNA ligases are classified into two families, according to their requirement for ATP or NAD^+ as a nucleotide cofactor. ATP-dependent DNA ligases are widely found in all three domains of life, whereas NAD^+-dependent DNA ligases exist mostly in Bacteria and are also exceptionally found in some halophilic archaea (Poidevin *et al.*, 2006) and eukaryotic viruses (Benarroch *et al.*, 2006). Most of our knowledge of DNA ligases has been obtained from *E. coli* for NAD^+-dependent enzymes and the bacteriophage T4 for ATP-dependent enzymes.

A common feature of the main replicative DNA ligases in Bacteria is that their ligation activity is NAD^+-dependent (Table 8.2). The thermophilic DNA ligases from *Thermus* species have been widely characterized *in vitro*, especially for applications to genetic manipulation, but not

Table 8.2 Distribution of DNA ligases in the three domains of life

Bacteria	Archaea	Eukarya
ATP-dependent	ATP-dependent	ATP-dependent
NAD^+-dependent	NAD^+-dependent	–

for the molecular mechanism of DNA replication. *T. thermophilus* was originally used for the detection of a point mutation in the β-globin gene by LCR (Barany, 1991). The amino acid sequences of the *Thermus* DNA ligases are highly homologous, with identities ranging from 85 to 98% (Tong *et al.*, 1999).

It was interesting that a gene encoding an eukaryotic ATP-dependent DNA ligase was originally found in the thermophilic archaeon, *Desulfolobus ambivalens* (Kletzin, 1992). This fact supported the proposal that the genetic information system in Archaea is similar to that in Eukarya, but differs from that in Bacteria. Subsequent identifications of the DNA ligases from archaeal organisms revealed that these enzymes mainly use ATP as a cofactor. However, this classification may not be so strict now (Table 8.2). The utilization of NAD^+, as well as ATP, as the cofactor has been observed in several DNA ligases from thermophiles, including *T. kodakarensis* (Nakatani *et al.*, 2000), *T. fumicolans* and *P. abyssi* (Rolland *et al.*, 2004), *Thermococcus* sp. NA1 (Kim *et al.*, 2006), and *T. acidophilum*, *Picrophilus torridus*, and *Ferroplasma acidophilum* (Ferrer et al., 2008). ATP is evidently preferable in all of the cases, but the dual co-factor specificity of these ligases *in vitro* is interesting, in terms of understanding the structures and functions of these DNA ligases. Another unusual co-factor specificity exists in the archaeal enzymes from *A. pernix* (Jeon and Ishikawa, 2003) and *Staphylothermus marinus* (Seo *et al.*, 2007), which use ADP as well as ATP. However, sufficient biochemical data have not been obtained, in terms of the cofactor specificity. It was pointed out that a thermostable *E. coli* adenylate kinase, which was present as a contaminant in the purified DNA ligases from the recombinant *E. coli* cell lysate, may convert ADP into ATP and AMP (Chen *et al.*, 2009). Actually, the DNA ligases from *P. horikoshii* (Keppetipola and Shuman, 2005) and *P. furiosus* (Kiyonari *et al.*, 2006) have a strict ATP preference. Further biochemical and structural analyses are required to resolve the issue of the dual co-factor specificity.

The physical and functional interactions between DNA ligase and PCNA from *P. furiosus* have been described. The stimulatory effect of PCNA on the DNA ligase activity was observed at a high salt concentration, at which the enzyme alone cannot bind to a nicked DNA substrate. Furthermore, the PCNA-binding site of in the DNA ligase is not the N- or C-terminal region, but an internal region of its peptide chain. This location of the PCNA-binding site is unusual, because the PIP-box generally resides in the N- or C-terminal region. The binding motif QKSFF, which was proposed as a shorter version of the PIP box, is actually looped out from the protein surface in the crystal structure (Kiyonari *et al.* 2006). The Gln and the first Phe residues in the QKSFF motif are especially important for the stable binding with PCNA. The location of the PCNA-binding site is conserved in the archaeal DNA ligases (Kiyonari *et al.*, 2007).

The human Lig I protein, complexed with DNA, was the first crystal structure of an ATP-dependent cellular DNA ligase (Pascal *et al.*, 2004). The crystal structure of Lig I (residues 233 to 919) in complex with a nicked, 5′ adenylated DNA intermediate revealed that the enzyme redirects the path of the double-stranded DNA, to expose the nick termini for the strand-joining reaction. The enzyme comprises the N-terminal DNA binding domain, the middle adenylation domain, and the C-terminal OB-fold domain. The N-terminal DNA-binding domain works to encircle the DNA substrate and stabilizes it in a distorted structure, positioning the catalytic core on the nick. The crystal structure of the full-length *P. furiosus* DNA ligase (PfuLig) was reported soon after that of human Lig I (Nishida *et al.*, 2006). The architecture of each domain in PfuLig resembles those in Lig I, but the domain arrangements strikingly differ between the two enzymes (Fig. 8.11). This domain rearrangement is probably derived from the helical extension of motif VI, which is conserved at the C-termini in the archaeal and eukaryotic DNA ligases. Both the shapes and electrostatic distributions are similar between

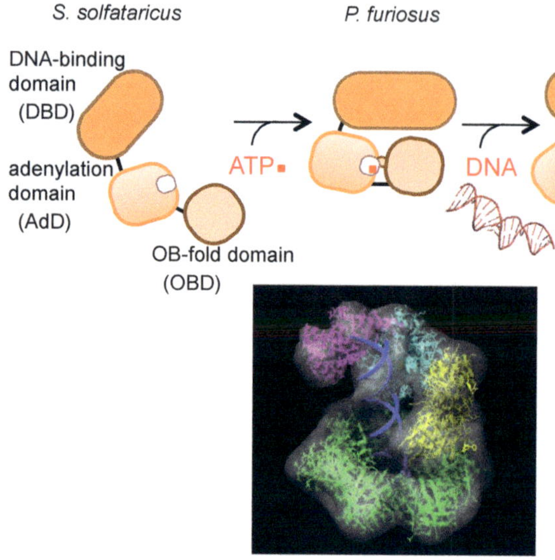

Figure 8.11 Mechanism of strand joining by DNA ligase. The process to capture the DNA strand and to perform ligation reaction by the ATP-dependent DNA ligases from Archaea and Eukarya are schematically shown. The DNA ligases, consisting of three distinct domains, flexibly change their conformations. Three crystal structures *S. solfataricus*, *P. furiosus*, and human support these conformational changes. Furthermore, the predicted process by DNA ligase with the help of PCNA is depicted at the bottom, which is a 3D model structure of the DNA ligase–PCNA–DNA complex obtained by an EM single particle analysis. The DNA ligase forms the crescent configuration, which embraces the DNA before ligation.

motif VI and the DNA substrate, and the DNA substrate in the open form of Lig I can be replaced by motif VI at the C-terminus in the closed form of PfuLig. The crystal structure of SsoLig from *S. solfataricus* represents the fully open structure, in which the three domains are highly extended (Fig. 8.11). The SsoLig–PCNA complex was also analysed by small-angle X-ray scattering (SAXS). The SsoLig bound to the PCNA ring still retains an open, extended conformation. The closed, ring-shaped conformation observed in the Lig I structure is probably the active form for catalysis, and the open-to-closed conformational change is important for ligation (Pascal *et al.*, 2006).

After the publication of these crystal structure analyses, the three-dimensional structure of the ternary complex, consisting of DNA ligase-PCNA-DNA, was obtained by EM single particle analysis (Mayanagi *et al.*, 2011b). The three domains of the crescent-shaped PfuLig, surrounding the central DNA duplex, were encircled by the closed PCNA ring. However, the relative orientations of the PfuLig domains were remarkably different from those of the crystal structures, suggesting a large domain rearrangement upon ternary complex formation. The new findings from this analysis are the PfuLig and PCNA contacts at two sites, the conventional PIP box and a novel second contact at the middle adenylation domain. It is also interesting that the DNA strand substantially tilts from the PCNA ring axis. Based on this structural model, a mechanism was proposed, in which the PCNA binding proteins, including DNA polymerase, ligase and flap endonuclease, are bound and released sequentially for Okazaki fragment maturation. In fact, most of the PCNA binding proteins share the same binding sites in the interdomain connecting loop (IDCL) and the C-terminal tail of the PCNA. The structural features exclude the possibility that the three proteins contact a single PCNA ring simultaneously, because DNA ligase occupies two of the three subunits of the PCNA trimer. In the case of the RFC–PCNA–DNA complex, which was obtained by the same EM analysis, RFC entirely covers the PCNA ring, thus blocking the access of other proteins (Miyata *et al.*, 2005). This ternary complex appears to favour a mechanism involving the sequential binding and release of replication factors, as schematically represented in Fig. 8.11.

In solution, a DNA ligase can adopt various conformations between the extended and closed forms, as described above. The ternary structure obtained by the EM analysis could be regarded as an intermediate state, just before DNA ligation.

The RNA portion and unannealed 5′-flap of the RNA/DNA primer strands synthesized by the primase, as described above, must be removed to join the Okazaki fragments into an intact continuous strand DNA (Fig. 8.10). FEN1 is mainly responsible for this task. Okazaki fragment maturation is highly coordinated with continuous DNA synthesis, and the interactions of DNA polymerase, FEN1, and DNA ligase with PCNA allow these enzymes to act sequentially during the maturation process, as described above.

FEN1, a structure-specific endonuclease, specifically recognizes double-stranded DNA with an unannealed 5′-flap (Liu *et al.*, 2004; Zheng *et al.*, 2011). In the eukaryotic Okazaki fragment processing system, 5′-flap DNA structures are formed by the strand displacement activity of Polδ, and Lig I seals the nick after the flapped DNA is cleaved by FEN1. These processing steps are facilitated by PCNA (Rossi *et al.*, 2006). The interactions between eukaryotic FEN1 and PCNA have been well characterized (Jónsson *et al.*, 1998; Warbrick, 2000), and the stimulatory effect of PCNA on the FEN1 activity was also shown (Tom *et al.*, 2000). An interesting crystal structure of the human FEN1–PCNA complex, in which three FEN1 molecules bind to each PCNA subunit of the trimer ring in different configurations, was reported (Sakurai *et al.*, 2005). This study supports a flip-flop transition mechanism, which enables the proteins to switch internally for different functions on the same DNA clamp, as described above.

Homologues of the eukaryotic FEN1 were found in Archaea (Shen *et al.*, 1998). The crystal structures of the archaeal FEN1s have been published (*M. jannaschii*, Hwang *et al.*, 1998; *P. furiosus*, Hosfield *et al.*, 1998; *P. horikoshii*, Matsui *et al.*, 2002; *A. fulgidus*, Chapados *et al.*, 2004; and *S. solfataricus*, Doré *et al.*, 2006). In addition, detailed biochemical studies were performed on *P. horikoshii* FEN1 (Matsui *et al.*, 1999, 2004). Thus, studies of the archaeal FEN1 proteins have provided important insights into the structural basis of the cleavage reaction of the flapped DNA. The flap endonuclease activity of *P. furiosus* FEN1 was stimulated by PCNA, and furthermore, this stimulatory effect was observed in the sequential actions of FEN1 and Lig *in vitro*. The quaternary complex of FEN1-Lig-PCNA-DNA was also isolated for the EM single particle analysis (Kiyonari *et al.*, unpublished). These studies provided a clearer image of the molecular mechanism. Recently, an EM analysis of the Okazakisome complex, containing *S. solfataricus* DNA polymerase, DNA ligase, and FEN1 on the DNA strand was published (Cannone *et al.*, 2015). This supports the tool belt model of Okazaki fragment maturation, in which the binding proteins are stably associated with PCNA throughout the process. The simultaneous binding of two different partners has also been demonstrated for the bacterial sliding clamp, although not for the Okazakisome (Indiani *et al.*, 2005). However, a recent report on human Okazaki fragment maturation suggested a mechanism involving the sequential switching of partners on the eukaryotic PCNA trimer, since simultaneous binding of multiple partners to PCNA does not provide significant functional advantages for PCNA-mediated Okazaki fragment maturation *in vitro* (Dovrat *et al.*, 2014). The molecular mechanism of Okazaki fragment maturation may be different between prokaryotes and eukaryotes.

Summary and perspectives

Research on the molecular mechanism of DNA replication has been a central theme since the dawn of molecular biology. The basic mechanism of DNA replication was determined from *E. coli* and the bacteriophages that infect *E. coli*. Archaeal organisms became popular in the total genome sequencing age, as described above, and most of the DNA replication proteins from the hyperthermophilic archaea are now equally understood by biochemical characterizations. In addition, studies on thermophilic bacteria and archaea are especially interesting, to understand the mechanisms by which cells live in extreme environmental conditions. Furthermore, it is also noteworthy that the proteins from the thermophilic organisms are more stable than those from mesophilic organisms, and they are especially advantageous for the structural and functional analyses of higher ordered complexes, such as the replisome. Studies on the higher ordered complexes, rather than single proteins, are essential for understanding each of the events involved in DNA metabolism, and the research on thermophiles will continuously contribute to the development and advancement of the DNA replication research field, as summarized in part in recent reviews (Beattie *et al.*, 2011; Ishino and Ishino, 2012).

In addition to basic research for understanding the DNA replication mechanism, DNA replication proteins from thermophiles are quite useful reagents for gene manipulations. Numerous enzymes have been commercialized and are utilized daily around the world. The most remarkable example is the thermostable DNA polymerases for PCR applications. Several DNA sequencing technologies, referred to as 'next-generation sequencing', have been developed (Shendure, 2008; Ansorge, 2009), and are now widely used for obtaining large quantities of data. These technologies apply DNA polymerases or DNA ligases from various sources, and thus these DNA replication enzymes are indispensable for the development of DNA manipulation technology. These facts prove that the progress in the basic research on DNA replication in thermophiles will promote the development of new technologies for genetic engineering.

References

Abe, Y., Jo, T., Matsuda, Y., Matsunaga, C., Katayama, T., and Ueda, T. (2007). Structure and function of DnaA N-terminal domains: specific sites and mechanisms in inter-DnaA interaction and in DnaB helicase loading on *oriC*. J. Biol. Chem. *282*, 17816–17827.

Akita, M., Adachi, A., Takemura, K., Yamagami, T., Matsunaga, F., and Ishino, Y. (2010). Cdc6/Orc1 from *Pyrococcus furiosus* may act as the origin recognition protein and Mcm helicase recruiter. Genes Cells, *15*, 537–552.

Ansorge, W.J. (2009). Next-generation DNA sequencing techniques. Nat. Biotechnol. *25*, 195–203.

Aparicio, O.M., Weinstein, D.M., and Bell, S.P. (1997). Components and dynamics of DNA replication complexes in *S. cerevisiae*: redistribution of MCM proteins and Cdc45p during S phase. Cell *91*, 59–69.

Bailey, S., Wing, R.A., and Steitz, T.A. (2006). The structure of *T. aquaticus* DNA polymerase III is distinct from eukaryotic replicative DNA polymerases. Cell *126*, 893–904.

Bailey, S., Eliason, W.K., and Steitz, T.A. (2007). Structure of hexameric DnaB helicase and its complex with a domain of DnaG primase. Science *318*, 459–463.

Barany, F. (1991). Genetic disease detection and DNA amplification using cloned thermostable ligase. Proc. Natl. Acad. Sci. U.S.A. *88*, 189–193.

Barry, E.R., and Bell, S.D. (2006). DNA replication in the archaea. Microbiol. Mol. Biol. Rev. *70*, 876–887.

Barcena, M., Ruiz, T., Donate, L.E., Brown, S.E., Dixon, N.E., Radermacher, M., and Carazo, J.M. (2001). The DnaB.DnaC complex: a structure based on dimers assembled around an occluded channel. EMBO J. *20*, 1462–1468.

Beattie, T.R., and Bell, S.D. (2011). Molecular machines in archaeal DNA replication. Curr. Opin. Chem. Biol, *15*, 614–619.

Bell, S.P., and Dutta, A. (2002). DNA replication in eukaryotic cells. Annu. Rev. Biochem. *71*, 333–374.

Benarroch, D., and Shuman, S. (2006). Characterization of mimivirus NAD+-dependent DNA ligase. Virology *353*, 133–143.

Berquist, B.R., DasSarma, P., and DasSarma, S. (2007). Essential and non-essential DNA replication genes in the model halophilic Archaeon, *Halobacterium* sp. NRC-1. BMC Genet. *8*, 31.

Bird, L.E., Pan, H., Soultanas, P., and Wigley, D.B. (2000). Mapping protein–protein interactions within a stable complex of DNA primase and DnaB helicase from *Bacillus stearothermophilus*. Biochemistry *39*, 171–182.

Biswas, S.B., Chen, P.H., and Biswas, E.E. (1994). Structure and function of *Escherichia coli* DnaB protein: role of the N-terminal domain in helicase activity. Biochemistry *33*, 11307–11314.

Biswas, E.E., and Biswas, S.B. (1999). Mechanism of DnaB helicase of *Escherichia coli*: structural domains involved in ATP hydrolysis, DNA binding, and oligomerization. Biochemistry *38*, 10919–10928.

Bocquier, A., Liu, L., Cann, I., Komori, K., Kohda, D., and Ishino, Y. (2001). Archaeal primase: bridging the gap between RNA and DNA polymerase. Curr. Biol. *11*, 452–456.

Bowman, G.D., O'Donnell, M., and Kuriyan, J. (2004). Structural analysis of a eukaryotic sliding DNA clamp-clamp loader complex. Nature *429*, 724–730.

Braithwaite, D.K., and Ito, J. (1993). Compilation, alignment, and phylogenetic relationships of DNA polymerases. Nucleic Acids Res. *21*, 787–802.

Brochier-Armanet, C., Boussau, B., Gribaldo, S., and Forterre, P. (2008). Mesophilic Crenarchaeota: proposal for a third archaeal phylum, the Thaumarchaeota. Nat. Rev. Microbiol. *6*, 245–252.

Cann, I., and Ishino, Y. (1999). Archaeal DNA replication: Identifying the pieces to solve a puzzle. Genetics *152*, 1249–1267.

Cann, I., Komori, K., Toh, H., Kanai, S., and Ishino, Y. (1998). A heterodimeric DNA polymerase: evidence that members of the euryarchaeota possess a novel DNA polymerase. Proc. Natl. Acad. Sci. U.S.A. *95*, 14250–14255.

Cann, I., Ishino, S., Hayashi, I., Komori, K., Toh, H., Morikawa, K., and Ishino, Y. (1999a). Functional interactions of a homolog of proliferating cell nuclear antigen with DNA polymerases in Archaea. J. Bacteriol. *181*, 6591–6599.

Cann, I., Ishino, S., Nomura, N., Sako, Y., and Ishino, Y. (1999b). Two family B DNA polymerases in Aeropyrum pernix, an obligate aerovic hyperthermophilic crenarchaeote. J. Bacteriol. *181*, 5984–5992.

Cann, I., Ishino, S., Yuasa, M., Daiyasu, H., Toh, H., and Ishino, Y. (2001). Biochemical analysis of the replication factor C from the hyperthermophilic archaeon *Pyrococcus furiosus*. J. Bacteriol. *183*, 2614–2623.

Cannone, G., Xu, Y., Beattie, T.R., Bell, S.D., and Spagnolo, L. (2015). The architecture of an Okazaki fragment-processing holoenzyme from the archaeon *Sulfolobus solfataricus*. Biochem. J. *465*, 239–245.

Capaldi, S.A., and Berger, J.M. (2004). Biochemical characterization of Cdc6/Orc1 binding to the replication origin of the euryarchaeon *Methanothermobacter thermoautotrophicus*. Nucleic Acids Res. *32*, 4821–4832.

Castrec, B., Laurent, S., Henneke, G., Flament, D., and Raffin, J.P. (2010). The glycine-rich motif of *Pyrococcus abyssi* DNA polymerase D is critical for protein stability. J. Mol. Biol. *396*, 840–848.

Chapados, B.R., Hosfield, D.J., Han, S., Qiu, J., Yelent, B., Shen, B., and Tainer, J.A. (2004). Structural basis for FEN-1 substrate specificity and PCNA-mediated activation in DNA replication and repair. Cell *116*, 39–50.

Chen, B., Sysoeva, T.A., Chowdhury, S., Guo, L., and Nixon, B.T. (2009). ADPase activity of recombinantly expressed thermotolerant ATPases may be caused by copurification of adenylate kinase of *Escherichia coli*. FEBS J. *276*, 807–815.

Chen, Y.H., Kocherginskaya, S.A., Lin, Y., Sriratana, B., Lagunas, A.M., Robbins, J.B., Mackie, R.I., and Cann, I.K. (2005). Biochemical and mutational analyses of a unique clamp loader complex in the archaeon *Methanosarcina acetivorans*. J. Biol. Chem. *280*, 41852–41863.

Chen, Y.H., Lin, Y., Yoshinaga, A., Chhotani, B., Lorenzini, J.L., Crofts, A.A., Mei, S., Mackie, R.I., Ishino, Y., and Cann, I.K.O. (2009). Molecular analyses of a three-subunit euryarchaeal clamp loader complex from *Methanosarcina acetivorans*. J. Bacteriol. *191*, 6539–6549.

Chen, Y.J., Yu, X., Kasiviswanathan, R., Shin, J.H., Kelman, Z., and Egelman, E.H. (2005). Structural polymorphism of *Methanothermobacter thermautotrophicus* MCM. J. Mol. Biol. *346*, 389–394.

Costa, A., van Duinen, G., Medagli, B., Chong, J., Sakakibara, N., Kelman, Z., Nair, S.K., Patwardhan, A., and Onesti, S. (2008). Cryo-electron microscopy reveals a novel DNA-binding site on the MCM helicase. EMBO J. *27*, 2250–2258.

Costa, A., Ilves, I., Tamberg, N., Petojevic, T., Nogales, E., Botchan, M.R., and Berger, J.M. (2011). The structural basis for MCM2-7 helicase activation by GINS and Cdc45. Nat. Struct. Mol. Biol. *18*, 471–477.

Costa, A., Hood, I.V., and Berger, J.M. (2013). Mechanisms for initiating cellular DNA replication. Annu. Rev. Biochem. *82*, 25–54.

Cubonová, L., Richardson, T., Burkhart, B.W., Kelman, Z., Connolly, B.A., Reeve, J.N., and Santangelo, T.J. (2013). Archaeal DNA polymerase D but not DNA polymerase B is required for genome replication in *Thermococcus kodakarensis*. J. Bacteriol. *195*, 2322–2328.

Daimon, K., Kawarabayasi, Y., Kikuchi, H., Sako, Y., and Ishino, Y. (2002). Three proliferating cell nuclear antigen-like proteins found in the hyperthermophilic archaeon *Aeropyrum pernix*: interactions with the two DNA polymerases. J. Bacteriol. *184*, 687–694.

De Falco, M., Fusco, A., DeFelice, M., Rossi, M., and Pisani, F.M. (2004). The DNA primase is activated by substrates containing a thymine-rich bubble and has a 3′-terminal nucleotidyl-transferase activity. Nucleic Acids Res. *32*, 5223–5230.

Dionne, I., Nookala, R.K., Jackson, S.P., Doherty, A.J., and Bell, S.D. (2003). A heterotrimeric PCNA in the hyperthermophilic archaeon *Sulfolobus solfataricus*. Mol. Cell *11*, 275–282.

Donovan, S., Harwood, J., Drury, L.S., and Diffley, J.F. (1997). Cdc6p-dependent loading of Mcm proteins onto pre-replicative chromatin in budding yeast Proc. Natl. Acad. Sci. U.S.A. *94*, 5611–5616.

Doré, A.S., Kilkenny, M.L., Jones, S.A., Oliver, A.W., Roe, S.M., Bell, S.D., and Pearl, L.H. (2006). Structure of an archaeal PCNA1–PCNA2-FEN1 complex: elucidating PCNA subunit and client enzyme specificity. Nucleic Acids Res. *34*, 4515–4526.

Dovrat, D., Stodola, J.L., Burgers, P.M., and Aharoni, A. (2014). Sequential switching of binding partners on PCNA during *in vitro* Okazaki fragment maturation. Proc. Natl. Acad. Sci. U.S.A. *111*, 14118–14123.

Duderstadt, K.E., and Berger, J.M. (2013). A structural framework for replication origin opening by AAA+ initiation factors. Curr. Opin. Struct. Biol. 23, 144–153.

Dueber, E.L., Corn, J.E., Bell, S.D., and Berger, J.M. (2007). Replication origin recognition and deformation by a heterodimeric archaeal Orc1 complex. Science *317*, 1210–1213.

Elkins, J.G., Podar, M., Graham, D.E., Makarova, K.S., Wolf, Y., Randau, L., Hedlund, B.P., Brochier-Armanet, C., Kunin, V., Anderson, I., *et al.* (2008). A korarchaeal genome reveals insights into the evolution of the Archaea. Proc. Natl. Acad. Sci. USA, *105*, 8102–8107.

Erzberger, J.P., and Berger, J.M. (2006). Evolutionary relationships and structural mechanisms of AAA+ proteins. Annu. Rev. Biophys. Biomol. Struct. 35, 93–114.

Erzberger, J.P., Pirruccello, M.M., and Berger, J.M. (2002). The structure of bacterial DnaA: implications for general mechanisms underlying DNA replication initiation. EMBO J. *21*, 4763–4773.

Evguenieva-Hackenberg, E., Walter, P., Hochleitner, E., Lottspeich, F., and Klug, G. (2003). An exosome-like complex in *Sulfolobus solfataricus*. EMBO Rep. *4*, 889–934.

Felczak, M.M., Simmons, L.A., and Kaguni, J.M. (2005). An essential tryptophan of Escherichia coli DnaA protein functions in oligomerization at the *E. coli* replication origin. J. Biol. Chem. *280*, 24627–24633.

Ferrer, M., Golyshina, O.V., Beloqui, A., Böttger, L.H., Andreu, J.M., Polaina, J., De Lacey, A.L., Trautwein, A.X., Timmis, K.N., and Golyshin, P.N. (2008). A purple acidophilic di-ferric DNA ligase from *Ferroplasma*. Proc. Natl. Acad. Sci. U.S.A. *105*, 8878–8883.

Franklin, M.C., Wang, J., and Steitz, T.A. (2001). Structure of the replicating complex of a pol α family DNA polymerase. Cell *105*, 657–667.

Fu, Y., Slaymaker, I.M., Wang, J.,Wang, G., and Chen, X.S. (2013). The 1.8-Å Crystal structure of the N-terminal domain of an Archaeal MCM as a right-handed filament. J. Mol. Biol. *426*, 1512–1523.

Fujikawa, N., Kurumizaka, H., Nureki, O., Terada, T., Shirouzu, M., Katayama, T., and Yokoyama, S. (2003). Structural basis of replication origin recognition by the DnaA protein. Nucleic Acids Res. *31*, 2077–2086.

Fukui, T., Atomi, H., Kanai, T., Matsumi, R., Fujiwara, S., and Imanaka, T. (2005). Complete genome sequence of the hyperthermophilic archaeon *Thermococcus kodakarensis* KOD1 and comparison with *Pyrococcus* genomes. Genome Res. *15*, 352–363.

Galletto, R., Rajendran, S., and Bujalowski, W. (2000). Interactions of nucleotide cofactors with the *Escherichia coli* replication factor DnaC protein Biochemistry *39*, 12959–12969.

Gaudier, M., Schuwirth, B.S., Westcott, S.L., and Wigley, D.B. (2007). Structural basis of DNA replication origin recognition by an ORC protein. Science *317*, 1213–1216.

Grainge, I., Scaife, S., and Wigley, D.B. (2003). Biochemical analysis of components of the pre-replication complex of Archaeoglobus fulgidus. Nucleic Acids Res. *31*, 4888–4898.

Grainge, I., Gaudier, M., Schuwirth, B.S., Westcott, S.L., Sandall, J., Atanassova, N., and Wigley, D.B. (2006). Biochemical analysis of a DNA replication origin in the archaeon Aeropyrum pernix. J. Mol. Biol. *363*, 355–369.

Grogan, D.W. (2000). The question of DNA repair in hyperthermophilic archaea. Trends Microbiol. *8*, 180–185.

Gueguen, Y., Rolland, J.L., Lecompte, O., Azam, P., Le Romancer, G., Flament, D., Raffin, J.P., and Dietrich, J. (2001). Characterization of two DNA polymerases from the hyperthermophilic euryarchaeon *Pyrococcus abyssi*. Eur. J. Biochem. *268*, 5961–5969.

Guo, C., Kosarek-Stancel, J.N., Tang, T.S., and Friedberg, E.C. (2009). Y-family DNA polymerases in mammalian cells. Cell Mol. Life. Sci. *66*, 2363–2381.

Haugland, G.T., Shin, J.H., Birkeland, N.K., and Kelman, Z. (2006). Stimulation of MCM helicase activity by a Cdc6 protein in the archaeon *Thermoplasma acidophilum*. Nucleic Acids Res. *34*, 6337–6344.

Haugland, G.T., Innselset, M., Madern, D., and Birkeland, N.K. (2008). Characterization of the Cdc6 Homologues from the Euryarchaeon *Thermoplasma acidophilum*. Open Biochem. J. 2, 129–134.

Hawkins, M., Malla, S., Blythe, M.J., Nieduszynski, C.A., and Allers, T. (2013). Accelerated growth in the absence of DNA replication origins. Nature *503*, 544–547.

Henneke, G., Flament, D., Hübscher, U., Querellou, J., and Raffin, J.P. (2005). The hyperthermophilic euryarchaeota *Pyrococcus abyssi* likely requires the two DNA polymerases D and B for DNA replication. J. Mol. Biol. *350*, 53–64.

Hosfield, D.J., Mol, C.D., Shen, B., and Tainer, J.A. (1998). Structure of the DNA repair and replication endonuclease and exonuclease FEN-1: coupling DNA and PCNA binding to FEN-1 activity. Cell *95*, 135–146.

Hu, J., Guo, L., Wu, K., Lang, S., and Huang, L. (2011). Template-dependent polymerization across discontinuous templates by the heterodimeric primase from the hyperthermophilic archaeon *Sulfolobus solfataricus*. Nucleic Acids Res. *40*, 3470–3483.

Hwang, K.Y., Baek, K., Kim, H.Y., and Cho, Y. (1998). The crystal structure of flap endonuclease-1 from *Methanococcus jannaschii*. Nat. Struct. Biol. *5*, 707–713.

Imamura, K., Fukunaga, K., Kawarabayasi, Y., and Ishino, Y. (2007). Specific interactions of three proliferating cell nuclear antigens with replication-related proteins in *Aeropyrum pernix*. Mol. Microbiol. *64*, 308–318.

Imamura, M., Uemori, T., Kato, I., and Ishino, Y. (1995). A non-α-like DNA polymerase from the hyperthermophilic archaeon *Pyrococcus furiosus*. Biol. Pharm. Bull. *18*, 1647–1652.

Indiani, C., McInerney, P., Georgescu, R., Goodman, M.F., and O'Donnell, M. (2005). A sliding clamp toolbelt binds high- and low-fidelity DNA polymerases simultaneously. Mol. Cell *19*, 805–815.

Ishino, S., and Ishino, Y. (2006). Conprehensive search for DNA polymerase in the hyperthermophilic archaeon, *Pyrococcus furiosus*. Nucleos. Nucleot. Nucl. *25*, 681–691.

Ishino, S., and Ishino, Y. (2012). Rapid progress of DNA replication studies in Archaea, the third domain of life. Sci. China. Life. Sci. *55*, 386–403.

Ishino, S., Oyama, T., Yuasa, M., Morikawa, K., and Ishino, Y. (2003). Mutational Analysis of *Pyrococcus furiosus* Replication Factor C based on the Three-Dimensional Structure. Extremophiles 7, 169–175.

Ishino, S., Fujino, S., Tomita, H., Ogino, H., Takao, K., Daiyasu, H., Kanai, T., Atomi, H., and Ishino, Y. (2011). Biochemical and genetical analyses of the three mcm genes from the hyperthermophilic archaeon, *Thermococcus kodakarensis*. Genes Cells *16*, 1176–1189.

Ishino, S., Kelman, L.M., Kelman, Z., and Ishino, Y. (2013). The archaeal DNA replication machinery: past, present and future. Genes Genet. Syst. *88*, 315–319.

Ishimi, Y. (1997). A DNA helicase activity is associated with an MCM4, -6, and -7 protein complex. J. Biol. Chem. 272, 24508–24513.

Ishino, Y., and Cann, I. (1998). The euryarchaeotes, a subdomain of archaea, survive on a single DNA polymerase: fact or farce? Genes Genet. Syst. 73, 323–336.

Ishino, Y., and Ishino, S. (2001). Novel DNA polymerases from Euryarchaeota. Meth. Enzymol. *334*, 249–260.

Ishino, Y., Tsurimoto, T., Ishino, S., and Cann, I.K. (2001). Functional interactions of an archaeal sliding clamp with mammalian clamp loader and DNA polymerase δ. Genes Cells *6*, 699–706.

Ito, N., Nureki, O., Shirouzu, M., Yokoyama, S., and Hanaoka, F. (2003). Crystal structure of the *Pyrococcus horikoshii* DNA primase–UTP complex: implications for the mechanism of primer synthesis. Genes Cells *8*, 913–923.

Ito, N., Matsui, I., and Matsui, E. (2007). Molecular basis for the subunit assembly of the primase from an archaeon *Pyrococcus horikoshii*. FEBS J. *274*, 1340–1351.

Ilves, I., Petojevic, T., Pesavento, J.J., and Botchan, M.R. (2010). Activation of the MCM2–7 helicase by association with Cdc45 and GINS proteins. Mol. Cell *37*, 247–258.

Indiani, C., and O'Donnell, M. (2006). The replication clamp-loading machine at work in the three domains of life. Nat. Rev. Mol. Cell Biol. 7, 751–761.

Iyer, L.M., Leipe, D.D., Koonin, E.V., and Aravind, L. (2004a). Evolutionary history and higher order classification of AAA+ ATPases. J. Struct. Biol. *146*, 11–31.

Iyer, L.M., Makarova, K.S., Koonin, E.V., and Aravind, L. (2004b). Comparative genomics of the FtsK-HerA superfamily of pumping ATPases: implications for the origins of chromosome segregation, cell division and viral capsid packaging. Nucleic Acids Res. *32*, 5260–5279.

Jacob, F., and Brenner, S.C.R. (1963). On the regulation of DNA synthesis in bacteria: the hypothesis of the replicon. Hebd. Seances Acad. Sci. *256*, 298–300.

Jeon, S.J., and Ishikawa, K. (2003). A novel ADP-dependent DNA ligase from *Aeropyrum pernix* K1. FEBS Lett. *550*, 69–73.

Jeruzalmi, D., O'Donnell, M., and Kuriyan, J. (2001). Crystal structure of the processivity clamp loader γ (gamma) complex of *E. coli* DNA polymerase III. Cell *106*, 429–441.

Jokela, M., Eskelinen, A., Pospiech, H., Rouvinen, J., and Syväoja, J.E. (2004). Characterization of the 3′ exonuclease subunit DP1 of *Methanococcus jannaschii* replicative DNA polymerase D. Nucleic Acids Res. *32*, 2430–2440.

Jónsson, Z.O., Hindges, R., and Hübscher, U. (1998). Regulation of DNA replication and repair proteins through interaction with the front side of proliferating cell nuclear antigen. EMBO J. *17*, 2412–2425.

Kaguni, J.M. (2006). DnaA: controlling the initiation of bacterial DNA replication and more. Annu. Rev. Microbiol. *60*, 351–375.

Kawakami, H., and Katayama, T. (2010). DnaA, ORC, and Cdc6: similarity beyond the domains of life and diversity. Biochem. Cell. Biol. *88*, 49–62.

Kelch, B.A., Makino, D.L., O'Donnell, M., and Kuriyan, J. (2011). The replication clamp-loading machine at work in the three domains of life. Science *334*, 1675–1680.

Kelman, Z., and White, M.F. (2005). Archaeal DNA replication and repair. Curr. Opin. Microbiol. *8*, 669–676.

Keppetipola, N., and Shuman, S. (2005). Characterization of a thermophilic ATP-dependent DNA ligase from the euryarchaeon *Pyrococcus horikoshii*. J. Bacteriol. *187*, 6902–6908.

Keyamura, K., Fujikawa, N., Ishida, T., Ozaki, S., Su'etsugu, M., Fujimitsu, K., Kagawa, W., Yokoyama, S., Kurumizaka, H., and Katayama, T. (2007). The interaction of DiaA and DnaA regulates the replication cycle in *E. coli* by directly promoting ATP DnaA-specific initiation complexes. Genes Dev. *21*, 2083–2099.

Kim, Y.J., Lee, H.S., Bae, S.S., Jeon, J.H., Yang, S.H., Lim, J.K., Kang, S.G., Kwon, S.T., and Lee, J.H. (2006). Cloning, expression, and characterization of a DNA ligase from a hyperthermophilic archaeon *Thermococcus* sp. Biotechnol. Lett. *28*, 401–407.

Kirk, B.W., and Kuchta, R.D. (1999). Arg304 of human DNA primase is a key contributor to catalysis and NTP binding: primase and the family X polymerases share significant sequence homology. Biochemistry *38*, 7727–7736.

Kiyonari, S., Takayama, K., Nishida, H., and Ishino, Y. (2006). Identification of a novel binding motif in *Pyrococcus furiosus* DNA ligase for the functional interaction with proliferating cell nuclear antigen. J. Biol. Chem. *281*, 28023–28032.

Kiyonari, S., Kamigochi, T., and Ishino, Y. (2007). A single amino acid substitution in the DNA-binding domain of *Aeropyrum pernix* DNA ligase impairs its interaction with proliferating cell nuclear antigen, Extremophiles *11*, 675–684.

Klemm, R.D., and Bell, S.P. (2001). ATP bound to the origin recognition complex is important for preRC formation. Proc. Natl. Acad. Sci. U.S.A. *98*, 8361–8367.

Kletzin, A. (1992). Molecular characterisation of a DNA ligase gene of the extremely thermophilic archaeon *Desulfurolobus ambivalens* shows close phylogenetic relationship to eukaryotic ligases. Nucleic Acids Res. *20*, 5389–5396.

Kobori, J.A., and Kornberg, A. (1982). The *Escherichia coli* dnaC gene product. III. Properties of the dnaB-dnaC protein complex J. Biol. Chem. 257, 13770–13775.

Kuba, Y., Ishino, S., Yamagami, T., Tokuhara, M., Kanai, T., Fujikane, R., Daiyasu, H., Atomi, H., and Ishino, Y. (2012). Comparative analyses of the two PCNAs from the hyperthermophilic archaeon, *Thermococcus lodakarensis*. Genes Cells *17*, 923–937.

Kubota, Y., Takase, Y., Komori, Y., Hashimoto, Y., Arata, T., Kamimura, Y., Araki, H., and Takisawa, H. (2003). A novel ring-like complex of Xenopus proteins essential for the initiation of DNA replication. Genes Dev. *17*, 1141–1152.

Krupovic, M., Gribaldo, S., Bamford, D.H., and Forterre, P. (2010). The evolutionary history of archaeal MCM helicases: a case study of vertical evolution combined with hitchhiking of mobile genetic elements. Mol. Biol. Evol. 27, 2716–2732.

Ladner, J.E., Pan, M., Hurwitz, J., and Kelman, Z. (2011). Crystal structures of two active proliferating cell nuclear antigens (PCNAs) encoded by *Thermococcus kodakaraensis*. Proc. Natl. Acad. Sci. U.S.A. *108*, 2711–2716.

Lao-Sirieix, S.H., and Bell, S.D. (2004). The heterodimeric primase of the hyperthermophilic archaeon *Sulfolobus solfataricus* possesses DNA and RNA primase, polymerase and 3′-terminal nucleotidyl transferase activities. J. Mol. Biol. *344*, 1251–1263.

Lao-Sirieix, S.H., Nookala, R.K., Roversi, P., Bell, S.D., and Pellegrini, L. (2005). Structure of the heterodimeric core primase. Nat. Struct. Mol. Biol. *12*, 1137–1144.

Le Breton, M., Henneke, G., Norais, C., Flament, D., Myllykallio, H., Querellou, J., and Raffin, J.P. (2007). The heterodimeric primase from the euryarchaeon *Pyrococcus abyssi*: a multifunctional enzyme for initiation and repair? J. Mol. Biol. *374*, 1172–1185.

Leipe, D.D., Aravind, L., and Koonin, E.V. (1999). Did DNA replication evolve twice independently? Nucleic Acids Res. *27*, 3389–3401.

Leipe, D.D., Aravind, L., Grishin, N.V., and Koonin, E.V. (2000). The bacterial replicative helicase DnaB evolved from a RecA duplication. Genome Res. *10*, 5–16.

Leipe, D.D., Koonin, E.V., and Aravind, L. (2003). Evolution and classification of P-loop kinases and related proteins. J. Mol. Biol. *333*, 781–815.

Levin, D.S., Bai, W., and Yao, N. (1997). An interaction between DNA ligase I and proliferating cell nuclear antigen: implications for Okazaki fragment synthesis and joining. Proc. Natl. Acad. Sci. U.S.A. *94*, 12863–12868.

Li, Z., Pan, M., Santangelo, T.J., Chemnitz, W., Yuan, W., Edwards, J.L., Hurwitz, J., Reeve, J.N., and Kelman, Z. (2011). A novel DNA nuclease is stimulated by association with the GINS complex. Nucleic Acids Res. *39*, 6114–6123.

Lipps, G. (2004). The replication protein of the *Sulfolobus islandicus* plasmid pRN1. Biochem. Soc. Trans. *32*, 240–244.

Lipps, G. (2009). Molecular biology of the pRN1 plasmid from *Sulfolobus islandicus*. Biochem. Soc. Trans. *37*, 42–45.

Lipps, G., Röther, S., Hart, C., and Krauss, G. (2003). A novel type of replicative enzyme harbouring ATPase, primase and DNA polymerase activity. EMBO J. *22*, 2516–2525.

Liu, L., Komori, K., Ishino, S., Bocquier, A.A., Cann, I.K., Kohda, D., and Ishino, Y. (2001). The archaeal DNA primase: biochemical characterization of the p41–p46 complex from *Pyrococcus furiosus*. J. Biol. Chem. *276*, 45484–45490.

Liu, Y., Kao, H.I., and Bambara, R.A. (2004). Flap endonuclease 1: a central component of DNA metabolism. Annu. Rev. Biochem. *73*, 589–615.

Lo, Y.H., Tsai, K.L., Sun, Y.J., Chen, W.T., Huang, C.Y., and Hsiao, C.D. (2009). The crystal structure of a replicative hexameric helicase DnaC and its complex with single-stranded DNA. Nucleic Acids Res. *37*, 804–814.

Lopez, P., Philippe, H., Myllykallio, H., and Forterre, P. (1999). Identification of putative chromosomal origins of replication in Archaea. Mol. Microbiol. *32*, 883–886.

Ludlam, A.V., McNatt, M.W., Carr, K.M., and Kaguni, J.M. (2001). Essential amino acids of Escherichia coli DnaC protein in an N-terminal domain interact with DnaB helicase J. Biol. Chem. *276*, 27345–27353.

Lundgren, M., Andersson, A., Chen, L., Nilsson, P., and Bernander, R. (2007). Three replication origins in *Sulfolobus* species: synchronous initiation of chromosome replication and asynchronous termination. Proc. Natl. Acad. Sci. U.S.A. *101*, 7046–7051.

Lyubimov, A.Y., Strycharska, M., and Berger, J.M. (2011). The nuts and bolts of ring-translocase structure and mechanism. Curr. Opin. Struct. Biol. *21*, 240–248.

McHenry, C.S. (2011). Bacterial replicases and related polymerases. Curr. Opin. Chem. Biol. *15*, 587–594.

Makarova, K.S., Wolf. Y.I., Mekhedov, S.L., Mirkin, B.G., and Koonin, E.V. (2005). Ancestral paralogs and pseudoparalogs and their role in the emergence of the eukaryotic cell. Nucleic Acids Res. *33*, 4626–4638.

Makarova, K.S., Koonin, E.V., and Kelman, Z. (2012). The CMG (CDC45/RecJ, MCM, GINS) complex is a conserved component of the DNA replication system in all archaea and eukaryotes. Biol. Direct 7, 7.

Makarova, K.S., Krupovic, M., and Koonin, E.V. (2014). Evolution of replicative DNA polymerases in archaea and their contributions to the eukaryotic replication machinery. Front. Microbiol. *5*, 354.

Marinsek, N., Barry, E.R., Makarova, K.S., Dionne, I., Koonin, E.V., and Bell, S.D. (2006). GINS, a central nexus in the archaeal DNA replication fork. EMBO Rep. *7*, 539–545.

Marszalek, J., and Kaguni, J.M. (1994). DnaA protein directs the binding of DnaB protein in initiation of DNA replication in *Escherichia coli*. J. Biol. Chem. *269*, 4883–4890.

Marszalek, J., Zhang, W., Hupp, T.R., Margulies, C., Carr, K.M., Cherry, S., and Kaguni, J.M. (1996). Domains of DnaA protein involved in interaction with DnaB protein, and in unwinding the *Escherichia coli* chromosomal origin. J. Biol. Chem. *271*, 18535–18542.

Matsui, E., Kawasaki, S., Ishida, H., Ishikawa, K., Kosugi, Y., Kikuchi, H., Kawarabayashi, Y., and Matsui, I. (1999). Thermostable flap endonuclease from the archaeon, *Pyrococcus horikoshii*, cleaves the replication fork-like structure endo/exonucleolytically. J. Biol. Chem. *274*, 18297–182309.

Matsui, E., Musti, K.V., Abe, J., Yamasaki, K., Matsui, I., and Harata, K. (2002). Molecular structure and novel DNA binding sites located in loops of flap endonuclease-1 from *Pyrococcus horikoshii*. J. Biol. Chem. *277*, 37840–37847.

Matsui, E., Nishio, M., Yokoyama, H., Harata, K., Damas, S., and Matsui, I. (2003). Distinct domain functions regulating *de novo* DNA synthesis of thermostable DNA primase from hyperthermophile *Pyrococcus horikoshii*. Biochemistry *42*, 14968–14976.

Matsui, E., Abe, J., Yokoyama, H., and Matsui, I. (2004). Aromatic residues located close to the active center are essential for the catalytic reaction of flap endonuclease-1 from hyperthermophilic archaeon *Pyrococcus horikoshii*. J. Biol. Chem. *279*, 16687–16696.

Matsumiya, S., Ishino, Y., and Morikawa, K. (2001). Crystal structure of an archaeal DNA sliding clamp: proliferating cell nuclear antigen from *Pyrococcus furiosus*. Protein Sci. *10*, 17–23.

Matsumiya, S., Ishino, S., Ishino, Y., and Morikawa, K. (2002). Physical interaction between proliferating cell nuclear antigen and replication factor C from *Pyrococcus furiosus*. Genes Cells *7*, 911–922.

Matsumiya, S., Ishino, S., Ishino, Y., and Morikaw, K. (2003). Intermolecular ion pairs maintain toroidal structure of *Pyrococcus furiosus* PCNA. Protein Sci. *12*, 823–831.

Matsunaga, F., Forterre, P., Ishino, Y., and Myllykallio, H. (2001). In vivo interactions of archaeal Cdc6/Orcl and minichromosome maintenance proteins with the replication origin. Proc. Natl. Acad. Sci. U.S.A. *98*, 11152–11157.

Matsunaga, F., Norais, C., Forterre, P., and Myllykallio, H. (2003). Identification of short 'eukaryotic' Okazaki fragments synthesized from a prokaryotic replication origin. EMBO Rep. *4*, 154–158.

Matsunaga, F., Glatigny, A., Mucchielli-Giorgi, M.H., Agier, N., Delacroix, H., Marisa, L., Durosay, P., Ishino, Y., Aggerbeck, L., and Forterre, P. (2007). Genomewide and Biochemical Analyses of DNA-binding activity of Cdc6/Orcl and Mcm proteins in *Pyrococcus* sp. Nucleic Acids Res. *35*, 3214–3222.

Matsunaga, F., Takemura, K., Akita, M., Adachi, A., Yamagami, T., and Ishino, Y. (2010). Localized melting of duplex DNA by Cdc6/Orcl at the DNA replication origin in the hyperthermophilic archaeon *Pyrococcus furiosus*. Extremophiles *14*, 21–31.

Mayanagi, K., Miyata, T., Oyama, T., Ishino, Y., and Morikawa, K. (2001). Three-dimensional electron microscopy of clamp loader small subunit from *Pyrococcus furiosus*. J. Struct. Biol, *134*, 35–45.

Mayanagi, K., Kiyonari, S., Nishida, H., Saito, M., Kohda, D., Ishino, Y., Shirai, T., and Morikawa, K. (2011a). Architecture of the DNA polymerase B-proliferating cell nuclear antigen (PCNA)-DNA ternary complex. Proc. Natl. Acad. Sci. U.S.A. *108*, 1845–1849.

Mayanagi, K., Kiyonari, S., Saito, M., Shirai, T., Ishino, Y., and Morikawa, K. (2011b). Mechanism of replication machinery assembly as revealed by the DNA ligase–PCNA–DNA complex architecture. Proc. Natl. Acad. Sci. U.S.A. *106*, 4647–4652.

Mendez, J., and Stillman, B. (2003). Perpetuating the double helix: molecular machines at eukaryotic DNA replication origins. Bioessays *25*, 1158–1167.

Messer, W. (2002). The bacterial replication initiator DnaA. DnaA and oriC, the bacterial mode to initiate DNA replication. FEMS Microbiol. Rev. *26*, 355–374.

Miyata, T., Suzuki, H., Oyama, T., Mayanagi, K., Ishino, Y., and Morikawa, K. (2005). Open clamp structure in the clamp-loading complex visualized by electron microscopic image analysis. Proc. Natl. Acad. Sci. U.S.A. *102*, 13795–13800.

Moldovan, G.-L., Pfander, B., and Jentsch, S. (2007). PCNA, the maestro of the replication fork. Cell *129*, 665–679.

Moyer, S.E., Lewis, P.W., and Botchan, M.R. (2006). Isolation of the Cdc45/Mcm2-7/GINS (CMG) complex, a candidate for the eukaryotic DNA replication fork helicase. Proc. Natl. Acad. Sci. U.S.A. *103*, 10236–10241.

Myllykallio, H., Lopez, P., López-García, P., Heilig, R., Saurin, W., Zivanovic, Y., Philippe, H., and Forterre, P. (2000). Bacterial mode of replication with eukaryotic-like machinery in a hyperthermophilic archaeon. Science *288*, 2212–2215.

Nakatani, M., Ezaki, S., Atomi, H., and Imanaka, T. (2000). A DNA ligase from a hyperthermophilic archaeon with unique cofactor specificity. J. Bacteriol. *182*, 6424–643.

Neuwald, A.F., Aravind, L., Spouge, J.L., and Koonin, E.V. (1999). AAA+: a class of chaperone-like ATPases associated with the assembly, operation, and disassembly of protein complexes. Genome Res. *9*, 27–43.

Nishida, H., Kiyonari, S., Ishino, Y., and Morikawa, K. (2006). The closed structure of an archaeal DNA ligase from *Pyrococcus furiosus*. J. Mol. Biol. *360*, 956–967.

Nishida, H., Mayanagi, K., Kiyonari, S., Sato, Y., Oyama, T., Ishino, Y., and Morikawa, K. (2009). Structural determinant for switching between the polymerase and exonuclease modes in the PCNA-replicative DNA polymerase complex. Proc. Natl. Acad. Sci. U.S.A. *106*, 20693–20698.

Norais, C., Hawkins, M., Hartman, A.L., Eisen, J.A., Myllykallio, H., and Allers, T. (2007). Genetic and physical mapping of DNA replication origins in *Haloferax volcanii*. PLoS Genet. *3*, e77.

Nozaki, S., and Ogawa, T. (2008). Determination of the minimum domain II size of *Escherichia coli* DnaA protein essential for cell viability. Microbiology *154*, 3379–3384.

Nunoura, T., Takaki, Y., Kakuta, J., Nishi, S., Sugahara, J., Kazama, H., Chee, G.J., Hattori, M., Kanai, A., Atomi, H., *et al.* (2011). Insights into the evolution of Archaea and eukaryotic protein modifier systems revealed by the genome of a novel archaeal group. Nucleic Acids Res. *39*, 3204–3223.

Ogino, H., Ishino, S., Mayanagi, K., Haugland, G.T., Birkeland, N.K., Yamagishi, A., and Ishino, Y. (2011). The GINS complex from the thermophilic archaeon, *Thermoplasma acidophilum* may function as a homotetramer in DNA replication. Extremophiles *15*, 529–539.

Ogino, H., Ishino, S., Haugland, G.T., Birkeland, N.K., Kohda, D., and Ishino, Y. (2014a). Activation of the MCM helicase from the thermophilic archaeon, *Thermoplasma acidophilum* by interactions with GINS and Cdc6-2. Extremophiles *18*, 915–924.

Ogino, H., Ishino, S., Oyama, T., Kohda, D., and Ishino, Y. (2014b). Disordered interdomain region of Gins is important for functional tetramer formation to stimulate MCM helicase in *Thermoplasma acidophilum*. Biosci. Biotechnol. Biochem. 79, 432–438.

Ogura, T., and Wilkinson, A.J. (2001). AAA+ superfamily ATPases: common structure—diverse function. Genes Cells *6*, 575–597.

Ohmori, H., Friedberg, E.C., Fuchs, R.P., Goodman, M.F., Hanaoka, F., Hinkle, D., Kunkel, T.A., Lawrence, C.W., Livneh, Z., Nohmi, T., *et al.* (2001). The Y-family of DNA polymerases. Mol. Cell *8*, 7–8.

Oyama, T., Ishino, Y., Cann, I., Ishino, S., and Morikawa, K. (2001). Atomic structure of the clamp loader small subunit from *Pyrococcus furiosus*. Mol. Cell *8*, 455–463.

Oyama, T., Ishino, S., Fujino, S., Ogino, H., Shirai, T., Mayanagi, K., Saito, M., Nagasawa, N., Ishino, Y., and Morikawa, K. (2011). Architectures of archaeal GINS complexes, essential DNA replication initiation factors. BMC Biol. *9*, 28.

Ozaki, S., and Katayama, T. (2009). DnaA structure, function, and dynamics in the initiation at the chromosomal origin. Plasmid *62*, 71–82.

Ozaki, S., Fujimitsu, K., Kurumizaka, H., and Katayama, T. (2006). The DnaA homolog of the hyperthermophilic eubacterium *Thermotoga maritima* forms an open complex with a minimal 149-bp origin region in an ATP-dependent manner. Genes Cells *11*, 425–438.

Pan, M., Kelman, L., and Kelman, Z. (2011a). The archaeal PCNA proteins. Biochem. Soc. Trans. *39*, 20–24.

Pan, M., Santangelo, T.J., Li, Z., Reeve, J.N., and Kelman, Z. (2011b). *Thermococcus kodakarensis* encodes three MCM homologs but only one is essential. Nucleic Acids Res. *39*, 9671–9680.

Pascal, J.M., O'Brien, P.J., Tomkinson, A.E., and Ellenberger, T. (2004). Human DNA ligase I completely encircles and partially unwinds nicked DNA. Nature *432*, 473–478.

Pascal, J.M., Tsodikov, O.V., Hura, G.L., Song, W., Cotner, E.A., Classen, S., Tomkinson, A.E., Tainer, J.A., and Ellenberger, T. (2006). A flexible interface between DNA ligase and PCNA supports conformational switching and efficient ligation of DNA. Mol. Cell *24*, 279–291.

Perler, F.B., Comb, D.G., Jack, W.E., Moran, L.S., Qiang, B., Kucera, R.B., Benner, J., Slatko, B.E., Nwankwo, D.O., Hempstead, S.K., *et al.* (1992). Intervening sequences in an Archaea DNA polymerase gene. Proc. Natl. Acad. Sci. U.S.A. *89*, 5577–5581.

Pelve, E.A., Lindås, A.C., Knöppel, A., Mira, A., and Bernander, R. (2012). Four chromosome replication origins in the archaeon Pyrobaculum calidifontis. Mol. Microbiol. *85*, 986–995.

Pisani, F.M., De Martino, C., and Rossi, M. (1992). A DNA polymerase from the archaeon Sulfolobus solfataricus shows sequence similarity to family B DNA polymerases. Nucleic Acids Res. *20*, 2711–2716.

Poidevin, L., and MacNeill, S.A. (2006). Biochemical characterisation of LigN, an NAD+-dependent DNA ligase from the halophilic euryarchaeon Haloferax volcanii that displays maximal *in vitro* activity at high salt concentrations. BMC Mol. Biol. *7*, 44.

Robinson, N.P., Dionne, I., Lundgren, M., Marsh, V.L., Bernander, R., and Bell, S.D. (2004). Identification of two origins of replication in the single chromosome of the archaeon *Sulfolobus solfataricus*. Cell *116*, 25–38.

Rolland. J.L., Gueguen, Y., Persillon, C., Masson, J.M., and Dietrich, J. (2004). Characterization of a thermophilic DNA ligase from the archaeon *Thermococcus fumicolans*. FEMS Microbiol. Lett. *236*, 267–273.

Rossi, M.L., and Bambara, R.A. (2006). Reconstituted Okazaki fragment processing indicates two pathways of primer removal. J. Biol. Chem. *281*, 26051–26061.

Roth, A., and Messer, W. (1995). The DNA binding domain of the initiator protein DnaA. EMBO J. *14*, 2106–2111.

Sakakibara, N., Kelman, L.M., and Kelman, Z. (2009). Unwinding the structure and function of the archaeal MCM helicase. Mol. Microbiol. *72*, 286–296.

Sakurai, S., Kitano, K., Yamaguchi, H., Hamada, K., Okada, K., Fukuda, K., Uchida, M., Ohtsuka, E., Morioka, H., and Hakoshima, T. (2005). Structural basis for recruitment of human flap endonuclease 1 to PCNA. EMBO J. *24*, 683–693.

Saluja, D., and Godson, G.N. (1995). Biochemical characterization of *Escherichia coli* temperature-sensitive *dnaB* mutants *dnaB8, dnaB252, dnaB70, dnaB43*, and *dnaB454*. J. Bacteriol. *177*, 1104–1111.

Sarmiento, F., Mrázek, J., and Whitman, W.B. (2013). Genome-scale analysis of gene function in the hydrogenotrophic methanogenic archaeon *Methanococcus maripaludis*. Proc. Natl. Acad. Sci. U.S.A. *110*, 4726–4731.

Sanchez-Pulido, L., and Ponting, C.P. (2011). Cdc45: the missing RecJ ortholog in eukaryotes? Bioinformatics *27*, 1885–1888.

Sato, T., Fukui, T., Atomi, H., and Imanaka, T. (2003). Targeted gene disruption by homologous recombination in the hyperthermophilic archaeon *Thermococcus kodakaraensis* KOD1. J. Bacteriol. *185*, 210–220.

Sato, T., Fukui, T., Atomi, H., and Imanaka, T. (2005). Improved and versatile transformation system allowing multiple genetic manipulations of the hyperthermophilic archaeon *Thermococcus kodakaraensis*. Appl. Environ. Microbiol. *71*, 3889–3899.

Schaper, S., Nardmann, J., Lüder, G., Lurz, R., Speck, C., and Messer, W. (2000). Identification of the chromosomal replication origin from *Thermus thermophilus* and its interaction with the replication initiator DnaA. J. Mol. Biol. *299*, 655–665.

Seitz, H., Weigel, C., and Messer, W. (2000). The interaction domains of the DnaA and DnaB replication proteins of Escherichia coli. Mol. Microbiol. *37*, 1270–1279.

Sekimizu, K., Bramhill, D., and Kornberg, A. (1987). ATP activates dnaA protein in initiating replication of plasmids bearing the origin of the E. coli chromosome. Cell *50*, 259–265.

Seo, M.S., Kim, Y.J., Choi, J.J., Lee, M.S., Kim, J.H., Lee, J.H., and Kwon, S.T. (2007). Cloning and expression of a DNA ligase from the hyperthermophilic archaeon *Staphylothermus marinus* and properties of the enzyme. J. Biotechnol. *128*, 519–530.

Shamoo, Y., and Steitz, T.A. (1999). Building a replisome from interacting pieces: sliding clamp complexed to a peptide from DNA polymerase and a polymerase editing complex. Cell *99*, 155–166.

Shen, B., Qiu, J., Hosfield, D., and Tainer, J.A. (1998). Flap endonuclease homologs in archaebacteria exist as independent proteins. Trends. Biochem. Sci. *23*, 171–173.

Shen, Y., Musti, K., Hiramoto, M., Kikuchi, H., Kawarabayashi, Y., and Matsui, I. (2001). Invariant Asp-1122 and Asp-1124 are essential residues for polymerization catalysis of family D DNA polymerase from *Pyrococcus horikoshii*. J. Biol. Chem. *276*, 27376–27383.

Shen, Y., Tang, X., and Matsui, I. (2003). Subunit interaction and regulation of activity through terminal domains of the family D DNA polymerase from *Pyrococcus horikoshii*. J. Biol. Chem. *278*, 21247–21257.

Shendure, J., and Ji, H. (2008). Next-generation DNA sequencing. Nat. Biotechnol. *26*, 1135–1145.

Shin, J.-H., Heo, G.Y., and Kelman, Z. (2008). The *Methanothermobacter thermautotrophicus* Cdc6-2 protein, the putative helicase loader, dissociates the minichromosome maintenance helicase. J. Bacteriol. *190*, 4091–4094.

Singleton, M.R., Dillingham, M.S., and Wigley, D.B. (2007). Structure and mechanism of helicases and nucleic acid translocases. Annu. Rev. Biochem. *76*, 23–50.

Slaymaker, I.M., Fu, Y., Toso, D.B., Ranatunga, N., Brewster, A., Forsburg, S.L., Zhou, Z.H., and Chen, X.S. (2013). Mini-chromosome maintenance complexes form a filament to remodel DNA structure and topology. Nucleic Acids Res. *41*, 3446–3456.

Soler, N., Marguet, E., Cortez, D., Desnoues, N., Keller, J., and van Tilbeurgh H. (2010). Two novel families of plasmids from hyperthermophilic archaea encoding new families of replication proteins. Nucleic Acids Res. *38*, 5088–5104.

Speck, C., and Messer, W. (2001). Mechanism of origin unwinding: sequential binding of DnaA to double- and single-stranded DNA. EMBO J. *20*, 1469–1476.

Speck, C., Chen, Z., Li, V., and Stillman, B. (2005). ATPase-dependent cooperative binding of ORC and Cdc6 to origin DNA. Nat. Struct. Mol. Biol. *12*, 965–971.

Strycharska, M.S., Arias-Palomo, E., Lyubimov, A.Y., Erzberger, J.P., O'Shea, V.L., Bustamante, C.J., and Berger, J.M. (2013). Nucleotide and partner-protein control of bacterial replicative helicase structure and function. Mol. Cell *52*, 844–854.

Stteter, K.O. (2013). A brief history of the discovery of hyperthermophilic life. Biochem. Soc. Trans. *41*, 416–420.

Sun, Y., Seo, M.S., Kim, J.H., Kim, Y.J., Kim, G.A., Lee, J.I., Lee, J.H., and Kwon, S.T. (2008). Novel DNA ligase with broad nucleotide cofactor specificity from the hyperthermophilic crenarchaeon *Sulfophobococcus zilligii*: influence of ancestral DNA ligase on cofactor utilization. Environ. Microbiol. *10*, 3212–3224.

Sutton, M.D., Carr, K.M., Vicente, M., and Kaguni, J.M. (1998). *Escherichia coli* DnaA protein. The N-terminal domain and loading of DnaB helicase at the *E. coli* chromosomal origin. J. Biol. Chem. 273, 34255–34262.

Tahirov, T.H., Makarova, K.S., Rogozin, I.B., Pavlov, Y.I., and Koonin, E.V. (2009). Evolution of DNA polymerases: an inactivated polymerase-exonuclease module in Pol ε and a chimeric origin of eukaryotic polymerases from two classes of archaeal ancestors. Biol. Direct *4*, 11.

Takayama, Y., Kamimura, Y., Okawa, M., Muramatsu, S., Sugino, A., and Araki, H. (2003). GINS, a novel multiprotein complex required for chromosomal DNA replication in budding yeast. Genes Dev. *17*, 1153–1165.

Tang, X.F., Shen, Y., Matsui, E., and Matsui, I. (2004). Domain topology of the DNA polymerase D complex from a hyperthermophilic archaeon *Pyrococcus horikoshii*. Biochemistry 43, 11818–11827.

Tom, S., Henricksen, L.A., and Bambara, R.A. (2000). Mechanism whereby proliferating cell nuclear antigen stimulates flap endonuclease 1. J. Biol. Chem. *275*, 10498–10505.

Tomkinson, A.E., Vijayakumar, S., Pascal, J.M., and Ellenberger, T. (2006). DNA ligases: structure, reaction mechanism, and function. Chem. Rev. *106*, 687–699.

Tong, J., Cao, W., and Barany, F. (1999). Biochemical properties of a high fidelity DNA ligase from *Thermus* species AK16D. Nucleic Acids Res. *27*, 788–794.

Tori, K., Kimizu, M., Ishino, S., and Ishino, Y. (2007). Both DNA polymerase BI and D from the hyperthermophilic archaeon, *Pyrococcus furiosus* bind to PCNA at the C-terminal PIP box motifs. J. Bacteriol. *189*, 5652–5657.

Uemori, T., Ishino, Y., Toh, H., Asada, K., and Kato, I. (1993). Organization and nucleotide sequence of the DNA polymerase gene from the archaeon *Pyrococcus furiosus*. Nucleic Acids Res., *21*, 259–265.

Uemori, T., Ishino, Y., Doi, H., and Kato, I. (1995). The hyperthermophilic archaeon Pyrodictium occultum has two alpha-like DNA polymerases. J. Bacteriol. *177*, 2164–2177.

Uemori, T., Sato, Y., Kato, I., Doi, H., and Ishino, Y. (1997). A novel DNA polymerase in the hyperthermophilic archaeon, *Pyrococcus furiosus*: gene cloning, expression, and characterization. Genes Cells 2, 499–512.

Vivona, J.B., and Kelman, Z. (2003). The diverse spectrum of sliding clamp interacting proteins. FEBS Lett. *546*, 167–172.

Walter, P., Klein, F., Lorentzen, E., Ilchmann, A., Klug, G., and Evguenieva-Hackenberg, E. (2006). Characterization of native and reconstituted exosome complexes from the hyperthermophilic archaeon *Sulfolobus solfataricus*. Mol. Microbiol. *62*, 1076–1089.

Walters, A.D., and Chong, J.P. (2010). An archaeal order with multiple minichromosome maintenance genes. Microbiology *156*, 1405–1414.

Wang, G., Klein, M.G., Tokonzaba, E., Zhang, Y., Holden, L.G., and Chen, X.S. (2008). The structure of a DnaB-family replicative helicase and its interactions with primase. Nat. Struct. Mol. Biol. *15*, 94–100.

Warbrick, E.M. (2000). The puzzle of PCNA's many partners. BioEssays *22*, 997–1006.

Watson, J.D., and Crick, F.H. (1953). Molecular structure of nucleic acids; a structure for deoxyribose nucleic acid. Nature *171*, 737–738.

Weigel, C., Schmidt, A., Seitz, H., Tüngler, D., Welzeck, M., and Messer, W. (1999). The N-terminus promotes oligomerization of the Escherichia coli initiator protein DnaA. Mol. Microbiol. *34*, 53–66.

Wickner, S., and Hurwitz, J. (1975). Interaction of *Escherichia coli dnaB* and *dnaC(D)* gene products *in vitro*. Proc. Natl. Acad. Sci. U.S.A. *72*, 921–925.

Wilkinson, A., Day, J., and Bowater, R. (2001). Bacterial DNA ligases. Mol. Microbiol. *40*, 1241–1248.

Williams, G.J., Johnson, K., Rudolf, J., McMahon, S.A., Carter, L., Oke, M., Liu, H., Taylor, G.L., White, M.F., and Naismith, J.H. (2006). Structure of the heterotrimeric PCNA from *Sulfolobus solfataricus*. Acta Crystallogr. Sect. F Struct. Biol. Cryst. Commun. *62*, 944–948.

Wu, K., Lai, X., Guo, X., Hu, J., Xiang, X., and Huang, L. (2007). Interplay between primase and replication factor C in the hyperthermophilic archaeon *Sulfolobus solfataricus*. Mol. Microbiol. *63*, 826–837.

Wu, Z., Liu, J., Yang, H., and Xiang, H. (1999). DNA replication origins in archaea. Front. Microbiol. *5*, 179.

Yoshimochi, T., Fujikane, R., Kawanami, M., Matsunaga, F., and Ishino, Y. (2008). The GINS complex from *Pyrococcus furiosus* stimulates the MCM helicase activity. J. Biol. Chem. *283*, 1601–1609.

Zheng, L., Jia, J., Finger, L.D., Guo, Z., Zer, C., and Shen, B. (2011). Functional regulation of FEN1 nuclease and its link to cancer. Nucleic Acids Res. *39*, 781–794.

Zillig, W., Kletzin, A., Schleper, C., Holz, I., Janekovic, D., Hain, J., Lanzendörfer, M., and Kristjansson, J.K. (1993). Screening for *Sulfolobales*, their plasmids and their viruses in Icelandic solfataras. Syst. Appl. Microbiol. *16*, 609–628.

Zuo, Z., Rodgers, C.J., Mikheikin, A.L., and Trakselis, M.A. (2010). Characterization of a functional DnaG-type primase in archaea: implications for a dual-primase system. J. Mol. Biol. *397*, 664–676.

Metabolic Engineering of Thermophiles for Biofuel Production

9

Ya-Jun Liu and Qiu Cui

Abstract

The severe energy crisis and environmental contamination require alternative means of fuel production, and biofuels produced from lignocellulose, a widely available and inexpensive substrate, are ideal substitutes. The major obstacle for the production of lignocellulosic biofuels is the conversion of the recalcitrant lignocellulose into fermentable sugars, although high-temperature conditions are beneficial to the deconstruction progress. Therefore, thermophiles, especially cellulolytic ones such as *Clostridium thermocellum*, are considered to be promising candidates for lignocellulosic biofuel production. Metabolic engineering is necessary because wild-type thermophilic candidates possess inherent disadvantages and do not meet the standards for industrialization, and several genetic manipulation tools have been developed to promote the targeted engineering and the application of thermophiles in the industrial production of biofuels. This chapter reviews the recent progress in the development of genetic tools for thermophiles and the metabolic engineering focused on the high production of lignocellulosic biofuels.

Introduction

With problems arising from climate change, greenhouse gas emissions, rapid population growth and an oil crisis, people have begun to realize the importance of developing renewable energy. The effective use and production of biofuels, one of the most important types of renewable energy, has become an increasingly hot topic and essential issue worldwide (Antoni *et al.*, 2007; Demain, 2009; Gronenberg *et al.*, 2013; Hoffert, 2010). Lignocellulosic biomass is the most abundantly available raw material on the Earth, and its sustainability and effective cost also make it an attractive feedstock for the production of second-generation biofuels. Thus, people have paid close attention to the utilization of lignocellulose materials on a global scale (de Souza, 2013; Demain, 2009; Gaykawad *et al.*, 2013; Hasunuma *et al.*, 2013; Lynd, 2010; Service, 2010).

Lignocellulosic biomass is composed of carbohydrate polymers, i.e. cellulose and hemicellulose, and an aromatic polymer (lignin). Cellulose is composed of six-carbon sugar monomers (i.e. glucose), while hemicellulose contains various five-carbon sugar monomers (e.g. xylose, arabinose, etc.). Both cellulose and hemicellulose are tightly bound to lignin, which makes the structure of lignocellulose resistant to deconstruction (Gowen and Fong, 2010; Hasunuma *et al.*, 2013; Olson *et al.*, 2012). In addition, although lignocellulosic materials from different sources have the same components, the specific compositions vary greatly with respect to component ratios (Gowen and Fong, 2010; Schwarz, 2001). Thus, the major bottleneck in the production of biofuels from lignocellulose is how to release soluble sugars efficiently from the recalcitrant substrate.

The conversion of cellulose to soluble sugars is mediated by multiple enzymes, including mainly endoglucanase, exoglucanase and β-glucosidase (Chandel *et al.*, 2012; Zhang *et al.*, 2006). The main intermediates and final products of the hydrolysis reactions are oligosaccharides and monosaccharides, which may inhibit lignocellulose degradation via feedback inhibition. Thus, successive fermentation of the resulting sugars is required along with the hydrolysis of lignocellulose to relieve the inhibitory effect (Gefen *et al.*, 2012; Kim *et al.*, 2013; Maki *et al.*, 2013). In contrast to separate enzymatic hydrolysis and fermentation (SHF), both consolidated bioprocessing (CBP) and simultaneous saccharification and fermentation (SSF) have advantages in combining the hydrolysis process with the fermentation process. CBP integrates the enzyme production, cellulose hydrolysis and fermentation in one step, which can further reduce the cost and simplify the technique compared with SSF. Thus, CBP has been considered the most promising strategy for biofuel production from lignocellulose, with outstanding potential (Kádár *et al.*, 2004; Lynd *et al.*, 2005; Olson *et al.*, 2012; Xu *et al.*, 2009).

High temperatures can promote lignocellulose deconstruction. In addition, thermal conditions reduce the risk of contamination and facilitate the recovery of volatile products (Frock and Kelly, 2012; Olson *et al.*, 2012; Taylor *et al.*, 2009). Thus, those thermophilic microorganisms and thermostable enzymes that can degrade lignocellulose under thermal conditions are of great interest (Anbar *et al.*, 2012; Gefen *et al.*, 2012; Xia *et al.*, 2013; Zhang *et al.*, 2006). However, because of their lower product yield, stress sensitivity or limited carbon sources (Lin and Xu, 2013; Taylor *et al.*, 2009), wild-type strains do not meet the standards for industrialization, which needs metabolically engineered thermophiles developed for higher yields of biofuels over a long period. The metabolic engineering of thermophiles makes slow process due to the lack of efficient thermo-techniques. Recently, several genetic tools have been developed for thermophiles, by using which, systematic engineering of cellulolytic and high-producing thermophiles has been performed, and notable progress has been made in enhancing biofuel production (Brown *et al.*, 2011; Chandrayan *et al.*, 2012; Cripps *et al.*, 2009; Mohr *et al.*, 2013; Olson, 2010; Taylor *et al.*, 2011). Here we review the recently developed genetic manipulation methods for thermophiles and their application in the production of biofuels from lignocellulose.

Metabolic engineering tools for thermophiles

The development of robust genetic systems is critical for the targeted engineering of thermophilic-based bioprocesses. The genetic system must be thermo-stable and should include a reliable DNA delivery system, a sensitive and specific screening system, a highly efficient gene targeting method, an effective endogenous/heterologous expression method, and reliable omics technology.

DNA transformation

Genetic manipulation of thermophilic microorganisms, especially Gram-positive cells, is generally considered more difficult than for mesophiles because (1) the cell envelope of thermophiles may inhibit efficient DNA delivery due to its low permeability; (2) competent cells made from thermophiles have low recovery efficiency; (3) plasmid backbones with thermostable replication origins and efficient selection markers are lacking; and (4) the available transformation protocols are limited (Averhoff, 2004; Lin and Xu, 2013; Taylor *et al.*, 2009, 2011). Traditional transformation methods include natural competency, protoplast transformation, conjugation, chemical treatment with $CaCl_2$ plus heat shock, ultrasound-based sonoporation and electrotransformation (Averhoff, 2004; de Grado *et al.*, 1999; Friedrich *et al.*, 2001; Koyama *et al.*, 1986; Lin *et al.*, 2010; Mohr *et al.*, 2013; Olson and Lynd, 2012b; Sato *et al.*, 2005; Waege *et al.*, 2010; Wan, 2013; Wu and Welker, 1989). Natural transformation has been observed in *Thermus thermophilus* strain HB27, an aerobic, Gram-negative extreme thermophile, and the transformation frequency ranges from 10^{-2} to 10^{-1} per μg DNA (Friedrich *et al.*, 2001; Koyama *et al.*, 1986). The chemical treatment plus heat shock method initially used for DNA transformation in mesophiles has also been applied to the transformation of thermophilic microorganisms such as *Pyrococcus furiosus* and *Thermococcus kodakaraensis* (Sato *et al.*, 2005; Waege *et al.*, 2010). Recently, Lin *et al.* (2010) developed an ultrasound-based sonoporation approach for *Thermoanaerobacter* sp. X514, and the transformation efficiency was 6×10^2 transformants/μg DNA.

The electroporation method is regarded as the most efficient and has the most broad adaptability among various species (Taylor *et al.*, 2011), particularly for cellulolytic thermophiles (Mohr *et al.*, 2013; Olson and Lynd, 2012b). However, the published electrotransformation method is often with low transformation efficiency or low stability and is not widely used. Cui *et al.* developed a novel pulse generation system MT01–3KV (3100 V, 100 A, 50 kHz) (Fig. 9.1) and anaerobic working station to obtain a higher transformation efficiency in both mesophilic and thermophilic *Clostridium* strains (Cui *et al.*, 2013; Mohr *et al.*, 2013). By using these custom-built facilities, the electroporation method, including electroporation parameters (i.e. voltage, frequency, duty ratio, time, and pulse duration), preparation of the competent cells, treatment with cell-wall-weakening agents, and plasmid backbones, was optimized for *Clostridium thermocellum*, and the transformation efficiency was increased to 10^5 CFU/μg DNA (unpublished data), which even met the requirement of homologous recombination (Lin *et al.*, 2010).

Vectors and screening systems

One of the major obstacles for gene targeting in thermophiles is the lack of proper vectors and suitable screening markers. Overexpression of foreign or native genes requires high-copy-number plasmids with thermostable and compact replication origins, while conditional or unconditional suicide plasmids are required for gene disruption, which needs higher transformation efficiency and more sensitive screening markers. Antibiotic-resistance genes, such as thermostable kanamycin, hygromycin, chloramphenicol, and bleomycin resistance genes, are usually used as screening markers (Lin and Xu, 2013). Functional screening markers are also widely applied. For instance, PyrF (i.e. orotidine 5-phosphate decarboxylase) is an essential enzyme in the pyrimidine biosynthesis and is a target of the antimetabolite 5-fluoroorotic acid (FOA). Thus, the targeted inactivation and ectopic expression of *pyrF* can be used as a selection or counterselection marker, and through use as a counterselection marker, the integrated antibiotic resistant gene can be removed after the genetic engineering (Cui *et al.*, 2013; Shaw *et al.*, 2011; Tripathi *et al.*, 2010). Green fluorescent proteins (GFP) are considered suitable reporters to detect the expression of foreign proteins, and thermostable GFP variants have been developed and used in thermophiles for protein tracing and identification (Cava *et al.*, 2008). Few green fluorescent reporter systems were established in obligate anaerobes because of the strict oxygen requirement for GFP-family proteins (Shaner *et al.*, 2007); however, most of the thermophilic microorganisms used for biofuel production are anaerobes. Recently, Hong *et al.* (2014) successfully expressed an oxygen-independent green fluorescent protein PpFbFP (Drepper *et al.*, 2007) in *C. thermocellum* and detected strong green fluorescence (Fig. 9.2, unpublished data). Therefore, PpFbFP may be used as a green fluorescent

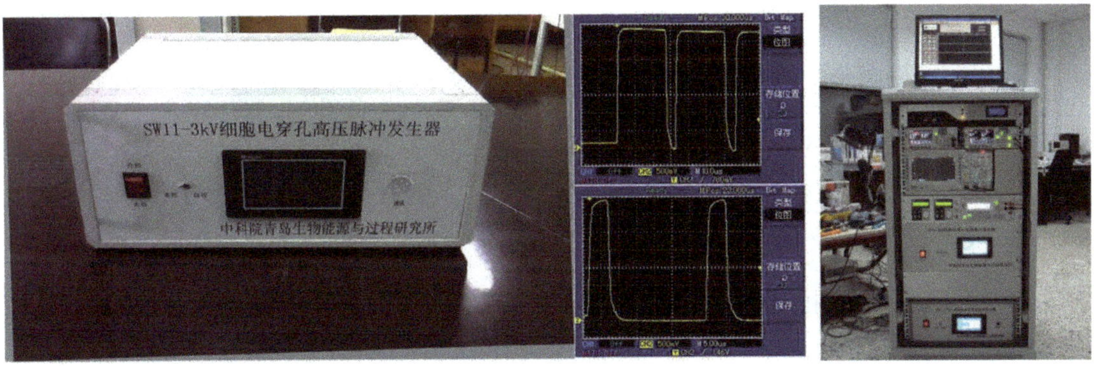

Figure 9.1 Custom pulse generation system MT01–3KV developed to obtain a higher transformation efficiency in *Clostridium* (Cui *et al.*, 2012, 2014; Hong *et al.*, 2014).

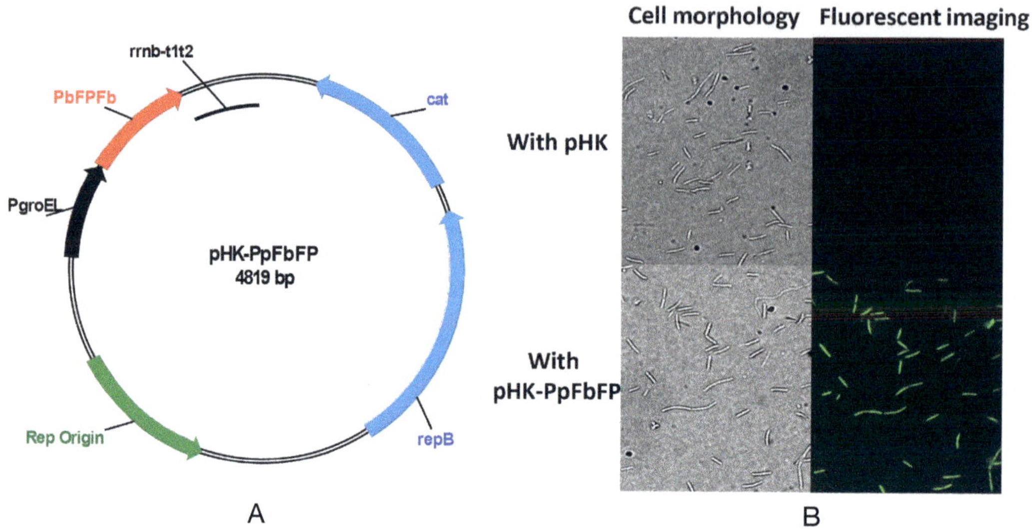

Figure 9.2 Construction of green fluorescent reporter system in *C. thermocellum* (unpublished data). (A) Schematic representation of the oxygen-independent green fluorescent protein PpFbFP expressing plasmid pHK-PpFbFP. (B) Cell morphology and fluorescent imaging of *C. thermocellum* transformed with plasmid pHK-PpFbFP, and pHK was used as the negative control.

reporter system for the expression, localization, movement and interaction of proteins in *C. thermocellum* and possibly other anaerobes.

Recently, thermostable plasmids carrying functional replication origins and screening gene markers have been reported for use in the genetic engineering of various thermophiles. For example, pIKM1-derived plasmids were used for gene deletion and expression in *Thermoanaerobacterium* (Mai and Wiegel, 2000) and *Thermoanaerobacter* strains (Lin *et al.*, 2010; Peng *et al.*, 2006), and p3CHPT and its derivatives were also applied in the genetic manipulation of *Thermoanaerobacter* strains (Yao and Mikkelsen, 2010). Furthermore, pUB110 and pUCG18-derived plasmids were used for integration and expression in *Geobacillus*, respectively (Cripps *et al.*, 2009; Taylor *et al.*, 2008), and pNW33N-derived plasmids were proven to produce high copy numbers in *C. thermocellum* and were suitable for both gene targeting via homologous recombination or the group II intron-based targeting method (Mohr *et al.*, 2013; Olson and Lynd, 2012a).

Targeted gene manipulation

Initially, gene targeting in thermophiles was performed via random mutation, such as transposon mutagenesis, due to the lack of DNA delivery methods (Taylor *et al.*, 2011). The Tn*916* transposon has been transferred to both *G. stearothermophilus* and *T. aquaticus*, and the integrations in chromosomal DNA have been identified (Natarajan and Oriel, 1991; Sen and Oriel, 1990). Zverlov *et al.* isolated six *C. thermocellum* mutants that had lost the ability to degrade microcrystalline cellulose. They detected a new transposable element IS*1447* inserted in the *cipA* gene of the chromosomal DNA (Zverlov *et al.*, 2008). However, random mutations require the construction of mutant libraries and reliable screening methods that commonly need tedious and long-term screening, therefore hindering high-throughput genetic manipulation. However, some gene mutations do not result in any apparent change in phenotype, thus requiring targeted gene manipulation.

Traditional directed gene manipulation methods are based on homologous recombination, signal transduction or Group II introns (Karberg *et al.*, 2001; Lang and Beatty, 2002; Stasiak, 1996), with homologous recombination being the most widely used method. Homologous recombination, also known as 'allelic exchange', is a type of genetic recombination in which similar nucleotide sequences are exchanged between the homologous arms of the plasmid and the target DNA sequences in the genomic DNA, facilitating in integration of the foreign DNA sequences carried by the plasmid into the chromosomal DNA target site. The efficiency of homologous recombination depends in part on the capability of the selection markers. Antibiotic resistance genes integrated into chromosomal DNA cannot be recycled, which limits systematic metabolic engineering. Heap *et al.* (2012) enabled the recycling of selection markers and successive disruption by using asymmetrical regions of homology to determine the order of homologous recombination events and by activating or inactivating counterselection markers that are coupled to the chromosome DNA, and the improved homologous recombination method was called allele coupled exchange (ACE).

Homologous recombination has been used in the genetic engineering of various thermophilic microorganisms. A series of *T. saccharolyticum* mutants have been constructed by targeted deletion to characterize xylan hydrolysis and xylose utilization in this organism (Podkaminer *et al.*, 2012) and to increase the ethanol or butanol production (Bhandiwad *et al.*, 2013; Shaw *et al.*, 2008). Cripps *et al.* (2009) constructed an engineered *G. thermoglucosidasius* strain with inactivated lactate dehydrogenase and pyruvate formate lyase and overexpressed pyruvate dehydrogenase. The directed knockout of lactate dehydrogenase was also performed in *T. mathranii* (Yao and Mikkelsen, 2010). Tripathi *et al.* (2010) utilized the homologous recombination method to construct a *pyrF*-disrupted mutant of *C. thermocellum* and a phosphotransacetylase-negative mutant with a disrupted acetate production pathway. A major exoglucanase, Cel48S, and a scaffoldin, CipA, from *C. thermocellum* have also been disrupted to verify their roles in cellulose degradation via the homologous recombination method (Olson *et al.*, 2013, 2010; Waller *et al.*, 2013).

On the basis of the homologous recombination mechanism, many genome editing methods, including the Cre/*lox*- or FLP/FRT-based recombination system, the transcription activator-like effector nucleases (TALENs) system, and the clustered regularly interspaced short palindromic repeats (CRISPR)/Cas system, have been recently developed to achieve high-throughput or scarless genetic manipulation (Bedell *et al.*, 2012; Pinkney *et al.*, 2012; Richter *et al.*, 2013). The TALEN and CRISPR/Cas systems employ engineered nucleases to create double-stranded breaks at desired sites in the target genes, which introduces an insertion or break through the endogenous processes of homologous recombination and non-homologous end-joining. These systems have received wide attention as efficient genome-scale engineering tools (Gaj *et al.*, 2013), and the CRISPR RNA processing in *C. thermocellum* has been characterized (Richter *et al.*, 2012). Although such approaches have not been applied to thermophiles to date, with the development of DNA transformation techniques, the improvement of screening systems, the engineering of thermophilic nucleases, and a deeper understanding of the interference mechanisms of the tools, it will be possible to exploit the potential of these tools in the genetic engineering of thermophilic microorganisms.

In addition to homologous recombination, the mobile group II intron-based targetron method is also a promising gene knock-out and knock-in technology (Liu *et al.*, 2015; Taylor *et al.*, 2011). The targetron is a gene-targeting vector derived from the mobile group II intron (Perutka *et al.*, 2004). It consists of an autocatalytic intron RNA and an intron-encoded reverse transcriptase, which use their combined activities to achieve highly efficient site-specific DNA integration, termed 'retrohoming', with readily programmable DNA target specificity (Lambowitz and Zimmerly, 2011) (Fig. 9.3). The most famous targetron method, ClosTron, is derived from a group II intron *Ll.LtrB* from *Lactococcus lactis* and has been used in the genetic engineering of various mesophilic *Clostridium* species (Cui *et al.*, 2012; Dong *et al.*, 2010; Heap *et al.*, 2007). Other group II introns, EcI5 from *Escherichia coli* and RmInt1 from *Sinorhizobium meliloti*, can also be used for targeted gene manipulation (Garcia-Rodriguez *et al.*, 2011; Zhuang *et al.*, 2009). However, these targetron methods are all derived from mesophilic group II introns and cannot be applied to thermophilic microorganisms.

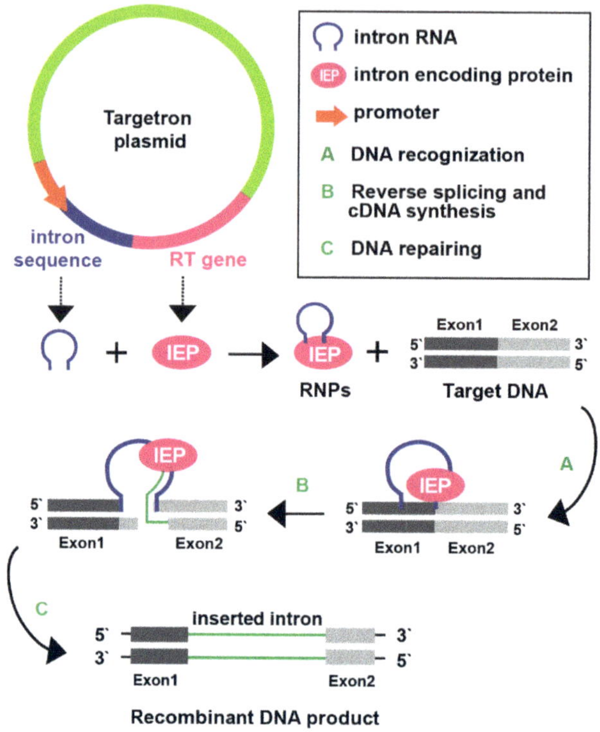

Figure 9.3 Thermotargetron method for gene targeting in thermophiles (Hong *et al.*, 2014). A targetron expression plasmid with a strong promoter is used to express a thermostable ~0.8-kb group II intron RNA (Tel3c) and a group II intron reverse transcriptase (Tel4c RT; denoted RT) from the thermophilic cyanobacterium *Thermosynechococcus elongatus* (Mohr *et al.*, 2010). The intron RNA is expressed as a precursor with short flanking exon sequences (denoted E1 and E2) and is spliced from this precursor with the assistance of the intron-encoded RT to yield a ribonucleoprotein (RNP) complex in which the RT is bound to the excised intron lariat RNA. Group II intron RNPs recognize DNA target sites primarily by base-pairing of sequence motifs in the intron RNA to the DNA target sequence, with only a small contribution from the intron-encoded RT. The intron RNA then uses its catalytic ('ribozyme') activity to insert into the top strand of the DNA target site between target exon sequences (E1′ and E2'), while the DNA endonuclease activity of the RT is used to cleave the bottom strand and the nicked DNA is used as a primer for reverse transcription of the inserted intron RNA. The resulting intron cDNA is integrated into the genome by host enzymes. Because the DNA target site is recognized mainly by base pairing of the intron RNA, the targetron can be programmed to insert into desired sites simply by modifying the base-pairing motifs in the intron RNA, taking into account the small number of target site nucleotide residues recognized by the RT. The inserted targetron contains multiple stop codons in all three reading frames, resulting in the expression of truncated proteins. Once inserted, the targetron cannot be spliced or mobilized in the absence of the group II intron RT expressed from the targetron plasmid, which is cured prior to the analysis of mutants. Targetron methods are described in detail in Mohr *et al.* (2013).

A thermotargetron system is constructed based on a mobile group II intron, Tel3c4c, from the thermophilic cyanobacterium *Thermosynechococcus elongates* (Mohr *et al.*, 2010) and has been used for targeted gene disruption in *C. thermocellum* (Mohr *et al.*, 2013) (Fig. 9.3). Six different chromosomal genes (*cipA*, *hfat*, *hyd*, *ldh*, *pta*, and *pyrF*) were inactivated, and high integration efficiencies were achieved (67–100%). Because the thermotargetron functions at high temperatures that promote DNA melting, it can recognize DNA target sequences almost entirely by base pairing of the intron RNA, with less contribution from the intron-encoded protein than for mesophilic targetrons. This feature increases the number of potential targetron-insertion sites while only moderately decreasing the DNA target specificity (Mohr *et al.*, 2013). Using the thermotargetron technique, eight *C. thermocellum* mutant strains have been further constructed to investigate the structure and function of important cellulosomal components (Hong *et al.*, 2014). In comparison, a total of seven *C. thermocellum* mutants constructed by homologous recombination methods have been reported, two for *cipA* and five for other genes (*ldh, pta, hpt, pyrF,* and *cel48S*) (Argyros *et al.*, 2011; Olson *et al.*, 2010; Tripathi *et al.*, 2010; Waller *et al.*, 2013). Thus, this novel technique can efficiently support high-throughput and systematic metabolic engineering of this important thermophilic microorganism for future industrial utilization of lignocellulosic biomass and the production of cellulosic ethanol and is also readily adaptable for gene targeting in other thermophiles. Furthermore, a targetron can also be used as a 'cargo' carrying target genes to insert at desired chromosomal locations (Frazier *et al.*, 2003) and can cooperate with other targeting methods (e.g. Cre/*lox*) for large scale genome engineering (Enyeart *et al.*, 2013).

Enzyme production

Different from microorganisms, enzymes can be advantageous as industrial catalysts as they usually tolerate the relatively harsh conditions of industrial processing and require less complex reaction conditions that are easier to control. Thermostable enzymes offer robust catalyst alternatives for biofuel production (Comfort *et al.*, 2004; Harris *et al.*, 2014; Turner *et al.*, 2007). Thus, thermostable enzymes have an obvious advantage in the conversion of biomass materials because high temperatures often promote better enzyme penetration and lignocellulose deconstruction (Harris *et al.*, 2014). Thermophilic enzymes used for biofuel production are mostly isolated or developed from thermophilic microorganisms (Harris *et al.*, 2014), especially hyperthermophiles (Turner *et al.*, 2007). Chandrayan *et al.* engineered the hyperthermophilic archaeon *P. furiosus* to overproduce the endogenous cytoplasmic hydrogenase SHI, which can be used for efficient *in vitro* biohydrogen production (Chandrayan *et al.*, 2012). An entire collection of cellulases, hemicellulases and other accessory proteins that are optimally active under the same conditions (high temperature and a complex substrate) to completely digest lignocellulosic biomass is of particular interest. However, the challenge will arise from how to achieve high-level expression of the proteins. Thermophiles usually grow slowly and produce a low amount of biomass, which severely limits enzyme production and their industrial-scale use for biomass degradation (Harris *et al.*, 2014). Fermentation conditions have been optimized to enhance the culture growth, but large-scale cultivation of thermophiles for enzyme production remains an obstacle due to the high cost and complex techniques (Schiraldi and De Rosa, 2002). Thus, heterologous expression of thermophilic enzymes in mesophilic microorganisms is a promising alternative method to reduce production costs, and candidate hosts include *Escherichia coli, Bacillus subtilis, Saccharomyces cerevisae, Pichia pastoris,* and *Trichoderma reesei* (Huang *et al.*, 2010; Moracci *et al.*, 1992; Ramchuran *et al.*, 2002, 2006; Zhang and Zhang, 2011). Protein thermostability can also be further enhanced by protein engineering or directed evolution (Anbar *et al.*, 2012; Lee *et al.*, 2011; Tang *et al.*, 2013; Zhang *et al.*, 2006).

Different codon usage or improper folding of the foreign proteins may result in low expression or low enzyme activity. In addition, some multi-enzyme complexes contain many components, and the heterologous expression and the *in vitro* assembly of the components may be difficult to achieve. For example, the cellulosome is a supermolecular machine that is excreted in the growth medium or attached to the cell wall of cellulolytic bacteria and consists of multiple structural scaffoldins and enzymatic subunits that interact with each other to efficiently degrade lignocellulose substrates (Fig. 9.4) (Bayer *et al.*, 1998; Demain *et al.*, 2005). The cellulosome systems of different cellulolytic bacteria have various structures and components (Bayer *et al.*, 2008). *C. thermocellum* produces a cellulosome with a molecular weight of ~100 MDa which may consist of more than 70 proteins (Hon-nami *et al.*, 1986; Riederer *et al.*, 2011; Zverlov *et al.*, 2005). Although great effort has been made on the construction of designer cellulosomes by the *in vitro* assembly of heterologously expressed cellulosomal proteins, the resulting cellulose hydrolysis activity is still much lower than the endogenous cellulosome of *C. thermocellum* (Borne *et al.*, 2013; Fierobe *et al.*,

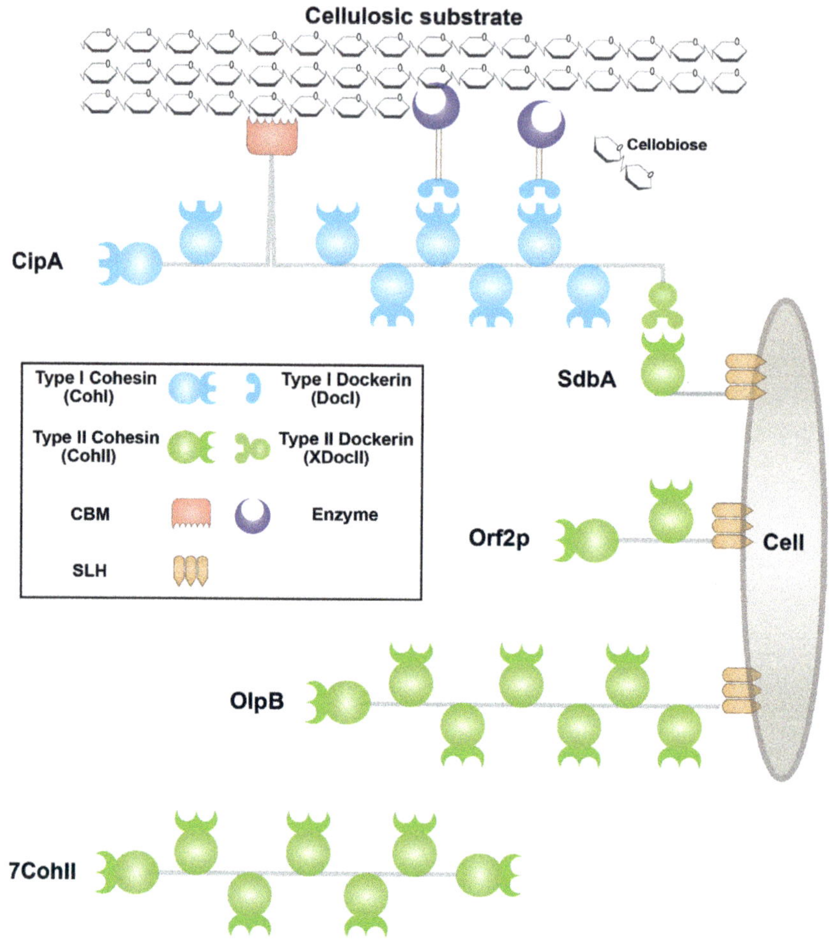

Figure 9.4 Schematic representation of structure and function of cellulosome from *C. thermocellum* (Hong *et al.*, 2014).

2002; Morais *et al.*, 2010, 2012). Thus, a better strategy for lignocellulose bioconversion may be to directly use cells from engineered cellulosome-producing microorganisms as catalysts via a CBP strategy.

Omics technology

Omics technologies for thermophiles have recently been developed to demonstrate the biological function of genes and their proteins in a complex intracellular system (Debnath *et al.*, 2010). For example, the genomic DNA of *C. thermocellum* wild-type strains ATCC27405 and DSM1313 has been sequenced and annotated in 2009 and 2011 (Feinberg *et al.*, 2011), respectively. The genome sequences of a cellulose adhesion-defective mutant strain AD2 and its ancestor *C. thermocellum* wild-type strain YS have also been analysed to show the role that a cellulosome plays in cellulose degradation (Brown *et al.*, 2012). Based on the available genomics data, transcriptomics, proteomics and metabolomics tools are used for deeper analysis of biological reactions under different conditions. For example, global transcriptome analyses of *C. thermocellum* have been performed using RNA-Seq or microarrays to investigate the intracellular response of *C. thermocellum* to different substrates and stresses at the genetic level (Raman *et al.*, 2011; Riederer *et al.*, 2011; Wilson *et al.*, 2013a,b). Dykstra *et al.* developed a proteomic quantitation method to measure the cellulosomal components of *C. thermocellum* simultaneously using multiple reaction monitoring mass spectrometry (MRM-MS)

(Dykstra *et al.*, 2014). Yang *et al.* analysed the transcriptomic, metabolomic and proteomic profiles of *C. thermocellum* ATCC27405 after ethanol stress and found that the largest response of *C. thermocellum* to ethanol shock treatment was related to nitrogen uptake and metabolism (Yang *et al.*, 2012). Zhu *et al.* compared the metabolomes of an ethanol-tolerant strain with the *C. thermocellum* wild-type strain and suggested that the cellodextrin synthesis may be a potential mechanism for stress resistance in *C. thermocellum* (Zhu *et al.*, 2013). Thus, omics technology, especially the combinatorial multilevel -omics, is considered an efficient approach to provide effective strategies for the metabolic engineering of thermophiles for focusing on the large-scale production of specific products, including biofuels.

Application of (engineered) thermophiles in biofuel production from lignocellulose

Lignocellulose materials such as straw, wood and rice husk are potential feedstocks for the production of renewable biofuels such as biogas, biohydrogen, bioethanol, biobutanol, and biodiesel. High-temperature conditions may facilitate the deconstruction and conversion of recalcitrant lignocellulose to release fermentable sugars, and therefore, the metabolic engineering of thermophiles for biofuel production is of interest worldwide (Frock and Kelly, 2012; Olson *et al.*, 2012; Taylor *et al.*, 2009). To date, thermophilic Clostridia, Bacilli and Archaea are the most well-studied for biofuel production (Argyros *et al.*, 2011; Henstra *et al.*, 2007; Kita *et al.*, 2012; Kotai *et al.*, 2013; Lu *et al.*, 2013; Mohr *et al.*, 2013; Waeonukul *et al.*, 2012), in which only *C. thermocellum* is able to degrade cellulose. Other thermophiles are used through SHF or fermented with *C. thermocellum* via SSF.

Hydrogen

Hydrogen is an attractive alternative fuel for the future because it has the advantage of being non-polluting, and its energy content is higher than methane (Logan *et al.*, 2002). Hydrogen can be produced from biomass via a thermochemical process. Biosyngas can be generated by gasification of lignocellulosic material, a thermal process achieved at high temperatures (>700°C) without combustion, and is rich in hydrogen and carbon monoxide. Hydrogen can be further obtained through steam reforming of the syngas (Bridgwater, 2003; Demirbas, 2007). In addition to thermal processing, biohydrogen can be produced through biophotolysis using solar light as energy and water as an electron donor, or through photofermentation of organic substrates from photosynthetic bacteria, which requires relatively restrictive growth conditions. In dark fermentation, biohydrogen is produced using renewable feedstocks as substrates in an anaerobic fermentation process, which may be a useful byproduct in future large-scale industrial fermentation (Bielen *et al.*, 2013; Singh *et al.*, 2013; Verhaart *et al.*, 2010).

The formation of hydrogen can be promoted by elevated temperatures, and many model thermophilic microorganisms that are able to produce biohydrogen efficiently have therefore been studied (Bielen *et al.*, 2013; Schut and Adams, 2009; Schut *et al.*, 2012; Soboh *et al.*, 2004; Verhaart *et al.*, 2010). *C. saccharolyticus* is one of the best choices for further research on hydrogen production from biomass because of its cellulolytic capability and favoured fermentation patterns (Bielen *et al.*, 2013). *C. saccharolyticus* can produce biohydrogen in a CBP operation. With untreated switchgrass (SG) as a carbon source, the H_2 yield of *C. saccharolyticus* is 11.2 mmol/g SG after 6 days of fermentation (Talluri *et al.*, 2013). *Thermotoga neapolitana* is a promising option as well because it can utilize organic wastes to produce hydrogen consistently with carbon dioxide as the only other prominent byproduct (Van Ooteghem *et al.*, 2002). Hyperthermophilic archaeon *P. furiosus* can also ferment a variety of sugar- or amino-acid-based oligomers and polymers, although cellulose and xylan cannot be used (Verhaart *et al.*, 2010). The hydrogen-producing capabilities of these thermophiles may be improved via metabolic engineering strategies, such as increasing the hydrogen yield and the enhancement of cell resistance to high-concentrations of hydrogen. The increase in the hydrogen yield can be achieved by redirecting metabolic fluxes or introducing non-native pathways in the cell. For example, the hydrogen production of *P. furiosus* was increased through overexpression of its native hydrogenase (Chandrayan *et al.*, 2012). The overexpression of a non-native [Fe]-hydrogenase in a non-hydrogen producing *E. coli* strain resulted in a high yield of hydrogen (Chittibabu *et al.*, 2006), but such a strategy has not been used in thermophiles to date.

Theoretically, only one-third of the energy content of the sugars derived from lignocellulose can be utilized in the conversion of biomass to biohydrogen via dark fermentation. Most sugars are converted into organic acids (e.g. ethanol, lactate, acetate) by anaerobic microorganisms. As a major component of biogas, methane can be produced from agricultural, municipal, and industrial wastes as sustainable energy via anaerobic digestion. The production process requires little energy but may yield high net energy, and the technology is scalable and globally applicable. In addition, methane production can be effectively applied to improve energy yields from other biofuel production processes, including bioethanol, biodiesel and biohydrogen production (Henstra *et al.*, 2007; Wall *et al.*, 2008). Thus, this method provides an alternative for saving energy and the carbon source by combining the production of biohydrogen and methane in a two-step biological process (Ljunggren and Zacchi, 2010; Zhu *et al.*, 2008).

Ethanol

Ethanol is the predominant biofuel produced via a microbial process (Antoni *et al.*, 2007). To produce bioethanol from lignocellulose, three biological events are involved, i.e. cellulase and hemicellulase production, cellulose and hemicellulose hydrolysis, and sugar and oligosaccharide fermentation, in which cellulolytic processes are essential. In simultaneous saccharification and fermentation (SSF) technology, free cellulases and hemicellulases are produced separately by specific microorganisms (e.g. *Bacillus*, *Trichoderma* and *Actinomyces*) and then directly added to fermentation reactors (typically utilizing yeasts such as *Saccharomyces cerevisiae*). This technology still requires a high cost for the production of expensive cellulase enzymes (over 50% of the total cost), and the total cost of SSF is more than four times higher than that of CBP (Gowen and Fong, 2010; Lynd *et al.*, 2005). CBP integrates the enzyme production, cellulose hydrolysis and fermentation in one step and has been proposed as a more ideal alternative for the production of lignocellulosic ethanol (Kádár *et al.*, 2004; Lynd *et al.*, 2005; Olson *et al.*, 2012; Xu *et al.*, 2009). CBP technology requires a microorganism or a co-culture that is capable of both efficient cellulose hydrolysis and fermentation.

To date, the ideal CBP microorganism is unknown, and two ways have been used to develop CBP organisms: engineering non-cellulosic microorganisms to become cellulolytic, or engineering cellulolytic microorganisms to produce ethanol (Argyros *et al.*, 2011; Kim *et al.*, 2013; Olson *et al.*, 2012; Shaw *et al.*, 2008; Yamada *et al.*, 2013; You *et al.*, 2012). To convert a non-cellulolytic microorganism into a cellulolytic one, heterologous expression of sufficient quantities of several types of cellulases and hemicellulases to support rapid growth and utilization of lignocelluloses will be the major challenge due to the limited genetic manipulation methods available for thermophiles compared with mesophiles (Olson *et al.*, 2012). Fungi *T. reesei* naturally produces high amounts of cellulase and is capable of ethanol production. Nevertheless, *T. reesei* is not ideal for industrial applications because its production rate of ethanol is low, and more importantly, it requires high energy for aeration and stirring during fermentation with solid cellulosic substrates (Fang and Xia, 2013; Xu *et al.*, 2009).

Thermophilic clostridia, such as *C. thermocellum*, can naturally utilize lignocellulose and produce ethanol and therefore is considered one of the most promising CBP candidates (Argyros *et al.*, 2011; Freier *et al.*, 1988; Lynd, 1996); however, the wild-type *C. thermocellum* strains possess inherent disadvantages and do not meet the requirements for industrial production. Thus, systematic engineering of *C. thermocellum* is needed to increase the ethanol yield by eliminating byproducts, improving the tolerance to ethanol, and introducing desirable metabolic pathways for assimilating most or all lignocellulose sugars (Xu *et al.*, 2009). Recently, the genome sequencing of *C. thermocellum* strains has been accomplished and a series of genetic tools have been developed for *C. thermocellum*, which facilitated the metabolic engineering of *C. thermocellum* as well as its co-culture with other thermophilic ethanol-producers to produce higher ethanol yields (Argyros *et al.*, 2011; Feinberg *et al.*, 2011; Mohr *et al.*, 2013; Olson, 2010; Olson *et al.*, 2010; Roberts *et al.*, 2010; Tripathi *et al.*, 2010).

As a strict anaerobe, *C. thermocellum* produces several fermentative end products including lactate, acetate and ethanol, which are all converted from pyruvate. To increase the production of ethanol, a range of targeted mutations aimed at eliminating lactate and acetate production has been performed to increase carbon flux from pyruvate to ethanol (Argyros *et al.*, 2011; Mohr *et al.*, 2013; Tripathi *et al.*, 2010). Argyros *et al.* constructed a *C. thermocellum* mutant with inactivated lactate dehydrogenase (Ldh) and phosphotransacetylase (Pta), whose acetate and lactate concentration in broth were below detection limits, and the ethanol yield was 4.2-fold higher than the wild-type strain (Argyros *et al.*, 2011). In the *ldh*- and *pta*-disrupted mutants, increased intracellular accumulation of pyruvate and extracellular secretion of amino acids was observed, which indicated that a disruption in fermentation end-product pathways may result in an intracellular redox imbalance (van der Veen *et al.*, 2013). Thus, genetic engineering of *C. thermocellum* on the pyruvate metabolic pathway and the recycling of redox power should be considered to further increase ethanol production (Mohr *et al.*, 2013).

The growth of *C. thermocellum* is inhibited by 2% (v/v) ethanol, although its cellulosome system is much less sensitive to ethanol (retains 17% and 55% activity in 2% and 8% ethanol, respectively) (Herrero and Gomez, 1980; Xu *et al.*, 2010). Compared to mesophiles, thermophilic microorganisms generally show less ethanol tolerance, presumably due to the higher permeability of solvents at higher temperatures. Timmons *et al.* suggested that ethanol-adapted *C. thermocellum* may increase membrane rigidity to resist ethanol (Timmons *et al.*, 2009), while Zhu *et al.* demonstrated that cellodextrin synthesis may be a potential mechanism for stress resistance in *C. thermocellum* (Zhu *et al.*, 2013). Alcohol dehydrogenase (Adh) was reported to play a key role in resisting ethanol stress. A mutated bifunctional acetaldehyde-CoA/alcohol dehydrogenase (AdhE) in *C. thermocellum* shifted its cofactor specificity completely from NADH-dependent to NADPH-dependent and endowed the mutant strain an ethanol-tolerant phenotype (Brown *et al.*, 2011). These findings encouraged targeted genetic engineering of *C. thermocellum* and other thermophiles to enhance the resistance to solvents.

Hemicellulose contributes 20–50% of various lignocellulose materials (Xia *et al.*, 2013). However, *C. thermocellum* cannot utilize pentose as a carbon source, which severely inhibits its economic potential as a biocatalyst in the CBP conversion of lignocellulose. According to the gene annotation, *C. thermocellum* harbours no enzymes catalysing the conversion of xylose to xylulose-5-phosphate to enter the pentose phosphate pathway, and the pentose phosphate pathway is also not complete. Thus, at least three genes encoding xylose isomerase, xylulose kinase and transaldehydase should be introduced and heterologously expressed to complement the xylose metabolic pathway in *C. thermocellum*, also considering the redox balance. Metabolic engineering focusing on xylose utilization is possible because genetic tools for thermophiles have been developed (Mohr *et al.*, 2013). Nevertheless, *C. thermocellum* can be co-cultured with other thermophilic microorganisms that are capable of pentose utilization based on the SSF technology that compliments its substrate-utilizing capability.

Thermoanaerobacter and *Thermoanaerobacterium* species are physiologically similar to the thermophilic Clostridia. Although they are non-cellulolytic, they are able to ferment both D-glucose and D-xylose into ethanol (Lynd *et al.*, 2005). Metabolic engineering of *Thermoanaerobacter* and *Thermoanaerobacterium* has been reported to enhance ethanol production. For example, Shaw *et al.* produced the engineered *T. saccharolyticum* strain ALK2 by knocking out genes involved in organic acid formation. ALK2 utilized glucose and xylose simultaneously and was able to produce ethanol as the only detectable organic product. Its homoethanologenic phenotype was stable for >150 generations in a continuous culture, and the maximum ethanol titre is 37 g/l (Shaw *et al.*, 2008). Another mutant of *T. saccharolyticum*, HKO7, was further engineered by deletion of the *hfs* gene cluster (encoding ferredoxin-linked hydrogenases) and the *ldh* gene (encoding a lactate dehydrogenase), and the ethanol production was increased by 44% compared with the wild-type strain (Shaw *et al.*, 2009). A mutant of *T. mathranii* has also been created by using a strategy similar to that used for *T. saccharolyticum* with the purpose of improving ethanol production, and the ethanol yield increased to 0.42 g/g glucose or xylose (Georgieva *et al.*, 2008). In the co-culture with an engineered *T. saccharolyticum* ethanologen, 38.1 g/l ethanol was produced from 92.2 g/l Avicel, which is the highest ethanol titre produced by a thermophilic, cellulolytic culture to date and is comparable to that of *S. cerevisiae* (Argyros *et al.*, 2011). *T. ethanolicus* was co-cultured with *C. thermocellum*, and the ethanol yield was increased 8- to 22-fold compared with the *C. thermocellum* pure culture (Fang, 2010). Other thermophiles can also be used for bioethanol production in addition to Thermoanaerobacterales. Xu and Tschirner (2011) used a co-culture consisting of *C. thermocellum* and *C. thermolacticum* to produce ethanol under alkaline conditions. Surprisingly, they detected that low levels of initially added ethanol had a stimulatory impact on the final ethanol production under selected conditions; however, high ethanol concentrations inhibited the fermentation process, and the highest ethanol yield was 75% (w/w) for microcrystallized cellulose and 90% (w/w) for xylose (Xu and Tschirner, 2011). Cripps *et al.* eliminated the lactate dehydrogenase and pyruvate formate lyase pathways in a *Geobacillus thermoglucosidasius* strain and then up-regulated its expression of pyruvate dehydrogenase to produce the engineered strain TM242. This mutant used various sugars as carbon sources and the ethanol yield was over 90% of the theoretical value (Cripps *et al.*, 2009).

Whether using CBP or SSF technology, lignocellulose hydrolysis is the most essential step in bioethanol production via lignocellulose hydrolysis and fermentation. Thus, although natural cellulolytic thermophiles still contain insufficiencies, great attention should be given and an effort made on metabolic engineering that can increase ethanol production.

Others

Recently, it has become possible to construct downstream fermentation pathways for biofuel production in thermophilic microorganisms. Thus, thermophiles are also better candidates to produce other biofuels in addition to hydrogen and ethanol.

Butanol has higher energy content than ethanol and is easily transported. The energy density and octane rating of butanol is comparable to gasoline and can be easily blended with gasoline or used 100% as a fuel without engine modification. Thus, biobutanol is an attractive candidate as a next-generation biofuel (Antoni *et al.*, 2007). A number of native mesophilic butanol producers, including *Clostridium* strains, have been engineered to increase the butanol yield (Atsumi *et al.*, 2008; Dusseaux et *al.*, 2013; Yu *et al.*, 2011), but thermophiles are not known to produce butanol endogenously, especially cellulolytic thermophiles. Despite the fact that *T. thermosaccharolyticum* contains the genes required for the butanol biosynthesis pathway, it produces butanol at very low level (Bhandiwad *et al.*, 2013). Bhandiwad *et al.* increased butanol production by 1.8-fold by overexpressing six native genes contributing to butanol synthesis in *T. thermosaccharolyticum*, but the yield remained low (5.1 mM, 0.04 g/g glucose) (Bhandiwad *et al.*, 2013). Moreover, *T. thermosaccharolyticum* has a low butanol tolerance, which limits the production of butanol (Green *et al.*, 2009). Green *et al.* selected a butanol-tolerant thermophilic Bacillaceae strain and then introduced the entire butanol biosynthesis pathway into the strain to make it capable of butanol production, increasing the yield to 0.4 g/g glucose (Green *et al.*, 2009).

Similar to butanol, isobutanol is a good choice for gasoline replacement. The construction of the isobutanol biosynthesis pathway has been successfully accomplished in *C. cellulolyticum*, a model mesophilic Clostridia strain, and the production of isobutanol was approximately 660 mg/l with microcrystalline cellulose as a carbon source (Higashide *et al.*, 2011). Therefore, it is possible to enable cellulolytic thermophiles (e.g. *C.*

thermocellum) to produce higher alcohols (i.e. alcohols with more than two carbons) by using lignocellulose as substrates via a CBP technology that introduces desirable biosynthesis pathways.

Biodiesel is defined as fatty acid mono-alkylesters and is produced from triacylglycerols. Due to the lack of oil feedstocks, much effort has been placed on increasing the cellular lipid content for biodiesel production in bacteria, algae and fungi/yeast (reviewed by Li *et al.*, 2008). Cellulolytic microorganisms usually cannot afford the high amount of acyl-CoA, ATP and reduced energy required for biodiesel production. However, the biosynthesis of lignocellulosic biodiesel can be achieved using an SSF technology because once lignocellulose is converted into fermentable sugars by thermophilic cellulose degraders, the fermentation for biodiesel production can be processed by heterotrophic oil-producing microorganisms (Chisti, 2007; Kalscheuer *et al.*, 2006; Shen *et al.*, 2007).

Conclusion

Biofuels are a promising next-generation energy alternative and are gradually supplementing fossil fuels such as oil and coal. Biofuel can be produced from renewable biomass in a chemical and/or a microbial process. Several steps such as pretreatment, hydrolysis, fermentation, and product separation are included in the production of lignocellulosic biofuels. The hydrolysis of lignocellulose is the essential step that converts the insoluble substrate into fermentable sugars and makes the recalcitrant substrate available for various microorganisms to produce a broad range of products. When considering the cost of production, cellulolytic thermophiles are better alternatives, while CBP technology is considered the most promising strategy for biofuel production from lignocellulose. Recently, biotechnology tools for thermophiles has been developed and allows for the genetic manipulation of biofuel producers, and the metabolic engineering of thermophiles has been preliminarily developed to overcome some of the inherent disadvantages for producing high yields of biofuels. However, thermophiles have rarely been used in the large-scale production of biofuels to date. Therefore, additional metabolic engineering is needed to further scale up the production of lignocellulosic biofuel using thermophiles. Studies on thermophilic biofuel producers should strive for not only an increase in biofuel yields but also the improvement of stress resistance, the stimulation of substrate utilization efficiency, and energy and reduced power coupling.

References

Anbar, M., Gul, O., Lamed, R., Sezerman, U.O., and Bayer, E.A. (2012). Improved thermostability of *Clostridium thermocellum* endoglucanase Cel8A by using consensus-guided mutagenesis. Appl. Environ. Microbiol. *78*, 3458–3464.

Antoni, D., Zverlov, V.V., and Schwarz, W.H. (2007). Biofuels from microbes. Appl. Microbiol. Biotechnol. *77*, 23–35.

Argyros, D.A., Tripathi, S.A., Barrett, T.F., Rogers, S.R., Feinberg, L.F., Olson, D.G., Foden, J.M., Miller, B.B., Lynd, L.R., Hogsett, D.A., *et al.* (2011). High ethanol titers from cellulose using metabolically engineered thermophilic, anaerobic microbes. Appl. Environ. Microbiol. *77*, 8288–8294.

Atsumi, S., Cann, A.F., Connor, M.R., Shen, C.R., Smith, K.M., Brynildsen, M.P., Chou, K.J., Hanai, T., and Liao, J.C. (2008). Metabolic engineering of *Escherichia coli* for 1-butanol production. Metab. Eng. *10*, 305–311.

Averhoff, B. (2004). DNA transport and natural transformation in mesophilic and thermophilic bacteria. J. Bioenerg. Biomembr. *36*, 25–33.

Bayer, E.A., Shimon, L.J., Shoham, Y., and Lamed, R. (1998). Cellulosomes-structure and ultrastructure. J. Struct. Biol. *124*, 221–234.

Bayer, E.A., Lamed, R., White, B.A., and Flint, H.J. (2008). From cellulosomes to cellulosomics. Chem. Rec. *8*, 364–377.

Bedell, V.M., Wang, Y., Campbell, J.M., Poshusta, T.L., Starker, C.G., Krug, R.G. 2nd, Tan, W., Penheiter, S.G., Ma, A.C., Leung, A.Y., *et al.* (2012). *In vivo* genome editing using a high-efficiency TALEN system. Nature *491*, 114–118.

Bhandiwad, A., Guseva, A., and Lynd, L. (2013). Metabolic Engineering of *Thermoanaerobacterium thermosaccharolyticum* for Increased n-Butanol Production. Adv. Microbiol. 3, 46–51.

Bielen, A., Verhaart, M., van der Oost, J., and Kengen, S. (2013). Biohydrogen production by the thermophilic bacterium *Caldicellulosiruptor saccharolyticus*: current status and perspectives. Life 3, 52–85.

Borne, R., Bayer, E.A., Pages, S., Perret, S., and Fierobe, H.P. (2013). Unraveling enzyme discrimination during cellulosome assembly independent of cohesin/dockerin affinity. FEBS J. *280*, 5764–5779.

Bridgwater, A.V. (2003). Renewable fuels and chemicals by thermal processing of biomass. Chem. Eng. J. *91*, 87–102.

Brown, S.D., Guss, A.M., Karpinets, T.V., Parks, J.M., Smolin, N., Yang, S., Land, M.L., Klingeman, D.M.,

Bhandiwad, A., Rodriguez, M. Jr., *et al.* (2011). Mutant alcohol dehydrogenase leads to improved ethanol tolerance in *Clostridium thermocellum*. Proc. Natl. Acad. Sci. U.S.A. *108*, 13752–13757.

Brown, S.D., Lamed, R., Morag, E., Borovok, I., Shoham, Y., Klingeman, D.M., Johnson, C.M., Yang, Z., Land, M.L., Utturkar, S.M., *et al.* (2012). Draft genome sequences for *Clostridium thermocellum* wild-type strain YS and derived cellulose adhesion-defective mutant strain AD2. J. Bacteriol. *194*, 3290–3291.

Cava, F., de Pedro, M.A., Blas-Galindo, E., Waldo, G.S., Westblade, L.F., and Berenguer, J. (2008). Expression and use of superfolder green fluorescent protein at high temperatures *In vivo*: a tool to study extreme thermophile biology. Environ. Microbiol. *10*, 605–613.

Chandel, A.K., Chandrasekhar, G., Silva, M.B., and Silverio da Silva, S. (2012). The realm of cellulases in biorefinery development. Crit. Rev. Biotechnol. *32*, 187–202.

Chandrayan, S.K., McTernan, P.M., Hopkins, R.C., Sun, J., Jenney, F.E., Jr., and Adams, M.W. (2012). Engineering hyperthermophilic archaeon *Pyrococcus furiosus* to overproduce its cytoplasmic [NiFe]-hydrogenase. J. Biol. Chem. *287*, 3257–3264.

Chisti, Y. (2007). Biodiesel from microalgae. Biotechnol. Adv. *25*, 294–306.

Chittibabu, G., Nath, K., and Das, D. (2006). Feasibility studies on the fermentative hydrogen production by recombinant *Escherichia coli* BL-21. Process Biochem. *41*, 682–688.

Comfort, D.A., Chhabra, S.R., Conners, S.B., Chou, C.-J., Epting, K.L., Johnson, M.R., Jones, K.L., Sehgal, A.C., and Kelly, R.M. (2004). Strategic biocatalysis with hyperthermophilic enzymes. Green Chem. *6*, 459.

Cripps, R.E., Eley, K., Leak, D.J., Rudd, B., Taylor, M., Todd, M., Boakes, S., Martin, S., and Atkinson, T. (2009). Metabolic engineering of *Geobacillus thermoglucosidasius* for high yield ethanol production. Metab. Eng. *11*, 398–408.

Cui, G.Z., Hong, W., Zhang, J., Li, W.L., Feng, Y., Liu, Y.J., and Cui, Q. (2012). Targeted gene engineering in *Clostridium cellulolyticum* H10 without methylation. J. Microbiol. Methods *89*, 201–208.

Cui, G.Z., Zhang, J., Hong, W., Xu, C., Feng, Y., Cui, Q., and Liu, Y.J. (2014). Improvement of ClosTron for successive gene disruption in *Clostridium cellulolyticum* using a *pyrF*-based screening system. Appl. Microbiol. Biotechnol. *98*, 313–323.

Debnath, M., Prasad, G.B.K.S., and Bisen, P.S. (2010). Molecular Diagnostics: Promises and Possibilities (Springer, Netherlands).

Demain, A.L. (2009). Biosolutions to the energy problem. J. Ind. Microbiol. Biotechnol. *36*, 319–332.

Demain, A.L., Newcomb, M., and Wu, J.H. (2005). Cellulase, clostridia, and ethanol. Microbiol. Mol. Biol. Rev. *69*, 124–154.

Demirbas, A. (2007). Progress and recent trends in biofuels. Prog. Energy Combust. Sci. *33*, 1–18.

Dong, H., Zhang, Y., Dai, Z., and Li, Y. (2010). Engineering *clostridium* strain to accept unmethylated DNA. PLoS One *5*, e9038.

Drepper, T., Eggert, T., Circolone, F., Heck, A., Krauss, U., Guterl, J.K., Wendorff, M., Losi, A., Gartner, W., and Jaeger, K.E. (2007). Reporter proteins for *In vivo* fluorescence without oxygen. Nat. Biotechnol. *25*, 443–445.

Dusseaux, S., Croux, C., Soucaille, P., and Meynial-Salles, I. (2013). Metabolic engineering of *Clostridium acetobutylicum* ATCC 824 for the high-yield production of a biofuel composed of an isopropanol/butanol/ethanol mixture. Metab. Eng. *18C*, 1–8.

Dykstra, A.B., St Brice, L., Rodriguez, M., Jr., Raman, B., Izquierdo, J., Cook, K.D., Lynd, L.R., and Hettich, R.L. (2014). Development of a multipoint quantitation method to simultaneously measure enzymatic and structural components of the *Clostridium thermocellum* cellulosome protein complex. J. Proteome Res. *13*, 692–701.

Enyeart, P.J., Chirieleison, S.M., Dao, M.N., Perutka, J., Quandt, E.M., Yao, J., Whitt, J.T., Keatinge-Clay, A.T., Lambowitz, A.M., and Ellington, A.D. (2013). Generalized bacterial genome editing using mobile group II introns and Cre-*lox*. Mol. Syst. Biol. *9*, 685.

Fang, H., and Xia, L. (2013). High activity cellulase production by recombinant *Trichoderma reesei* ZU-02 with the enhanced cellobiohydrolase production. Bioresour. Technol. *144*, 693–697.

Fang, Z. (2010). Enhanced role of the co-culture of thermophilic anaerobic bacteria on cellulosic ethanol. Environ. Sci. (Chinese) *31*, 1059–1065.

Feinberg, L., Foden, J., Barrett, T., Davenport, K.W., Bruce, D., Detter, C., Tapia, R., Han, C., Lapidus, A., Lucas, S., *et al.* (2011). Complete genome sequence of the cellulolytic thermophile *Clostridium thermocellum* DSM1313. J. Bacteriol. *193*, 2906–2907.

Fierobe, H.P., Bayer, E.A., Tardif, C., Czjzek, M., Mechaly, A., Belaich, A., Lamed, R., Shoham, Y., and Belaich, J.P. (2002). Degradation of cellulose substrates by cellulosome chimeras. Substrate targeting versus proximity of enzyme components. J. Biol. Chem. *277*, 49621–49630.

Frazier, C.L., San Filippo, J., Lambowitz, A.M., and Mills, D.A. (2003). Genetic manipulation of *Lactococcus lactis* by using targeted group II introns: generation of stable insertions without selection. Appl. Environ. Microbiol. *69*, 1121–1128.

Freier, D., Mothershed, C.P., and Wiegel, J. (1988). Characterization of *Clostridium thermocellum* Jw20. Appl. Environ. Microbiol. *54*, 204–211.

Friedrich, A., Hartsch, T., and Averhoff, B. (2001). Natural transformation in mesophilic and thermophilic bacteria: identification and characterization of novel, closely related competence genes in *Acinetobacter* sp. strain BD413 and *Thermus thermophilus* HB27. Appl. Environ. Microbiol. *67*, 3140–3148.

Frock, A.D., and Kelly, R.M. (2012). Extreme thermophiles: moving beyond single-enzyme biocatalysis. Curr. Opin. Chem. Eng. *1*, 363–372.

Gaj, T., Gersbach, C.A., and Barbas, C.F. 3rd (2013). ZFN, TALEN, and CRISPR/Cas-based methods for genome engineering. Trends Biotechnol. *31*, 397–405.

Garcia-Rodriguez, F.M., Barrientos-Duran, A., Diaz-Prado, V., Fernandez-Lopez, M., and Toro, N. (2011). Use of RmInt1, a group IIB intron lacking the intron-encoded protein endonuclease domain, in gene targeting. Appl. Environ. Microbiol. *77*, 854–861.

Gaykawad, S.S., Zha, Y., Punt, P.J., van Groenestijn, J.W., van der Wielen, L.A., and Straathof, A.J. (2013). Pervaporation of ethanol from lignocellulosic fermentation broth. Bioresour. Technol. *129*, 469–476.

Gefen, G., Anbar, M., Morag, E., Lamed, R., and Bayer, E.A. (2012). Enhanced cellulose degradation by targeted integration of a cohesin-fused beta-glucosidase into the *Clostridium thermocellum* cellulosome. Proc. Natl. Acad. Sci. U.S.A. *109*, 10298–10303.

Georgieva, T.I., Mikkelsen, M.J., and Ahring, B.K. (2008). Ethanol production from wet-exploded wheat straw hydrolysate by thermophilic anaerobic bacterium *Thermoanaerobacter* BG1L1 in a continuous immobilized reactor. Appl. Biochem. Biotechnol. *145*, 99–110.

Gowen, C.M., and Fong, S.S. (2010). Exploring biodiversity for cellulosic biofuel production. Chem. Biodivers. *7*, 1086–1097.

de Grado, M., Castan, P., and Berenguer, J. (1999). A high-transformation-efficiency cloning vector for *Thermus thermophilus*. Plasmid *42*, 241–245.

Green, E.G., Jenkinson, E.G., and Crow, M.G. (2009). Production of butanol. International Patent Application. WO/2009/122192.

Gronenberg, L.S., Marcheschi, R.J., and Liao, J.C. (2013). Next generation biofuel engineering in prokaryotes. Curr. Opin. Chem. Biol. *17*, 462–471.

Harris, P.V., Xu, F., Kreel, N.E., Kang, C., and Fukuyama, S. (2014). New enzyme insights drive advances in commercial ethanol production. Curr. Opin. Chem. Biol. *19*, 162–170.

Hasunuma, T., Okazaki, F., Okai, N., Hara, K.Y., Ishii, J., and Kondo, A. (2013). A review of enzymes and microbes for lignocellulosic biorefinery and the possibility of their application to consolidated bioprocessing technology. Bioresour. Technol. *135*, 513–522.

Heap, J.T., Pennington, O.J., Cartman, S.T., Carter, G.P., and Minton, N.P. (2007). The ClosTron: a universal gene knock-out system for the genus *Clostridium*. J. Microbiol. Methods *70*, 452–464.

Heap, J.T., Ehsaan, M., Cooksley, C.M., Ng, Y.K., Cartman, S.T., Winzer, K., and Minton, N.P. (2012). Integration of DNA into bacterial chromosomes from plasmids without a counter-selection marker. Nucleic Acids Res. *40*, e59.

Henstra, A.M., Sipma, J., Rinzema, A., and Stams, A.J. (2007). Microbiology of synthesis gas fermentation for biofuel production. Curr. Opin. Biotechnol. *18*, 200–206.

Herrero, A.A., and Gomez, R.F. (1980). Development of ethanol tolerance in *Clostridium thermocellum* – effect of growth temperature. Appl. Environ. Microbiol. *40*, 571–577.

Higashide, W., Li, Y., Yang, Y., and Liao, J.C. (2011). Metabolic engineering of *Clostridium cellulolyticum* for isobutanol production from cellulose. Appl. Environ. Microbiol. *77*, 2727–2733.

Hoffert, M.I. (2010). Farewell to fossil fuels? Science *329*, 1292–1294.

Hon-nami, K., Coughlan, M., Hon-nami, H., and Ljungdahl, L. (1986). Separation and characterization of the complexes constituting the cellulolytic enzyme system of *Clostridium thermocellum*. Arch. Microbiol. *145*, 13–19.

Hong, W., Zhang, J., Feng, Y., Mohr, G., Lambowitz, A.M., Cui, G.-Z., Liu, Y.-J., and Cui, Q. (2014). The contribution of cellulosomal scaffoldins to cellulose hydrolysis by *Clostridium thermocellum* analyzed by using thermotargetrons. Biotechnol. Biofuels 7, 80.

Huang, Y.C., Chen, G.H., Chen, Y.F., Chen, W.L., and Yang, C.H. (2010). Heterologous expression of thermostable acetylxylan esterase gene from *Thermobifida fusca* and its synergistic action with xylanase for the production of xylooligosaccharides. Biochem. Biophys. Res. Commun. *400*, 718–723.

Kádár, Z., Szengyel, Z., and Réczey, K. (2004). Simultaneous saccharification and fermentation (SSF) of industrial wastes for the production of ethanol. Ind. Crop. Prod. *20*, 103–110.

Kalscheuer, R., Stolting, T., and Steinbuchel, A. (2006). Microdiesel: *Escherichia coli* engineered for fuel production. Microbiology *152*, 2529–2536.

Karberg, M., Guo, H., Zhong, J., Coon, R., Perutka, J., and Lambowitz, A.M. (2001). Group II introns as controllable gene targeting vectors for genetic manipulation of bacteria. Nat. Biotech. *19*, 1162–1167.

Kim, S., Baek, S.H., Lee, K., and Hahn, J.S. (2013). Cellulosic ethanol production using a yeast consortium displaying a minicellulosome and beta-glucosidase. Microb. Cell Fact. *12*, 14.

Kita, A., Iwasaki, Y., Sakai, S., Okuto, S., Takaoka, K., Suzuki, T., Yano, S., Sawayama, S., Tajima, T., Kato, J., *et al.* (2012). Development of genetic transformation and heterologous expression system in carboxydotrophic thermophilic acetogen *Moorella thermoacetica*. J. Biosci. Bioeng. *115*, 347–352.

Kotai, L., Szepvolgyi, J., Szilagyi, M., Zhibin, L., Baiquan, C., Sharma, V., and K, P. (2013). Biobutanol from renewable agricultural and lignocellulose resources and its perspectives as alternative of liquid fuels. In Liquid, Gaseous and Solid Biofuels – Conversion Techniques. Fang, Z., ed. (InTech, Rijeka, Croatia).

Koyama, Y., Hoshino, T., Tomizuka, N., and Furukawa, K. (1986). Genetic transformation of the extreme thermophile *Thermus thermophilus* and of other Thermus spp. J. Bacteriol. *166*, 338–340.

Lambowitz, A.M., and Zimmerly, S. (2011). Group II Introns: Mobile Ribozymes that Invade DNA. Cold Spring Harb. Perspect. Biol. 3, a003616.

Lang, A.S., and Beatty, J.T. (2002). A bacterial signal transduction system controls genetic exchange and motility. J. Bacteriol. *184*, 913–918.

Lee, H.L., Chang, C.K., Teng, K.H., and Liang, P.H. (2011). Construction and characterization of different fusion proteins between cellulases and beta-glucosidase to improve glucose production and thermostability. Bioresour. Technol. *102*, 3973–3976.

Li, Q., Du, W., and Liu, D. (2008). Perspectives of microbial oils for biodiesel production. Appl. Microbiol. Biotechnol. *80*, 749–756.

Lin, L., and Xu, J. (2013). Dissecting and engineering metabolic and regulatory networks of thermophilic bacteria for biofuel production. Biotechnol. Adv. *31*, 827–837.

Lin, L., Song, H., Ji, Y., He, Z., Pu, Y., Zhou, J., and Xu, J. (2010). Ultrasound-mediated DNA transformation in thermophilic Gram-positive anaerobes. PLoS One *5*, e12582.

Liu, Y.J., Zhang, J., Cui, G.Z., and Cui, Q. (2015). Current progress of targetron technology: development, improvement and application in metabolic engineering. Biotechnol. J. *10*. In press.

Ljunggren, M., and Zacchi, G. (2010). Techno-economic analysis of a two-step biological process producing hydrogen and methane. Bioresour. Technol. *101*, 7780–7788.

Logan, B.E., Oh, S.E., Kim, I.S., and Van Ginkel, S. (2002). Biological hydrogen production measured in batch anaerobic respirometers. Environ. Sci. Technol. *36*, 2530–2535.

Lu, Y., Li, N., Yuan, X., Hua, B., Wang, J., Ishii, M., Igarashi, Y., and Cui, Z. (2013). Enhancing the cellulose-degrading activity of cellulolytic bacteria CTL-6 (*Clostridium thermocellum*) by co-culture with non-cellulolytic bacteria W2–10 (Geobacillus sp.). Appl. Biochem. Biotechnol. *171*, 1578–1588.

Lynd, L.R. (1996). Overreview and evaluation of fuel ethanol from cellulosic biomass: technology, economics, the environment, and policy. Annu. Rev. Energy Environ. *21*, 403–465.

Lynd, L.R. (2010). Make way for ethanol. Science 33, 1176–1177.

Lynd, L.R., van Zyl, W.H., McBride, J.E., and Laser, M. (2005). Consolidated bioprocessing of cellulosic biomass: an update. Curr. Opin. Biotechnol. *16*, 577–583.

Mai, V., and Wiegel, J. (2000). Advances in development of a genetic system for *Thermoanaerobacterium* spp.: expression of genes encoding hydrolytic enzymes, development of a second shuttle vector, and integration of genes into the chromosome. Appl. Environ. Microbiol. *66*, 4817–4821.

Maki, M.L., Armstrong, L., Leung, K.T., and Qin, W. (2013). Increased expression of beta-glucosidase A in *Clostridium thermocellum* 27405 significantly increases cellulase activity. Bioengineered *4*, 15–20.

Mohr, G., Ghanem, E., and Lambowitz, A.M. (2010). Mechanisms used for genomic proliferation by thermophilic group II introns. PLoS Biol. *8*, e1000391.

Mohr, G., Hong, W., Zhang, J., Cui, G.-Z., Yang, Y., Cui, Q., Liu, Y.-J., and Lambowitz, A.M. (2013). A targetron system for gene targeting in thermophiles and its application in *Clostridium thermocellum*. PLoS One *8*, e69032.

Moracci, M., La Volpe, A., Pulitzer, J.F., Rossi, M., and Ciaramella, M. (1992). Expression of the thermostable beta-galactosidase gene from the archaebacterium *Sulfolobus solfataricus* in *Saccharomyces cerevisiae* and characterization of a new inducible promoter for heterologous expression. J. Bacteriol. *174*, 873–882.

Morais, S., Barak, Y., Caspi, J., Hadar, Y., Lamed, R., Shoham, Y., Wilson, D.B., and Bayer, E.A. (2010). Cellulase-xylanase synergy in designer cellulosomes for enhanced degradation of a complex cellulosic substrate. mBio *1*, e00285.

Morais, S., Morag, E., Barak, Y., Goldman, D., Hadar, Y., Lamed, R., Shoham, Y., Wilson, D.B., and Bayer, E.A. (2012). Deconstruction of lignocellulose into soluble sugars by native and designer cellulosomes. mBio 3, e00508–00512.

Natarajan, M.R., and Oriel, P. (1991). Conjugal transfer of recombinant transposon Tn916 from *Escherichia coli* to *Bacillus stearothermophilus*. Plasmid 26, 67–73.

Olson, D.G. (2010). Electrotransformation of Gram-positive, anaerobic, thermophilic bacteria. United States Patent Application. 20110281362.

Olson, D.G., and Lynd, L.R. (2012a). Computational design and characterization of a temperature-sensitive plasmid replicon for Gram-positive thermophiles. J. Biol. Eng. *6*, 5.

Olson, D.G., and Lynd, L.R. (2012b). Transformation of *Clostridium thermocellum* by electroporation. Methods Enzymol. *510*, 317–330.

Olson, D.G., Tripathi, S.A., Giannone, R.J., Lo, J., Caiazza, N.C., Hogsett, D.A., Hettich, R.L., Guss, A.M., Dubrovsky, G., and Lynd, L.R. (2010). Deletion of the Cel48S cellulase from *Clostridium thermocellum*. Proc. Natl. Acad. Sci. U.S.A. *107*, 17727–17732.

Olson, D.G., McBride, J.E., Shaw, A.J., and Lynd, L.R. (2012). Recent progress in consolidated bioprocessing. Curr. Opin. Biotechnol. 23, 396–405.

Olson, D.G., Giannone, R.J., Hettich, R.L., and Lynd, L.R. (2013). Role of the CipA scaffoldin protein in cellulose solubilization, as determined by targeted gene deletion and complementation in *Clostridium thermocellum*. J. Bacteriol. *195*, 733–739.

Peng, H., Fu, B., Mao, Z., and Shao, W. (2006). Electrotransformation of *Thermoanaerobacter ethanolicus* JW200. Biotechnol. Lett. *28*, 1913–1917.

Perutka, J., Wang, W., Goerlitz, D., and Lambowitz, A.M. (2004). Use of computer-designed group II introns to disrupt *Escherichia coli* DExH/D-box protein and DNA helicase genes. J. Mol. Biol. *336*, 421–439.

Pinkney, J.N., Zawadzki, P., Mazuryk, J., Arciszewska, L.K., Sherratt, D.J., and Kapanidis, A.N. (2012). Capturing reaction paths and intermediates in Cre-*loxP* recombination using single-molecule fluorescence. Proc. Natl. Acad. Sci. U.S.A. *109*, 20871–20876.

Podkaminer, K.K., Guss, A.M., Trajano, H.L., Hogsett, D.A., and Lynd, L.R. (2012). Characterization of xylan utilization and discovery of a new endoxylanase in *Thermoanaerobacterium saccharolyticum* through targeted gene deletions. Appl. Environ. Microbiol. *78*, 8441–8447.

Raman, B., McKeown, C.K., Rodriguez, M., Jr., Brown, S.D., and Mielenz, J.R. (2011). Transcriptomic analysis of *Clostridium thermocellum* ATCC 27405 cellulose fermentation. BMC Microbiol. *11*, 134.

Ramchuran, S.O., Nordberg Karlsson, E., Velut, S., De Mare, L., Hagander, P., and Holst, O. (2002). Production of heterologous thermostable glycoside hydrolases and the presence of host-cell proteases in substrate limited fed-batch cultures of *Escherichia coli* BL21(DE3). Appl. Microbiol. Biotechnol. *60*, 408–416.

Ramchuran, S.O., Vargas, V.A., Hatti-Kaul, R., and Karlsson, E.N. (2006). Production of a lipolytic enzyme originating from *Bacillus halodurans* LBB2 in the methylotrophic yeast *Pichia pastoris*. Appl. Microbiol. Biotechnol. *71*, 463–472.

Richter, H., Zoephel, J., Schermuly, J., Maticzka, D., Backofen, R., and Randau, L. (2012). Characterization of CRISPR RNA processing in *Clostridium thermocellum* and Methanococcus maripaludis. Nucleic Acids Res. *40*, 9887–9896.

Richter, H., Randau, L., and Plagens, A. (2013). Exploiting CRISPR/Cas: interference mechanisms and applications. Int. J. Mol. Sci. *14*, 14518–14531.

Riederer, A., Takasuka, T.E., Makino, S., Stevenson, D.M., Bukhman, Y.V., Elsen, N.L., and Fox, B.G. (2011). Global gene expression patterns in *Clostridium thermocellum* as determined by microarray analysis of chemostat cultures on cellulose or cellobiose. Appl. Environ. Microbiol. *77*, 1243–1253.

Roberts, S.B., Gowen, C.M., Brooks, J.P., and Fong, S.S. (2010). Genome-scale metabolic analysis of *Clostridium thermocellum* for bioethanol production. BMC Syst. Biol. *4*, 31.

Sato, T., Fukui, T., Atomi, H., and Imanaka, T. (2005). Improved and versatile transformation system allowing multiple genetic manipulations of the hyperthermophilic archaeon *Thermococcus kodakaraensis*. Appl. Environ. Microbiol. *71*, 3889–3899.

Schiraldi, C., and De Rosa, M. (2002). The production of biocatalysts and biomolecules from extremophiles. Trends Biotechnol. *20*, 515–521.

Schut, G.J., and Adams, M.W. (2009). The iron-hydrogenase of *Thermotoga maritima* utilizes ferredoxin and NADH synergistically: a new perspective on anaerobic hydrogen production. J. Bacteriol. *191*, 4451–4457.

Schut, G.J., Nixon, W.J., Lipscomb, G.L., Scott, R.A., and Adams, M.W. (2012). Mutational Analyses of the enzymes involved in the metabolism of hydrogen by the hyperthermophilic archaeon *Pyrococcus furiosus*. Front. Microbiol. *3*, 163.

Schwarz, W.H. (2001). The cellulosome and cellulose degradation by anaerobic bacteria. Appl. Microbiol. Biotechnol. *56*, 634–649.

Sen, S., and Oriel, P. (1990). Transfer of transposon Tn916 from *Bacillus subtilis* to Thermus aquaticus. FEMS Microbiol. Lett. *55*, 131–134.

Service, R.F. (2010). Is there a road ahead for cellulosic ethanol? Science *329*, 784–785.

Shaner, N.C., Patterson, G.H., and Davidson, M.W. (2007). Advances in fluorescent protein technology. J. Cell Sci. *120*, 4247–4260.

Shaw, A.J., Podkaminer, K.K., Desai, S.G., Bardsley, J.S., Rogers, S.R., Thorne, P.G., Hogsett, D.A., and Lynd, L.R. (2008). Metabolic engineering of a thermophilic bacterium to produce ethanol at high yield. Proc. Natl. Acad. Sci. U.S.A. *105*, 13769–13774.

Shaw, A.J., Hogsett, D.A., and Lynd, L.R. (2009). Identification of the [FeFe]-hydrogenase responsible for hydrogen generation in *Thermoanaerobacterium saccharolyticum* and demonstration of increased ethanol yield via hydrogenase knockout. J. Bacteriol. *191*, 6457–6464.

Shaw, A.J., Covalla, S.F., Hogsett, D.A., and Herring, C.D. (2011). Marker removal system for *Thermoanaerobacterium saccharolyticum* and development of a markerless ethanologen. Appl. Environ. Microbiol. *77*, 2534–2536.

Shen, J.-j., Li, F.-c., Yang, Q.-l., Feng, D., Qin, S., and Zhao, Z. (2007). Fermentation of Spartina anglica acid hydrolysate by *Trichosporon cutaneum* for microbial lipid production. Mar. Sci. (Qingdao Chinese Edition) *31*, 38.

Singh, V., Mani, I., Chaudhary, D.K., and Dhar, P.K. (2013). Metabolic engineering of biosynthetic pathway for production of renewable biofuels. Appl. Biochem. Biotechnol. *172*, 1158–1171.

Soboh, B., Linder, D., and Hedderich, R. (2004). A multisubunit membrane-bound [NiFe] hydrogenase and an NADH-dependent Fe-only hydrogenase in the fermenting bacterium *Thermoanaerobacter tengcongensis*. Microbiology *150*, 2451–2463.

de Souza, W.R. (2013). Microbial degradation of lignocellulosic biomass. In Sustainable Degradation of Lignocellulosic Biomass – Techniques, Applications and Commercialization, Chandel, A.K., and Silvério da Silva, S., eds. (InTech, Rijeka, Croatia).

Stasiak, A. (1996). Getting down to the core of homologous recombination. Science *272*, 828–829.

Talluri, S., Raj, S.M., and Christopher, L.P. (2013). Consolidated bioprocessing of untreated switchgrass to hydrogen by the extreme thermophile *Caldicellulosiruptor saccharolyticus* DSM 8903. Bioresour. Technol. *139*, 272–279.

Tang, C.D., Li, J.F., Wei, X.H., Min, R., Gao, S.J., Wang, J.Q., Yin, X., and Wu, M.C. (2013). Fusing a carbohydrate-binding module into the *Aspergillus usamii* beta-mannanase to improve its thermostability and cellulose-binding capacity by in silico design. PLoS One *8*, e64766.

Taylor, M., Tuffin, M., Burton, S., Eley, K., and Cowan, D. (2008). Microbial responses to solvent and alcohol stress. Biotechnol. J. 3, 1388–1397.

Taylor, M.P., Eley, K.L., Martin, S., Tuffin, M.I., Burton, S.G., and Cowan, D.A. (2009). *Thermophilic ethanologenesis*: future prospects for second-generation bioethanol production. Trends Biotechnol. *27*, 398–405.

Taylor, M.P., van Zyl, L., Tuffin, I.M., Leak, D.J., and Cowan, D.A. (2011). Genetic tool development underpins recent advances in thermophilic whole-cell biocatalysts. Microb. Biotechnol. *4*, 438–448.

Timmons, M.D., Knutson, B.L., Nokes, S.E., Strobel, H.J., and Lynn, B.C. (2009). Analysis of composition and structure of *Clostridium thermocellum* membranes from wild-type and ethanol-adapted strains. Appl. Microbiol. Biotechnol. *82*, 929–939.

Tripathi, S.A., Olson, D.G., Argyros, D.A., Miller, B.B., Barrett, T.F., Murphy, D.M., McCool, J.D., Warner, A.K., Rajgarhia, V.B., Lynd, L.R., *et al.* (2010). Development of *pyrF*-based genetic system for targeted gene deletion in *Clostridium thermocellum* and creation of a pta mutant. Appl. Environ. Microbiol. *76*, 6591–6599.

Turner, P., Mamo, G., and Karlsson, E.N. (2007). Potential and utilization of thermophiles and thermostable enzymes in biorefining. Microb. Cell Fact. *6*, 9.

Van Ooteghem, S.A., Beer, S.K., and Yue, P.C. (2002). Hydrogen production by the thermophilic bacterium *Thermotoga neapolitana*. Appl. Biochem. Biotechnol. *98–100*, 177–189.

van der Veen, D., Lo, J., Brown, S.D., Johnson, C.M., Tschaplinski, T.J., Martin, M., Engle, N.L., van den Berg, R.A., Argyros, A.D., Caiazza, N.C., *et al.* (2013). Characterization of *Clostridium thermocellum* strains with disrupted fermentation end-product pathways. J. Ind. Microbiol. Biotechnol. *40*, 725–734.

Verhaart, M.R., Bielen, A.A., van der Oost, J., Stams, A.J., and Kengen, S.W. (2010). Hydrogen production by hyperthermophilic and extremely thermophilic bacteria and archaea: mechanisms for reductant disposal. Environ. Technol. *31*, 993–1003.

Waege, I., Schmid, G., Thumann, S., Thomm, M., and Hausner, W. (2010). Shuttle vector-based transformation system for *Pyrococcus furiosus*. Appl. Environ. Microbiol. *76*, 3308–3313.

Waeonukul, R., Kosugi, A., Tachaapaikoon, C., Pason, P., Ratanakhanokchai, K., Prawitwong, P., Deng, L., Saito, M., and Mori, Y. (2012). Efficient saccharification of ammonia soaked rice straw by combination of *Clostridium thermocellum* cellulosome and *Thermoanaerobacter brockii* beta-glucosidase. Bioresour. Technol. *107*, 352–357.

Wall, J.D., Harwood, C.S., and Demain, A.L. (2008). Bioenergy (ASM Press, Washington, USA).

Waller, B.H., Olson, D.G., Currie, D.H., Guss, A.M., and Lynd, L.R. (2013). Exchange of type II dockerin-containing subunits of the *Clostridium thermocellum* cellulosome as revealed by SNAP-tags. FEMS Microbiol. Lett. *338*, 46–53.

Wan, H.W. (2013). Transformation of the thermophilic bacterium, *Geobacillus debilis*, by conjugation with the mesophilic bacterium, *Escherichia coli* (Department of Microbiology, The University of Manitoba, Canada).

Wilson, C.M., Rodriguez, M., Jr., Johnson, C.M., Martin, S.L., Chu, T.M., Wolfinger, R.D., Hauser, L.J., Land, M.L., Klingeman, D.M., Syed, M.H., *et al.* (2013a). Global transcriptome analysis of *Clostridium thermocellum* ATCC 27405 during growth on dilute acid pretreated Populus and switchgrass. Biotechnol. Biofuels 6, 179.

Wilson, C.M., Yang, S., Rodriguez, M., Jr., Ma, Q., Johnson, C.M., Dice, L., Xu, Y., and Brown, S.D. (2013b). *Clostridium thermocellum* transcriptomic profiles after exposure to furfural or heat stress. Biotechnol. Biofuels *6*, 131.

Wu, L.J., and Welker, N.E. (1989). Protoplast transformation of *Bacillus stearothermophilus* NUB36 by plasmid DNA. J. Gen. Microbiol. *135*, 1315–1324.

Xia, Y., Fang, H.H.P., and Zhang, T. (2013). Recent studies on thermophilic anaerobic bioconversion of lignocellulosic biomass. RSC Adv. *2*, 15528–15542.

Xu, C., Qin, Y., Li, Y., Ji, Y., Huang, J., Song, H., and Xu, J. (2010). Factors influencing cellulosome activity in consolidated bioprocessing of cellulosic ethanol. Bioresour. Technol. *101*, 9560–9569.

Xu, L., and Tschirner, U. (2011). Improved ethanol production from various carbohydrates through anaerobic thermophilic co-culture. Bioresour. Technol. *102*, 10065–10071.

Xu, Q., Singh, A., and Himmel, M.E. (2009). Perspectives and new directions for the production of bioethanol using consolidated bioprocessing of lignocellulose. Curr. Opin. Biotechnol. *20*, 364–371.

Yamada, R., Nakatani, Y., Ogino, C., and Kondo, A. (2013). Efficient direct ethanol production from cellulose by cellulase- and cellodextrin transporter-co-expressing *Saccharomyces cerevisiae*. AMB Express 3, 34.

Yang, S., Giannone, R.J., Dice, L., Yang, Z.K., Engle, N.L., Tschaplinski, T.J., Hettich, R.L., and Brown, S.D. (2012). *Clostridium thermocellum* ATCC27405 transcriptomic, metabolomic and proteomic profiles after ethanol stress. BMC Genomics *13*, 336.

Yao, S., and Mikkelsen, M.J. (2010). Metabolic engineering to improve ethanol production in *Thermoanaerobacter mathranii*. Appl. Microbiol. Biotechnol. *88*, 199–208.

You, C., Zhang, X.Z., Sathitsuksanoh, N., Lynd, L.R., and Zhang, Y.H. (2012). Enhanced microbial utilization of recalcitrant cellulose by an *ex vivo* cellulosome–microbe complex. Appl. Environ. Microbiol. *78*, 1437–1444.

Yu, M., Zhang, Y., Tang, I.C., and Yang, S.T. (2011). Metabolic engineering of *Clostridium tyrobutyricum* for n-butanol production. Metab. Eng. *13*, 373–382.

Zhang, X.Z., and Zhang, Y.H. (2011). Simple, fast and high-efficiency transformation system for directed evolution of cellulase in *Bacillus subtilis*. Microb. Biotechnol. *4*, 98–105.

Zhang, Y.H., Himmel, M.E., and Mielenz, J.R. (2006). Outlook for cellulase improvement: screening and selection strategies. Biotechnol. Adv. *24*, 452–481.

Zhu, H., Stadnyk, A., Beland, M., and Seto, P. (2008). Co-production of hydrogen and methane from potato waste using a two-stage anaerobic digestion process. Bioresour. Technol. *99*, 5078–5084.

Zhu, X., Cui, J., Feng, Y., Fa, Y., Zhang, J., and Cui, Q. (2013). Metabolic adaption of ethanol-tolerant *Clostridium thermocellum*. PLoS One *8*, e70631.

Zhuang, F., Karberg, M., Perutka, J., and Lambowitz, A.M. (2009). *EcI5*, a group IIB intron with high retrohoming frequency: DNA target site recognition and use in gene targeting. RNA *15*, 432–449.

Zverlov, V.V., Kellermann, J., and Schwarz, W.H. (2005). Functional subgenomics of *Clostridium thermocellum* cellulosomal genes: identification of the major catalytic components in the extracellular complex and detection of three new enzymes. Proteomics 5, 3646–3653.

Zverlov, V.V., Klupp, M., Krauss, J., and Schwarz, W.H. (2008). Mutations in the scaffoldin gene, cipA, of *Clostridium thermocellum* with impaired cellulosome formation and cellulose hydrolysis: insertions of a new transposable element, IS1447, and implications for cellulase synergism on crystalline cellulose. J. Bacteriol. *190*, 4321–4327.

Thermophilic Viruses and Their Association with Thermophiles

10

Wakao Fukuda and Tadayuki Imanaka

Abstract

Viruses are small infectious particles containing own genome which consists of either DNA or RNA. Until now, a lot of viruses, which can infect organisms in the three domains, *Eukaryote, Bacteria* and *Archaea,* have been found from all over the world. Among them, a few per cent of these known viruses can infect thermophilic archaea and bacteria. Viruses of thermophiles are classified into over 10 families and virus particles (virions) of them show wide variety of shapes and size. Some viruses are lytic, but the others are non-lytic. The wide diversity of virions from thermophiles is interesting to get new insights of the viral world.

Viruses are commonly known as the cause of serious disease. In response, adaptive immune systems, named CRISPR (clustered regularly interspaced short palindromic repeat sequences) system, protect most of archaea from infections of extrachromosomal elements (ECEs). On the other hand, viruses can contribute to the evolutions of lives. The ECEs can transfer beneficial genes for adaptation to environment from one organism to another. Intriguingly, the regions formed by integration of viruses (provirus regions) also have been observed in some genomes of extremophiles, but some of these provirus regions lost the ability for virion-formation. The archaeal strains lacking virus-like region have shown growth defects, indicating that virus-like regions help cell growth in the certain conditions. Although virus infected to ancestor in ancient days might include virulence, virus-like regions seem to contribute to adaptation in certain environment at the present day.

Introduction

From the sequence comparisons of 16S and 18S ribosomal RNA genes, all living organisms on our planet can be divided into three domains, *Bacteria, Eukarya* and *Archaea* (Woese *et al.,* 1990). Regardless of domains, all living organisms have genome DNA as the genetic material, and can grow, metabolize, and maintain homeostasis (Alberts *et al.,* 2007b). In performing a life activity, they use their own machinery for DNA replication, transcription and translation, and they can provide their own source of metabolic energy. On the other hand, viruses are small infectious particles containing own genome which consists of either DNA or RNA, and most viruses infect specific host cell (Alberts *et al.,* 2007c). Viruses replicate only inside the host cell, and they do not have their own machinery for DNA replication, transcription, translation and metabolisms. Therefore, viruses are distinguished from normal living organisms, even though they resemble living organisms in some respects.

A lot of viruses have been found from all over the world, and viruses can infect organisms in the three domains, *Eukaryote, Bacteria* and *Archaea* (Pietilä *et al.,* 2014). A single virus particle (a virion) consists of genome packaged in a protein coat, which in some viruses is further enclosed by a lipid envelope (Alberts *et al.,* 2007c). The virus genome consists of either single-stranded or double-stranded nucleic acid which consists of either DNA or RNA. Virus particles (virions) show wide variety of shapes and size; bacilliform, filamentous, helical, icosahedral, pleomorphic, prolate, spherical, spindle-shaped, complex (head-tailed shape with a tail) and so on (Pietilä *et al.*, 2014). Physical appearances of virions are various, and therefore they cannot be classified by their relatedness into a single phylogenetic tree (Alberts *et al.*, 2007c).

Viruses invade into the host cell, and replicate themselves inside the host. Consequently, infections of viruses often cause serious diseases. On the other hand, extrachromosomal elements (ECEs), such as viruses and plasmid, may be able to transfer a mechanism from one organism to another. Therefore, viruses can carry beneficial genes for adaptation to environment, which has contributed to the evolutions of lives. Because of not only phylogenetic purpose but also progress for molecular biology, studies on ECEs have long been progressed strenuously. Until now, various ECEs have been found not only from mesophiles but also from extremophiles (Table 10.1) (Pietilä *et al.*, 2014). On the other hand, the regions formed by integration of ECEs have been observed in some archaeal genomes (She *et al.*, 2007; Mochizuki *et al.*, 2011; Gorlas *et al.*, 2012). Such putative provirus regions are found in many archaea, whereas archaeal viruses are

Table 10.1 Archaeal viruses (Pietilä *et al.*, 2014; Prangishvili, 2007)

Classification	Virus	Genome	Morphotype
Crenarchaeal virus			
Ampullaviridae	*Acidianus bottle-shaped virus* (ABV)	dsDNA, linear	Bottle-shaped
Bicaudaviridae	*Acidianus two-tailed virus* (ATV)	dsDNA, circular	Spindle-shaped, with two long tails
Clavaviridae	*Aeropyrum pernix bacilliform virus 1* (APBV1)	dsDNA, circular	Bacilliform
Fuselloviridae	*Sulfolobus spindle-shaped virus 1* (SSV1)	dsDNA, circular	Spindle-shaped
	Sulfolobus spindle-shaped virus 6 (SSV6)		
Globuloviridae	*Pyrobaculum spherical virus* (PSV)	dsDNA, linear	Spherical, with helical core
Guttaviridae	*Sulfolobus newzealandicus droplet-shaped virus* (SNDV)	dsDNA, circular	Droplet-shaped
	Aeropyrum pernix ovoid virus 1 (APOV1)		Ovoid-shaped
Lipothrixviridae	*Thermoproteus tenax virus 1* (TTV1)	dsDNA, linear	Filamentous shaped
	Sulfolobus islandicus filamentous virus (SIFV)		
	Acidianus filamentous virus 1 (AFV1)		
	Acidianus filamentous virus 2 (AFV2)		
Rudiviridae	*Sulfolobus islandicus rod-shaped virus 2* (SIRV2)	dsDNA, linear	Rod-shaped
Spiraviridae	*Aeropyrum coil-shaped virus* (ACV)	ssDNA, circular	Coil-shaped
Turriviridae	*Sulfolobus turreted icosahedral virus* (STIV)	dsDNA, circular	Tailless icosahedral
Euryarchaeal virus			
Myoviridae	*Halorubrum sodomense tailed virus 2* (HSTV-2)	dsDNA, linear	Head-tailed, with a long non-contractile tail
Podoviridae	*Halorubrum sodomense tailed virus 1* (HSTV-1)	dsDNA, linear	Head-tailed, with a short tail
Siphoviridae	*Haloarcula vallismortis tailed virus 1* (HVTV-1)	dsDNA, linear	Head-tailed, with a long contractile tail
Salterprovirus[a]	*His 1 virus*	dsDNA, linear	Spindle-shaped, with a short tail
Unclassified	*Pyrococcus abyssi virus 1* (PAV1)	dsDNA, circular	Spindle-shaped, with a short tail
	Thermococcus prieurii virus 1 (TPV1)		

[a]Family was not assigned.

mainly found from hyperthermophilic archaea and halophilic archaea (Pietilä *et al.*, 2014). In general, thermophiles are divided into thermophilic archaea and thermophilic bacteria. In this chapter, viruses and putative provirus regions of archaea are mainly introduced.

Archaeal viruses

It is well known that eukaryotic viruses and bacteriophages show diverse morphotypes (Alberts *et al.*, 2007c). Moreover, archaeal viruses, which have been found from various thermophilic archaea including both *Crenarchaeota* and *Euryarchaeota*, show further diverse and unusual morphotypes (Fig. 10.1 and Table 10.1) (Pietilä *et al.*, 2014). The data of taxonomic classification of viruses are available on the website of International Committee on Taxonomy of Viruses (ICTV, http://www.ictvonline.org/). From the virus taxonomy database released by ICTV in 2013, archaeal viruses are classified into over 10 unique families. The families *Lipothrixviridae* and *Rudiviridae* belong to the same order, *Ligamenvirales*, but most of virus families from hyperthermophilic archaea are not assigned to an order. Most of crenarchaeal viruses show no relatives among bacteriophages. On the other hand, the order *Caudovirales* consists of bacteriophage and euryarchaeal viruses which infect halophilea and/or methanogens (families *Myoviridae*, *Podoviridae* and *Siphoviridae*). A family of viruses named *Sphaerolipoviridae* has been proposed recently, and the family *Sphaerolipoviridae* is divided into three genera; *Alphasphaerolipovirus*, *Betasphaerolipovirus* and *Gammasphaerolipovirus*. *Alphasphaerolipovirus* (*Haloarcula hispanica* viruses SH1, PH1 and HHIV-2) and *Betasphaerolipovirus* (*Natrinema* virus SNJ1) infect halophilic archaea which are not hyperthermophile. However, intriguingly,

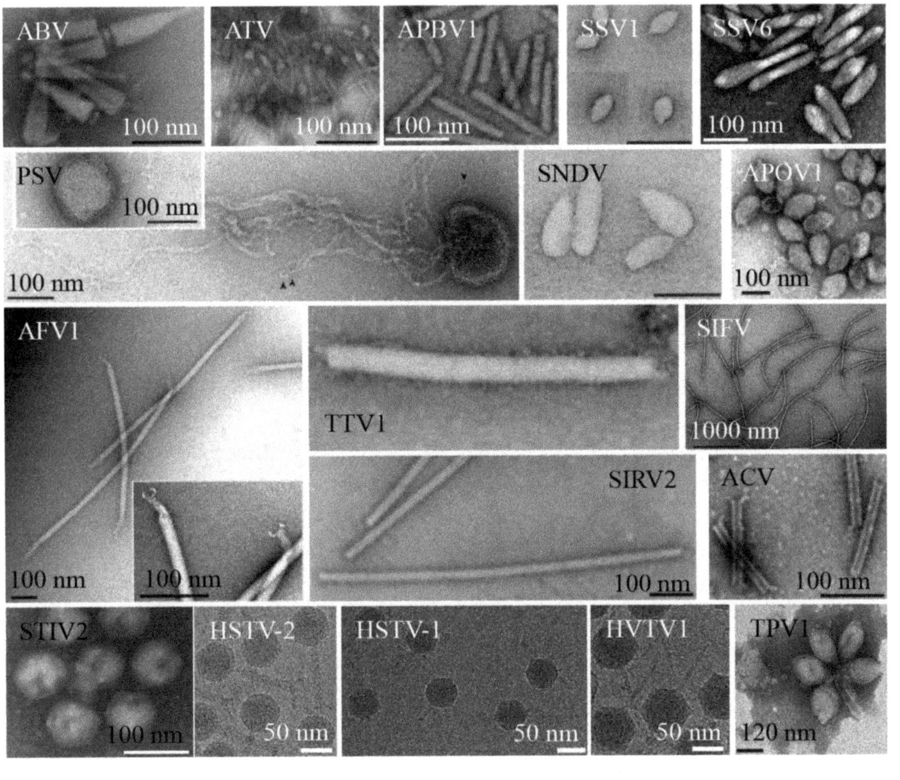

Figure 10.1 Electron micrographs of virions of indicated viruses. Crenarchaeal viruses: ABV, ATV, APBV1, SSV6, PSV, APOV1, SIRV2, ACV, STIV2, (Prangishvili, 2013), SIFV (Arnold *et al.*, 2000), SSV1, SNDV, TTV1 (Prangishvili, 2003), and AFV (Bettstetter *et al.*, 2003). Euryarchaeal viruses: HSTV-2, HVTV1 (Pietilä *et al.*, 2013b), HSTV-1 (Pietilä *et al.*, 2014), and TPV1 (Gorlas *et al.*, 2012).

Gammasphaerolipovirus infects thermophilic bacteria, *Thermus thermophilus* (Pawlowski *et al.*, 2014). The relationship between halophilic archaeal virus and bacteriophage suggests the gene transfer, and it is interesting to know when and where there was the association between halophilic archaea and bacteria (Prangishvili *et al.*, 2006a; Atanasova *et al.*, 2012). Although genomes of all known archaeal viruses are double- or single-stranded DNA, metagenome analysis on the segments from high-temperature acidic hot springs in Yellowstone National Park in the United States suggested the presence of archaeal RNA virus (Bolduc *et al.*, 2012). The features of each family are described below.

Family *Lipothrixviridae* (order *Ligamenvirales*)

The order *Ligamenvirales* are divided into two families, *Lipothrixviridae* and *Rudiviridae*. Viruses in the family *Lipothrixviridae* infect hyperthermophilic crenarchaeota; *Thermoproteus tenax* [*Thermoproteus tenax virus 1* (TTV1)], *Sulfolobus islandicus* [*Sulfolobus islandicus filamentous virus* (SIRV)] and *Acidianus filamentous* [*Acidianus filamentous virus 1* (AFV1)] (Prangishvili, 2007; Pietilä *et al.*, 2014). The family *Lipothrixviridae* is divided into four genera; *Alphalipothrixvirus* (TTV1), *Betalipothrixvirus* (SIRV), *Gammalipothrixvirus* (AFV1) and *Deltalipothrixvirus* (AFV2). Morphotype of virions is flexible filament (0.4–2 μm in length and 20–40 nm in width), but structure of virion termini are diverse (Fig. 10.1) (Zillig *et al.*, 1998). For example, both ends of AFV1 show claw-like structures, and the ends of SIRV virion are tapered and mop-like structure (Zillig *et al.*, 1998; Bettstetter *et al.*, 2003). The genomes of the viruses are linear double-stranded DNA (15.9–56 kbp). Most virions in this family are enveloped, and envelopes contain virus-encoded protein and lipid. However, no lipids were found in the virions of *Deltalipothrixviruses*. Integration of viral genome has not been observed. However, fragments of *Alphalipothrixvirus* DNA have been detected in the host chromosome (Prangishvili, 2007).

Family *Rudiviridae* (order *Ligamenvirales*)

Viruses in the family *Rudiviridae* belong to the order *Ligamenvirales*, and infect hyperthermophilic archaea: *S. islandicus* [*Sulfolobus islandicus rod-shaped virus 1* and 2 (SIRV1 and SIRV2)] and *Acidianus* spp. [*Acidianus rod-shaped virus 1* (ARV1)] (Prangishvili, 2007). Virions tube like superhelical rigid particles formed by the DNA and a single basic, glycosylated DNA-binding protein (0.6–0.9 μm in length and ca. 20 nm in width) (Fig. 10.1) (Prangishvili, 2007). Virions are unenveloped, and no lipid component could be detected in virione. Genomes of the viruses are linear double-stranded DNA (25.9–35 kbp). Growth retardation is occurred by infection of virus. Extrusions of virions occur without cell lysis, and no decrease of cell density or formation of cell debris is observed during growth of infected cells.

Family *Fuselloviridae*

Viruses in the family *Fuselloviridae* mainly infect *Sulfolobus* spp. The family *Fuselloviridae* is divided into two genera; *Alphafusellovirus* [*Sulfolobus spindle-shaped virus 1*, 2 (SSV1, SSV2)] and *Betafusellovirus* [SSV6, and *Acidianus spindle-shaped virus 1*] (ICTV database released in 2013). Natural host of SSV1 and SSV2 are *S. shibatae* and *S. islandicus*, respectively, and both SSV1 and SSV2 can also infect *S. solfataricus* (Prangishvili, 2007). Genomes of the viruses are circular double-stranded DNA. Virions are enveloped and spindle shaped, a short tail is observed at one of the two pointed ends (Fig. 10.1).

Although infected cells show growth retardation, replication and release of virions do not cause cell lysis. Replication of virions can be induced by UV-radiation (Stedman *et al.*, 1999). In the infected cells, viral genome could integrate into host genome. Although hosts of fuselloviruses are *Crenarchaeota*, the regions formed by integration of fuselloviruses at the locus of tRNA gene have been observed in genomes of *Euryarchaeota* (She *et al.*, 2002; She *et al.*, 2007; Mochizuki *et al.*, 2011; Gorlas *et al.*, 2012).

Family *Guttaviridae*

Viruses in the family *Guttaviridae* divided into two genus in the database provided by ICTV in 2013; *Alphaguttavirus* [*Sulfolobus newzealandicus droplet shaped virus* (SVDV)] and *Betaguttavirus* [*Aeropyrum pernix ovoid virus 1* (APOV1)]. Morphotypes of two genera differ as the name of a virus shows; virion of SVDV is droplet-shaped, and virion of APOV1 ovoid-shaped (Fig. 10.1) (Prangishvili, 2007; Mochizuki *et al.*, 2011). Genomes of the viruses are circular double-stranded DNA (SNDV, ca. 20 kb in length; APOV, 13,769 bp in length). Cell lysis does not observed at the time of release of virions. APOV1 exist as a putative provirus in the genome of *A. pernix*, and virions are found in the culture after continuous cultivation of *A. pernix* during without extensive aeration (Mochizuki *et al.*, 2011).

Family *Globuloviridae*

Viruses in the family *Globuloviridae* consist of one genus (*Globulovirus*) with two species [*Pyrobaculum spherical virus* (PSV) and *Thermoproteus tenax spherical virus 1* (TTSV1)]. Virions are enveloped and spherical shaped (100 nm in diameter for PSV) (Fig. 10.1) (Prangishvili, 2007). On the virion surface, a variable number of protrusions are observed. Genomes of the viruses are linear double-stranded DNA (PSV, 28,337bp in length; TTSV1, 20,993 bp in length). Replication and release of virions do not cause cell lysis, nor cause significant growth retardation (Prangishvili, 2007).

Family *Bicaudaviridae*

Viruses in the family *Bicaudaviridae* consist of one genus (*Bicaudavirus*) with one species [*Acidianus two-tailed virus* (ATV)]. Virions show two conformations; one is lemon-shaped virion, and the other is lemon-shaped with two tails at each pointed ends of the virion (maximum length is ca. 1 μm) (Fig. 10.1). Genome of ATV is circular double-stranded DNA and contains 62,730 bp in length (Prangishvili, 2007). ATV is a typical temperate virus, and viral genome is able to integrate into host genome. If infected cells were grown at the optimal growth temperature of 85°C, neither a decrease in cell density nor the formation of cell debris was observed. However, the lysogenic cycle could be interrupted by stress conditions, such as cold-shock, UV irradiation, treatment of mitomycin C (Prangishvili *et al.*, 2006b).

Family *Ampullaviridae*

Viruses in the family *Ampullaviridae* consist of one genus (*Ampullavirus*) with one species [*Acidianus bottle-shaped virus* (ABV)]. Virions are enveloped and look like a bottle (230 nm in length, 75 nm in width at the broad end, and 4 nm in width at the pointed end) (Fig. 10.1). Genome of ABV is linear double-stranded DNA and contains 23,794 bp in length (Prangishvili, 2007). Virions are released at the stationary phase, and lysis of host cells is not observed. Although growth rate of infected cells are longer than uninfected cell, the final cell densities are in the same level.

Family *Clavaviridae*

Viruses in the family *Clavaviridae* consist of one genus (*Clavavirus*) with one species [*Aeropyrum pernix bacilliform virus 1* (APBV1)]. Virions of APBV1 are rigid bacilliform (140 nm in length and 20 nm in width) (Fig. 10.1). Genome of APBV1 is circular double-stranded DNA. Size of APBV1 genome is very small (5278 bp in length), and the genome codes 14 putative genes (Mochizuki *et al.*, 2010; Pietilä *et al.*, 2014). Infection of APBV1 caused neither retardation of host cell growth nor lysis of host cell, and integration of viral genome into host genome was not observed (Mochizuki *et al.*, 2010).

Family *Spiraviridae*

Viruses in the family *Spiraviridae* consist of one genus (*Alphaspiravirus*) with one species [*Aeropyrum coil-shaped virus* (ACV)]. Virions of ACV are non-enveloped and cylindrical shaped (Fig. 10.1), and helix-forming filament is over 4000 nm in length (28 nm in diameter) (Mochizuki *et al.*, 2012; Prangishvili, 2013). Genome of ACV is circular single-stranded DNA (24,893 nt), and the genome contains 57 predicted open reading flames (ORFs) (Mochizuki *et al.*, 2012). The capsid protein binds to the genome and this nucleoprotein filament is condensed into the helical coil of the virion forms. ACV was discovered and isolated from a culture of *A. pernix*. ACV cannot be replicated in any of the available strains of *A. pernix*, and therefore ACV is propagated in the original *A. pernix* culture.

Family *Turriviridae*

Viruses in the family *Turriviridae* consist of one genus (*Alphaturrivirus*); *Sulfolobus turreted icosahedral virus* (STIV) and STIV2. Virions of STIV and STIV2 are encapsidated, and show tailless icosahedral structure with turrets (Fig. 10.1) (Rice *et al.*, 2004; Happonen *et al.*, 2010; Prangishvili, 2013). The structural organization of capsid is similar to other viruses such as human adenovirus, the algal virus PBCV-1 and bacteriophage PRD1 (Rice *et al.*, 2004). Genomes of STIV and STIV2 are circular double-stranded DNA, and have 17,663 bp including 36 predicted ORFs (Rice *et al.*, 2004; Maaty *et al.*, 2006). STIV and STIV2 are lytic virus (Brumfield *et al.*, 2009).

Euryarchaeal viruses in the order *Caudovirales*

The *Caudovirales* are known as the order of tailed bacteriophage, and the order *Caudovirales* are classified by tail morphology into three families, *Myoviridae* (a long non-contractile tail), *Podoviridae* (a short tail) and *Siphoviridae* (a long contractile tail) (Pietilä *et al.*, 2013b, 2014). For example, one of the most well known bacteriophages, *Enterobacteria phage λ*, belongs to the family *Siphoviridae*, and *Enterobacteria phage* T4 belongs to the family *Myoviridae*. In addition, bacteriophage belonging to *Myoviridae* and *Siphoviridae* infect thermophilic bacteria, *Thermus* spp. (Yu *et al.*, 2006). Meanwhile, some viruses of *Caudovirales* infect Euryarchaeota, either extremely halophiles or methanogens. About 40 head-tailed viruses have been isolated from haloarchaea, and *Halobacterium phage phi H* (family *Myoviridae*, genus *Phihlikevirus*) and *Methanobacterium* phage psiM1 (family *Siphoviridae*, genus *Psimunalikevirus*) are listed in ICTV database released in 2013 (Pietilä *et al.*, 2014). In addition, *Haloarcula vallismortis tailed virus 1* (HVTV-1), *Haloarcula sinaiiensis tailed virus 1* and 2 (HSTV-1 and HSTV-2) are belong to *Siphoviridae*, *Podoviridae* and *Myoviridae*, respectively (Pietilä *et al.*, 2013b, 2014).

Virions of *Caudovirales* show a 'head and tail' morphology, and the head contains the double-strand linear DNA genome (Fig. 10.1). Haloarchaeal viruses often lost infectivity at low salinity, and recover infectivity at high salinity (Pietilä *et al.*, 2013c, 2014).

Genus *Salterprovirus*

His1 virus is classified as the genus *Salterprovirus*, but family was not assigned (Pietilä *et al.*, 2014). Virion of His1 virus shows a spindle-shaped morphology such as SSV1 virus, and therefore, His1 virus was originally thought to be closely related to the SSV1 virus. However, there are some differences between His1 virus and spindle-shaped viruses. For example, the His1 virus genome is linear dsDNA, whereas SSV1 (Family *Fuselloviridae*) genome is circular dsDNA. Furthermore, nucleotide or predicted amino acid sequence of His1 virus shows no similarity to those of SSV1. Especially, His1 genome encodes no integrase, unlike the majority of spindle-shaped temperate viruses. Although it was reported by Bath *et al.* (2006) that His1 virus was lytic, Pietilä *et al.* (2013a) reported that His1 virus was clearly non-lytic. Although growth retardation of the infected cells was observed at the stationary phase, cell lysis was not observed (Pietilä *et al.*, 2013a).

Unclassified viruses of *Thermococcales*

Although most of known hyperthermophilic archaeal viruses infect crenarchaea, a few viruses which infect hyperthermophilic euryarchaea were found: *Thermococcus prieurii* virus 1 (TPV1) and *Pyrococcus abyssi* virus 1 (PAV1) (Geslin *et al.*, 2007; Gorlas *et al.*, 2012). Virions of both viruses are spindle-shaped, with a short tail (Fig. 10.1). Similar to fusselloviruses such as SSV1, virions of TPV1 and PAV1 are spindle-shaped and genomes of these viruses are circular double-stranded DNA. However, they show no significant DNA sequence similarity to fuselloviruses, and remain unclassified (Pietilä *et al.*, 2014).

Effect of infection

Virus particles are replicate inside the host cell by using the machinery of host cell, and then virion are released from host cell. In this life cycle of viruses, viruses regulate intracellular vital activity of host organisms by modifying the transcription and translation machineries to propagate in the host cells. It has been reported that the genes associated with DNA replication and repair have been up-regulated by the lytic virus, STIV in the family *Turriviridae* (Ortmann *et al.*, 2008; Walther *et al.*, 2010). Viruses can be released from the host cell by lysis, and enveloped viruses are released by budding (without lysis). As the results, it is well known that various viruses cause serious diseases in the case of animals and plants (Alberts *et al.*, 2007c). In the case of bacteriophage, their life cycles are divided into two types; one is a lytic cycle (virulent phage) and another is a lysogenic cycle (temperate phages). In the lysogenic cycle, the viral genome is integrated by genetic recombination into a specific site in the host genome. Similarly, as described above, both type of viruses (lytic and non-lytic) are known in the case of archaeal viruses. In addition, the regions formed by integration of extrachromosomal DNA have been observed in some archaeal genomes (see the next section) (She *et al.*, 2007; Mochizuki *et al.*, 2011; Gorlas *et al.*, 2012).

Defence mechanism from viral infection

Fundamentally, infection of a virus is harmful for the host cell. In response, adaptive immune systems protect us from infections of pathogens in our bodies (Alberts *et al.*, 2007a). In the case of most archaea (approximately 90%), including hyperthermophiles, and many bacteria (approximately 40%), clustered, regularly interspaced short palindromic repeats (CRISPRs) protects them from the infections of viruses and plasmids (Grissa *et al.*, 2007; Hale *et al.*, 2008; Lillestol *et al.*, 2009; Horvath and Barrangou, 2010; Marraffini and Sontheimer, 2010; Elmore *et al.*, 2013; Su *et al.*, 2013). CRISPR interference is an adaptive immune system to reject the invading DNA molecules by the similar system to RNA interference (RNAi) found in eukaryote. CRISPR systems mainly consist of CRISPR-associated (*cas*) genes, leader sequences and repeat-spacer array (Fig. 10.2) (Horvath and Barrangou, 2010). The spacer sequences (21–72 bp) correspond to the DNA sequence of virus or plasmid which

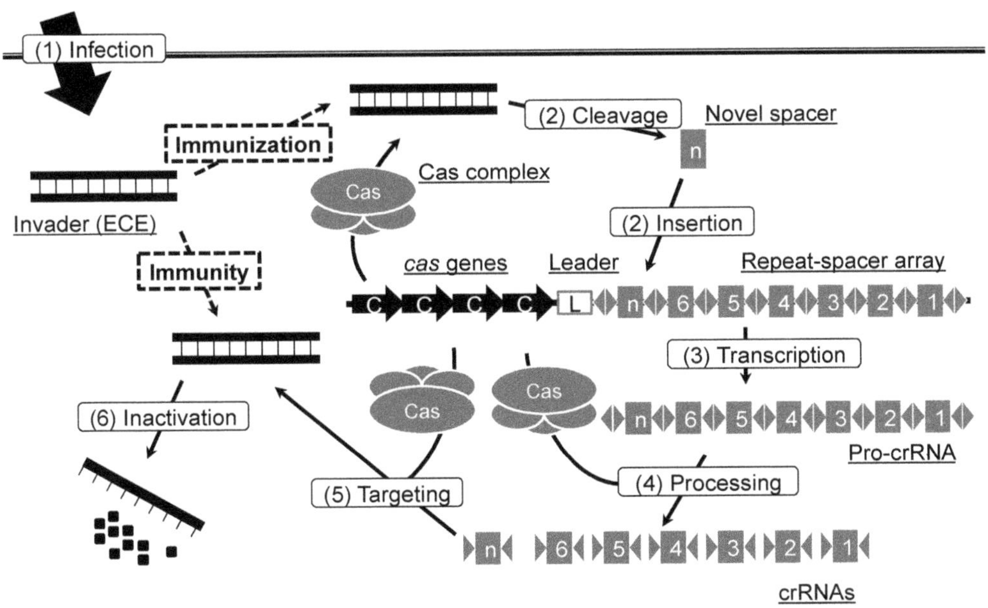

Figure 10.2 Outline of the mechanism for CRISPR system (Horvath and Barrangou, 2010). (1) Invading exogenous DNA from ECE. (2) Integration of a novel repeat-spacer unit at the leader end of the CRISPR locus by a Cas complex. (3) Transcription of the CRISPR repeat-spacer array (production of pre-crRNA). (4) Maturation of pre-crRNA (production of crRNAs). (5) Targeting of invading nucleic acid by the corresponding crRNA. (6) Inactivation of exogenous DNA. Arrows labelled C, *cas* genes; Diamonds, repeat sequences; rectangles, spacers; rectangles labelled L, CRISPR leader.

previously encountered. Gene *cas* encodes a combination of proteins such as nucleases, helicases, polymerases, and polynucleotide-binding proteins. When virus or plasmid invades into the cell, Cas complex cleavage DNA of invader. Then, the DNA fragment is inserted into CRISPR region as a novel repeat-spacer unit. The cluster of repeat-spacer unit is transcribed into a pre-crRNA, and pre-crRNA is processed into mature crRNAs by Cas complex. When the virus or plasmid, which have been encountered previously, invades into the cell, crRNAs with Cas complex interfere with the corresponding invading nucleic acid. As a result, the invading nucleic acid is inactivated. As described above, CRISPR region can acquire new repeat-spacer units which are derived from invader DNA. CRISPR system behaves as an 'adaptive immune systems' of procaryote.

Provirus regions

Although infection of a virus is generally harmful, viruses can transfer beneficial machineries from other organism. Until now, whole genome analyses have been performed on the various microorganisms, and it has become clear that the genome sizes vary according to the microorganism (Koonin and Wolf, 2008). The gene contents of prokaryotic genome have been modified during the process of evolution. New genes have been acquired by gene genesis (duplication and divergence of single genes) and horizontal gene transfer (HGT), and excessive genes have been lost (Kunin and Ouzounis, 2003, Koonin and Wolf, 2008). Especially, HGT provides a mechanism for transferring a number of genes from one organism to another via ECEs, such as virus (bacteriophage) and plasmid, enabling a rapid adaptation of microbes to a new environment (Canchaya *et al.*, 2003). Although these ECEs replicates only inside the host cell, they may carry beneficial genes for adaptation to environment. When the archaeal genomes were analysed, the regions formed by integration of ECEs at the locus of tRNA gene have been observed in some genomes of extremophiles (Fig. 10.3) (She *et al.*, 2002; Wiezer and Merkl, 2005; She *et al.*, 2007; Mochizuki *et al.*, 2011; Gorlas *et al.*, 2012). It is known that putative proviruses in the genome of the hyperthermophilic archaeon *A. pernix* form virions (APSV1 and APOV1) under special conditions (Mochizuki *et al.*, 2011), but some of these provirus regions lost the ability for virion-formation. On the basis of their sequences and functions, archaeal integrases can be classified into three major categories, the Xer type, the pNOB8 type, and the SSV type. However, the property and gene content of integrated region district differ from each other even in the same type of integrated region (She *et al.*, 2007). It is well known that genome contents are subject to modification by gene deletion and degradation in the course of evolution. As a result, only acquired genes will be retained and ameliorated.

In the case of hyperthermophilic archaeon, *Thermococcus kodakarensis*, four virus-like integrated elements, TKV1–TKV4, have been identified in the *T. kodakarensis* genome (Fukui *et al.*, 2005). About 30–40 ORFs are found on the each region, and some of them are annotated as fragment of integrase, transcriptional regulator,

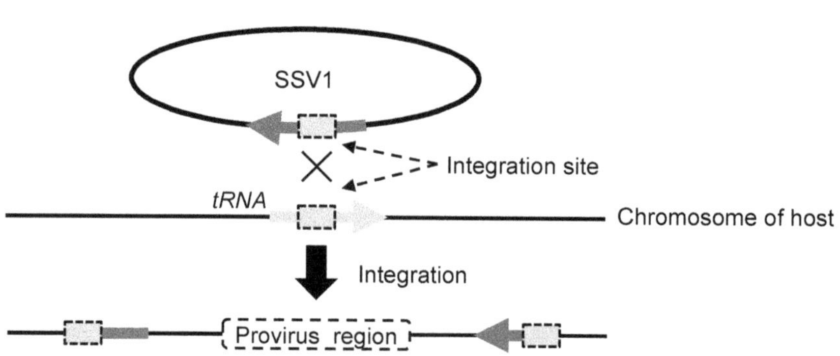

Figure 10.3 Integration of virus genome into host genome.

DNA polymerase sliding clamp, DNA replication licensing factor, and so on (Tagashira *et al.*, 2013). However, 75% of them are annotated as hypothetical proteins. In the regions of TKV1 through TKV3, genes corresponding to the N-terminal region of the integrase (homologous to integrase from SSV1) are overlapped with downstream half of $tRNA^{Val}$, $tRNA^{Glu}$, and $tRNA^{Arg}$, respectively (Fukui *et al.*, 2005). However, only a few other homologues of SSV1 components were found in the four virus-like regions. The ORFs of TKV1 show sequence identities against ORFs of a plasmid from *Thermococcus* sp. strain AMT11 (pAMT11) (Gonnet *et al.*, 2011), and the ORFs of TKV2 and TKV3 show sequence identities against ORFs of a plasmid from *Thermococcus* sp. strain 26–2 (pT26–2) (Soler *et al.*, 2010). On the other hand, orientation of the attachment sites in TKV4 is the same as that of $tRNA^{Leu}$, and the sequence of TKV4 is distinct from TKV1, TKV2, and TKV3 (Fukui *et al.*, 2005). TKV4 shows sequence identities to MVV which is an archaeal provirus integrated into *Methanococcus voltae* A3 genome, and it has been reported that TKV4 and MMV belong to the PRD1-adenovirus of Euryarchaeota (Krupovic and Bamford, 2008). The PRD1-adenovirus lineage spans all three domains of life; *Enterobacteria phage PRD1* belongs to family *Tectiviridae*, genus *Tectivirus*, and adenoviruses are member of the family *Adenoviridae* which infect animals (Benson *et al.*, 1999; Benson *et al.*, 2004; Khayat *et al.*, 2005).

In the case of the genus *Thermococcus*, virus-like particles containing DNA have been found in the culture of *T. gammatolerans* (Soler *et al.*, 2008). On the other hand, formation of virus-like particles has been observed in the culture of *T. kodakarensis*, but any DNA has not been obtained from the virus-like particles (Soler *et al.*, 2008). To understand the function of four virus-like integrated elements in *T. kodakarensis*, four virus-like regions were removed from *T. kodakarensis* genome (Tagashira *et al.*, 2013). As the result of growth property analyses, mutants lacking each of the four virus-like regions exhibited decrease in the cell concentration and/or less growth rates compared to growth of parental strain (KU216), when the *T. kodakarensis* strains were grown at optimal growth temperature (85°C) in nutrient-rich medium. These results indicated that the genes in virus-like regions stimulated the cell growth under the observed growth condition. In contrast, infections of the provirus isolated from the *Thermococcus prieurii* (TPV1) retarded the growth of several *Thermococcus* species including *T. kodakarensis* (Gorlas *et al.*, 2012). Although infections of the virus, which was the origin of each virus-like region in *T. kodakarensis*, might be harmful for the cell growth at the time, the remaining virus like regions are helpful for the cell growth of *T. kodakarensis*. As the result of transcriptome analyses, genes involved in amino acid, energy or nucleotide metabolisms, and transport systems were up- or down-regulated in the cells of mutant strains. Interestingly, a decrease in transcriptional levels of glutamine synthetase (TK1796) gene (*Tk*-glnA) was observed in the cells of four mutant strains. Growths of TKV1 disrupted strain and TKV4 disrupted strain have shown no difference compared with that of KU216 by the addition of glutamate or glutamine, and the result suggested that TKV1 and TKV4 contributed to supply of amino acids to the cell.

The strains lacking virus-like region have shown growth defects, indicating that virus-like regions help cell growth in the observed medium. In addition, transcriptome analyses revealed that genes concerning various carbon metabolisms and transport systems were down-regulated in the cells of mutant strains. Although virus infected to ancestor of *T. kodakarensis* in ancient days might include virulence, virus-like regions seem to contribute to adaptation in certain environment for *T. kodakarensis* at the present day.

Bacteriophage from thermophilic bacteria

As described above, various archaeal viruses infecting extremophiles have been reported. In addition bacteriophages also have been isolated from thermophilic bacteria such as *Bacillus* (*Geobacillus*), *Thermus* (optimum growth temperature, about 70–75°C) and *Meiothermus* (optimum growth temperature, about 50–65°C) (Da Costa *et al.*, 2005; Nobre and Da Costa, 2005; Liu *et al.*, 2006; Yu *et al.*, 2006; Lin *et al.*, 2011; Pawlowski *et al.*, 2014). Viruses belonging to the families

Myoviridae, Siphoviridae, Tectiviridae, Inoviridae and *Sphaerolipoviridae* infect *Thermus* spp. (Yu *et al.*, 2006; Pawlowski *et al.*, 2014). *Myoviridae* and *Siphoviridae* are assigned to the Order *Caudovirales* (see also 1. 11 in this chapter). The genomes of these viruses are double stranded DNA. φfYS40 (*Myoviridae*), P23-45 and P74-26 (*Siphoviridae*) are known as bacteriophages of *T. thermophilus*, and genomes of these 3 viruses have been sequenced (Naryshkina *et al.*, 2006; Minakhin *et al.*, 2008). A bacteriophage infecting *Meiothermus* sp. TG17 was isolated and named MMP17 (*Meiothermus Myoviridae phage 17*) (Lin *et al.*, 2011). MMP17 is a typical myovirus with an icosahedral head (42 nm in diameter) and a tail (120 nm in length and 17 nm in width), and the capsid and tail l size are smaller than those of φfYS40 (head size, 125 nm in diameter; tail size, 178 nm × 27 nm).

Tectiviridae and *Inoviridae* are not assigned to an order in the ITCV database released in 2013. *Thermus* tectiviruses and inoviruses were isolated from alkaline hot springs (Yu *et al.*, 2006). Type strain of *Tectivirus* is *Enterobacteria phage* PRD1, and *Tectivirus Thermus phage P37–14* has been listed in the ICTV database (Abrescia *et al.*, 2004). Virion of *Tectivirus* is non-enveloped and icosahedral shaped without tail, and the genomes of them are double stranded DNA. Type strain of *Inovirus* is well known *Enterobacteria phage* M13 with genome of single stranded DNA. Virion of *Thermus* inovirus PH75 are filamentous shaped (~6 nm × 910 nm) (Overman *et al.*, 2005).

Sphaerolipoviridae is a recently proposed as a new family of viruses (Pawlowski *et al.*, 2014). Intriguingly, the two genera of family *Sphaerolipoviridae* infect halophilic archaea (*Alphasphaerolipovirus* and *Betasphaerolipovirus*), and the other infects thermophilic bacteria, *T. thermophilus* (*Gammasphaerolipovirus*) (Pawlowski *et al.*, 2014). The virions of *Sphaerolipoviridae* show icosahedral, tailless shape with an internal lipid membrane located between the protein capsid and the dsDNA genome. Genome DNA of *Alphasphaerolipovirus* is circular with inverted terminal repeats and terminal proteins, but those of *Betasphaerolipovirus* and *Gammasphaerolipovirus* are linear. *Gammasphaerolipovirus* comprises two members, lytic phage P23-77 and temperate phage IN93. In addition to these viruses, provirus regions (*Gammasphaerolipoprovirus*) were found in the *Thermus* and *Meiothermus* genomes.

Conclusion

Viruses populate almost all ecosystems on Earth, and It has been known that viruses can infect organisms in the all three domains, *Eukaryote, Bacteria* and *Archaea* (Lawrence *et al.*, 2009). On the one hand infection of a virus is harmful for host cell. On the other viruses can provide mechanisms for the rapid adaptation of microbes to a new environment. In fact, as described above, various viruses case cell lysis and/or growth retardation, and provirus regions have been found on the genome of many organisms. Furthermore, because of not only phylogenetic and epidemiologic purpose but progress for molecular biology, studies on viruses have long been progressed strenuously. However, physical appearances and architecture of virions are various, and therefore they cannot be classified by their relatedness into a single phylogenetic tree (Alberts *et al.*, 2007c). Especially, the virions of hyperthermophilic archaea show great variety of appearances (Fig. 10.1). On the other hand, the number of the found archaeal viruses is a small number compared with the number of whole. It is expected that research of 'unique' archaeal viruses lead further understandings of the viruses.

References

Abrescia, N.G.A., Cockburn, J.J.B., Grimes, J.M., Sutton, G.C., Diprose, J.M., Butcher, S.J., Fuller, S.D., Martín, C.S., Burnett, R.M., Stuart, D.I., *et al.* (2004). Insights into assembly from structural analysis of bacteriophage PRD1. Nature 432, 68–74.

Alberts, B., Johnson, A., Lewis, J., Raff, M., and Roberts, K. (2007a). The adaptive immune system. In Molecular Biology of the Cell, 5th edn (Garland Science, New York), pp. 1539–1601.

Alberts, B., Johnson, A., Lewis, J., Raff, M., and Roberts, K. (2007b). Cells and genomes. In Molecular Biology of the Cell, 5th edn. (Garland Science, New York), pp. 1–44.

Alberts, B., Johnson, A., Lewis, J., Raff, M., and Roberts, K. (2007c). Pathogens, infection, and innate immunity. In Molecular Biology of the Cell, 5th edn (Garland Science, New York), pp. 1485–1538.

Arnold, H.P., Zillig, W., Ziese, U., Holz, I., Crosby, M., Utterback, T., Weidmann, J.F., Kristjansond, J.K., Klenkc, H.P., Nelsonc, K.E., and Fraser, C.M. (2000). A novel lipothrixvirus, SIFV, of the extremely thermophilic crenarchaeon *Sulfolobus*. Virology *267*, 252–266.

Atanasova, N.S., Roine, E., Oren, A., Bamford, D.H., and Oksanen, H.M. (2012). Global network of specific virus–host interactions in hypersaline environments. Environ. Microbiol. *14*, 426–440.

Bath, C., Cukalac, T., Porter, K., and Dyall-Smith, M.L. (2006). His1 and His2 are distantly related, spindle-shaped haloviruses belonging to the novel virus group, *Salterprovirus*. Virology *350*, 228–239.

Benson, S.D., Bamford, J.K., Bamford, D.H., and Burnett, R.M. (1999). Viral evolution revealed by bacteriophage PRD1 and human adenovirus coat protein structures. Cell *98*, 825–833.

Benson, S.D., Bamford, J.K.H., Bamford, D.H., and Burnett, R.M. (2004). Does common architecture reveal a viral lineage spanning all three domains of life? Mol. Cell *16*, 673–685.

Bettstetter, M., Peng, X., Garrett, R.A., and Prangishvili, D. (2003). AFV1, a novel virus infecting hyperthermophilic archaea of the genus *Acidianus*. Virology *315*, 68–79.

Bolduc, B., Shaughnessy, D.P., Wolf, Y.I., Koonin, E.V., Roberto, F.F., and Young, M. (2012). Identification of novel positive-strand RNA viruses by metagenomic analysis of archaea-dominated Yellowstone hot springs. J. Virol. *86*, 5562–5573.

Brumfield, S.K., Ortmann, A.C., Ruigrok, V., Suci, P., Douglas, T., and Young, M.J. (2009). Particle assembly and ultrastructural features associated with replication of the lytic archaeal virus *sulfolobus* turreted icosahedral virus. J. Virol. *83*, 5964–5970.

Canchaya, C., Fournous, G., Chibani-Chennoufi, S., Dillmann, M.L., and Brussow, H. (2003). Phage as agents of lateral gene transfer. Curr. Opin. Microbiol. *6*, 417–424.

Da Costa, M.S., Nobre, M.F., and Rainey, F.A. (2005). Genus I. *Thermus* Brock and Freeze 1969, 295[AL] emend. Nobre, Trüper and da Costa 1996b, 605. In Bergey's Manual of Systematic Bacteriology, 2nd edn, Brenner, D.J., Kreig, N.R., Staley, J.T., and Garrity, G.M., eds. (Springer, New York), vol. 1, pp. 404–414.

Elmore, J.R., Yokooji, Y., Sato, T., Olson, S., Glover, C.V. 3rd, Graveley, B.R., Atomib, H., Ternsa, R.M., and Terns, M.P. (2013). Programmable plasmid interference by the CRISPR-Cas system in *Thermococcus kodakarensis*. RNA Biol. *10*, 828–840.

Fukui, T., Atomi, H., Kanai, T., Matsumi, R., Fujiwara, S., and Imanaka, T. (2005). Complete genome sequence of the hyperthermophilic archaeon *Thermococcus kodakaraensis* KOD1 and comparison with *Pyrococcus* genomes. Genome Res. *15*, 352–363.

Geslin, C., Gaillard, M., Flament, D., Rouault, K., Le Romancer, M., Prieur, D., and Erauso, G. (2007). Analysis of the first genome of a hyperthermophilic marine virus-like particle, PAV1, isolated from *Pyrococcus abyssi*. J. Bacteriol. *189*, 4510–4519.

Gonnet, M., Erauso, G., Prieur, D., and Le Romancer, M. (2011). pAMT11, a novel plasmid isolated from a *Thermococcus* sp. strain closely related to the virus-like integrated element TKV1 of the *Thermococcus kodakaraensis* genome. Res. Microbiol. *162*, 132–143.

Gorlas, A., Koonin, E.V., Bienvenu, N., Prieur, D., and Geslin, C. (2012). TPV1, the first virus isolated from the hyperthermophilic genus *Thermococcus*. Environ. Microbiol. *14*, 503–516.

Grissa, I., Vergnaud, G., and Pourcel, C. (2007). The CRISPRdb database and tools to display CRISPRs and to generate dictionaries of spacers and repeats. BMC Bioinformatics *8*, 172.

Hale, C., Kleppe, K., Terns, R.M., and Terns, M.P. (2008). Prokaryotic silencing (psi)RNAs in *Pyrococcus furiosus*. RNA *14*, 2572–2579.

Happonen, L.J., Redder, P., Peng, X., Reigstad, L.J., Prangishvili, D., and Butcher, S.J. (2010). Familial relationships in hyperthermo- and acidophilic archaeal viruses. J. Virol. *84*, 4747–4754.

Horvath, P., and Barrangou, R. (2010). CRISPR/Cas, the immune system of bacteria and archaea. Science *327*, 167–170.

Khayat, R., Tang, L., Larson, E.T., Lawrence, C.M., Young, M., and Johnson, J.E. (2005). Structure of an archaeal virus capsid protein reveals a common ancestry to eukaryotic and bacterial viruses. Proc. Natl. Acad. Sci. U.S.A. *102*, 18944–18949.

Koonin, E.V., and Wolf, Y.I. (2008). Genomics of bacteria and archaea: the emerging dynamic view of the prokaryotic world. Nucleic Acids Res. *36*, 6688–6719.

Krupovic, M., and Bamford, D.H. (2008). Archaeal proviruses TKV4 and MVV extend the PRD1-adenovirus lineage to the phylum *Euryarchaeota*. Virology *375*, 292–300.

Kunin, V., and Ouzounis, C.A. (2003). The balance of driving forces during genome evolution in prokaryotes. Genome Res. *13*, 1589–1594.

Lawrence, C.M., Menon, S., Eilers, B.J., Bothner, B., Khayat, R., Douglas, T., and Young, M.J. (2009). Structural and functional studies of archaeal viruses. J. Biol. Chem. *284*, 12599–12603.

Lillestol, R.K., Shah, S.A., Brugger, K., Redder, P., Phan, H., Christiansen, J., and Garrett, R.A. (2009). CRISPR families of the crenarchaeal genus *Sulfolobus*: bidirectional transcription and dynamic properties. Mol. Microbiol. *72*, 259–272.

Lin, L., Han, J., Ji, X., Hong, W., Huang, L., and Wei, Y. (2011). Isolation and characterization of a new bacteriophage MMP17 from *Meiothermus*. Extremophiles *15*, 253–258.

Liu, B., Wu, S., Song, Q., Zhang, X., and Xie, L. (2006). Two novel bacteriophages of thermophilic bacteria isolated from deep-sea hydrothermal fields. Curr. Microbiol. *53*, 163–166.

Maaty, W.S., Ortmann, A.C., Dlakic, M., Schulstad, K., Hilmer, J.K., Liepold, L., Weidenheft, B., Khayat, R., Douglas, T., Young, M.J., and Bothner, B. (2006). Characterization of the archaeal thermophile *Sulfolobus* turreted icosahedral virus validates an evolutionary link among double-stranded DNA viruses from all domains of life. J. Virol. *80*, 7625–7635.

Marraffini, L.A., and Sontheimer, E.J. (2010). CRISPR interference: RNA-directed adaptive immunity in bacteria and archaea. Nat. Rev. Genet. *11*, 181–190.

Minakhin, L., Goel, M., Berdygulova, Z., Ramanculov, E., Florens, L., Glazko, G., Karamychev, V.N., Slesarev,

A.I., Kozyavkin, S.A., Khromov, I., *et al.* (2008). Genome comparison and proteomic characterization of Thermus thermophilus bacteriophages P23–45 and P74–26: siphoviruses with triplex-forming sequences and the longest known tails. J. Mol. Biol. *378*, 468–480.

Mochizuki, T., Yoshida, T., Tanaka, R., Forterre, P., Sako, Y., and Prangishvili, D. (2010). Diversity of viruses of the hyperthermophilic archaeal genus *Aeropyrum*, and isolation of the *Aeropyrum pernix* bacilliform virus 1, APBV1, the first representative of the family *Clavaviridae*. Virology *402*, 347–354.

Mochizuki, T., Sako, Y., and Prangishvili, D. (2011). Provirus induction in hyperthermophilic archaea: characterization of *Aeropyrum pernix* spindle-shaped virus 1 and *Aeropyrum pernix* ovoid virus 1. J. Bacteriol. *193*, 5412–5419.

Mochizuki, T., Krupovic, M., Pehau-Arnaudet, G., Sako, Y., Forterre, P., and Prangishvili, D. (2012). Archaeal virus with exceptional virion architecture and the largest single-stranded DNA genome. Proc. Natl. Acad. Sci. U.S.A. *109*, 13386–13391.

Naryshkina, T., Liu, J., Florens, L., Swanson, S.K., Pavlov, A.R., Pavlova, N.V., Inman, R., Minakhin, L., Kozyavkin, S.A., Washburn, M., *et al.* (2006). *Thermus thermophilus* bacteriophage φYS40 genome and proteomic characterization of virions. J. Mol. Biol. *364*, 667–677.

Nobre, M.F., and Da Costa, M.S. (2005). Genus II. *Meiothermus* Nobre, Trüper and da Costa 1996b, 605^{VP}. In Bergey's Manual of Systematic Bacteriology, 2nd edn, Brenner, D.J., Kreig, N.R., Staley, J.T., and Garrity, G.M., eds. (Springer, New York), vol. 1, pp. 414–424.

Ortmann, A.C., Brumfield, S.K., Walther, J., McInnerney, K., Brouns, S.J., van de Werken, H.J., Bothner, B., Douglas, T., van de Oost, J., and Young, M.J. (2008). Transcriptome analysis of infection of the archaeon *Sulfolobus solfataricus* with *Sulfolobus* turreted icosahedral virus. J. Virol. *82*, 4874–4883.

Overman, S.A., Bondre, P., Maiti, N.C., and Thomas, G.J. Jr. (2005). Structural characterization of the filamentous bacteriophage PH75 from Thermus thermophilus by Raman and UV-resonance Raman spectroscopy. Biochemistry *44*, 3091–3100.

Pawlowski, A., Rissanen, I., Bamford, J.K., Krupovic, M., and Jalasvuori, M. (2014). *Gammasphaerolipovirus*, a newly proposed bacteriophage genus, unifies viruses of halophilic archaea and thermophilic bacteria within the novel family *Sphaerolipoviridae*. Arch. Virol. *159*, 1541–1554.

Pietilä, M.K., Atanasova, N.S., Oksanen, H.M., and Bamford, D.H. (2013a). Modified coat protein forms the flexible spindle-shaped virion of haloarchaeal virus His1. Environ. Microbiol. *15*, 1674–1686.

Pietilä, M.K., Laurinmaki, P., Russell, D.A., Ko, C.C., Jacobs-Sera, D., Butcher, S.J., Bamforda, D.H., and Hendrix, R.W. (2013b). Insights into head-tailed viruses infecting extremely halophilic archaea. J. Virol. *87*, 3248–3260.

Pietilä, M.K., Laurinmaki, P., Russell, D.A., Ko, C.C., Jacobs-Sera, D., Hendrix, R.W., Bamford, D.H., and Butcher, S.J. (2013c). Structure of the archaeal head-tailed virus HSTV-1 completes the HK97 fold story. Proc. Natl. Acad. Sci. U.S.A. *110*, 10604–10609.

Pietilä, M.K., Demina, T.A., Atanasova, N.S., Oksanen, H.M., and Bamford, D.H. (2014). Archaeal viruses and bacteriophages: comparisons and contrasts. Trends Microbiol. *22*, 334–344.

Prangishvili, D. (2003). Evolutionary insights from studies on viruses of hyperthermophilic archaea. Res. Microbiol. *154*, 289–294.

Prangishvili, D. (2007). Families of DNA viruses infecting hyperthermophilic crenarchaea. In Archaea: Evolution, Physiology, and Molecular Biology, Garrett, R.A., and Klenk, H.P., eds. (Wiley-Blackwell, Singapore), pp. 60–74.

Prangishvili, D. (2013). The wonderful world of archaeal viruses. Annu. Rev. Microbiol. *67*, 565–585.

Prangishvili, D., Forterre, P., and Garrett, R.A. (2006a). Viruses of the Archaea: a unifying view. Nat. Rev. Microbiol. *4*, 837–848.

Prangishvili, D., Vestergaard, G., Haring, M., Aramayo, R., Basta, T., Rachel, R., and Garrett, R.A. (2006b). Structural and genomic properties of the hyperthermophilic archaeal virus ATV with an extracellular stage of the reproductive cycle. J. Mol. Biol. *359*, 1203–1216.

Rice, G., Tang, L., Stedman, K., Roberto, F., Spuhler, J., Gillitzer, E., Johnson, J.E., Douglas, T., and Young, M. (2004). The structure of a thermophilic archaeal virus shows a double-stranded DNA viral capsid type that spans all domains of life. Proc. Natl. Acad. Sci. U.S.A. *101*, 7716–7720.

She, Q., Brugger, K., and Chen, L. (2002). Archaeal integrative genetic elements and their impact on genome evolution. Res. Microbiol. *153*, 325–332.

She, Q., Zhu, H., and Xiang, X. (2007). Integration mechanisms: possible role in genome evolution. In Archaea: Evolution, Physiology, and Molecular Biology. Garrett, R.A., and Klenk, H.P., eds. (Wiley-Blackwell, Singapore), pp. 113–123.

Soler, N., Marguet, E., Verbavatz, J.M., and Forterre, P. (2008). Virus-like vesicles and extracellular DNA produced by hyperthermophilic archaea of the order *Thermococcales*. Res. Microbiol. *159*, 390–399.

Soler, N., Marguet, E., Cortez, D., Desnoues, N., Keller, J., van Tilbeurgh, H., Sezonov, G., and Forterre, P. (2010). Two novel families of plasmids from hyperthermophilic archaea encoding new families of replication proteins. Nucleic Acids Res. *38*, 5088–5104.

Stedman, K.M., Schleper, C., Rumpf, E., and Zillig, W. (1999). Genetic requirements for the function of the archaeal virus SSV1 in *Sulfolobus solfataricus*: construction and testing of viral shuttle vectors. Genetics *152*, 1397–1405.

Su, A.A., Tripp, V., and Randau, L. (2013). RNA-Seq analyses reveal the order of tRNA processing events and the maturation of C/D box and CRISPR RNAs in the hyperthermophile Methanopyrus kandleri. Nucleic Acids Res. *41*, 6250–6258.

Tagashira, K., Fukuda, W., Matsubara, M., Kanai, T., Atomi, H., and Imanaka, T. (2013). Genetic studies on the virus-like regions in the genome of hyperthermophilic archaeon, *Thermococcus kodakarensis*. Extremophiles *17*, 153–160.

Walther, J., Sierocinski, P., and van der Oost, J. (2010). Hot transcriptomics. Archaea *2010*, 897585.

Wiezer, A., and Merkl, R. (2005). A comparative categorization of gene flux in diverse microbial species. Genomics *86*, 462–475.

Woese, C.R., Kandler, O., and Wheelis, M.L. (1990). Towards a natural system of organisms: proposal for the domains Archaea, Bacteria, and Eucarya. Proc. Natl. Acad. Sci. U.S.A. *87*, 4576–4579.

Yu, M.X., Slater, M.R., and Ackermann, H.W. (2006). Isolation and characterization of *Thermus* bacteriophages. Arch. Virol. *151*, 663–679.

Zillig, W., Arnold, H.P., Holz, I., Prangishvili, D., Schweier, A., Stedman, K., She, Q., Phan, H., Garrett, R., and Kristjansson, J.K. (1998). Genetic elements in the extremely thermophilic archaeon *Sulfolobus*. Extremophiles 2, 131–140.

Index

Note: Where terms appear in tables, page references are in **bold**; where terms appear in figures, page references are in *italic*.

N

O

P

R

S

T

V

Z

www.ingramcontent.com/pod-product-compliance
Lightning Source LLC
LaVergne TN
LVHW071337210926
848486LV00003B/165
* 9 7 8 1 9 1 0 1 9 0 1 3 5 *